3-21-97

NMR SPECTROSCOPY

D1120417

NMR SPECTROSCOPY

Basic Principles, Concepts, and Applications in Chemistry

Harald Günther
University of Siegen, Siegen, Germany

Second Edition

Translated by Harald Günther

JOHN WILEY & SONS
Chichester · New York · Brisbane · Toronto · Singapore

First Published in 1992 © Georg Thieme Verlag,
Stuttgart under the title *NMR-Spektroscopie:
Grundlagen, Konzepte und Anwendungen der Protonen-
und Kohlenstoff-13 Kernresonanz-Spektroscopie in der
Chemie* by Harald Günther

English language translation copyright © 1995 by John Wiley & Sons Ltd,
Baffins Lane, Chichester,
West Sussex PO19 1UD, England

Telephone: National Chichester (01243) 779777
International +44 1243 779777

Reprinted June 1996

Other Wiley Editorial Offices

John Wiley & Sons, Inc., 605 Third Avenue,
New York, NY 10158-0012, USA

Jacaranda Wiley Ltd, 33 Park Road, Milton,
Queensland 4064, Australia

John Wiley & Sons (Canada) Ltd, 22 Worcester Road,
Rexdale, Ontario M9W 1L1, Canada

John Wiley & Sons (SEA) Pte Ltd, 37 Jalan Pemimpin #05-04,
Block B, Union Industrial Building, Singapore 2057

Library of Congress Cataloging-in-Publication Data
Günther, H. (Harald)
   [NMR Spektroskopie.   English]
   NMR spectroscopy : basic principles, concepts, and applications in
chemistry   /   H. Günther. — 2nd ed.
      p.   cm.
   Includes bibliographical references and index.
   ISBN 0-471-95199-4 : — ISBN 0-471-95201-X (pbk.)
   1. Nuclear magnetic resonance spectroscopy. I. Title.
QD96.N8G8313   1994
543'.0877—dc20   94-23084
                                                                      CIP

British Library Cataloguing in Publication Data

A catalogue record for this book is available from the British Library

ISBN 0 471 95199 4 (cloth)
ISBN 0 471 95201 X (paper)

Typeset in 10/12pt Times by Techset Composition Ltd, Salisbury, Wilts
Printed and bound in Great Britain by Biddles Ltd, Guildford and King's Lynn

CONTENTS

PREFACE

During the last 20 years we have witnessed an unprecedented development of new techniques in the field of nuclear magnetic resonance spectroscopy, which has served not only to maintain but even strengthen its position as the most important spectroscopic tool in chemistry. Aside from the construction of cryomagnets, the potential of Fourier transform spectroscopy and the introduction of two-dimensional methods form the basis for these achievements. Their importance is underlined by the fact that the Nobel Prize in Chemistry was awarded in 1991 to the Swiss physical chemist Richard R. Ernst, who introduced FT methods to n.m.r. and pioneered the development of multidimensional n.m.r. spectroscopy.

The present edition of *NMR Spectroscopy* takes this situation into account and includes these new areas with an introduction into the basic principles and the demonstration of practical applications. The first six chapters contain the material important for n.m.r. beginners as well as for users who are primarily interested in the relations between n.m.r. parameters and chemical structure. The physical principles of n.m.r. in general and Fourier transform n.m.r. and pulse spectroscopy in particular are introduced in Chapter 7, while Chapter 8 is devoted to two-dimensional experiments. This also includes a short treatment of the product operator formalism as well as the application of phase cycles and pulsed field gradients. Chapter 9 treats dynamic phenomena and Chapter 10 collects selected topics with a discussion of the nuclear Overhauser effect, rotating frame experiments, the CIDNP effect, n.m.r. of paramagnetic materials, n.m.r. in liquid crystals, and high-resolution solid state n.m.r., as well as a short introduction into applications in biology and medicine. A separate Chapter on carbon-13 n.m.r. and an Appendix with data collections and the mathematical formalism of a number of theoretical aspects concludes the volume.

Of course a number of important topics, for example n.m.r. of nuclei like ^{15}N or ^{31}P, could not be included for reasons of space. Nevertheless it is hoped that the present text serves its purpose to introduce students to modern n.m.r. and its applications in chemistry and related sciences, and helps to guide them from their first contacts with this fascinating tool to more advanced uses during graduation.

Apart from those persons whose help was acknowledged in the first edition, I am indebted to a number of coworkers who contributed new figures to the present edition and to a number of colleagues for their advice. In particular I would like to thank Professors E. R. Andrew, R. Benn, P. Joseph-Nathan, and A. Maercker as well as Drs G. Englert, W. Holzer, and H. Uzar for supplying me with original figures from their work. From experiments done in our own laboratory new figures were kindly provided by Dr H. Hausmann and Drs W. Andres, P. Bast, K. Bergander, M.

Ebener, O. Eppers, W. Frankmölle, G. von Fircks, Th. Fox, M. Kreutz, H.-E. Mons, D. Moskau, P. Schmitt, D. Schmalz and J. R. Wesener. Finally, I would like to express my gratitude to the publisher and the people in the production department for their cooperation.

Siegen, January 1995   H. Günther

INTRODUCTION

Of the important spectroscopic aids that are at the disposal of the chemist for use in structure elucidation, nuclear magnetic resonance (n.m.r.) spectroscopy is relatively recent. In 1945 two groups of physicists working independently—Purcell, Torrey and Pound at Harvard University and Bloch, Hansen and Packard at Stanford University—first succeeded in observing the phenomenon of nuclear magnetic resonance in solids and liquids. At the beginning of the 1950s, after an extraordinarily short time, the phenomenon was called upon for the first time in the solution of a chemical problem. Since that time its importance has steadily increased, a situation highlighted in 1991 by the award of the Nobel Prize in Chemistry to the Swiss physical chemist Richard R. Ernst from the ETH Zürich for his outstanding contributions to the development of experimental n.m.r. techniques.

The physical foundation of nuclear magnetic resonance spectroscopy lies in the magnetic properties of atomic nuclei. The interaction of the nuclear magnetic moment of a nucleus with an external magnetic field, B_0, leads, according to the rules of quantum mechanics, to a nuclear energy level diagram, because the magnetic energy of the nucleus is restricted to certain discrete values E_i, the so-called *eigenvalues*. Associated with the eigenvalues are the *eigenstates*, which are the only states in which an elementary particle can exist. They are also called *stationary states*. Through a high-frequency transmitter, transitions between eigenstates within the energy level diagram can be stimulated. The absorption of energy can be detected, amplified and recorded as a spectral line, the so-called *resonance signal* (Figure 1).

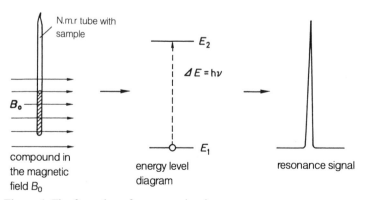

Figure 1 The formation of an n.m.r. signal

In this way a *spectrum* can be generated for a compound containing atoms whose nuclei have non-zero magnetic moments. Among these are the proton, ^1H, the fluorine nucleus, ^{19}F, the nitrogen isotopes, ^{14}N and ^{15}N, and many others of chemical interest. However, the carbon nucleus, ^{12}C, which is so important in organic chemistry, has, like all other nuclei with even mass and atomic number, no magnetic moment. Therefore, n.m.r. studies with carbon are limited to the isotope ^{13}C, which has a natural abundance of only 1.1%.

In order to illustrate a *nuclear magnetic resonance spectrum* and its essential characteristics, the proton n.m.r. spectrum of ethyl formate is reproduced in Figure 2. The spectrum was measured at a magnetic field of 1.4 tesla with a radiofrequency of 60 MHz. In addition to the resonance signals observed at different frequencies, it shows a step curve produced by a built-in electronic *integrator*. The heights of the steps are proportional to the areas under the corresponding spectral lines.

The following points should be noted:

1. Several resonance signals are found for various protons in a molecule. These arise because the protons reside in different chemical environments. The resonance signals are separated by a so-called *chemical shift*.

2. The area under a resonance signal is proportional to the number of protons that give rise to the signal. It can be measured by *integration*.

3. Not all spectral lines are simple, i.e. singlets. For some, characteristic splitting patterns are followed, forming triplets or quartets. This splitting is the result of *spin–spin coupling*—a magnetic interaction between different nuclei.

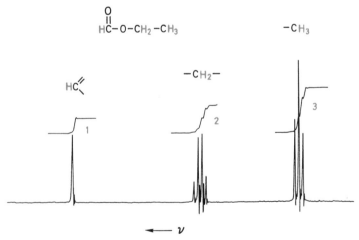

Figure 2 ^1H n.m.r. spectrum of ethyl formate

Empirically determined correlations between the spectral parameters, *chemical shift* and *spin–spin coupling*, on the one hand and the structure of chemical compounds on the other form the basis for the application of proton and, in general, nuclear magnetic resonance to the *structure determination* of unknown samples. In this respect the nuclear magnetic moment has proved itself to be a very sensitive probe with which one can gather extensive information. Thus, with the aid of the chemical shift it leads to knowledge of the chemical environment in which the nucleus responsible for a signal is situated, and the integration of the spectrum allows one to draw conclusions concerning the relative numbers of nuclei present. Spin–spin coupling makes it possible to define the positional relationship between the nuclei since the magnitude of this interaction—*the coupling constant (J)*—depends upon the number and type of bonds separating them. The multiplicity of the resonance signals and the intensity distribution within the multiplet are, moreover, in simple cases as illustrated by the ethyl group of ethyl formate, clearly dependent upon the number of protons on the neighbouring group.

A further possible application of nuclear magnetic resonance spectroscopy is based on the observation that the n.m.r. spectra of many compounds are temperature dependent. One encounters such a case with dimethylformamide, the spectrum of which shows a doublet for the resonance of the methyl protons at 40 °C while at 160 °C a singlet is observed (Figure 3).

The cause of this different behaviour at the two temperatures is the high barrier to rotation about the carbonyl carbon–nitrogen bond (88 kJ mol^{-1}) which possesses

a

some double bond character as illustrated by the resonance form (a). The two methyl groups therefore have a relatively long life-time in different chemical environments, namely *cis* or *trans* to the carbonyl oxygen, and this makes it possible to record separate resonances. At higher temperatures the rate of internal rotation is increased and a frequent interconversion of methyl groups between chemically different positions results, so that we are obviously no longer able to differentiate between them.

It follows that for a number of molecules the line shape of the n.m.r. signal is dependent upon *dynamic processes* and the *rates* of such processes can be studied with the aid of nuclear magnetic resonance spectroscopy. What is even more significant is that one can study fast reversible reactions that cannot be followed by means of classical kinetic methods. Thus, the progress achieved in the fields of fluxional molecules and conformational analysis would have been unimaginable without n.m.r spectroscopy.

Nuclear magnetic resonance spectroscopy is also used successfully for the study

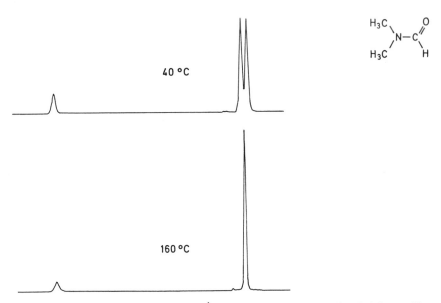

Figure 3 Temperature dependence of the ^{1}H n.m.r. spectrum of N,N-dimethyl formamide

of reaction mechanisms in all branches of chemistry. In these experiments, magnetic isotopes of hydrogen, carbon, or nitrogen (^{2}H, ^{13}C, ^{15}N), to name only a few elements, can be used in labelling experiments that are devised to follow the fate of a particular atom during the reaction of interest. Labelling with radioactive carbon, ^{14}C, can be replaced today in many cases by labelling experiments with the stable but n.m.r. active carbon isotope ^{13}C. Only where the highest sensitivity is indispensable does the use of the radiocarbon method still prevail.

For radical reactions a technique has become important which was initiated by the discovery that when the thermal decomposition of dibenzoyl peroxide in cyclohexanone was carried out directly in an n.m.r. spectrometer, the observed resonance signal of the benzene protons first appeared as an emission line rather than the expected absorption line. Figure 4 illustrates the results of this experiment as a function of time, t, and one can see that with the passage of time the initially produced emission signal decays and becomes a weak and finally a strong absorption signal.

$$C_6H_5-\underset{\underset{O}{\|}}{C}-O-O-\underset{\underset{O}{\|}}{C}-C_6H_5 \xrightarrow{\Delta,+2H^\bullet} 2\,C_6H_6 + 2\,CO_2$$

Such an observation can be considered as an indication of the transitory existence of radical pairs. The effect, known as *chemically induced dynamic nuclear polarization* (CIDNP), has been studied intensively.

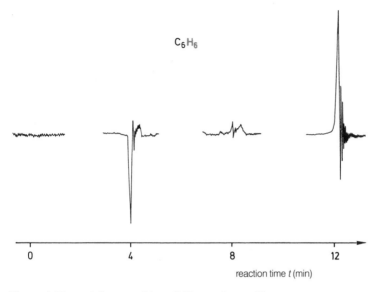

Figure 4 Thermal decomposition of dibenzoyl peroxide

The various aspects of the application of nuclear magnetic resonance to problems of inorganic, organic, and physical chemistry are supplemented by a remarkable variety of experimental techniques that lend a special position to n.m.r. spectroscopy in comparison with other spectroscopic methods. In addition to the versatile physics of the n.m.r. experiment, the variety of magnetic nuclei that are of significance to chemistry also contributes to this situation.

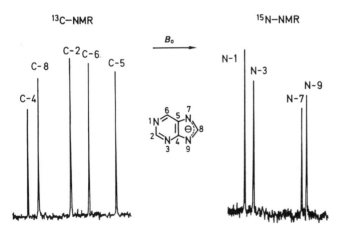

Figure 5 Carbon-13 and nitrogen-15 n.m.r. spectra of purine anion

In the fields of organic chemistry and biochemistry, ^{13}C n.m.r. plays a major role, but n.m.r. investigations of ^{19}F, ^{15}N and ^{31}P nuclei also yield valuable informations. As is demonstrated in Figure 5 with the ^{13}C and ^{15}N n.m.r. spectra of purine anion, the chemical shifts of these nuclei are sensitive to the chemical structure. Taking in addition information from proton n.m.r., each position in the molecule is labelled with a reporter that provides data about bonding, structure, and reactivity.

For inorganic chemistry a large number of metal nuclei are of interest and have become available for n.m.r. experiments due to the rapid development of the experimental techniques (Figure 6). Since nearly all elements of the Periodic Table contain a stable isotope with a magnetic moment, a large area is accessible for n.m.r. investigations, even if the natural abundance of many of these isotopes is rather small.

Another innovation of general importance is *high resolution n.m.r. spectroscopy of solids* which opened up new areas of structural research in inorganic and organic chemistry. Fast sample rotation and magnetization transfer from sensitive to insensitive nuclei—methods known as *magic angle spinning (MAS)* and *cross polarisation (CP)*—provide the basis for the measurement of chemical shifts and the study of dynamic processes even in solids.

The most spectacular development, which has revolutionized practically all branches of n.m.r. spectroscopy, is without doubt the concept of *two-dimensional (2D) n.m.r. spectroscopy*. Special techniques of impulse spectroscopy allow the recording of n.m.r. spectra with two independent frequency dimensions F_1 and F_2. The signals of such a 2D spectrum are characterized by a frequency pair f_1, f_2. In some experiments, the frequency axis F_2 only contains chemical shifts, while F_1 only contains spin–spin coupling constants. Both parameters are, therefore, separated by the 2D n.m.r. experiment. For practical purposes spectra with chemical shift data on both frequency axes are the most important because they

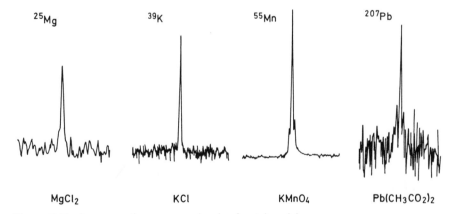

Figure 6 Nuclear magnetic resonance signals of metal nuclei

allow a shift correlation between resonance frequencies of different nuclei and in this way a spectral assignment. One distinguishes homo- and heteronuclear shift correlations because F_1 and F_2 can contain frequencies of the same nuclide, e.g. a proton n.m.r. spectrum, or of different nuclides, e.g. proton signals in F_1 and carbon-13 signals in F_2.

A homonuclear two-dimensional shift correlation, a so-called *COSY spectrum* (from *correlated spectroscopy*) is shown in Figure 7 for the protons of ethylformate. The new and important aspect is the observation of *cross peaks* which appear in addition to the normal spectrum recorded on the diagonal. Cross peaks, which have coordinates $F_1 \neq F_2$, indicate a spin–spin coupling between the nuclei that leads to the respective diagonal signals, which have the coordinates $F_1 = F_2$. The *contour*

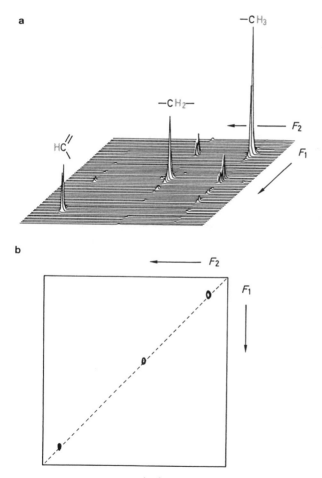

Figure 7   Two-dimensional ^{1}H,^{1}H COSY spectrum of ethylformate with the axes F_1 and F_2 and diagonal and cross peaks (red); (a) stacked plot; (b) contour plot

diagram shown in Figure 7b gives a particularly clear demonstration of the characteristic cross-peak positions.

COSY spectroscopy is important for the analysis of complex spectra with intensive signal overlap, where coupled nuclei cannot be recognized any more on the basis of simple multiplet structures. Other 2D n.m.r. spectra show cross peaks resulting from non-scalar interactions between nuclei which are close in space or which participate in a chemical exchange process. In this way information about atomic distances or the mechanism of intramolecular dynamic processes becomes available. Two-dimensional n.m.r. thus paved the way for the successful investigation of the structures of complex molecules like natural products and biopolymers from the field of proteins or nucleic acids. These examples may suffice to underline the enormous potential of the concept of 2D n.m.r. spectroscopy.

In summary, this short overview shows that n.m.r. spectroscopy is an indispensable tool for all branches of chemistry. Apart from that, the method has its place in other sciences like physics and, more recently, biology. Furthermore, with the development of the n.m.r. imaging techniques, the n.m.r. experiment has found new applications as an important diagnostic tool even in medicine.

UNITS

The *Système International* (SI), based on the metre, kilogram, second and ampere, is now accepted for all units of physico-chemical quantities. Accordingly, SI units have generally been used in the present text. In chemistry, however, the old centimetre, gram, second (CGS) system is still in use and, of course, older textbooks and research papers employed this system. It seems, therefore, necessary to point out some of the main changes that occur when SI units are used:

1. For the magnetic field we use the symbol B, the magnetic induction field or magnetic flux density. The former use of H is incorrect, since this symbolizes the magnetic field intensity. The SI unit for the magnetic induction field is the tesla (T), 10^4 times the electromagnetic unit, the gauss (G).

2. The SI unit for energy is the joule (J), and this replaces the calorie. Accordingly, activation energies are now given in kJ mol^{-1}, entropies in J K^{-1} mol^{-1} (4.184 times the numerical values in kcal mol^{-1} or cal K^{-1} mol^{-1}, respectively).

3. The SI system uses rationalized equations. In these, the factors 2π or 4π appear where expected on geometrical grounds, i.e. if the equation refers to situations where circular or spherical symmetry is involved.

4. The permeability of free space, μ_0, often appears explicitly in SI equations. Its value is $4\pi \times 10^{-7}$, and its dimension kg m s^{-2} A^{-2}.

1 THE PHYSICAL BASIS OF THE NUCLEAR MAGNETIC RESONANCE EXPERIMENT. PART I

In this chapter, an elementary presentation of the nuclear magnetic resonance experiment as applied to the proton is given and should suffice for the empirical and chemically routine application of the method, and as a preparation for the material in Chapters 2–6. A more detailed treatment of the physical principles is given in Chapter 7.

1. The quantum mechanical model for the isolated proton

The magnetic properties of the atomic nucleus form the basis of nuclear magnetic resonance spectroscopy. We know from nuclear physics that several nuclei, among them the proton, possess *angular momentum, P,* which in turn is responsible for the fact that these nuclei also exhibit a *magnetic moment, μ*. The two quantities are related through the expression

$$\mu = \gamma P \tag{1.1}$$

where γ, the *magnetogyric ratio*, is a constant characteristic of the particular nucleus.

According to quantum theory, angular momentum and nuclear magnetic moment are *quantized*, a fact that cannot be explained by arguments based on classical physics. The allowed values or *eigenvalues* of the maximum component of the angular momentum in the z direction of an arbitrarily chosen Cartesian coordinate system are measured in units of $\hbar$ ($h/2\pi$) and are defined by the relation

$$P_z = \hbar m_I \qquad (1.2)$$

where m_I is the *magnetic quantum number* which characterizes the corresponding *stationary* or *eigenstates* of the nucleus. According to the *quantum condition*

$$m_I = I, I - 1, I - 2, \ldots, -I \qquad (1.3)$$

these magnetic quantum numbers are related to the *spin quantum number, I,* of the respective nucleus. The total number of possible eigenstates or energy levels is thus equal to $2I + 1$.

The proton has a spin quantum number $I = \frac{1}{2}$ and the z component of its angular momentum, which is often called nuclear *spin**, is according to equation (1.2) given by

$$P_z = \pm \hbar I \qquad (1.4)$$

Consequently, the proton can exist in only two spin states, which are characterized by the magnetic quantum numbers $m_I = \frac{1}{2}$ and $m_I = -\frac{1}{2}$. Further, for the magnitude of the magnetic moment in the z direction, we have

$$\mu = \gamma \hbar m_I = \pm \gamma \hbar I = \pm \gamma \hbar / 2 \qquad (1.5)$$

The proton can therefore be pictured as a *magnetic dipole*, the z component, μ_z, of which can have a parallel or anti-parallel orientation with respect to the positive z direction of the coordinate system. Thus the direction of the vector μ is quantized (Figure 1.1a).

By analogy with the relations that describe the magnetic properties of electrons, equation (1.5) is sometimes expressed in the form

$$\mu_z = g_N \mu_N m_I \qquad (1.6)$$

Here g_N is the nuclear g factor which is characteristic of the ratio of the nuclear charge to the nuclear mass and μ_N is the nuclear magneton, the value of which can be calculated from

$$\mu_N = eh/4\pi m_P c \qquad (1.7)$$

where e is the electronic charge, h is Planck's constant, m_P is the mass of the proton, and c is the velocity of light, to 5.0505×10^{-27} JT^{-1}. From equations (1.5) and (1.6) it follows that $\gamma \hbar = g_N \mu_N$.

In quantum mechanics, an atomic system is described by means of *wave functions* that are solutions of the well known Schrödinger equation. For the purpose of the following discussion we introduce *eigenfunctions* α and β for the proton corresponding to the $m_I = \frac{1}{2}$ state and $m_I = -\frac{1}{2}$ state, respectively. In Chapter 5 we shall consider and describe in more detail the properties of these functions, since

---

* The eigenstates or stationary states of nuclei are therefore also called *spin states*.

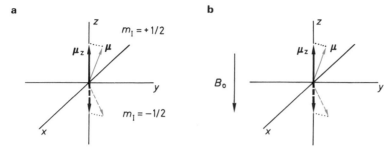

Figure 1.1   Nuclear magnetic moments in the magnetic field B_o

through them the energy of a spin system in a magnetic field can be determined. For the present they serve simply to identify the energy levels of the proton.

The α and β states for the nuclei of *spin* $\frac{1}{2}$ have the same energy, i.e. they are *degenerate*. Only in a static magnetic field B_0 is this degeneracy destroyed as a result of the interaction of the nuclear moment μ with B_0. When the direction of B_0 coincides with the z axis, as in Figure 1.1b, an energy difference

$$\Delta E = 2\mu_z B_0 \tag{1.8}$$

for the two spin states results, since the potential energy of a magnetic dipole in the field B_0 is $\mu_z B_0$ or $-\mu_z B_0$, depending on its orientation (Figure 1.2). The energy separation between the states produced in this way is proportional to the strength of the field B_0. It provides the necessary condition for the observation of a spectral line and thus forms the basis of the nuclear magnetic resonance experiment.

Because of its lower energy, the β state will be preferentially occupied and, according to the Bohr frequency condition, $\Delta E = h\nu$, we need an energy quantum

$$h\nu_0 = 2\mu_z B_0 = \gamma \hbar B_0 \tag{1.9}$$

or radiation of frequency

$$\nu_0 = \gamma B_0 \quad \text{or (with } \omega = 2\pi\nu) \; \omega_0 = \gamma B_0 \tag{1.10}$$

in order to stimulate a transition to the state of higher energy. Equation (1.10) describes the so-called *resonance condition*, where the radiation frequency exactly matches the energy gap. The n.m.r. signal corresponds to the arrow in Figure 1.2 and ν_0, the *Larmor frequency*, according to equation (1.10), varies with the strength of the B_0-field employed in the experiment. For protons with $\gamma_H = 2.675 \times 10^8 \; T^{-1} \; s^{-1}$ a field of 1.4 T yields $\nu_0 = 60$ MHz, which corresponds to a wavelength, λ, of 5 m, typical for radiowaves at the ultra-highfrequency end of the radiowave region. Routine spectrometers today operate at ^1H frequencies of 100 or 200 MHz with magnetic fields of 2.3 or 4.6 T, respectively.

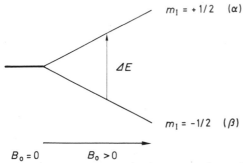

Figure 1.2   Energy separation between nuclear spin states

2. The experimental demonstration of quantized angular momentum and of the resonance equation (1.10)

It seems appropriate to mention here two experiments of outstanding significance that verify the existence of nuclear magnetic moments and illustrate their behaviour in a magnetic field as described above. They are the Stern–Gerlach experiment and the molecular beam experiment of Rabi.

Stern and Gerlach passed a stream of silver atoms, which, in the ground state, possess a total angular momentum of $\frac{1}{2}$, through a non-homogeneous magnetic field and found two discrete spots on the photographic plate used as a detector (Figure 1.3). The splitting of the beam of atoms is a direct consequence and a striking experimental documentation of the quantum nature of the magnetic energy of atoms. The magnetic moment of the individual silver atoms could be oriented either parallel or antiparallel to the external magnetic field, that is, an atom in the magnetic field would be either paramagnetic or diamagnetic. Paramagnetic and diamagnetic particles are, however, affected differently in a non-homogeneous magnetic field (Figure 1.3b). Because of the different field strengths at the dipole ends (illustrated by the density of the lines of force in the figure), one end of the dipole will be attracted or repelled more strongly than the other, resulting in a net accelerating force on the particle. If all orientations of the atomic moments relative to the magnetic field were allowed, as expected on the basis of classical theory, the experiment should yield a smear of silver atoms along a horizontal line. The observation of only two spots immediately tells us that only two distinct orientations, i.e. two discrete values of magnetic energy, exist.

The quantization of magnetic energy demonstrated in this experiment is the result of the splitting of electronic states but it is also valid for nuclear spin states. This was demonstrated by the experiments of Rabi and his co-workers, who investigated the behaviour of molecular beams in an apparatus illustrated schematically in Figure 1.4. Only molecules for which the total electronic magnetic moment was zero were used

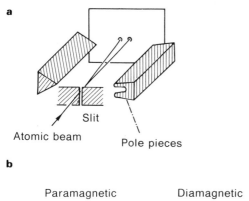

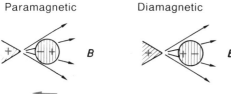

Figure 1.3 (a) Schematic representation of the Stern-Gerlach experiment. (b) The behavior of paramagnetic and diamagnetic particles in an inhomogeneous magnetic field; the arrows indicate the direction of motion

in these experiments so that any observable magnetic effect had to be ascribed to the magnetic properties of the nuclei.

In the experiment a molecular beam is directed obliquely between the pole pieces of a magnet A, which produces a non-homogeneous magnetic field and, as described above for the Stern–Gerlach experiment, the beam is split in two. Only the paramagnetic molecules following that path *a* pass through the slit into the homogeneous field of magnet B and they are finally focused by the magnetic field of

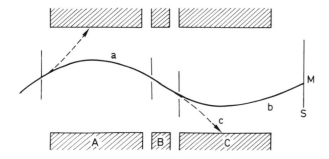

Figure 1.4 The principle of the experimental procedure for the detection of the resonance condition according to Rabi

magnet C, the inhomogeneity of which is exactly opposite to that of A. The screen S serves as a detector with which one can measure the intensity of the molecular beam focused at M. Now, if one irradiates the molecular beam in the region between the pole pieces of magnet B with radiofrequency radiation, there results at a particular frequency, depending upon the field strength of the magnet B, a sharp decrease in the intensity of the molecular beam at M. At that frequency/field strength ratio the resonance condition 1.10 is met, the orientation of part of the nuclear magnetic moments changes through the absorption of energy, and these diamagnetic particles are diverted by the effect of the non-homogeneous field C along path c rather than proceeding along path b at the detector M.

3. The nuclear magnetic resonance experiment on compact matter and the principle of the nuclear magnetic resonance spectrometer

The significance of the experiments of the Bloch and Purcell groups mentioned in the Introduction is that there the nuclear magnetic resonance experiment was performed for the first time on compact matter. In both solids and liquids magnetic nuclei are distributed between their energy states. For a very large number of protons, as for example exists in a macroscopic sample of hydrogen-containing material, the distribution of protons between the ground state and an excited state is given by the Boltzmann relation

$$\frac{N_\alpha}{N_\beta} = \exp\left(\frac{-\Delta E}{kT}\right) = \exp\left(\frac{-\gamma h B_0}{2\pi kT}\right) \approx 1 - \frac{\gamma h B_0}{2\pi kT} \qquad (1.11)$$

where N_β and N_α are the numbers of nuclei in the ground and excited states, respectively, ΔE is the energy difference between them, k is the Boltzmann constant, and T is the absolute temperature. Since ΔE in the above case is very small, the number of nuclei in the lower state at equilibrium is only slightly larger than the number of nuclei in the higher state ($N_\beta \gtrsim N_\alpha$). At a field strength of 1.4 T and room temperature, ΔE for a proton is about 0.021 J mol^{-1} and the population excess in the lower state, which determines the *probability* of a transition and in this way the *sensitivity* of the experiment amounts to only 0.001%. Relatively weak signals have thus to be detected in nuclear magnetic resonance spectroscopy.

The most important parts of the n.m.r. spectrometer are the magnet and the radio frequency (r.f.) transmitter and detector. The compound to be investigated is contained in a sample tube—a glass tube of approx. 15 cm length and 5 or 10 mm in diameter—in the external magnetic field B_0. A radio frequency coil in the probehead yields the r.f. radiation and the stimulated signal is detected either through the same coil or through a separate coil (single coil or cross coil type spectrometer). After amplification and transmission of the signal to an x, y plotter, the spectrum can be recorded and the resonance frequencies measured (Figure 1.5).

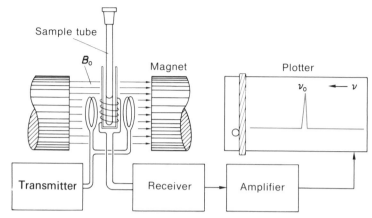

Figure 1.5 Schematic diagram of an n.m.r. spectrometer with separate transmitter and detector coil (cross coil arrangement)

Two independent experimental techniques are available today for the realization of an n.m.r. experiment: *CW* and *FT n.m.r. spectroscopy*. The basic principles of both methods will be discussed shortly with the help of Figure 1.6. The physical foundations will be treated in more detail in Chapter 7.

The acronym CW is used for *continuous wave* and means that during the recording of an n.m.r. spectrum the frequency v of a weak r.f. transmitter is varied continuously. The vector of the macroscopic magnetization of the sample, M (Figure 1.6a), which is the resultant of the individual nuclear magnetic moments μ_z defined in Figure 1.1, deviates during resonance ($v = \gamma B_0$) from its position on the z-axis through the combined action of the r.f. field and the magnetic field B_0 (Figure 1.6b). This creates an x, y component, the so-called *transverse magnetization*, which induces a signal in the detector coil L_E. The signal amplitude corresponds to the stationary state between nuclear excitation and relaxation, i.e. the return of the individual nuclei to the ground state. After resonance ($v > \gamma B_0$), the vector M returns to its position on the z-axis.

In the expression *FT n.m.r. spectroscopy* the letters FT stand for *Fourier transformation*. Here, nuclear excitation is achieved through an r.f. pulse, a strong radio frequency field (c. 50 Watt) of short duration (typically 10–50 μs). The vector M will also be turned away from the z-axis in the direction of the y-axis. However, after the pulse, r.f. radiation ceases and only the magnetic field B_0 acts upon M which starts a precession around the z-axis with the Larmor frequency characteristic of the particular nucleus (Figure 1.6d). The time signal induced in the detector coil through the precessional motion of the x, y component of M, $S(t)$, fades away through relaxation (Figure 1.6e). Its Fourier transformation yields the frequency signal $S(v)$, which is identical with the CW signal. This procedure for the measurement of n.m.r. spectra is also known as *pulse Fourier transform (PFT) n.m.r. spectroscopy*.

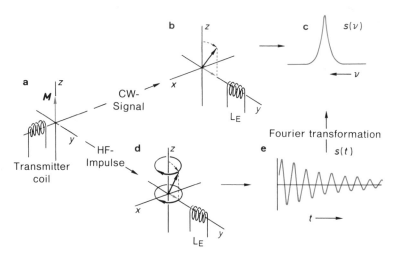

Figure 1.6 N.m.r. signal generation in CW and FT spectroscopy

Despite the fact that the FT method yields the spectrum only via the time signal *S(t)*, it has, if compared to the CW method, a very important advantage. As we shall show in more detail in Chapter 7, the complete n.m.r. spectrum can be excited with a single pulse and the corresponding time signal recorded within one second. On the other hand, a CW spectrometer needs 250 s for recording the spectrum since every signal has to be measured separately. This difference in measuring time was the main factor which led to a complete replacement of CW by PFT n.m.r. spectrometers.

Thus, the nuclear magnetic resonance spectrometer possesses all the elements that we encounter in optical spectrometers: *radiation source, sample cell*, and *detector*. We must, however, point out a few significant differences.

One is that the sample must be in a strong magnetic field before energy can be absorbed. Another difference is that a prism for scanning the spectrum is not needed because the radiation generated by the CW transmitter is monochromatic. In the FT experiment on the other hand, the radiation is polychromatic because a radio-frequency field generated by an r.f. pulse has a broad frequency spectrum, but the analysis of the receiver signal and its separation into individual resonance signals is achieved via Fourier transformation. A powerful computer is thus an important part of an FT n.m.r. spectrometer.

Moreover, in n.m.r. spectroscopy there is a dramatic difference between the spectra of *solids* on the one hand and those of *liquids or solutions* on the other. In solids the rigid orientation of the nuclei with respect to the external magnetic field B_0 as well as to their neighbouring nuclei leads to mechanisms which cause severe line broadening. For example, the variation, ΔB, in the local magnetic field caused by the mutual interaction of two nuclear magnetic moments separated by a distance r, oriented at an angle θ to the direction of B_0, is given by

$$\Delta B = \pm \tfrac{3}{2}\mu(3\cos^2\theta - 1)r^{-3}\frac{\mu_0}{4\pi} \qquad (1.12)$$

where μ_0 is the permeability in free space. In a solid, the magnetic field therefore varies from place to place and the spectra of solids are characterized by lines which are several kHz wide and generally not easily analysed.

In a liquid the factor $3\cos^2\theta - 1$ becomes zero because of the random thermal translational and rotational motions of the molecules, a fact that can easily be derived if the time average over $3\cos^2\theta - 1$ is replaced by the average obtained from $\Sigma_{x,y,z}(3\cos^2\theta_{x,y,z} - 1)/3$. The *dipolar interactions* between nuclei therefore cancel. Only in this situation high-resolution spectra with discrete resonance signals and line widths smaller than 1 Hz result. One speaks, therefore, of *high-resolution* n.m.r. spectroscopy.

It is interesting to note, however, that the dipolar interaction between nuclei also cancels, according to equation (1.12), if $\theta = 54.7°$. Thus, if one very rapidly rotates the solid under examination, mostly a crystalline powder deposited in a so-called *rotor*, around an axis that forms the 'magic' angle of 54.7 with the direction of the external field, one can eliminate the perturbing interaction since all lines connecting magnetic moments would have the angle $\theta = 54.7°$ as an *average* value. This technique—called magic angle spinning (MAS)—forms the basis for a new branch of n.m.r. spectroscopy treated in more detail in Chapter 10: *high-resolution n.m.r. of solids*.

Up to now we have concentrated our discussion of the n.m.r. experiment on the process by which energy is absorbed. The equilibrium distribution of nuclei between spin states expressed in equation (1.11) presupposes, however, that excited nuclei can return to the lower spin state since, otherwise, the population difference in the two states would tend to zero and the system would be *saturated*. The phenomenon by which the energy of the excited state is transferred to the environment as thermal energy and the nucleus reverts to the lower spin state is called *relaxation*, mentioned already above in connection with our description of signal generation. We will consider this phenomenon in detail in Chapter 7. For the moment we will keep in mind that relaxation is as vital as absorption for the success of an n.m.r. experiment.

As it was inferred above, the resonance frequency for protons lies in the region of radiowaves. In the electromagnetic spectrum, then, and in the series of well known spectrometric methods, nuclear magnetic resonance spectroscopy takes its place at the long wave-length end (Figure 1.7).

Experimentally the resonance condition (1.10) can be met in two ways if a CW experiment is performed: at fixed field strength the frequency can be varied (*frequency sweep*) or the field strength can be varied while the stimulating radiation is maintained at a constant frequency (*field sweep*). In the FT experiment there is no alternative because pulse excitation always occurs at a constant field. All spectra are reproduced in such a way that the frequency increases from right to left.

In addition to the electromagnets that have already been mentioned, permanent magnets are also employed in n.m.r. spectrometers. With these their greater sensitivity to environmental influences (temperature variations and background magnetic fields) is a disadvantage: however, one can dispense with the water-cooling accessory that is required with electromagnets to dissipate the heat they generate.

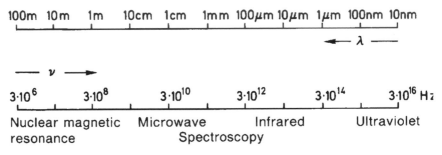

Figure 1.7   Electromagnetic spectrum

Both types of magnets yield a maximum magnetic field strength of 2.3 T. Therefore, so-called *cryo-* or *superconducting magnets* are used today in the majority of n.m.r. spectrometers. Here the magnetic field is generated by a superconducting coil which is held at the temperature of liquid helium. With this technique one can achieve much higher magnetic fields—up to 17.5 T—and, following equation 1.11, the sensitivity of the n.m.r. experiment can be considerably improved. Further advantages of these types of spectrometers will become apparent later. Since the cooling of the superconducting coil requires liquid nitrogen and liquid helium, the costs of maintenance for these instruments are higher than for instruments employing permanent magnets. They are, however, less than those for instruments with electromagnets, since due to an improved dewar technique the helium consumption is rather low (typically ca. 0.5 1/24 h or less).

4. The magnetic properties of other nuclei

As we have mentioned in the Introduction, not all nuclei possess magnetic moments. Furthermore, in the case of nuclei heavier than the proton, spin quantum numbers greater than $\frac{1}{2}$ are possible. The spin states of such nuclei are characterized, according to equation 1.2, by the magnetic quantum numbers $m_I = I, I - 1, I - 2, \ldots, -I$ and the energy level diagram for the deuteron ($I = 1$), for example, has the appearance illustrated in Figure 1.8.

Generally, for nuclei with even mass and atomic numbers, the even–even nuclei, $I = 0$ and for all other nuclei $I \geqq \frac{1}{2}$. I is an integral multiple of 1 for even–odd nuclei and for odd–odd and odd–even nuclei it is an integral multiple of $\frac{1}{2}$. The nuclei important to organic chemists are listed in Table 1.1, with their properties that are relevant to nuclear magnetic resonance. In particular, in addition to the n.m.r. spectroscopy of the proton, the spectra of such nuclei as ^{19}F, ^{13}C, ^{15}N, ^{29}Si, and ^{31}P have been extensively investigated.

One can further see from Table 1.1 that all nuclei with $I > \frac{1}{2}$ possess a *nuclear quadrupole moment* as a result of non-spherical distribution of nuclear charge. These nuclei can, therefore, interact with electrical field gradients in the environment—

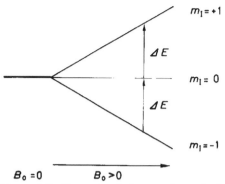

Figure 1.8   Energy levels for a nucleus of spin quantum number $I = 1$

especially those due to the electron shell in the molecule in which the nucleus is situated—and these interactions are of significance in relaxation phenomena. Further, because of the aforementioned interaction, even in the absence of an externally applied magnetic field, such nuclei possess spin states of different energies between which transitions can be stimulated. The stimulation and detection of these transitions is known as *nuclear quadrupole spectroscopy.*

The *sensitivity* of a nucleus to investigation by an n.m.r. experiment depends on the magnitude of its magnetic moment μ, which determines the energy difference

Table 1.1   Nuclear properties of nuclei important for the n.m.r. spectroscopy of organic compounds

Nucleus	Spin quantum number, I	Magnetic moment, μ (units of μ_N)	Magnetogyric ratio, γ (10^8 rad T^{-1} s^{-1})	Resonance frequency ν_0 (MHz at a field of 1 T)	Relative sensitivity at constant field	Natural abundance (%)	Quadrupole moment, Q ($e \times 10^{-28}$ m^2)
^{1}H	$\frac{1}{2}$	2.79277	2.675	42.577	1.000	99.98	0.003
^{2}H	1	0.85735	0.411	6.536	0.009	0.0156	0.111
^{10}B	3	1.8007	0.288	4.575	0.02	18.83	0.036
^{11}B	$\frac{3}{2}$	2.6880	0.858	13.660	0.165	81.17	—
^{13}C	$\frac{1}{2}$	0.70216	0.673	10.705	0.016	1.108	0.02
^{14}N	1	0.40369	0.193	3.076	0.001	99.635	—
^{15}N	$\frac{1}{2}$	−0.28298	−0.271	4.315	0.001	0.365	−0.004
^{17}O	$\frac{5}{2}$	−1.8930	−0.363	5.772	0.029	0.037	—
^{19}F	$\frac{1}{2}$	2.6273	2.517	40.055	0.834	100.0	—
^{29}Si	$\frac{1}{2}$	−0.55492	−0.531	8.460	0.079	4.70	—
^{31}P	$\frac{1}{2}$	1.1316	1.083	17.235	0.066	100.0	—

between the nuclear spin states and therefore, following equation (1.11), the population excess in the lower energy state. It can be shown that at constant field the signal strength should be proportional to

$$\frac{I+1}{I^2}\mu^3 B_0^2 \qquad (1.13)$$

an expression that also demonstrates the importance of strong fields, B_0. In practice the factor $B_0^{3/2}$ is found. In addition the *natural abundance* is a critical factor. The n.m.r. spectroscopy of ^{13}C and ^{15}N was thus severely hampered in early years by the low concentration of these nuclei in molecules with natural isotopic distribution and it is only since the recent introduction of the *Fourier transform technique* that this problem has been overcome.

Aside from those nuclei listed in Table 1.1 a large number of other nuclei have been detected by n.m.r. In practice, nearly all elements of the Periodic Table have an isotope which is n.m.r. active and the resonance of which can be measured. Because of the sensitivity problem mentioned already above, but also because of the sometimes large quadrupole moments for nuclei with $I > \frac{1}{2}$ which lead to fast relaxation and thus a shortening of the life time in the excited state, such measurements are not routine in all cases.

We can summarize the foregoing discussion of n.m.r. spectroscopy as follows. The nuclear magnetic resonance experiment allows us to record the resonance signals of magnetic nuclei. Each type of nucleus is thereby characterized by its

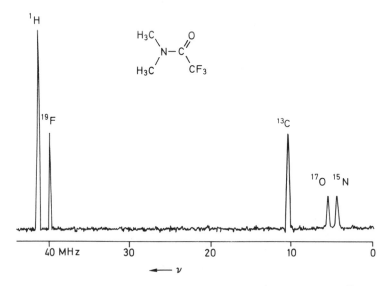

Figure 1.9 Hypothetical nuclear magnetic resonance spectrum of [^{15}N]dimethyltrifluor-oacetamide with the magnetic nuclei ^{15}N, ^{17}O, ^{13}C, ^{19}F, and ^{1}H in a field of 1.0 T

individual resonance frequency. In a theoretical experiment for a hypothetical molecule such as $[^{15}N]$dimethyltrifluoroacetamide we might therefore expect the spectrum shown in Figure 1.9. For the purpose of simplifying the illustration, all nuclei were assumed to have the same sensitivity and the same natural abundance.

2 THE PROTON MAGNETIC RESONANCE SPECTRA OF ORGANIC MOLECULES

Up to now we have been concerned with the magnetic resonance of a single nucleus and with explaining the physical basis of an n.m.r. experiment. We will now turn our attention to the nuclear magnetic resonance spectra of organic molecules and in so doing will encounter two new phenomena: the *chemical shift of the resonance frequency* and the *spin–spin coupling*. These two phenomena form the foundation for the application of nuclear magnetic resonance spectroscopy in chemistry and related disciplines. They will be treated in the following sections.

1. The chemical shift

The hypothetical spectrum of [^{15}N]dimethyltrifluoroacetamide presented at the end of Chapter 1 may have suggested that n.m.r. spectroscopic is employed for the detection of magnetically different nuclei in a compound. For at least two reasons this is not the case. Firstly, experimental considerations make such an application

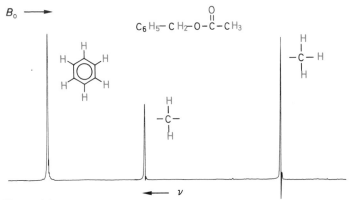

Figure 2.1

difficult, if not impossible, since conditions and techniques must be modified to measure the resonance frequencies of different nuclei. Secondly, the elemental composition of organic compounds can be determined far more easily and accurately by other techniques such as elemental analysis or mass spectrometry.

The significance of n.m.r. spectroscopy in chemistry is therefore not based on its ability to differentiate between elements, but on its ability to distinguish a particular nucleus with respect to its environment in the molecule. That is, one finds that the resonance frequency of an individual nucleus is influenced by the distribution of electrons in the chemical bonds of the molecule. The value of the resonance frequency of a particular nucleus is therefore *dependent upon molecular structure*.

Using the proton to demonstrate this, means that a compound such as benzyl acetate, for example, will produce three different n.m.r. signals, one each for the protons of the phenyl, methylene, and methyl groups (Figure 2.1). This effect, produced by the different chemical environments of the protons in the molecule, is known as the chemical shift of the resonance frequency or more simply as the *chemical shift*. Thus, with an applied magnetic field of 1.4 T, the proton resonances of a molecule do not occur at $v_0 = 60$ MHz but rather at $v_0 \pm \Delta v$, where Δv in general is less than 1 kHz. Other magnetic nuclei are affected similarly and this phenomenon forms the basis of applied n.m.r. spectroscopy.

In a first analysis, the chemical shift is caused by the electrons of the C–H bond in which the proton is involved. The applied magnetic field, B_0, induces circulations in the electron cloud surrounding the nucleus such that, following Lenz's law, a magnetic moment μ, opposed to B_0, is produced (Figure 2.2). Thus, at the proton the applied field strength, B_0, does not prevail or, in other words, the local field at the nucleus is smaller than the applied field. This effect corresponds to a magnetic shielding of the nucleus that reduces B_0 by an amount equal to σB_0 where σ is known as the *shielding or screening constant* of the particular proton:

$$B_{\text{local}} = B_0(1 - \sigma) \qquad (2.1)$$

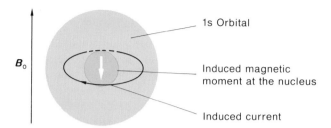

1s Orbital

Induced magnetic
moment at the nucleus

Induced current

Figure 2.2 Shielding of a proton in a magnetic field

The shielding constant, σ, is proportional to the electron density of the 1s orbital of the hydrogen atom and σB_0 is the magnitude of the secondary field induced at the proton. This magnetic shielding has the effect that a higher external field is required to meet the resonance condition in an experiment in which the field is varied, while when operating in the frequency sweep mode at a constant field, B_0, the resonance condition is met at a lower frequency than might be expected.

In the case of an unperturbed spherical electron distribution as exists, for example, in a hydrogen atom, the induced movement of charge leads to a pure diamagnetic effect. The shielding constant, σ, may then be calculated by the Lamb formula (equation 2.2) from the electron density $\rho(r)$ around the nucleus:

$$\sigma = \frac{\mu_0 e^2}{3m_e} \int_0^\infty r\rho(r)\mathrm{d}r \tag{2.2}$$

$\rho(r)$ itself is a function of the distance r from the nucleus and the other terms are well known constants.

In molecules the situation is more complex, for here one must consider the electronic circulation within the entire molecule. In these cases it can be shown that the perturbation of the spherical symmetry of electron distribution caused by the presence of other nuclei reduces the diamagnetic effect. This diminution can be treated as corresponding to a paramagnetic moment that strengthens the external field B_0. The value of σ in molecules then corresponds to the sum of diamagnetic and paramagnetic components of the induced electronic motion:

$$\sigma = \sigma_{\text{dia}} + \sigma_{\text{para}} \tag{2.3}$$

Theoretical calculations of chemical shifts rely on advanced quantum chemical methods and are often restricted to small molecules. A practical approach for the treatment of the chemical shift in cases that are of immediate interest to the chemist is possible if the shielding is separated into contributions of individual atoms and groups which, in turn, can be evaluated with the assistance of simple models. These contributions may be represented by a third term and from equation (2.3) it follows that

$$\sigma = \sigma_{\text{dia}}^{\text{local}} + \sigma_{\text{para}}^{\text{local}} + \sigma' \tag{2.4}$$

where $\sigma_{\text{dia}}^{\text{local}}$ and $\sigma_{\text{para}}^{\text{local}}$ are now the *local diamagnetic* and *local paramagnetic* contributions to the shielding constant of the respective nucleus.

For protons, $\sigma_{\text{dia}}^{\text{local}}$ and σ' are of primary significance since theoretical calculations show that strong paramagnetic effects arise only for heavier nuclei where energetically low-lying atomic orbitals are available and a mixing of ground state and excited state wave functions takes place through the external magnetic field. Thus, in the case of fluorine, for example, the paramagnetic contribution to the chemical shift is essentially the result of the availability of low-lying p orbitals.

This introduction to the expected dependence of the shielding constant on chemical structure will be considerably expanded in Chapter 4. Here we simply assert that for protons in molecules σ is always *positive*. For example, the semi-empirically determined shielding constant of the protons in a hydrogen molecule is 26.6×10^{-6} and for the local magnetic field at the nucleus the relation $B_{\text{local}} < B_0$ holds. These nuclei are shielded and their resonance signals are recorded at higher field strengths (or at lower frequencies) than that of an isolated proton. For individual compounds, protons of certain groups are designated as *shielded* or *deshielded* relative to a reference line if their absorptions occur at higher or lower field. This can be seen in the spectrum of benzyl acetate (Figure 2.1), where relative to the methylene protons the methyl protons are shielded and the aromatic protons are deshielded. In cases such as this one often speaks of a *diamagnetic* or a *paramagnetic* shift, respectively, of the resonance signals.

1.1   CHEMICAL SHIFT MEASUREMENTS

The discussion of the differential shielding of individual protons assumes that a system of measurement for the chemical shift has been established. In order to specify the position of a resonance signal in an n.m.r. spectrum it would, in principle, be possible to measure the strength of the external field, B_0, or the resonance frequency, v, at which the resonance line of interest appears. These parameters are, however, unsuited for the characterization of the chemical shift because n.m.r. spectrometers operate at different B_0 fields (for example 1.4, 2.3, or 5.2 T) and according to equation (1.10) the resonance frequency varies with field strength. Furthermore, an absolute determination of the field strength or the resonance frequency is technically difficult to obtain.

Hence one is obliged to measure the position of the resonance signal *relative* to that of a *reference compound* or *standard*. In proton n.m.r. the compound used under normal circumstances is *tetramethylsilane (TMS)*, the twelve protons of which give a sharp signal that is always recorded simultaneously with the spectrum of the sample under investigation. Thus, benzyl acetate, in a spectrometer employing a frequency of 60 MHz, gives rise to the spectrum shown in Figure 2.3a.

The next step involves a *calibration* which in a CW experiment is accomplished by modulating the reference signal with a known frequency, for example 500.0 Hz. A *side band* can be recorded within this interval against which the calibration of the spectrum can be determined. In this fashion we obtain the relative separations of the

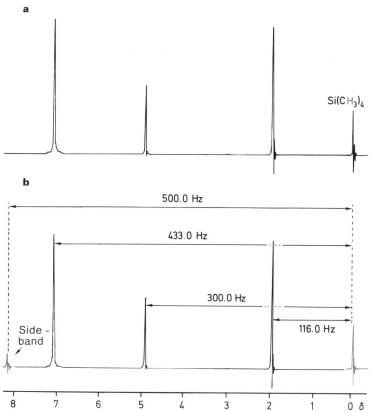

Figure 2.3 (a) Proton magnetic resonance spectrum of benzyl acetate in the presence of tetramethylsilane as an internal standard. (b) Proton magnetic resonance spectrum of benzyl acetate with side band callibration; 60 MHz

proton n.m.r. signals of benzyl acetate from the reference signal in units of frequency (Hz); (Figure 2.3b). In an FT n.m.r. experiment the line frequencies are directly available after data processing.

As frequencies in the order of Hz or kHz are easily measured with considerable precision, the determination of the chemical shift is straightforward. Nevertheless, the data thus obtained have the disadvantage that their values, according to equation (1.10), are field dependent. Recording of the spectrum at 2.3 T with a radio-frequency of 100 MHz would yield signals at 721, 500, and 193 Hz for the three groups of protons in benzyl acetate. Therefore, a dimensionless quantity has been introduced for the chemical shift that is defined as follows:

$$\delta = \frac{\nu_{\text{substance}} - \nu_{\text{reference}}}{\nu_0} \qquad (2.5)$$

Here, ν_0 is the operating frequency of the spectrometer employed (for example 60 MHz) and as units for the δ *scale* one uses parts per million (ppm). Thus, for the proton resonances of benzyl acetate δ values of 7.21, 5.00, and 1.93 ppm are found, regardless of whether the spectrum is measured at 1.4 T and 60 MHz or 2.3 T and 100 MHz. The relations between the frequency scale in Hz and the δ-scale in ppm are again illustrated in the following diagram for two typical spectrometer frequencies:

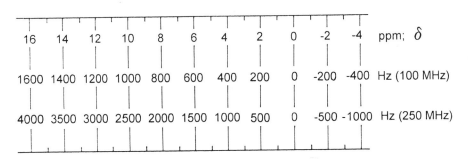

The absence of an absolute energy scale makes the comparison of n.m.r. spectra difficult if agreement cannot be reached upon a universal reference. The previously mentioned tetramethylsilane fulfils the requirements that such a substance must meet. The TMS signal is an intense singlet, the chemical shift of which is different from most of the other proton resonances so that the superposition of a sample resonance signal with that of TMS is seldom observed. The substance is essentially chemically inert and it can easily be removed from the sample after recording the spectrum. The δ scale of proton magnetic resonance is thus based on this reference compound. Furthermore, TMS is also used as reference in carbon-13 n.m.r. spectroscopy, where it gives rise to a ^{13}C n.m.r. signal at high field.

The recording of spectra with modern spectrometers is, in fact, highly automated. Either pre-calibrated chart paper can be used or the calibration is directly provided while plotting the spectra. Of course, spectral regions of interest can be recorded in great detail with a large scale, e.g. 1 Hz/cm or more.

In the original literature, δ values are often given which, for technical reasons, refer to other reference compounds such as cyclohexane, methylene chloride, or benzene. The need to employ a different reference becomes obvious when the sample being investigated has a resonance signal that is superimposed on the TMS signal. In such cases one usually labels the δ value with an identifying subscript, for example, $\delta_{C_6H_6}$. This kind of notation is especially important in the n.m.r. spectroscopy of heavier nuclei where general agreement on a universal standard has in many cases not been reached.

Individual contributions to the chemical shift of proton resonances caused by substituents will be symbolized in this text as $\Delta\sigma$ (ppm) for they represent changes in the shielding constant. A positive sign signifies an increase and a negative sign a decrease in the shielding.

1.2   THE INTEGRATION OF THE SPECTRUM

With the establishment of the δ scale, we are now able to assign the protons in a particular structural element to a definite region in the spectrum. Before proceeding, one more property of the n.m.r. spectrum that may have already been noticed in reference to Figure 2.1 should be mentioned. The signals have different *intensities*. A more detailed examination shows that the area under a resonance signal is proportional to the number of protons that gives rise to that signal. An electronic *integrator* built into the spectrometer, automatically produces the step curve shown in Figure 2.4. The relative heights of the steps in the curve indicate a proton ratio in benzyl acetate of 5:2:3. This measurement yields valuable and often crucial additional information.

An important point to note here is the fact that only the *relative number* of protons can be determined by integration. Thus, were it not for the chemical shift differences, ethyl formate and diethyl malonate, to choose only one example, would have identical spectra:

$$HCOOCH_2CH_3 \qquad\qquad \begin{array}{c} H \\ \\ H \end{array}\!\!\!C\!\!\!\begin{array}{c} COOCH_2CH_3 \\ \\ COOCH_2CH_3 \end{array}$$

The integration of resonance signals finds important *applications* in *analytical chemistry*, where it is employed to determine the constitution of mixtures or the percentage of an impurity present. For example, the mass m_A of component A in a mixture can be determined if an amount m_B of a known substance B is added to a

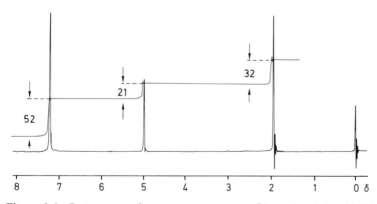

Figure 2.4   Proton magnetic resonance spectrum of benzyl acetate with integration; step heights indicated in millimetres

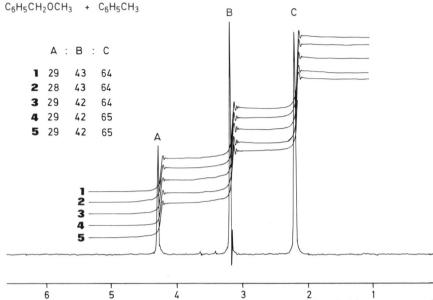

$C_6H_5CH_2OCH_3$ + $C_6H_5CH_3$

A	:	B	:	C
1	29	43	64
2	28	43	64
3	29	42	64
4	29	42	65
5	29	42	65

Figure 2.5 Proton magnetic resonance spectrum of a mixture of methyl benzyl ether and toluene in the region from δ 0 to 6

weighed sample of the mixture and signals assigned to A and B are integrated. In this case, the mass of A in the sample is given by

$$m_A = m_B \cdot \frac{N_B}{N_A} \cdot \frac{A_A}{A_B} \cdot \frac{M_A}{M_B} \tag{2.6}$$

where N is the number of protons responsible for the signals chosen, A is the area under the signals and M_A and M_B are the molecular weights of A and B, respectively.

In order to illustrate the application of the integration of n.m.r. signals, Figure 2.5 shows the spectrum of a mixture of methylbenzyl ether and toluene in a molar ratio of 1:1.491 (as determined by weighing). From the integration of the two methyl signals at δ 2.2 and 3.2, one obtains, as an average of five integrations, a molar ratio of 1:1.519 $\pm$ 0.030. The correct value lies within the error limit and the magnitude of the error (2%) is typical of the magnitude of errors in this type of determination (2–4%). Figure 2.5 also shows that, because of differences in line widths in the methyl resonances B and C, the *signal height* is not an accurate measure of the number of protons responsible for a signal.

Exercise 2.1 Determine the molar ratio of methyl benzyl ether to toluene by reference to the integration given in Figure 2.5 for the methylene protons of methyl benzyl ether and the methyl portons of toluene.

1.3   THE DEPENDENCE OF THE RESONANCE FREQUENCY ON STRUCTURE — A GENERAL SURVEY

Thanks to the chemical shift, nuclear magnetic resonance spectroscopy yields important data which, like the group frequencies in infrared spectroscopy, are used for the determination of the structures of unknown substances. The *characteristic absorption ranges* for the most important types of protons that are present in organic molecules are indicated in Figure 2.6.

The following general statements can be made. For *aliphatic* C–H bonds the shielding decreases in the series $CH_3 > CH_2 > CH$. While the protons of methyl groups at saturated centres absorb at δ 0.9, the resonance for the protons of cyclohexane occurs at δ 1.4. An exception is observed in the case of cyclopropane, the protons of which absorb at δ 0.22. For *olefinic* protons, the resonances lie in the region from δ 4.0 to 6.5, and only in special instances, such as with compounds like acrolein ($CH_2 = CHCHO$), below δ 6.5. The resonance signals of protons in *aromatic* molecules occur in a characteristic region between δ 7.0 and 9.0. Although sp^2-hybridized bonds are present, as in the olefins, an additional deshielding obviously exists here. One observes the opposite effect in acetylenes where the *acetylenic* proton absorbs at δ 2.9. Electronegative elements such as nitrogen, oxygen, and halogens produce paramagnetic shifts of the resonances of neighbouring protons and neighbouring multiple bonds have the same effect. The resonance signals of *aldehyde* and *carbocylic acid* protons are found at very low field.

The n.m.r. signals of the protons of OH, NH, NH_2, and CO_2H groups deserve special consideration for their position is strongly dependent upon concentration, temperature, and the solvent employed. For example, the OH signal is observed at δ 1.4 in purified methanol and at δ 4.0 in an impure sample. As will be explained in greater detail later, the cause of this difference lies in the ability of the OH group to form hydrogen bonds. The presence of traces of acid or water promotes exchange processes and these result in a shift of the proton resonance frequency. In addition, the shape of the resonance signal can also be changed by this chemical exchange and protons of COOH, NH_2, and NH groups in particular very frequently show broad resonance signals that sometimes are hidden in the noise (Figure 2.7). To complete this preliminary survey, the δ values of certain protons in a series of small organic molecules that are characteristic representatives of several classes of compounds are collected in Table 2.1.

Exercise 2.2   The proton resonance spectrum of a mixture of acetone, methylene chloride and chloroform is integrated and results in step heights of 10, 18, and 36 mm for the signals at δ 7.27, 5.30, and 2.17, respectively. In what molar ratio are the three substances present?

Exercise 2.3   Figure 2.8 shows the integrated proton n.m.r. spectrum of a mixture of toluene, benzene, and methylene chloride. Assign the resonances with the help of Table 2.1 and determine the molar ratio of the three compounds.

24

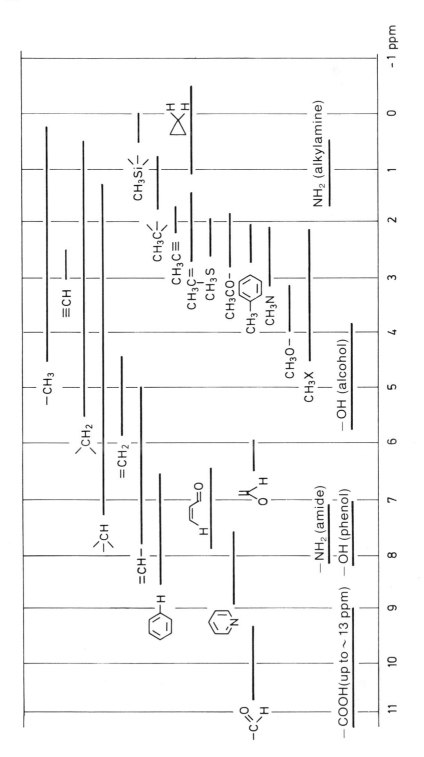

Figure 2.6 δ-Scale of chemical shifts of proton resonances in organic compounds

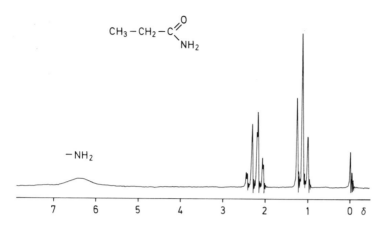

Figure 2.7 Proton magnetic resonance spectrum of propionamide

Table 2.1 Proton resonances (δ values) of selected organic compounds

		δ
Cyclopropane	C_3H_6	0.22
Ethane	CH_3-CH_3	0.88
Ethylene	$CH_2=CH_2$	5.84
Acetylene	$HC\equiv CH$	2.88
Benzene	C_6H_6	7.27
Propene	$CH_2=CH-CH_3$	1.71
Propyne	$CH\equiv C-CH_3$	1.80
Acetone	$CH_3-CO-CH_3$	2.17
Cyclohexane	C_6H_{12}	1.44
Methyl chloride	CH_3Cl	3.10
Methylene chloride	CH_2Cl_2	5.30
Chloroform	$CHCl_3$	7.27
Ethanol	CH_3CH_2OH	1.22
	CH_3CH_2OH	3.70
	CH_3CH_2OH	2.58
Acetic acid	CH_3-COOH	2.10
	CH_3-COOH	8.63
Acetaldehyde	CH_3-CHO	2.20
	CH_3-CHO	9.80
Diethyl ether	$(CH_3CH_2)_2O$	1.16
	$(CH_3CH_2)_2O$	3.36
Ethyl acetate	$CH_3COOCH_2CH_3$	2.03
	$CH_3COOCH_2CH_3$	1.25
	$CH_3COOCH_2CH_3$	4.12
Trimethylamine	$N(CH_3)_3$	2.12
Triethylamine	$N(CH_2CH_3)_3$	2.42
Toluene	$C_6H_5-CH_3$	2.32
Benzaldehyde	C_6H_5-CHO	9.96

Exercise 2.4 For an unknown substance, signals are observed at δ 2.32 and 7.10 (area ratio 3:5) at a radiofrequency of 100 MHz. How large would the chemical shift difference (in Hz) be if the spectrum of the same substance were measured with a 60-MHz spectrometer? What compound fits these data (the molecular formula of C_7H_8)?

Exercise 2.5 What differences would be expected in the n.m.r. spectra of the following pairs of isomers with respect to the position and areas under the resonance signals?

$$\textbf{a}\quad \underset{\underset{CH_3}{|}}{\overset{\overset{CH_3}{|}}{H_3C-C}}-O-CH_2-C_6H_5 \qquad\qquad \textbf{a'}\quad H_3C-O-\underset{\underset{CH_3}{|}}{\overset{\overset{CH_3}{|}}{C}}-C_6H_4-CH_3$$

$$\textbf{b}\quad \underset{\underset{CH_3}{|}}{\overset{\overset{CH_3}{|}}{H_3C-C}}-C\equiv C-CH_3 \qquad\qquad \textbf{b'}\quad \underset{\underset{CH_3}{|}}{\overset{\overset{CH_3}{|}}{H_3C-C}}-CH_2-C\equiv C-H$$

Exercise 2.6 Using the proton magnetic resonance spectra in Figure 2.9, determine the structure of compounds a–j. In addition to the molecular formula of each compound the relative areas of the signals are given. Note: the solution of the problems is facilitated if one first determines the double bond equivalents each compound. For the compound $C_aH_b(O_c)$ the double bond equivalents (DBE) are given by

$$DBE = \frac{(2a+2)-b}{2}$$

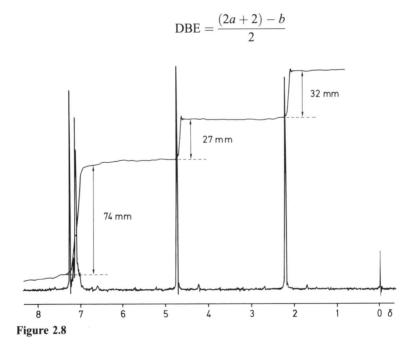

Figure 2.8

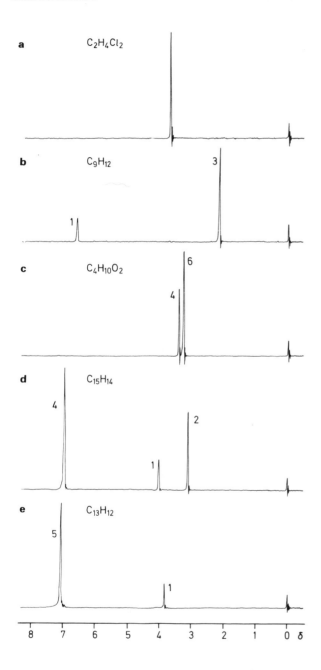

Figure 2.9 (*continued overleaf*)

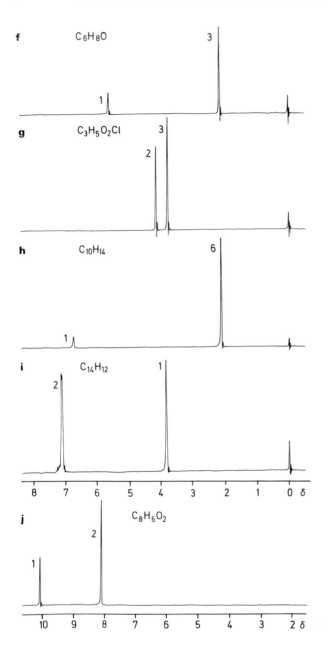

Figure 2.9 (*continued*)

2. Spin–spin coupling

In comparing the spectrum of benzyl acetate with that of ethyl formate, as is done in Figure 2.10, we notice not only a difference in the position of the resonance signals but also a difference in the *multiplicity* of the signals. In one case singlets are observed for both the methyl and methylene protons, and in the other the same protons give rise to a triplet and a quartet, respectively, each with a rather distinct intensity distribution. The cause of this fine structure is *spin–spin coupling*. It is brought about by a magnetic interaction between individual protons that is not transmitted through space but rather by the bonding electron through which the protons are indirectly connected.

As Figure 2.11 shows schematically for the hydrogen fluoride molecule and the protons of a methylene group, the magnetic moment of nucleus A causes a weak magnetic polarization of the bonding electrons that is transmitted by way of the overlapping orbitals to nucleus X. As a consequence, depending on the spin state of A, the external field at X is either augmented or diminished. That is, the magnitude

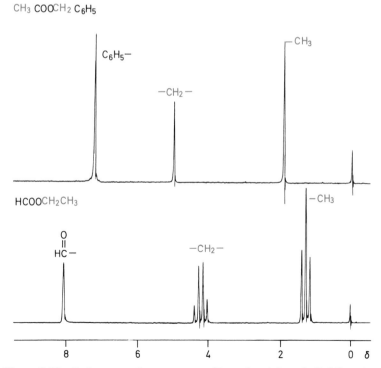

Figure 2.10   Proton magnetic resonances of benzyl acetate and ethyl formate

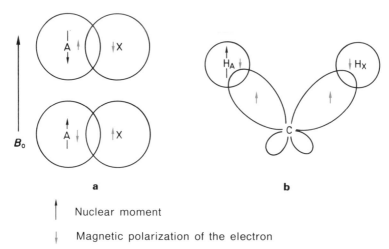

↑ Nuclear moment

↓ Magnetic polarization of the electron

Figure 2.11 Schematic representation of the nuclear spin–spin interaction through the bonding electrons (a) in the HF molecule and (b) in a CH_2 group. The low-energy state corresponds to the antiparallel arrangement of nuclear and magnetic moments. For the transmission of magnetic polarization the Pauli principle and Hundi's rule apply. Thus the magnetic polarization for orbitals at different centres, i.e. those involved in a bond, is antiparallel, while that for degenerate orbitals at the same centre, for example the hybrid orbitals at a carbon atom, is parallel

of the local magnetic field responsible for the resonance frequency nucleus X varies and the n.m.r. signal is split into a doublet. The same is true for nucleus A. Because the two spin states of A are almost equally probable, the lines of the doublet have the same intensity. In the following, we will refer to nuclei between which a spin–spin interaction exists as a *spin system*.

The energy, E, of the spin–spin interaction between two nuclei A and X is proportional to the scalar product of their nuclear magnetic moments μ_A and μ_B and can be expressed with the help of equation (1.5) as

$$E = J_{AX}\mathbf{I_A}\mathbf{I_X} \qquad (2.7)$$

where $\mathbf{I_A}$ and $\mathbf{I_X}$ are the so-called *nuclear spin vectors* of both nuclei and J_{AX} is known as the *scalar coupling constant* between them. Following equation 1.5, it can also be seen that the coupling is proportional to the product of the magnetogyric ratios of the coupled nuclei. To eliminate this dependence a *reduced coupling constant*, K_{AX}, defined by

$$K_{AX} = \frac{1}{\gamma_A\gamma_X} \cdot J_{AX} \qquad (2.8)$$

is used sometimes if the magnitudes of coupling constants between different nuclei are compared or in order to compensate for the negative sign introduced by negative magnetogyric ratios.

An important consequence of equation (2.7) is that the energy of the coupling and consequently the coupling constant, unlike the chemical shift, is independent of the strength of the external magnetic field. These constants are therefore expressed in frequency units (Hz). Because the coupling constant is independent of the spectrometer frequency, the separation between two lines in a spectrum can be identified as a coupling constant by measurements made at different B_0 fields. In the case of a spin–spin coupling the line splitting (in Hz) remains the same while in the case of chemical shifts it is changed.

In order to reinforce our understanding, the phenomenon of line splitting caused by spin–spin coupling will be explained further by reference to the energy level diagram of a two-spin system. First, we obtain four different states for two nuclei in an external field B_0 in the absence of spin–spin coupling, $(J=0)$. That is, both nuclear moments can be oriented either parallel or antiparallel to B_0 and one can be parallel and the other antiparallel and *vice versa* (Figure 2.12a). The transitions to be expected for the A nucleus (A_1 and A_2) that we wish to consider here have the same energy and consequently only one resonance line is observed.

In the presence of spin–spin coupling $(J > 0)$ the eigenvalues of the spin system are, as a consequence of the coupling, either stabilized or destabilized according to

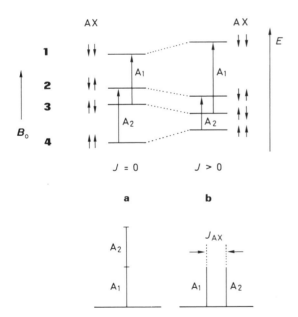

Figure 2.12 Nuclear magnetic energy level diagram for a two-spin system: (a) without spin–spin coupling; (b) with spin–spin coupling. The antiparallel orientation of the nuclear moment was chosen as the-low energy state so that the eigenvalues (2) and (3) are stabilized, while eigenvalues (1) and (4) are destabilized. For the sake of clarity only the lines of the nucleus are shown.

the relative orientation of the nuclear moments. Following convention, we will choose, the antiparallel arrangement of the nuclear moments as the low-energy state and thus the energy level diagram in Figure 2.12b results. It can be seen that the transitions A_1 and A_2 are no longer of equal energy and this has the effect of splitting the spectral line into a doublet.

A *quantitative* treatment of spin–spin coupling in the case of an AX system, that is for two nuclei i and j, is rather straightforward, since as will be shown later, the eigenvalues of a spin system can be calculated using the simple relation

$$E(\text{Hz}) = \sum_i v_i\, m_I(i) + \sum\sum_{i<j} J_{ij}\, m_I(i) m_I(j) \tag{2.9}$$

We obtain, with the resonance frequencies v_A and v_X and the coupling constant J_{AX} (all quantities in Hz):

(1) $E_1 = \frac{1}{2}v_A + \frac{1}{2}v_X + \frac{1}{4}J_{AX}$

(2) $E_2 = \frac{1}{2}v_A - \frac{1}{2}v_X - \frac{1}{4}J_{AX}$

(3) $E_3 = -\frac{1}{2}v_A + \frac{1}{2}v_X - \frac{1}{4}J_{AX}$

(4) $E_4 = -\frac{1}{2}v_A - \frac{1}{2}v_X + \frac{1}{4}J_{AX}$

The destabilization or stabilization of the eigenstates of the AX system as a result of spin–spin coupling thus amounts to $\pm\frac{1}{4}J_{AX}$, depending upon whether the orientations of the two nuclear moments are parallel or antiparallel. For the frequency of the spectral lines, by application of the *selection rule* $\Delta m_T = \pm 1$ (where $m_T = \sum_i m_I(i)$ is the *total spin* of the eigenstate under consideration), one obtains:

(3) $\to$(1): $E_1 - E_2 = v_A + \frac{1}{2}J_{AX}$

(4) $\to$(2): $E_2 - E_4 = v_A - \frac{1}{2}J_{AX}$

(2) $\to$(1): $E_1 - E_2 = v_X + \frac{1}{2}J_{AX}$

(4) $\to$(3): $E_3 - E_4 = v_X - \frac{1}{2}J_{AX}$

so that the line splitting at v_A and v_X equals exactly J_{AX}.

By definition the coupling constant is *positive* when the low-energy state has an antiparallel arrangement of nuclear moments as shown in Figure 2.12. When the reverse is true that coupling constant is *negative*. Thus, for the spin–spin coupling of the two protons of a methylene group J is negative since here the low-energy spin pairing leads to a parallel arrangement of the nuclear moments. The generalization into a rule that the coupling constant for spin–spin interaction over an even number of bonds is negative and over an odd number of bonds is positive has exceptions, as will be shown later.

The fact that spin–spin coupling is transmitted through chemical bonds makes the coupling constant J a sensitive parameter for the types of bonds involved and for their spatial orientation in the molecule. Before we discuss the relationship between the magnitude of the H–H coupling constants and the structure of organic molecules in more detail, the simple rules that are used for the interpretation of the splitting patterns will be explained. These rules, called *first-order rules*, are conceived as limiting cases of the quantum mechanical analysis of nuclear magnetic resonance spectra as we will treat them in Chapter 4. Their validity is therefore restricted and in connection with their introduction we should discuss a few points that will enable us to define the limits of their applicability.

2.1   SIMPLE RULES FOR THE INTERPRETATION OF MULTIPLET STRUCTURES

Let us consider once again the ethyl group in the spectrum of ethyl formate (Figure 2.10, p. 29). On the basis of the intensities of the signals the triplet and the quartet can be assigned to the methyl and the methylene groups, respectively. The number of lines in each group, that is, their *multiplicity*, is larger by just one than the number of protons in the neighbouring group. This can be understood if we consider the possible combinations of the magnetic quantum numbers, $m_I(i)$, of the protons of each group. We shall now employ the wave functions α and β for the characterization of the two possible spin states the individual protons and thus the following scheme results:

CH$_2$ group	m_T	CH$_3$ group			m_T
$\alpha\alpha$	$+1$	$\alpha\alpha\alpha$			$+\frac{3}{2}$
$\alpha\beta$ $\beta\alpha$	0	$\alpha\alpha\beta$	$\alpha\beta\alpha$	$\beta\alpha\alpha$	$+\frac{1}{2}$
$\beta\beta$	-1	$\alpha\beta\beta$	$\beta\alpha\beta$	$\beta\beta\alpha$	$-\frac{1}{2}$
		$\beta\beta\beta$			$-\frac{3}{2}$

The individual combinations are distinguished by means of their *total spins, m_T*, that characterize the magnetic properties of the group of nuclei under consideration.

The fact that the three protons of the methyl group can exist in four different magnetic states leads, by analogy with the analysis made above for the two spin system, to the observed quartet for the resonance of the methylene protons in which the 1:3:3:1 intensity distribution is a result of the relative probabilities of the different spin combinations.

Completely analogous considerations apply for the structure of the methyl proton resonance. Their generalization leads to the following rules that can easily be verified with the aid of equation (2.10):

1. For nuclei with spin quantum number $I=\frac{1}{2}$ the *multiplicity of the splitting* equals $n+1$, where n is the number of nuclei in the neighbouring group. If another neighbouring group is present the protons of which have a chemical shift different from that of the protons in the first group, the effect of the second group must be considered separately. The sequence in which the effects of the protons in neighbouring groups are considered is immaterial. Thus, if a nucleus H_M has two chemically different neighbouring nuclei H_A and H_X, the signal for H_M would be

split into a doublet of doublets. A triplet would be observed only if J_{AM} and J_{AX} were by chance identical.

2. The *line separations* (in Hz) correspond to the coupling constants between the nuclei under consideration.

3. The *relative intensities* within a multiplet are given by the coefficients of the binomial expansion

$$1 : \frac{n}{1} : \frac{n(n-1)}{2 \cdot 1} : \frac{n(n-1)(n-2)}{3 \cdot 2 \cdot 1} \cdots$$

They also can be read directly from the Pascal triangle

```
n = 0                    1
    1                  1   1
    2                1   2   1
    3              1   3   3   1
    4            1   4   6   4   1
    5          1   5  10  10   5   1
    6        1   6  15  20  15   6   1
```

4. The *magnitude* of the *spin–spin coupling* between protons in general decreases as the number of bonds between the coupled nuclei increases. The coupling constant is finally reduced to the order of magnitude of the natural line width so that a splitting is no longer observed.

5. The *splitting patterns* are independent of the *signs* of the *coupling constants*. These must be determined by other means and we will come back to this later.

The fourth rule becomes clear through a comparison of the spectra of ethyl formate and benzyl acetate made in Figure 2.11. The interaction between the CH_2 and the CH_3 protons is transmitted in one case through *three* bonds and in the other through *five* bonds. It is seen to be too small in benzyl acetate to produce a splitting.

The splitting patterns of a series of alkyl groups provide a demonstration of the application of the above rules. The splitting pattern of the ethyl group has already been discussed. In order to expand the series, Figure 2.13 shows the characteristic patterns of a few other groups.

As is apparent from the foregoing discussion, the coupling constant, J, is determined in the spectrum by measuring the separation of adjacent lines in the multiplet under consideration. The observed splitting must then also be found in the multiplet of the neighbouring group of protons. This is illustrated in Figure 2.14 for the three aromatic protons of 2,4-dinitrophenol. We use the rule of thumb that the magnitude of J decreases as the number of bonds between the coupled nuclei increases. This leads to the result that the assignments made should be such that $J_{ac} < J_{ab} < J_{bc}$.

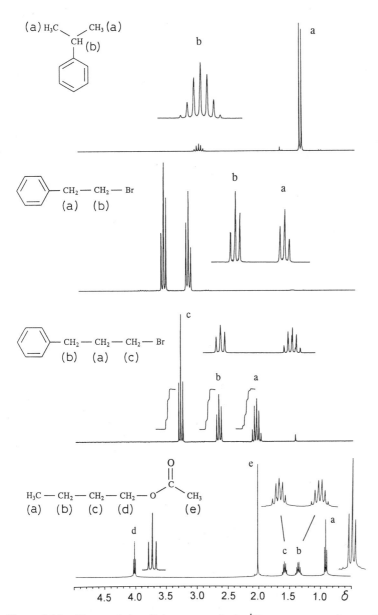

Figure 2.13 Characteristic splitting patterns in the ^{1}H n.m.r. spectra of some alkyl groups at 200 MHz (upper three spectra) and 400 MHz (lowest spectrum); the spectrum of 1-phenyl-3-bromopropane demonstrates that the determination of the proton number from signal heights instead of integration can be misleading

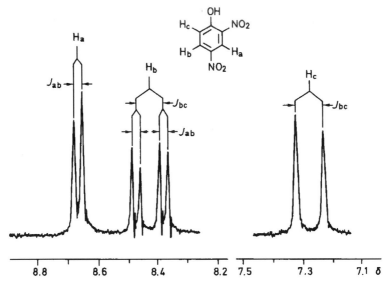

Figure 2.14   Signal splitting due to spin–spin coupling in the 100 MHz proton magnetic resonance spectrum of 2,4-dinitrophenol. One finds J_{bc} = 9.1 Hz and J_{ab} = 2.8 Hz. J_{ac} is not observed (rule 4)

2.2 SPIN–SPIN COUPLING WITH OTHER NUCLEI

2.2.1 Nuclei of Spin $I=\frac{1}{2}$

The aspects of spin–spin coupling discussed above are also valid for other nuclei of spin $I=\frac{1}{2}$. For example, many first order splittings are found in the ^{19}F n.m.r. spectra of organic fluoro compounds or in the ^{31}P n.m.r. spectra of organic phosphorus compounds. In addition, to stay with these examples, fluoro and phosphorus organic compounds possess mixed spin systems composed of protons and ^{19}F and ^{31}P nuclei, respectively. ^{1}H,^{19}F and ^{1}H,^{31}P coupling constants then lead to line splittings in the ^{1}H,^{19}F and ^{31}P n.m.r. spectra of these molecule (Figure 2.15).

In contrast to ^{19}F or ^{31}P nuclei, which both have a natural abundance of 100%, other nuclei with the spin quantum number $\frac{1}{2}$ are present in organic or organometallic molecules only with a few percent. These nuclei are known as *rare nuclei*. Prominent examples are the ^{13}C nucleus (1.1%), the ^{29}Si nucleus (4.7%) as well as several metal nuclei such as ^{199}Hg (16.9%), the cadmium isotopes ^{111}Cd and ^{113}Cd (12.9 and 12.3%, respectively), but also tin (^{117}Sn, 7.7%; ^{199}Sn, 8.8%) or platinum (^{195}Pt, 33.7%). Coupling of these nuclei to protons is observed only for those molecules which contain the n.m.r. active isotope. The corresponding resonance signals are, therefore, in most cases considerably less intensive than that of the residual molecules which contain the corresponding n.m.r inactive isotopes. They can be observed in the ^{1}H n.m.r. spectrum of the particular compound as so-called *satellite*

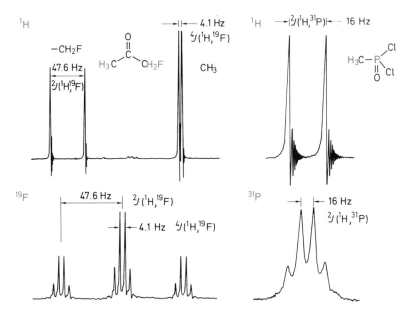

Figure 2.15 ^{1}H n.m.r. and ^{19}F as well as ^{31}P n.m.r. spectra of fluoroacetone and methylphosphorusdichloride, respectively (measuring frequencies 60, 75.3 and 32.3 MHz)

lines. Figure 2.16 shows as an example the proton resonance spectrum of bis(trimethylsilyl)mercury.

If the natural abundance of an n.m.r. active nuclide falls below 1%, the observation of satellite spectra becomes difficult for sensitivity reasons. In these cases isotopic enrichment is necessary to observe the coupling constant of interest.

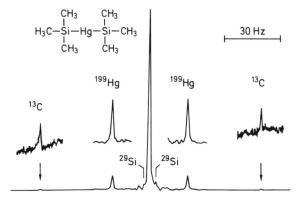

Figure 2.16 Satellite lines in the ^{1}H n.m.r. spectrum of bis(trimethylsilyl)mercury: ^{13}C (1.1%), $^1J(^{13}C,^1H)$ = 119.6 Hz; ^{199}Hg (16.9%), $^3J(^{199}Hg,^1H)$ = 40.7 Hz; ^{29}Si (4.7%), $^2J(^{29}Si,^1H)$ = 6.6Hz

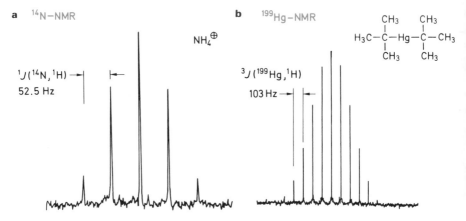

Figure 2.17 ^{14}N n.m.r. spectrum of ammonium ion (a) and ^{199}Hg n.m.r. spectrum di-t-butylmercury (b); measuring frequencey 4.33 and 14.3 MHz, respectively. In figure 2.17b eight of the 19 expected lines are not visible because of their low intensity

The importance of ^{13}C satellites in ^{1}H n.m.r. spectra for spectral analysis will be discussed in Chapter 6.

In the spectra of the rare isotopes, on the other hand, the line splitting that arises from coupling to the protons of the particular compound is always clearly visible. The splitting pattern corresponds in many cases to the first order rules. As examples we show in Figure 2.17 the ^{14}N n.m.r. signal of ammonium ion and the ^{199}Hg resonance in di-t-butylmercury. The ^{13}C n.m.r. spectrum of norbornane discussed later shows similar results (Figure 2.24, p. 47).

2.2.2 Nuclei of Spin $I > \frac{1}{2}$

In cases in which spin–spin coupling involves a nucleus that has a spin quantum number I greater than $\frac{1}{2}$, the multiplicity and the intensity distribution of the splitting pattern deviates from the rules given above. For example, a neighbouring deuteron $(I = 1)$ splits a proton signal into a triplet the lines of which have equal intensities. This follows from the fact that the possible orientations of the deuteron's spin relative to an external field, namely $m_I = +1$, 0 and -1, are in practice equally probable. In general the multiplicity of an n.m.r. signal caused by n neighbouring nuclei is given by $2nI + 1$. Examples for this rule are found in Exercise 2.9.

Exercise 2.7 What multiplicity and intensity distribution should be expected according to the first-order rules for the nuclei designated a, b, c, and d in the compounds **1–6**? Consider coupling over only two or three bonds.

1 CH_3-OH
    a   b

4 $CF_3-CHF-CH_3$
    a   bc   d

2 $(CH_3)_3C-CH_2Br$
    a   b

5 $(CH_3)_2CH-O-CH_2-CH_3$
    a   b   c   d

3 $CH_3-CHCl-CH_2-O-CH_3$
    a   b   c   d

6 $CHDCl_2$
    ab

2.3 THE LIMITS OF THE SIMPLE SPLITTING RULES

2.3.1 *The Notion of Magnetic Equivalence*

As we have already mentioned, a few qualifying remarks are necessary concerning the validity of the first-order rules for the analysis of the fine structure of nuclear magnetic resonance signals. The explanation often given leads one to the erroneous assumption that no spin–spin coupling occurs between protons within a group, for example the three protons of a methyl group, because there is no indication of coupling between these protons in the spectrum. Therefore, we want to introduce a rule here that will be substantiated in detail later. It states: *The spin–spin coupling between magnetically equivalent nuclei does not appear in the spectrum.* By *magnetically equivalent* we mean that all of the nuclei under consideration possess the same resonance frequency and only one characteristic spin–spin interaction with the nuclei of a neighbouring group. Nuclei with the same resonance frequency are called *isochronous*. They are usually also chemically equivalent, that is, they have identical chemical environments. Chemically equivalent nuclei are, however, not necessarily magnetically equivalent.

The protons of a methyl group are magnetically equivalent since, as a consequence of the rotation about the C–C bond, all three protons have the same time-averaged chemical environment and therefore the same resonance frequencies. The coupling constant to the protons of a neighbouring CH_2 or CH group is likewise necessarily identical for each of the three protons, as the three conformations a, b, and c are of equal energy and therefore equally populated. The geometric relation

Exercise 2.8. Figure 2.18 (page 40) shows a series of splitting patterns for three different protons or proton groups that we shall identify as A, M and X. Determine the coupling constants J_{AM}, J_{AX}, and J_{MX} and also the number of protons in each group by reference to the multiplicity and intensity distribution of the signals.

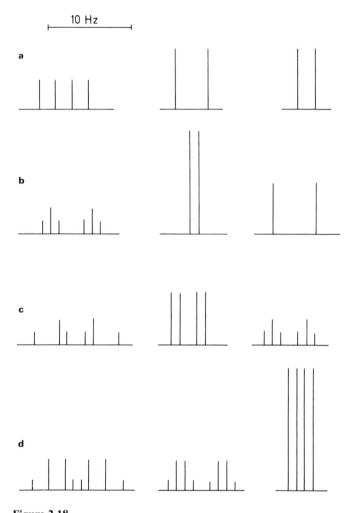

Figure 2.18

Exercise 2.9. Figure 2.19 (page 41) shows the proton resonance spectra of NaBH$_4$ and C$_6$H$_5$CHD$_2$ and also those of the ammonium ions ^{14}NH$_4^+$ and ^{15}NH$_4^+$. Assign the spectra, explain the multiplicity and indicate the value of the coupling constants.

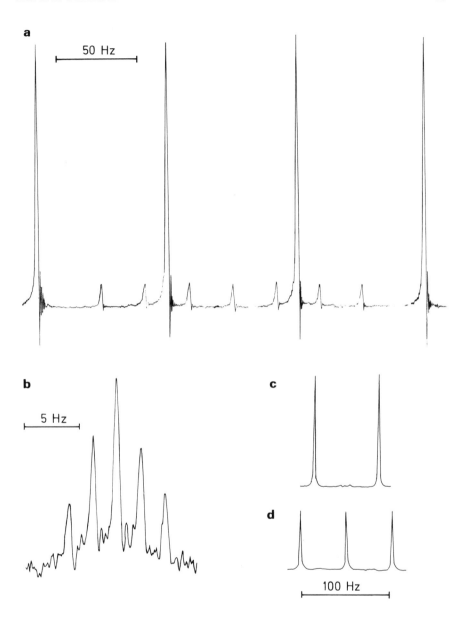

Figure 2.19

between the individual methyl protons and their neighbours, which determines the magnitude of the coupling constant, thus becomes identical for each of the three methyl protons. The same arguments apply for the nine protons of a t-butyl group. With this group, however, it is possible in special cases to reduce the rate of rotation about the bond to the next carbon by cooling to low temperatures to such an extent that chemically different methyl groups can be distinguished in the n.m.r. spectrum (Figure 2.20).

In other classes of compounds the magnetic non-equivalence of protons is often suggested by the structure. The protons in 1,1-difluoroethylene (**1**) as well as the two fluorine nuclei are chemically equivalent but magnetically non-equivalent for two different coupling constants are observed. In such a case $J_{13} \neq J_{14}$ or $J_{cis} \neq J_{trans}$.

Similarly, there are two non-identical coupling constants, J_{13} and J_{14}, for the protons in furan (**2**). The α and β protons are therefore magnetically non-equivalent. In difluoromethane (**3**), on the other hand, the protons, as well as the fluorine nuclei, are magnetically equivalent ($J_{13} = J_{14} = J_{23} = J_{24}$).

In the customary notation for spin systems these different properties of nuclei are taken into account. As we shall explain later in detail (p. 192), the magnetic nuclei in 1,1-difluoromethylene and in furan represent AA'XX' systems while those in difluoromethane are classified as an A_2X_2 system.

In this context it is of significance that the first-order rules formulated above for the explanation of the line splittings in nuclear magnetic resonance spectra apply only for groups of magnetically equivalent nuclei. If the nuclei in a group are not

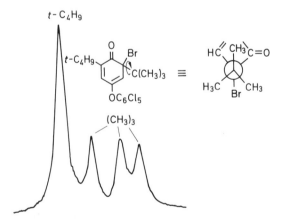

Figure 2.20   Proton absorption of the chemically non-equivalent methyl groups of a t-butyl group in 6-bromo-4-pentachlorophenoxy-2,4-di-t-butylcyclohex-2,4-dien-1-one at −90 °C and 220 MHz (after Ref. 1). The spin-spin coupling between the methyl protons, because of it's small value, leads only to a line broadening

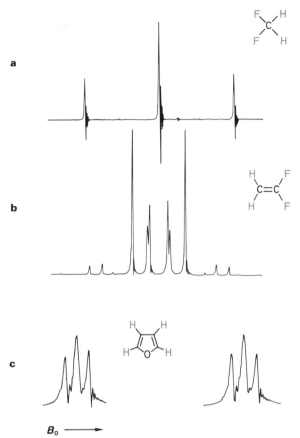

Figure 2.21 Proton magnetic resonance spectrum of (a) difluoromethane, (b) 1,1-difluoroethylene (after Ref. 2), and (c) furan

magnetically equivalent, individual coupling constants cannot be taken from the spectrum and the chemical shift can be determined from the centre of the multiplet with sufficient accuracy without analysis only if the relation $J/v_0\delta < 0.1$ holds (cf. Section 2.3.2, p. 44). This is made readily apparent through the comparison of the spectra of difluoromethane and 1,1-difluoroethylene. In the former (Figure 2.21a) we observe the expected triplet for the proton resonance while in the latter (Figure 2.21b) a complicated splitting pattern results. In this case the coupling constants can be determined only after a precise analytical procedure which will be considered in Chapter 5. In addition, we should emphasize that the presence of simple splitting patterns are not always a guarantee that the first-order rules can be applied. For example, in the n.m.r. spectrum of furan one finds two triplets centred at δ 6.37 and 7.42 (Figure 2.21c) which, as we have already explained, may not be interpreted as indicating that the two α protons are equally coupled to the two β protons and that the coupling constant, J, corresponds to the 3.2 Hz that can be determined from the line separation. We shall return to the treatment of such *deceptively simple* cases in Chapter 5.

2.3.2 The Significance of the Ratio $J/v_0\delta$

A further important restriction that must be applied to the first-order rules states that they can be employed with confidence only where the chemical shift difference, $v_0\delta$ (Hz), of the individual groups of magnetically equivalent nuclei is large compared with the coupling constant connecting these groups. We then speak of a *first-order spectrum*. If the magnitude of $v_0\delta$ is of the same order of magnitude as the coupling constant, more lines are observed in the spectrum than would be expected according to the first-order rules (Figure 2.22). It should be noted that the intensity distribution in the lines of the two groups of signals is also dramatically affected. The intensity of the lines nearest to the multiplet of the neighbouring group is greatly enhanced while that of the other lines decreases. This is called the *roof effect*.

This effect has considerable diagnostic value in the assignment of coupling constants for it indicates whether the resonance of neighbouring protons coupled with a particular group lie at a higher or lower field than that of the group under consideration.

Increased multiplicity and altered intensity distribution are therefore indications of *spectra of higher order* which must be analysed by more exact methods. In applying

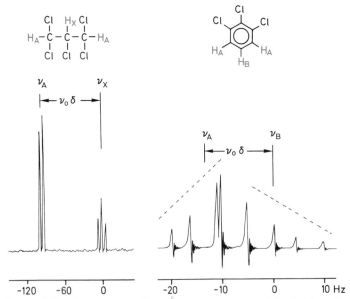

Figure 2.22 Proton magnetic resonance spectrum of 1,1,2,3,3-pentachloropropane and 1,2,3-trichlorobenzene. The ratio $J/v_0\delta$ in the first case is 0.06 and in the second it is 0.7; the number of lines increases from five to eight

the first-order rules in the determination of chemical shifts and coupling constants one obtains only approximate values, the errors become larger and larger as the ratio $J/v_0\delta$ increases. However, because of the field dependence of $v_0\delta$, complicated spectra can be simplified by using high B_0-fields from superconductive magnets.

Finally, it should be pointed out that, unlike first-order spectra, all spectra of higher order depend upon the *relative signs* of the coupling constants (if spin systems of more than two nuclei are involved). While the relative signs of the coupling constants in a higher order spectrum can be assigned through a detailed analysis of the spectrum, this can be accomplished in a first order spectrum only by performing additional experiments which we shall consider in Chapters 7 and 8. The

Table 2.2 Typical values of H-H coupling constants in organic compounds

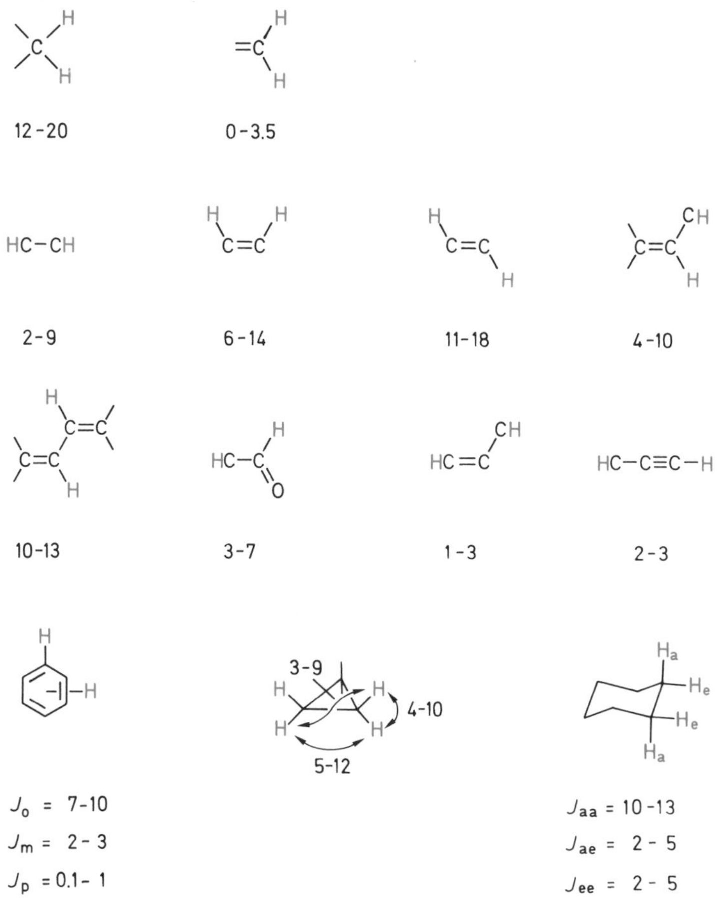

12-20   0-3.5

HC-CH

2-9   6-14   11-18   4-10

10-13   3-7   1-3   2-3

J_o = 7-10 J_{aa} = 10-13
J_m = 2-3 J_{ae} = 2-5
J_p = 0.1-1 J_{ee} = 2-5

3-9
4-10
5-12

determination of the *absolute* signs of the coupling constants has been possible in only a few cases. The signs of these parameters, the consideration of which is essential for a meaningful discussion of the experimental data, are given in Chapter 4. They are based, in most cases, on the determination of the sign relative to that of the one-bond coupling between a ^{13}C nucleus and a proton which is positive.

2.4   SPIN–SPIN DECOUPLING

As we shall discuss in detail in Chapter 6, spin–spin coupling yields important information about molecular structure. The interpretation of n.m.r. spectra, however, is complicated through line splittings. In cases of higher order spectra situations arise where by direct inspection even the chemical shifts can be determined only approximately. The possibility of simplifying complicated spectra by eliminating scalar spin–spin coupling experimentally is, therefore, of considerable practical importance. The techniques used for this purpose are known as *double resonance experiments*. They will be discussed in more detail in Chapter 7. Because of its importance for the simplification of n.m.r. spectra, however, and since it is easily

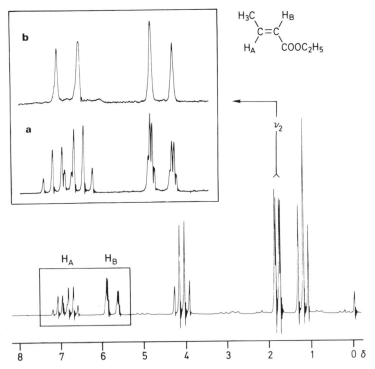

Figure 2.23 Spin–spin decoupling experiment with *trans*-ethylcrotonate; (a) single resonance spectrum of the olefinic protons; (b) double resonance spectrum with $\nu_2 = \nu_{CH_3}$

performed even with routine spectrometers, we shall describe the *spin–spin decoupling experiment* here.

Spin–spin decoupling rests on the application of a second radio frequency source S_2 in addition to the transmitter S_1 used for the detection of the spectrum. Consider a spin system of two nuclei A and X with a scalar coupling $J(A,X)$. If we observe the A nucleus with the transmitter S_1 while irradiating the X nucleus with the transmitter S_2, the scalar coupling between A and X vanishes. As we shall justify in Chapter 7, the application of S_2 to the X resonance leads to a situation where the nuclear magnetic moment μ_x points along the y-axis of the coordinate system (Figure 1.1). Since the vector μ_A still points along the z-axis, the scalar product of μ_A and μ_X and thus the energy of spin–spin coupling (equation (2.8)) will become zero. As a consequence, the line splitting for the A nucleus disappears.

Two examples may suffice to illustrate these facts. Figure 2.23 shows a spin–spin decoupling experiment with *trans*-ethylcrotonate. The resonances of the olefinic protons H_A and H_B are extensively split due to spin–spin coupling with the protons of the methyl group (Figure 2.23a). Upon irradiation of the methyl resonance with a second r.f. source, the spectrum simplifies to the expected AB system (Figure

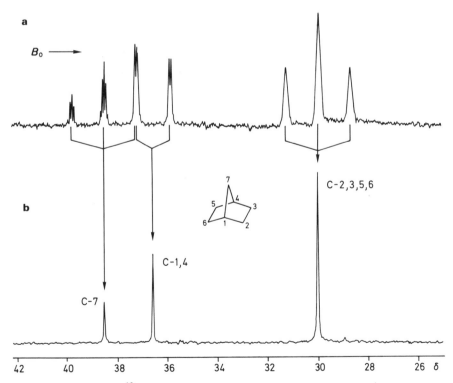

Figure 2.24   100 MHz ^{13}C n.m.r. spectrum of norbornane without and with ^{1}H-decoupling ((a) and (b), respectively); the δ scale refers to the ^{13}C resonance of tetramethylsilane

2.23b). Since in this experiment the irradiated and the observed nucleus are from the same nuclide, the experiment is known as *homonuclear* decoupling experiment.

Figure 2.24 shows an example for a *heteronuclear* decoupling experiment with the ^{13}C n.m.r. spectrum of norbornane, where the multiplets of the various ^{13}C resonances result from $^{13}C,^{1}H$ coupling. However, if we record the ^{13}C n.m.r. spectrum in the frequency range $v(^{13}C)$ (in this particular experiment 100 MHz) while irradiating at the same time the proton resonances in the frequency range $v(^{1}H)$ (here 400 MHz), all couplings disappear. The ^{13}C resonances are then recorded as singlets which can be assigned immediately on the basis of their relative intensities. Because all proton resonances of norbornane were irradiated at once, the technique is called *broadband decoupling*.

2.5   THE STRUCTURE DEPENDENCE OF THE SPIN–SPIN COUPLING—A GENERAL SURVEY

We want to conclude this discussion, as we did that of the chemical shift, with a general survey of the dependence of H–H coupling on structure. For this purpose the most important types of spin–spin couplings and the range of coupling constants are summarized in Table 2.2 (p. 45). It can be seen from these data that the magnitude of coupling constants for protons in general lies in the range of 5–17 Hz and that these parameters are very sensitive to the geometry of the coupling path, i.e. the stereochemistry of the molecule.

The spin–spin interaction of olefinic protons offers a typical example. For a pair of isomers J_{trans} is always larger than J_{cis}. Similarly, we see that in cyclohexane $J_{aa} > J_{ee}$. On the other hand, we observe that $J_{trans} < J_{cis}$ in cyclopropane. In addition to coupling through a carbon skeleton, spin–spin interaction can be transmitted through heteroatoms in groups of the type H–C–O–H and H–C–N–H. *Long-range coupling* over more than three bonds is generally observed in unsaturated systems in which the π-electrons prove to be effective transmitters of magnetic information.

Finally we note that the coupling constants of many molecules are influenced by conformational equilibria and other dynamic processes. The observed data are then average values which are formed on the basis of the mole fractions p, from the data of the individual conformers. In the simplest case of an equilibrium between two conformations A and B the equation

$$J_{exp} = p_A J_A + p_B J_B$$

holds. An analogous equation exists for chemical shifts. These relations are discussed in more detail in Chapter 9.

Exercise 2.10. Figure 2.25 presents a series of proton magnetic resonance spectra with significant signal splitting as the result of spin–spin coupling; frequency 60 MHz.

1. Determine the structures of compounds (a)–(d).
2. Assign the signals in the spectrum of vinyl acetate (e) to the appropriate protons by means of determining the coupling constants.

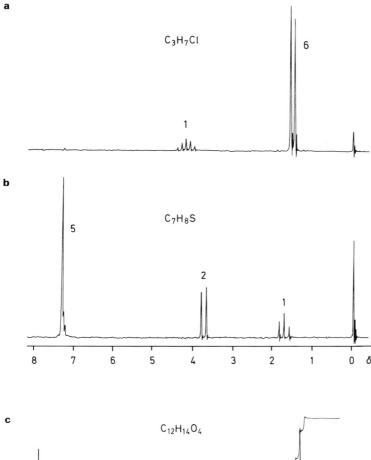

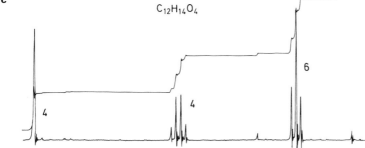

Figure 2.25

3. Analyse the absorption of the olefinic protons in crotonaldehyde (f) and determine and assign the coupling constants. Which proton absorbs at lower field?

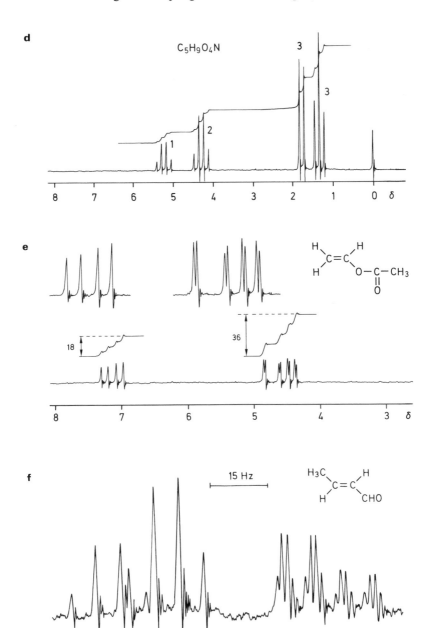

Figure 2.25 *(continued)*

Exercise 2.11. Figure 2.26 shows the 90-MHz spectrum of the H^a and H^b protons by pyrimidine. Assign the resonances and determine the coupling constants.

Exercise 2.12. Figure 2.27 shows the olefinic resonances of thujic ester. Assign the resonances to appropriate protons and determine the vicinal coupling constants.

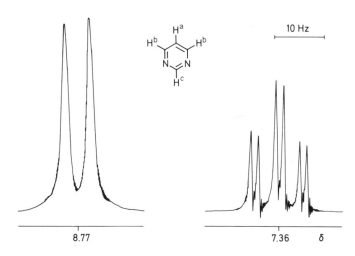

Figure 2.26

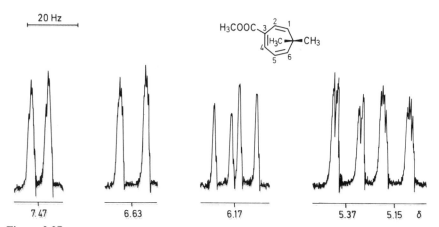

Figure 2.27

Exercise 2.13. Figure 2.28 shows the 80 MHz ^{1}H n.m.r spectrum of 2,4-dinitrophenyl-2-pyridylsulphide. Assign the protons to the structural formula given and estimate the coupling constants.

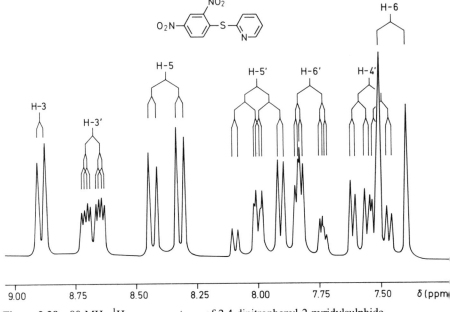

Figure 2.28 80 MHz ^{1}H n.m.r. spectrum of 2,4-dinitrophenyl-2-pyridylsulphide

3. References

1. H. Kessler, *Angew. Chem.,* **82**, 237 (1970); *Angew. Chem. Int. Ed. Engl.,* **9**, 219 (1970).
2. E. D. Becker, *High-resolution NMR*, 2nd Ed., Academic Press, New York 1980, p. 91.

3 EXPERIMENTAL ASPECTS OF NUCLEAR MAGNETIC RESONANCE SPECTROSCOPY

Having introduced the fundamental parameters of high-resolution nuclear magnetic resonance spectroscopy in the preceding chapter, we shall now examine more closely the experimental aspects of the technique.

1. Sample preparation and sample tubes

Sample preparation in n.m.r. spectroscopy is extremely simple. For ^{1}H n.m.r. obviously *solvents* that have no protons are preferred, but this limitation is not serious because of the ready availability of deuterated compounds. The most widely used solvent is deuterochloroform, $CDCl_3$. For poorly soluble samples a series of other deuterated solvents is available: dimethyl sulphoxide-d_6, acetone-d_6, acetonitrile-d_3, benzene-d_6, and D_2O. Table 3.1 summarizes the most frequently used solvents and their properties, which show, as a result of isotope efects, small differences between deuterated and non-deuterated compounds.

Another aspect of solvent selection arises in connection with the fact that in some cases the compound to be investigated is prepared in the n.m.r. tube by the reaction of a starting material with the solvent or with a third component. Protonated ketones are obtained in this way in trifluoroacetic acid, while simple carbonium ions were directly observed for the first time when aliphatic fluorides were reacted with antimony(V) fluoride in sulphur dioxide resulting in the formation of the

Table 3.1 Properties of solvents used for NMR spectroscopy

Solvent	Formula	Boiling point (°C)	Melting point (°C)	Magnetic volume suscepti-bility, $\chi_v \times 10^6$	$\delta(^1H)$	$\delta(^{13}C)$
Carbon tetrachloride	CCl_4	76.8	−22.8	−0.684	—	96.0
Carbon disulphide	CS_2	46.3	−111.5	−0.681	—	192.8
Chloroform	$CHCl_3$	61.3	−63.5	−0.733	7.27	77.2
Chloroform-d	$CDCl_3$	60	−64.1	—	7.26[a]	77.0
Diethylether-d_{10}	$C_4D_{10}O$	—	—	—	—	—
Dimethyl sulphoxide	CH_3SOCH_3	189	8	−0.609	2.58	—
Dimethyl sulphoxide-d_6	CD_3SOCD_3	190	20.2	—	2.5[a]	39.5
Acetonitrile	CH_3CN	82	−45.7	−0.486	1.95	—
Acetonitrile-d_3	CD_3CN	79	−42	—	—	1.3; 118.1
Acetone	CH_3COCH_3	56.5	−95	−0.460	2.17	—
Acetone-d_6	CD_3COCD_3	55	−94.5	—	†2.05	30.5; 205.1
Methylene chloride	CD_2Cl_2	40.1	−96.7	−0.733	5.35	—
Methylene chloride-d_2	CD_2Cl_2	39	−97	—	5.31[a]	53.7
1,4-Dioxane	$C_4H_8O_2$	101.5	12	−0.589	3.68	—
1,4-Dioxane-d_8	$C_4D_8O_2$	100	11	—	3.53[a]	66.3
Cyclohexane	C_6H_{12}	81.4	6.5	−0.631	1.43	27.5
Cyclohexane-d_{12}	C_6D_{12}	—	—	—	1.38	26.0
Benzene	C_6H_6	80.1	5.5	−0.626	7.27	—
Benzene-d_6	C_6D_6	79	7.6	—	7.16[a]	128.7
Pyridine	C_5H_5N	115.3	−42	−0.612	6.9–8.5	—
Pyridin-d_5	C_5D_5N	114	−41	—	—	—
Water	H_2O	100.0	0.0	−0.721	4.8	—
Heavy water	D_2O	101.4	3.8	−0.719	4.72[a]	—
Tetrahydrofuran-d_8	C_4D_8O	64	−108.5	—	1.72[a]; 3.57[a]	25.3; 67.2
Sulphur dioxide	SO_2	−10.0	−75.5	−0.812	—	—

[a] Residual ^1H n.m.r. signal of partially deuterated material

corresponding hexafluoroantimonate:

$$R{-}F \ + \ SbF_5 \longrightarrow R^{\oplus}[SbF_6]^{\ominus}$$

Similar results can be obtained with the so-called 'magic acid', a mixture of fluorosulphonic acid and antimony(V) fluoride. It is obvious that in such cases the choice of solvent is considerably more than a routine decision and that the n.m.r. tube has served in these and other instances as a very useful 'reaction flask' in organic chemistry.

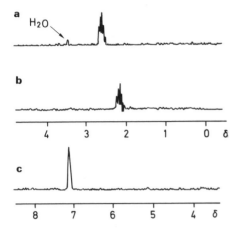

Figure 3.1 N.m.r. absorptions of partially deuterated 'impurities' in (a) dimethyl sulphoxide-d_6, (b) acetone-d_6, and (c) benzene-d_6. Dimethyl sulphoxide is hygroscopic and therefore most samples of it are contaminated with a trace of water. For the fine structure of the absorption, compare Exercise 2.9 (p. 40)

It should be noted that the small fractions (usually less than 0.5%) of only partially deuterated solvent molecules give rise to low-intensity signals in the spectrum. Those produced by a few solvents are illustrated in Figure 3.1. The spin–spin splitting apparent in the cases of dimethyl sulphoxide and acetone arises through the presence of the CHD_2 groups, that is, through the interaction of the proton with the two deuterons the total spin of which can take the m_T values of $+2$, $+1$, 0, -1, and -2 in the statistical ratio of 1:2:3:2:1. Except for a small high-field shift which originates from a deuterium induced isotope effect on the ^1H chemical shift, the proton resonance signals of partially deuterated solvents have the same δ values as those of the nondeuterated compounds.

To prepare a sample for measurement, about 5–10 mg or 10 μl of the substance is placed in the *sample cell* or *n.m.r. tube*, a cylindrical glass tube of about 17 cm in length and 5 mm o.d. (Figure 3.2a) and dissolved by the addition of about 0.5 ml of solvent. The solution should fill the tube then to a height of 3–4 cm. Finally, the *standard* or *reference compound*, usually tetramethylsilane, is added. Since this compound, because of its low boiling point (26 °C) and its high vapour pressure, is difficult to handle in small amounts it is advantageous to have 5% (v/v) solutions of TMS in the most frequently used solvents on hand.

As with the choice of solvent, the choice of the standard depends upon the substance to be investigated. For example, in the case of TMS for cyclopropane derivatives and especially for silyl compounds a superposition of resonance and reference signals can result. In these cases cyclohexane, methylene chloride (dichloro-methane), or benzene can be used as a reference. In aqueous systems, 1,4-dioxane or t-butanol are applicable but especially advantageous is the sodium salt of

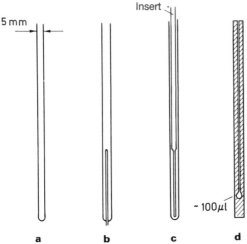

Figure 3.2   N.m.r. sample tubes: (a) for measurement with an internal standard; (b) and (c) with capillaries for measurements with an external standard; (d) for microsamples

the partially deuterated 3-trimethylsilylpropionic acid (**4**), the ^{1}H n.m.r. signal of which in aqueous and methanolic solutions is found at exactly $\delta = 0.00$.

$$(CH_3)_3Si-CD_2-CD_2-COONa$$

4

Generally, each substance the protons of which give rise to a signal of sufficient intensity, is suitable as a reference. In this connection it is reasonable to wonder to what extent measurements made with different reference compounds can be compared with respect to chemical shifts of the resonance signals. In the following section a few fundamental remarks are made concerning this subject.

2. Internal and external standards; solvent effects

Following the classical electromagnetic equations the magnetic flux density B in a substance exposed to an external magnetic field consists of two terms:

$$B = \mu_0(H + M) \tag{3.1}$$

where H is the field strength of the applied field, M is the magnetization induced in the substance, and μ_0 is the permeability, a constant equal to $4\pi \times 10^{-7}$ kg m s^{-2} A^{-2}. The magnetization M is in turn dependent upon the external field strength according to

$$M = \chi_v H \tag{3.2}$$

where χ_v is a dimensionless constant, the volume susceptibility, that is characteristic of the material. For diamagnetic substances χ_v is negative and in general is independent of the temperature. In the case of n.m.r. measurements the field strength that exists within the n.m.r. tube is influenced by the magnetic susceptibility of the solvent. For this reason an *internal standard* is employed. That is, the reference substance and the sample are contained in the same solution so that both are exposed to the same magnetic environment and corrections of the experimental results are not necessary. On the other hand, with an *external standard*, when the reference substance is contained in a coaxial capillary separated from the volume that contains the sample (Figure 3.2b, c), the field strengths that exist in the capillary and the sample solution are different owing to the different volume susceptibilities. The chemical shifts recorded must then be corrected. For a cylindrical n.m.r. tube the axis of which is parallel to the direction of the magnetic field, a situation met for cryomagnets, the correction is given by

$$\delta_{cor.} = \delta_{exp.} - \frac{4\pi}{3}(\chi_v^{Standard} - \chi_v^{Sample}) \times 10^6. \tag{3.3}$$

For iron magnets, where the sample cell is oriented perpendicular to the axis of the magnetic field, the factor $+2\pi/3$ must be used. In the case of dilute samples, the volume susceptibility of the sample solution can be approximated as being equal to that of the solvent. Differences in δ values caused by the susceptibility effect can be as much as 1 ppm.

It must be emphasized that while, with the aid of the internal standard, the susceptibility correction can be obviated, specific interactions between the solvent and the reference substance cannot be avoided. When, for example, chloroform is used as a reference substance in benzene-d_6 as solvent, to consider an extreme case, the resonance signal of cyclohexane (concentration 20% v/v) is recorded at -4.96 ppm. With carbon tetrachloride as solvent and using the same reference the chemical shift is -5.80 ppm. The difference of 0.84 ppm between the two measurements arises from the fact that chloroform associates with benzene in such a way that the chloroform proton is specifically shielded (cf. p. 98). If we now try to determine the 'δ value' for cyclohexane on the basis of the above measurements by reference to the value for chloroform on the δ_{TMS} scale, we obtain values of 2.31 (7.27–4.96) or 1.47 (7.27–5.80) ppm. Thus, only the measurement with carbon tetrachloride as solvent gives an acceptable result (cf. Table 2.1).

This example shows that measurements conducted in different solvents or with different reference compounds lead to equivalent results only when there are no *specific interactions* between the solvent and the reference substance or the sample. Solvent-standard combinations for which specific interactions of this kind are known or expected should therefore be avoided. On the other hand, association effects can also be advantageous, since interactions of this type often lead to changes in the relative chemical shifts that influence the appearance of the spectrum. Furthermore, in addition to solvents, such changes can be caused by a simple concentration dependence of the resonance frequencies. For example, with a particular concentration of benzyl alcohol in acetone the resonance of the CH_2OH group appears as a singlet, while in pure benzyl alcohol the expected AB_2 system is

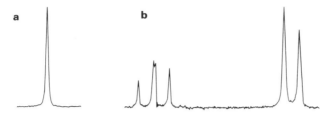

Figure 3.3   Concentration dependance of the CH_2OH group absorbsion of benzyl alcohol in acetone as solvent. (a) Acetone to benzyl alcohol ratio=7:5(v/v); (b) pure benzyl alcohol

observed (Figure 3.3). In the first case the determination of the coupling constant $J(CH_2OH)$ is not possible. Thus, the concentration dependence of the chemical shift can be used to increase the information obtainable from the spectrum.

Especially in steroid chemistry, specific *solvent effects* have been systematically studied and used to advantage. Particularly valuable for this purpose is benzene because of its high magnetic anisotropy and its tendency to form specific complexes with the solute. If benzene is used instead of chloroform, the proton resonance signals of individual methyl groups in steroids can often be differentiated. Figure 3.4 provides an example with the spectra of 3-oxo-4,4-dimethyl-5-α-androstane. Acetone, because of its dipole moment, is also suitable for the production of specific solvent effects.

The solvent dependence of spin–spin coupling is, in general, less marked than that of the resonance frequency, but in polar solvents variations in even this n.m.r. parameter have been observed. Thus, in formaldoxime **(5)** and in 1-chloro-2-

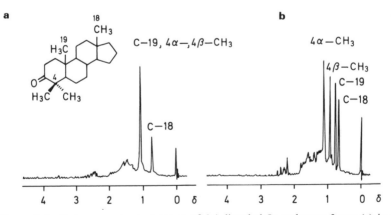

Figure 3.4   Proton resonance spectrum of 4,4-dimethyl-5α-androstan-3-one (a) in deutero-chloroform and (b) in benzene (after Ref. 1)

ethoxyethylene (**6**), to take two examples, variations in the geminal and vinical coupling constants over two and three bonds, respectively, have been measured in the range of 7.6–9.9 and 4.2–6.3 Hz.

Also noteworthy is the property of dimethyl sulphoxide to slow down the exchange of protons of OH as well as NH and NH_2 groups. This solvent is therefore used to advantage when spin–spin interactions of the type $H-C-O-H$, $H-C-C-O-H$ or $H-C-N-H$ are to be investigated. These coupling constants, which depend upon the stereochemistry of the bonds involved, cannot be determined in the presence of rapid proton exchange (cf. Chapter 9).

5   **6**

3. Tuning the spectrometer

Nuclear magnetic resonance spectrometers available today have been so simplified in their operation that the measurement of a spectrum can be accomplished with a few adjustments that are clearly described in the instrument instruction manual. Therefore, only a few general remarks need be added here. As mentioned already in Chapter 1, two different experimental techniques are principally available in n.m.r. spectroscopy: the *CW* and the *Fourier transform (FT) method*. For various reasons, the FT method has by now completely replaced the CW technique and practically all spectrometers used in chemical research laboratories are of the Fourier transform type. Only few CW instruments are still in use for ^{1}H n.m.r. measurements. Because the experimental procedures of the FT technique will be introduced in more detail in Chapter 7, in the following section only general aspects and a few points which are important for CW measurements will be discussed.

During all experiments fast rotation of the sample tube around its long axis is achieved by means of a small air turbine (cf. Figures 3.2 and 1.5). This has the effect of improving the field homogeneity because, as a result of the macroscopic movement of the sample, the individual nuclei are exposed to a time-averaged value of the external field B_0, the magnitude of which varies within certain limits over the sample volume. As Figure 3.5 demonstrates for the resonance signal of tetramethylsilane, this experimental trick is indispensable in obtaining sharp resonance signals. As a result of spinning the sample tube *spinning side bands* appear on both sides of the principal signal and at equal distance from it. The difference in frequency between the central signal and the side bands is equal to the rotational frequency of the sample cell, so that at higher spinning frequencies the

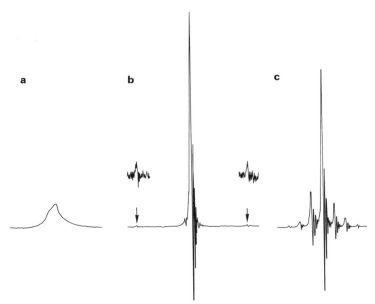

Figure 3.5 The effect of spinning the sample tube on the shape of the resonance signal: (a) without spinning; (b) with spinning, spinning side bands magnified; (c) with slow spinning (*c*.10 rps)

side bands move away from the central band and, in so doing, decrease in intensity. By correctly adjusting the spectrometer and sufficiently rapid spinning of the sample the intensity of the side bands becomes so low that a perturbation of the spectrum through this effect is not observed.

Since the chemical shifts are field dependent, it is important that the field/frequency ratio of the particular spectrometer is constant during measurement. For this purpose a *lock channel* is installed. The continuous recording of a reference signal allows one to detect even very small changes in the magnetic field strength and corrections for a possible field drift can be applied. Modern spectrometers use a *heterolock* system with the ^{2}H n.m.r. signal of the deuterated solvents as *lock signal*.

An important condition for recording well resolved n.m.r. spectra is the homogeneity of the external magnetic field B_0. Field homogeneity can be optimized using the reference signal in CW spectrometers and the ^{2}H lock signal in FT spectrometers which is recorded in the CW mode and displayed on the screen. With instruments that are regularly checked it is generally sufficient to optimize the *y*-gradient of the B_0 field if the instrument has an iron magnet, as is usually the case for CW instruments. With superconducting magnets of FT spectrometers, because of the different orientation of the sample cell (see above and Chapter 10), the *z*-gradient must be adjusted. Good field homogeneity is documented by the wiggles which follow the CW signal. The longer the train of wiggles, the better the homogeneity of the field (Figure 3.6a). In FT n.m.r. a similar phenomenon can be observed for the time signal S(t) (cf. Figure 1.6, p. 8), but not for the frequency signal $S(v)$ which is obtained after Fourier transformation. The signal shape should be symmetric if the sweep direction is changed (Figure 3.6b). For Fourier transform spectra the signal shape can be improved if the time signal $S(t)$ is subjected to certain mathematical operations before Fourier transformation. These aspects will be considered in Chapter 7.

In cases where great precision is required, the second pair of lines in the spectrum of 1,2-*dichlorobenzene* can be used to improve further the performance of the spectrometer. For a well adjusted instrument, the splitting here should approach the base line (Figure 3.6c). Thus, the better the *resolution* of a spectrometer, the smaller the frequency difference between the two resonance signals that can still be separately recorded or *resolved*.

A widely used test for the n.m.r. line shape provided by the spectrometer is the *hump test*, which is performed with the ^{1}H n.m.r. signal of chloroform (Figure 3.7).

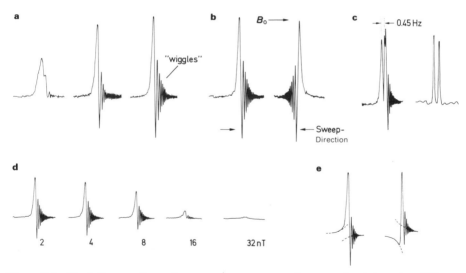

Figure 3.6 The influence of some instrument parameters on the n.m.r. signal shape in a CW experiment: (a) ^{1}H n.m.r. signal of a TMS with varying adjustment of the y-gradients; (b) symmetric signals for both sweep directions; (c) resolution test with the second signal pair in the ^{1}H n.m.r. spectrum of 1,2-dichlorobenzene: CW recording at 60 MHz (left), FT recording at 80 MHz (right). The resolution amounts to 0.45 Hz. In the CW spectrum the second small splitting is not real. It results from the wiggles of the first signal which are superimposed with the second signal. If the sweep direction is reversed, this effect is observed at the other line. At higher field strengths the spectrum changes due to the increase of chemical shifts and the resolution test is performed with a different line pair; (d) different r.f. frequency power in a CW experiment; for a value of 32 nT the resonance is practically saturated, i.e. $N_\alpha = N_\beta$ (see p. 6); (e) incorrect signal phase

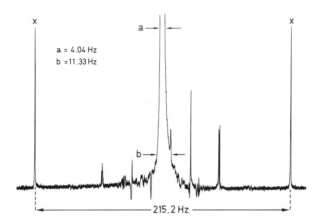

Figure 3.7 Humptest with the chloroform ^{1}H n.m.r. signal on a 400 MHz FT n.m.r. spectrometer with superconducting magnet. The ^{13}C satellites (x) belong to 1.1% $^{13}CHCl_3$. The splitting is caused by the $^{13}C,^{1}H$ coupling constants over one bond. The typical signal widths which characterize the line shape are indicated

For this purpose the signal is measured with high sensitivity in order to detect the ^{13}C satellite lines (see p. 214ff.). The line width of the central signal is then measured at the height of the satellites and at the base. The intensity of the rotational side bands should be smaller than that of the ^{13}C satellites.

After locking the spectrometer one can start the experiment. For CW spectrometers, the reference signal must be adjusted to the $\delta = 0$ value and the amplitude of the transmitter as well as the detector phase adjustment must be optimized (Figure 3.6d,e). For FT measurements, the pulse angle, the receiver gain and parameters that determine the spectral width and resolution (to be discussed in detail in Chapter 7) are chosen by the operator.

After recording the spectrum it is *integrated*. For very exact integrations it is necessary to determine the average of several individual measurements. Further, the integration is unreliable if the detector phase is incorrectly adjusted or if the signals to be integrated vary considerably in their relative intensities. Thus, the number of methylene groups in an aldehyde of the general structure $CH_3-(CH_2)_n-CHO$ could not be determined with certainty if $n > 20$. Finally, for overlapping signals a separate integration is not possible with the standard equipment. In these cases a curve analyser may be used if the line shape of both signals is known.

Since frequency measurements are of central importance for n.m.r. spectroscopy an independent check of the precision of the spectrometer is desirable. This is most easily done with the standard sample of chloroform and tetramethylsilane in $CDCl_3$. The difference of the resonance signals amounts to 7.27 ppm or the equivalent in Hz. The calibration of the spectra in Hz or ppm can be provided by modern spectrometers directly with the plotter. As the most important experimental parameter the measuring frequency, which determines the relation Hz/ppm, should always be given. Recommendations for recording n.m.r. spectra in standard fashion are given in the Appendix on p. 540ff.

4. Increasing the sensitivity

Compared to other spectroscopic methods, as for instance u.v. or e.s.r., n.m.r is relatively insensitive because of the small energy difference between ground and excited states. Therefore, an important goal of new experimental developments was always to minimize this disadvantage by improving the sensitivity. In particular, the n.m.r. spectroscopy of insensitive nuclei like ^{13}C or ^{15}N could only be expected to become routine methods if improvements in this direction were made.

If we look at the Boltzmann distribution of the nuclear spins introduced on p. 6, we find that increasing field strength should result in a higher population of the lower energy level. Theory predicts that the signal intensity should be proportional to B_0^2, in practical measurements one finds the factor $B_0^{3/2}$. Accordingly, the signal-to-noise ratio, as defined for the quartet of the methylene protons in ethylbenzene in Figure 3.8, improves from ca. 25:1 to ca. 50:1 if we compare a 60 MHz spectrometer with a 100 MHz spectrometer. This is equivalent to changing the field strength from 1.4 to 2.3 T. With the 400 MHz spectrometer and B_0 field of 9.2 T a ratio of 800:1 results.

These relations were the driving force behind the development of superconducting magnets which in the meantime have reached the considerable field strength of 17.5 T. This corresponds to a ^1H resonance frequency of 750 MHz. Aside from increasing technological difficulties in the construction of such high-field magnets and the rising operational costs, this development is limited by the fact that even small molecules start to orient in high magnetic fields and the spectra will become much more complicated (see Chapter 10).

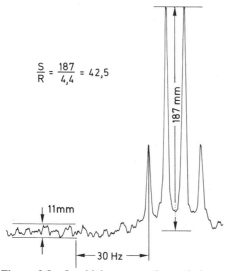

$$\frac{S}{R} = \frac{187}{4,4} = 42,5$$

187 mm

11mm

← 30 Hz →

Figure 3.8 Sensitivity test on the methylene quartet of 1% solution of ethylbenzene. The signal to noise ratio (*S/N*) is defined as the quotient of the average signal height, *S*, and the average noise level, *N*, which is determined by the relation N = noise height/2.5

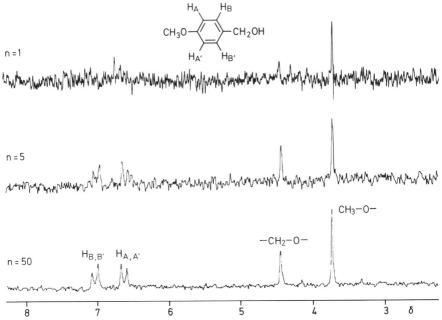

Figure 3.9 Spectrum accumulation of CW spectra by means of a CAT (computer of averaged transients) accessory for a 0.2% solution of 4-methoxybenzyl alcohol in carbon tetrachloride. The hydroxyl proton resonance could not be detected (cf. p. 23); n = number of scans; sweep time for a single scan 250 s

A further increase in sensitivity can be attained by increasing the observation time, t, for, according to theory, the electronic noise increases proportionally to $\sqrt{t}$ while the intensity of a coherent signal increases proportionally to t. The application of this principle became practical after minicomputers were available and one was able to *digitize* the n.m.r. spectrum. The spectral region is then divided into a number of channels (*words* in computer language) so that during the measurement a corresponding number of data points can be stored. Repeated recording allowed the summation of 50 or more individual CW spectra. Since signals originating from random noise vary in their intensity and, more important, their sign, whereas a true n.m.r. absorption always gives a positive response, the signal to noise ratio is improved. According to the above mentioned correlation between the observation time, t, and intensity, the improvement is equal to $\sqrt{n}$, where n is the number of times the spectrum is swept (Figure 3.9). As one can easily see, however, improving the sensitivity by spectral accumulation of CW spectra requires long measurement times. For example, for the third spectrum in Figure 3.9 already 3.5 hours were needed for an improvement of the signal-to-noise-ratio by a factor of 7. Very often much higher improvement factors are necessary, which leads to extremely time consuming experiments. For instance, the ^{13}C nucleus, especially important for organic chemistry, is less sensitive for the n.m.r. experiment than the proton by a

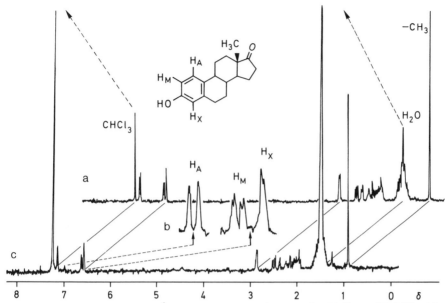

Figure 3.10 Fourier transform ^{1}H n.m.r. spectrum of estrone in CDCl$_3$ at 400 MHz; (a) concentration 3.7×10^{-3} molar (weight 0.5 mg), measuring time 20 s; (b) concentration 3.7×10^{-4} molar, measuring time 25 min. The signals of the residual CHCl$_3$ molecules in the solvent (0.04%) and the water traces disturb the spectrum; (c) enlarged signals of the aromatic protons from spectrum (b); one recognizes the ortho- and meta-^{1}H,^{1}H coupling which allows an assignment (see also the substituent increments for the $\delta(^1$H) values on p. 107)

factor of 5800, as calculated from its low natural abundance and its small magnetic moment (see Table 1.1 pp. 11). In order to remove this disadvantage 5800^2 scans would be necessary. Accordingly, sensitivity improvement by spectral accumulation became only attractive after the introduction of Fourier transform spectroscopy where the FID is recorded within 1 s or less, leading to a situation where several hundreds or thousands of transients can be accumulated in a relatively short period of time. Furthermore, additional methods to improve the sensitivity are available and the recording of a typical ^{13}C n.m.r. spectrum today needs hardly more than 30 min or 1 h measuring time. For sensitive nuclei we have a situation where already 1 mg of a compound or even less is sufficient for recording a ^{1}H n.m.r. spectrum (Figure 3.10). Concentrations which were reserved before for u.v. spectroscopy (10^{-5} molar) are now also within the reach of n.m.r. Highly purified solvents have to be used for such measurements in order to avoid disturbing signals from impurities of low concentration or the residual protons of the solvent molecules.

5. The measurement of spectra at different temperatures

For a number of reasons it is desirable that n.m.r. spectra can be recorded at different temperatures. The main application of variable-temperature n.m.r. spectroscopy is in the area of temperature-dependent n.m.r. line shapes, where information about rate processes, usually involving intramolecular dynamics, is obtained. This field will be covered extensively in Chapter 9. In addition, it is also possible to detect unstable intermediates at low temperatures, while on the other hand the solubility of a poorly soluble compound can be improved at elevated temperatures.

The spectrometers available today are routinely equipped with a *variable-temperature probe* permitting experiments between -150 and $+200$ °C. To accomplish this, nitrogen is usually run through the probe after being brought to the desired temperature in a cooling chamber containing liquid nitrogen or with an electric heater. The temperature of the gas flow can be checked by using a thermocouple and regulated automatically. Other cooling systems for low-temperature studies use the gas stream boiled off from a reservoir of liquid nitrogen to cool the sample chamber. The temperature is then controlled simply by varying the flow-rate.

Ideally, the *temperature measurement* should take place within the sample tube itself, but for technical reasons this is not always possible. In practice the temperature is determined by placing the thermocouple in the nitrogen stream directly below the n.m.r. tube or by recording the spectra of standard samples of methanol or ethylene glycol both before and after the spectrum of the sample under investigation is measured. With these compounds the chemical shift differences. Δv (in Hz), between the resonances of the CH_3 or the CH_2 protons, respectively, and the OH proton, are temperature dependent. Precise measurements have led to the following relationships, where v_0 is the spectrometer frequency used in MHz:

Low temperature, methanol:

$$175 - 330 \text{ K}; \ T = 403.0 - (29.46/v_0)|\Delta v| - (23.832/v_0^2)|\Delta v|^2 \qquad (3.4)$$

High temperature, ethylene glycol:

$$310 - 410 \text{ K}; \ T = 466.0 - (101.64/v_0)|\Delta v| \qquad (3.5)$$

These equations result in errors of only about ± 0.5 °C for pure aerated samples of the two substances, in which the line splitting due to spin–spin coupling has been eliminated by the addition of a trace (0.03% v/v) of concentrated hydrochloric acid. Linear correlations are given in the original literature [2] for certain sections of the low temperature region. If deuterated methanol, CD_3OD (99.8%), is used, the signals of the residual CD_2H and OH groups, which belong to different molecules, yield the following equation with an error of ± 0.7 K (Ref. 3):

$$180 - 300 \text{ K}; \ T = 398.7 - (26.94/v_0)|\Delta v| - (24.436/v_0^2)|\Delta v|^2 \qquad (3.6)$$

If the signals of the methanol and the ethylene glycol do not interfere with the spectrum of the sample under investigation, the most accurate temperature measurement in the region between -100 and $+140$ °C is obtained by use of the n.m.r. tube illustrated in Figure 3.2c, in which the capillary is filled with methanol or the ethylene glycol. This has the advantage that temperature and spectrum are measured simultaneously.

The *choice of solvent* presents special problems in variable-temperature measurements. At high temperatures dimethyl sulphoxide, hexachlorobutadiene, decalin, and nitrobenzene have been used successfully. Of course, the highly volatile tetramethylsilane must be replaced in these experiments with a different reference substance. Cyclosilane-d_{18} (7) with a boiling point of 208 °C and a singlet at δ 0.327 appears to be suitable. At low temperatures, acetone-d_6 and carbon disulphide, perhaps mixed with chloroform, can be used down to about -100 °C. Below -100 °C, fluorinated hydrocarbons such as trifluorobromomethane and difluorodichloromethane are usually used. Dimethyl ether and carbon oxysulphide are also suitable because of their low freezing points (-138.5 and -138 °C, respectively). Frequently only mixtures of several components lead to satisfactory results.

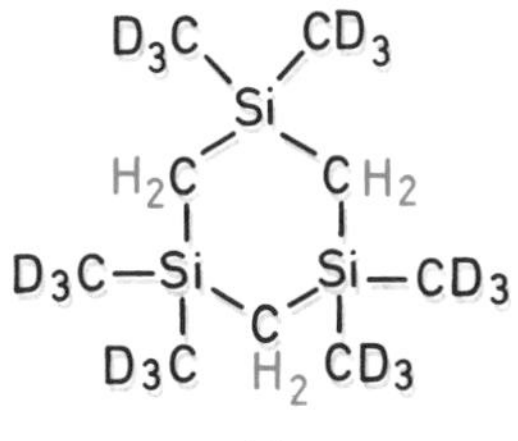

7

6. References

1. N. S. Bhacca and D. H. Williams, *Application of NMR Spectroscopy in Organic Chemistry*, Holden-Day, San Francisco, 1965.
2. A. L. Van Geet, *Anal. Chem.* **42**, 679 (1970); **40**, 2227 (1968).
3. E. W. Hansen, *Anal. Chem.* **57**, 2993 (1985).

Recommended Reading

Review Articles

(a) P. Laszlo, Solvent Ects and Nuclear Magnetic Resonance in F 5, **3**, 231 (1967).
(b) J. Ronayne and D. H. Williams, Solvent Ects in Proton Magnetic Resonance Spectroscopy, in F2, **2**, 83 (1969).
(c) J. L. Deutsch and S. M. Poling, The Determination of Paramagnetic Susceptibility by NMR, *J. Chem. Educ.*, **46**, 167 (1969).
(d) J. Löliger and R. Scheffold, Paramagnetic Moment Measurement by NMR, *J. Chem. Educ.*, **49**, 646 (1972).

4 CHEMICAL SHIFT AND SPIN–SPIN COUPLING AS FUNCTIONS OF STRUCTURE

In Chapter 2 it became clear that the dependence of proton resonance frequencies and spin–spin coupling on chemical structure leads to an abundance of important information that is of both theoretical and practical interest. The rapid development of nuclear magnetic resonance doubtlessly has been due in great part to the fact that it was recognized very early as a method of great utility in the solution of one of the central problems of chemical research—the determination of structure. Each new measurement provided data that proved to be characteristic for a particular class of compounds or structural units. A large number of empirical correlations between n.m.r. parameters and structure were discovered in this way. The abundance of experimental results also whetted interest in and advanced the understanding of the theoretical basis of these correlations so that now the majority of these effects can be satisfactorily explained. The models developed for these interpretations are described in this chapter.

1. The origins of proton chemical shifts

Having provided a general summary of the characteristic absorption regions of the most important proton types in organic compounds in Figure 2.6 (p. 24), we shall now discuss, in terms of the approach introduced on p. 17, the individual contributions of different structural elements to the chemical shift. Since for protons the previously mentioned local paramagnetic contribution to the screening constant is negligible, because of the large energy gap between the 1s and the 2p orbitals, we can confine ourselves at the outset to the consideration of two effects. They are

1. The *local diamagnetic* contribution of the electron cloud around the proton under consideration ($\sigma_{\text{dia}}^{\text{local}}$), and

2. The effect of *neighbouring* atoms and groups in the molecule (σ').

Thus, within this approximation the influence of substituents and neighbouring atoms is two-fold: firstly, they will effect $\sigma_{\text{dia}}^{\text{local}}$ through changes of the electron density at the proton caused by inductive and mesomeric mechanisms, and secondly, electron circulations induced by the external field B_0 within these neighbouring atoms and groups will give rise to magnetic moments, i.e. secondary fields that change B_{local} at the proton. In addition, electric field and van-der-Waals effects may be considered, and also the influence of the surrounding medium. Consequently, any change in proton screening may be expressed as a sum of several terms:

$$\Delta\sigma = \Delta\sigma_{\text{dia}}^{\text{local}} + \Delta\sigma_{\text{magn}} + \Delta\sigma_{\text{el}} + \Delta\sigma_{\text{W}} + \Delta\sigma_{\text{med}} \qquad (4.1)$$

where the last four contributions stand for the magnetic, electric field, van der Waals, and medium effects.

1.1   THE INFLUENCE OF THE ELECTRON DENSITY AT THE PROTON

As we have already mentioned, the diamagnetic contribution to the shielding of a nucleus may be calculated from the Lamb formula (equation 2.2), which is strictly applicable, however, only in the case of spherical symmetry, i.e. for the neutral hydrogen atom. Here a value of 17.8 ppm results for σ_{dia}. If inductive effects present in a molecule reduce the electron density in the hydrogen 1s orbital, deshielding is expected. Thus, the screening constants in the hydrogen halides not unexpectedly

fall in the order HF<HCl<HBr<HI, which is, however, also affected by the
magnetic properties of the halide atoms.

1.2   THE INFLUENCE OF THE ELECTRON DENSITY AT
NEIGHBOURING CARBON ATOMS

In organic compounds protons are not usually bonded directly to electronegative
elements. Nevertheless, their influence has far reaching effects through the carbon
skeleton of a compound and the charge density at the neighbouring carbon atom
becomes a determining factor for the resonance frequency of a proton. In order to
illustrate the expected relationship between the proton chemical shift and the
electronegativity of the substituents, Figure 4.1 shows results for the methyl halides.
The δ values found are consistent with the decreasing electronegativity of the
halogens in the order F>Cl>Br>I.

	CH_3F	CH_3Cl	CH_3Br	CH_3I	CH_3H
$\delta(CH_3)$	4.13	2.84	2.45	1.98	0.13
E(Pauling)	4.0	3.0	2.8	2.5	2.1

A similar correlation between the resonance frequency of the methyl protons and
the polarity of the C–X bond exists for other methyl derivatives of the type CH_3X
and Figure 4.2 provides a general summary of the phenomenon. As is evident, the
increased shielding of the protons in the series X = Hg, Sn, Cd, Zn, Al, Mg, and Li
parallels the growing ionic character of the corresponding metal–carbon bonds
which according to Pauling is (in the above order) 9, 12, 15, 18, 22, 35, and 43%.

Finally, the magnitude of the inductive effect and its propagation through the C–C
bond framework is clearly illustrated with the spectrum of nitropropane (Figure 4.3,
p. 73). Here, $\Delta\delta$ values of 3.45, 0.72, and 0.12 ppm are found for the a-, b-, and c-
protons, respectively, if the proton resonance frequencies propane (δ_{CH_3} 0.91, δ_{CH_2}
1.33) are used as a reference.

Linear relations such as those shown in Figure 4.1 suggest that n.m.r. data can be
used as a measure of electronegativities. Appropriate equations have indeed been
proposed, but they must be used with caution because additional effects usually play
an important role in determining proton resonance frequencies.

The significance of such other sources of proton shielding is demonstrated by
observations made on the *ethyl* halides. In Figure 4.1b,c the expected change in the
resonance frequency—low-field shift with increasing electronegativity—is noted for
the methylene protons while the reverse trend is observed for the methyl protons. In
this case, owing to geometrical factors to be discussed later, the magnetic anisotropy
of the C–H and C–X bonds is important and the magnetic contribution to the
shielding constant, $\Delta\sigma_{magn}$, dominates.

In unsaturated compounds in which the carbon of the C–H bond under
consideration has a positive or a negative partial charge, shifts to lower and higher
field, respectively, are observed. When the proton resonances of tropylium cation

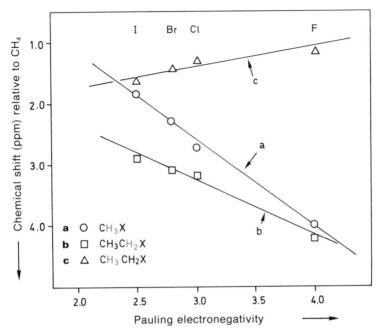

Figure 4.1 Correlation between the chemical shifts of alkyl halide protons and the electronegativity of the halogens

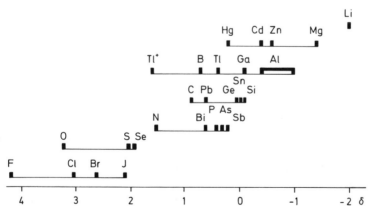

Figure 4.2 Proton resonances of the methyl derivatives of the representative elements (after Ref. 1)

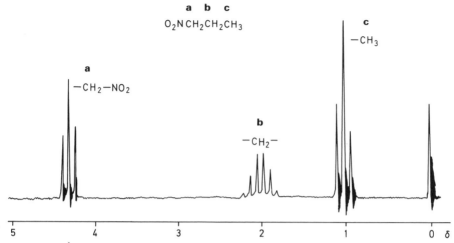

Figure 4.3 ^{1}H n.m.r. spectrum of nitropropane

and cyclopentadienyl anion are compared with that of benzene, $\Delta\sigma$ values of -1.90 and $+1.90$ ppm result, the signs and magnitudes of which reflect the charge deficiency and the charge excess, respectively, of $\frac{1}{7}$ and $\frac{1}{5}$ of an electron per C atom.

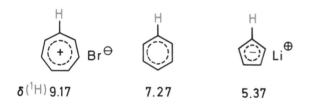

This observation led to the discovery of a linear correlation between the π-electron density and the chemical shift of the protons in these compounds that, as is shown in Figure 4.4, also extends to other aromatic ions. From these data the empirical relation

$$\Delta\sigma = 10.0 \, \Delta\rho \qquad (4.2)$$

has been developed, where $\Delta\sigma$ is the change in the shielding constant and $\Delta\rho$ is the change in the π-electron density relative to benzene.

Physically, this effect can be interpreted as the influence of the electric field of the partial charge residing in the $2p_z$ orbital of the carbon atom on the electron cloud of the carbon–hydrogen bond. It results in a shift of the electrons either towards the carbon or towards the hydrogen and the proton is either deshielded or shielded, respectively. We shall return to this model later.

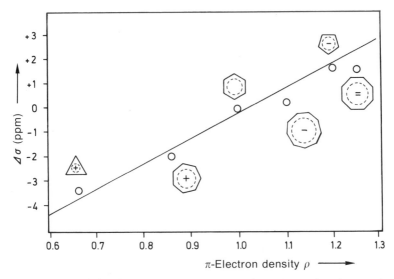

Figure 4.4 Relation between proton resonance frequency and π-electron density at the carbon atoms in aromatic ions. The regression analysis of the points shown yields $\Delta\sigma = 9.54\Delta\rho$ with a standard deviation of 0.65 ppm and a correlation of coefficient of 0.993

The π-electron density at the corresponding carbon atom is also of significant importance for the resonance frequency of protons in *substituted benzenes*, notably for those in positions *ortho* and *para* to the substituents. Here too a linear correlation exists between charge density changes, $\Delta\rho$, obtained from Hückel MO calculations, and the changes in the proton shielding constant, $\Delta\sigma$ (Figure 4.5). In this case the proportionality constant is 12.7.

Pronounced charge density effects for ^1H chemical shifts are observed in monosubstituted olefines, where strong substituents can lead to a considerable polarization of the π-bond. As shown by the mesomeric structures **a** and **b**, the $-M$ and the $+M$-effect result in deshielding and shielding for the β-proton, respectively:

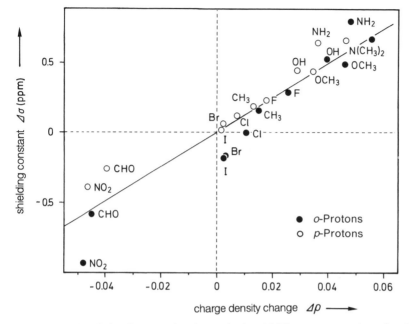

Figure 4.5   Correlation between the change in the shielding constants, $\Delta\sigma$, of *ortho-* and *para*-protons in monosubstituted benzenes and the corresponding electron density change $\Delta\rho = \rho(C_6H_5X) - 1.00$, relative to benzene (after Ref. 2)

A comparison of the resonance frequencies in methylvinylketone and methylvinyl ether with the $\delta(^1H)$ value of ethane documents this effect most clearly:

It is, therefore, not surprising that olefinic protons in some cases are stronger shielded than protons at saturated carbon atoms, as the spectrum of the bicyclic lactone shown in Figure 4.6 convincingly demonstrates.

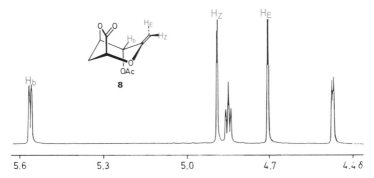

Figure 4.6 Partial 200 MHz ^{1}H n.m.r. spectrum of a bicyclic lactone with strong shielding of the olefinic protons H_E and H_Z (after Ref. 3)

Analogous observations regarding charge density effects are made for saturated compounds and in Table 4.1 some representative data for carbanions and carbocations are summarized. It can be seen that the protons in carbocations are strongly deshielded, but the influence of the positive charge in saturated systems rapidly decreases with the distance from the charge. The data for the dimethyl carbocation and the dipropylcyclo-propenylium ion illustrate this effect. From the results for the allyl cation it can be concluded that, contrary to predictions of simple resonance theory or the Hückel MO model, the central carbon atom must bear considerable positive charge. In the allyl anion, to the extent that it exists in allylmagnesium bromide, it is seen that the negative charge is concentrated primarily at the terminal carbon atoms, doubtlessly a consequence of electron repulsion. The saturated carbanions show strong shielding effects only for the α-protons with resonances at high field from the tetramethylsilane signal. On the other hand, extensive delocalization of charge is again indicated by the δ values of the aromatic protons in benzyllithium and phenyldimethyl carbenium ion, where the *ortho-* and *para*-protons are affected most. Noteworthy in this context is the comparison of the carbenium ion results with the δ-values of trimethylanilinium ion, where only a $+I$ effect is operative. While the resonance of the *meta*-protons is influenced least in the former, a substituent effect decreasing in the order *ortho>meta>para* is clearly recognized in the trimethylanilinium ion.

In order to conclude our discussion, we mention that the chemical shifts in molecules like carboxylic or amino acids may show a strong dependence on pH which can be used for pK_a value determinations. For fast proton exchange an average spectrum is observed that results from the protonated and deprotonated species that are in equilibrium. Titration can then be studied by n.m.r. spectroscopy as shown in Figure 4.7 for alanine. Here, with increasing pH the doublet of the methylproton resonance moves to higher field (Figure 4.7b). The inversion point of the titration curve (Figure 4.7a) yields a pK_a value of 9.6 because at a concentration ratio $[A^-]/[HA] = 1$ for anion and acid $pH = pK_a$.

Table 4.1 Proton resonances in carbocations and carbanions

For more precise evaluations the use of the Henderson–Haselbalch equation is to be preferred. The ratio $[A^-]/[HA]$ is then replaced by the quotient $(\delta_{max}-\delta)/(\delta-\delta_{min})$ (equation 4.3); δ_{max} and δ_{min} are the ^1H frequencies under acidic and basic conditions, respectively, which are independent of pH, and δ is the pH dependent shift:

$$pH = pK_a + \log\frac{\delta_{max} - \delta}{\delta - \delta_{min}} \qquad (4.3)$$

If the logarithmic term of equation (4.3) is plotted ag.   ‑traight line re⸱
and the pK_a value is obtained at the intersection with th⸱

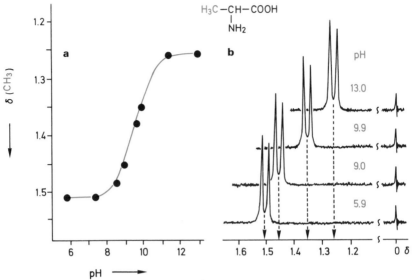

Figure 4.7 pH Dependence of the methyl ^{1}H resonance in alanine; (a) titration curve from a complete measurement series in H_2O; (b) methyl ^{1}H signal at selected pH values

---

Exercise 4.1 Calculate the δ values of the protons in (a) anisole and (b) the triphenylmethyl cation on the basis of the given π-electron densities using equation (4.2) and δ(benzene) = 7.27 as reference point. Also use 12.7 as the proportionality factor in (a) and compare the results with the experimental data in Table 4.6 (p. 105).

1.000 1.020

1.018 ⟨benzene ring⟩—O—CH₃

a

0.94 0.95

0.81 ⟨benzene ring⟩—C⊕(C₆H₅)(C₆H₅)

b

1.3 THE INFLUENCE OF INDUCED MAGNETIC MOMENTS OF NEIGHBOURING ATOMS AND BONDS

The diamagnetic shielding of a proton by its 1s electron density is relatively small compared with the shielding of nuclei of heavier atoms that have filled inner shells. Therefore, additional effects such as the influence of magnetic dipoles at neighbouring atoms or groups, that alter the local magnetic field responsible for the resonance frequency, are much more significant in determining the shift of the proton resonance than that of heavier nuclei.

Let us first consider a *diatomic* molecule AB. Through the external field B_0 a magnetic moment μ_A, that we consider as a localized point dipole at the centre of A, is induced at A; μ_A is proportional to the magnetic susceptibility, χ_A, of A and can be broken down in a Cartesian coordinate system into its components $\mu_A(x)$, $\mu_A(y)$, and $\mu_A(z)$. Its contribution to the shielding of the nucleus B is given by

$$\Delta\sigma \frac{1}{4\pi} \sum_{i=x,y,z} \chi_A^i (1 - 3\cos^2\theta_i)/R^3 \qquad (4.4)$$

where θ is the angle between the direction of $\mu_A(xyz)$ and the A–B bond axis and R is the distance between the centre of A and the nucleus B.

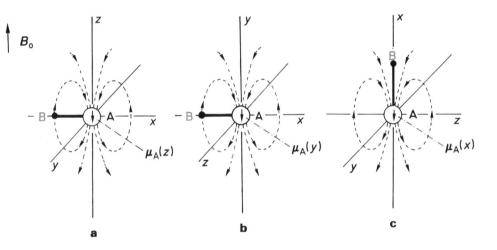

In the geometric arrangement (a) the secondary field at B is parallel to B_0. The induced field thus augments the externally applied field and the resonance of B appears at lower field, i.e. B is deshielded. The same situation develops when the direction of B_0 and the y-axis of the molecular coordinate system coincide (b). However, in arrangement (c) the induced field at B is opposed to B_0 and shielding results. In solution, the molecules of the sample are undergoing rapid rotation and averaging takes place. According to the factor $1-3\cos^2\theta$ in equation (4.3) the resulting net effect is zero as long as the components $\chi_A(x)$, $\chi_A(y)$ and $\chi_A(z)$ of the susceptibility χ_A have the same values in which case the group A is said to be *magnetically isotropic*. When this is not the case, A possesses a *magnetic anisotropy* $\Delta\chi$ which, according to its orientation, can effect a paramagnetic or a diamagnetic shift of the resonance frequency of the nucleus B, By means of the relation

$$\Delta\chi = \chi_\parallel - \chi_\perp \qquad (4.5)$$

the anisotropy of a group with an axis of symmetry is defined as the difference between the susceptibilities parallel and perpendicular to the axis of symmetry. The magnetic contribution to the chemical shift of individual protons can then be determined by the McConnell equation:

$$\Delta\sigma = \tfrac{1}{3}\Delta\chi(1 - 3\cos^2\theta)/4\pi R^3 \qquad (4.6)$$

if the magnitude and sign of the magnetic anisotropy $\Delta\chi$ of a group with axial symmetry is known. For example, for two points in the vicinity of a C–C single bond that have a distance $R = 0.3$ nm from the centre of the bond the following results are obtained with a value of $\Delta\chi_{C-C} = 140 \times 10^{-36}$ m^3 per molecule ($\theta = 0°$ and $90°$, respectively):

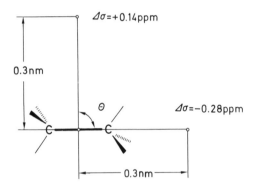

Generally, the results of equation (4.6) are graphically represented by shielding cones for the various groups, whose nodal plane ($\Delta\sigma = 0$) is fixed at the 'magic angle' of $54.7°$.

An experimental verification of the proposed shielding behaviour for the C–C single bond is found in the case of cyclohexane. Here, a difference of about 0.5 ppm exists between the resonance frequencies of the axial and equatorial protons that can be measured at low temperature where the rate of chair–chair interconversion (cf. p. 363) is slow on the n.m.r. time scale. Obviously H_a is more strongly shielded than H_e.

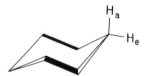

This differential influence on the resonance is an important aid in conformational analysis. As the example of α- and β-methoxygalactose (**8a** and **8b**, respectively) shows, there is even an effect for axial and equatorial methoxy groups.

It is of significance for proton chemical shifts that almost all chemical bonds are magnetically anisotropic and the circulations of electrons induced by the external

8a   **8b**

magnetic field therefore yield important contributions to the proton shielding constants.

Of the *multiple bonds* the $C=C$ and the $C=O$ double bonds and the $C\equiv C$ and $C\equiv N$ triple bonds possess particularly strong magnetic anisotropies. The special resonance position observed for the protons of acetylene mentioned at the outset is explained on this basis. According to Figure 4.8, these protons lie in the shielding region of the electron circulation induced by the magnetic field, that is the magnetic anisotropy is negative, contrary to that for a single bond. Also, a comparison of the resonances of the methyl protons in toluene and in propyne (Table 2.1) demonstrates the increased shielding near the $C\equiv C$ bond axis. On the other hand, in the regions alongside the triple bond deshielding is expected. An experimental verification of this effect is found in 4-ethynylphenanthrene (**9**), where the chemical shift of H(5) is 1.71 ppm downfield from the resonance of the same proton in phenanthrene itself. Similar considerations apply to the nitrile group.

The magnetic properties of acetylene are further illuminated if one considers the case in which the bond axis and the direction of B_0 are perpendicular (Figure 4.8b). The π-electron circulation is now hindered, a situation that leads to a paramagnetic moment in the centre of the bond. Again the protons are shielded.

For quantitative calculations with equation (4.6), apart from the geometrical data, one must know the $\Delta\chi$ values. Those used frequently for axial symmetric groups like C–H, $C=C$ and $C\equiv C$ bonds ($\Delta\chi_{C-H} = 90$, $\Delta\chi_{C-C} = 140$, $\Delta\chi_{C\equiv C} = -340 \cdot 10^{-36}$ m^3/molecule) have proven valuable in practical cases but cannot be regarded as completely secured. Clearly, the experimental determination of such bond increments is difficult. Furthermore, in view of the approximations used for deriving equation (4.6), the precision of the calculated results should not be

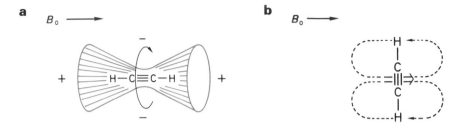

Figure 4.8   Schematic representation of the magnetic anisotropic effect of the triple bond

overestimated. For many applications, however, already the sign of the resonance shifts to be expected can be of significance.

With the $C=C$ and $C=O$ double bonds the situation is more complicated as a consequence of the fact that these groups have lost cylindrical symmetry. Experience has shown, however, that to a good approximation the shielding effects of both groups as well as those of the nitro group can be represented by the diagrams given in Figure 4.9.

Thus, for compounds **10** and **11**, in which the protons of interest are situated above the double bond system, the resonance frequencies are shifted toward higher field strength. In 1,3,5-cycloheptatriene (**12**) the resonances of the methylene protons can be determined at low temperature (cf. p. 365). In this case the quasi-axial proton is more strongly shielded by the C3–C4 double bond. In the region of the nodal surface between shielding and deshielding reliable predictions concerning the influence of a double bond on the proton resonance frequency are difficult to make. This results on the one hand from the approximations inherent in equation (4.5) and on the other hand from the uncertainty with which the geometry of the molecule under investigation is usually known. For example, the resonance of the *syn* proton in norbornene (**13**) is found at lower field than that of the *anti* proton, while in substituted norbornenes this order is reversed, as the example of the isomeric pair **13a**/**13b** shows.

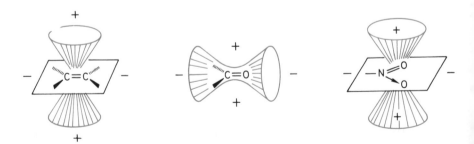

Figure 4.9   Schematic representation of the magnetic anisotropic effect of the carbon–carbon double bond, the carbonyl group, and the nitro group

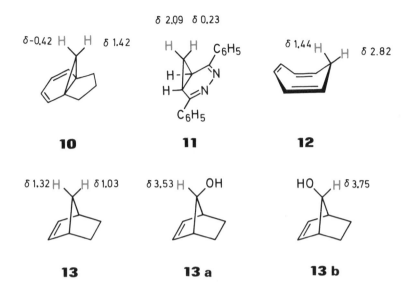

10   11   12

13   13 a   13 b

The resonance of olefinic protons that are shifted toward lower field compared with proton resonances in saturated hydrocarbons is an indication of the extent of the deshielding region of the C=C double bond, but the different hybridization of the carbon may also be important. The paramagnetic shift of the central protons in 1,3-butadiene (14) that exists almost completely in the planar *s-trans* conformation as well as the low position of the vinyl proton resonance in 1,1,2,5,6,6-hexamethyl-1,3,5-hexatriene (15) are, however, in good agreement with the predictions.

14   15

The particularly low field absorption of the *aldehyde proton* resonance is the result of the combined electronic and magnetic effects. Possibly the dipole moment of this group plays an additional rôle (cf. Section 1.6). Also, compounds 16 and 17 distinctly show the deshielding effect of the C = O function on neighbouring protons which lie in the nodal region of the π-bond.

In α,β-*unsaturated ketones* and *aldehydes* the participation of resonance structures such as 18b is of special significance. As a result, the chemical shift is dominated by electronic effects and the β-protons are strongly deshielded. In the case of malonic anhydride mesomeric and anisotropic effects work in concert and the resonance

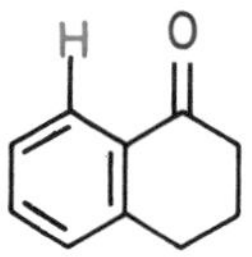

$\Delta\sigma \sim -0.7$ ppm

16

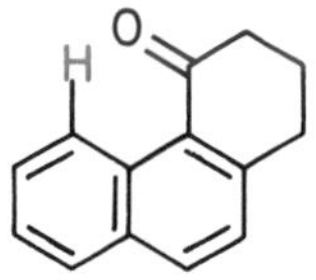

$\Delta\sigma \sim -1.8$ ppm

17

18 a   **18 b**

frequency of the olefinic protons lies at especially low field. In diethyl fumarate (**20**) and cyclopentenone (**21**), similar situations are found. In contrast, the olefinic protons of diethyl malonate (**22**) are shielded because here the *cis*-position of the carbethoxy groups distorts the coplanar arrangement of the π-system, thereby reducing the deshielding caused by the mesomeric effect and the diamagnetic anisotropy of the $C = O$ group.

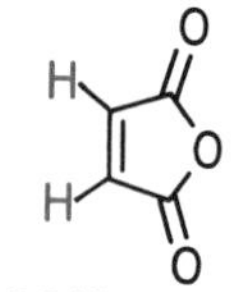

δ 7.10

19

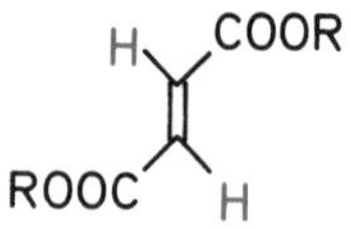

δ 6.83

20

δ 6.10

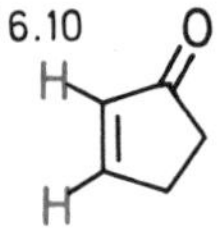

δ 7.71

21

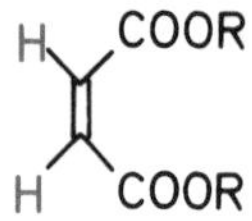

δ 6.28

22

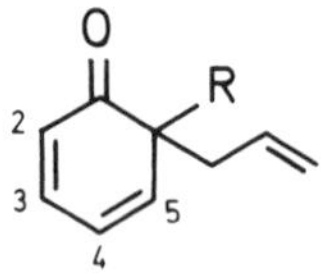

	ρ	δ
2	1.007	5.93
3	0.969	7.07
4	1.004	6.28
5	0.986	6.38

23

Finally, electronic effects again dominate in the case of 2,4-cyclohexadiene-1-one (**23**), where the proton resonance frequencies follow the calculated π-electron densities. Thus the proton H-2 at the carbon atom with the largest charge density absorbs at the highest field, while H-3 bonded to the carbon with the lowest electron density is strongly deshielded. This brief analysis shows that, as a rule, only the consideration of all factors that are responsible for the variation of the shielding constants allows for a satisfactory interpretation of the experimental findings.

Exercise 4.2 Explain on the basis of the effects just discussed the chemical shifts of the olefinic protons in the olefinic compounds **a–f**

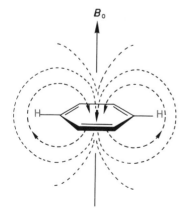

1.4 THE RING CURRENT EFFECT IN CYCLIC CONJUGATED π SYSTEMS

A special case to which reference has already been made is encountered with the proton resonance of benzene. As will be shown in this section, the reduced shielding of aromatic protons as compared to olefinic protons is caused by electronic circulations that cover the entire molecule. In terms of a simple model, an aromatic molecule can be visualized as a current loop where the π-electrons are free to move on a circle formed by the σ framework. If these compounds are subjected to the external magnetic field B_0, a diamagnetic ring current is induced. The secondary field resulting from this current can then be approximated by the field of a dipole opposed to B_0 and centred in the middle of the ring (Figure 4.10). As a result, protons in the molecular plane and outside the ring are deshielded. Conversely, protons in the region above or below the plane of the ring are strongly shielded.

Figure 4.10 Secondary magnetic field of a benzene ring

This approach, that originated with Pauling, was first formulated quantitatively for proton magnetic resonance by Pople. If the benzene ring is considered as a circular wire perpendicular to the direction of the field B_0, the π-electrons move around the σ framework with the Larmor frequency

$$\omega = \frac{\mu_0}{4\pi} \frac{eB_0}{2m_e} \tag{4.7}$$

where e is the charge of an electron, B_0 the external field strength, m_e the mass of an electron and μ_0 the permeability of free space. In a system with six electrons the current intensity is

$$i = \frac{3e\omega}{\pi} = \frac{\mu_0}{4\pi} \frac{3e^2 B_0}{2\pi m_e} \tag{4.8}$$

The magnetic properties of this ring current are now approximated as due to a magnetic dipole at the centre of the ring the magnitude of which is given by

$$\mu = i\pi r^2 \tag{4.9}$$

or, in combination with equation (4.8), by

$$\mu = \frac{\mu_0}{4\pi} \frac{3e^2 B_0 r^2}{2m_e} \tag{4.10}$$

where r is the radius of the ring. The secondary magnetic field of this dipole at a proton located in a distance R from the centre of the ring is μ/R^3 and, according to equation (4.1)

$$B' = -B_0 \sigma = \frac{\mu_0}{4\pi} \frac{3e^2 B_0 r^2}{2m_e R^3} \tag{4.11}$$

The contribution to the shielding constant is then given by

$$\Delta\sigma = -\frac{\mu_0}{4\pi} \frac{e^2 r^2}{2m_e R^3} \tag{4.12}$$

where the statistical factor of 1/3 is introduced to account for the situation in which the plane of the ring is oriented parallel to B_0 and no ring current is induced.

With the known values for e, m_e and μ_0 as well as the data $r = 0.14$ nm and $R = 0.25$ nm, $\Delta\sigma = -1.77 \times 10^{-6}$ or $\Delta\delta = +1.77$ ppm. This result compares well with the shift difference between the resonance frequencies observed above for the olefinic protons of 1,3-cyclohexadiene (δ 5.8) and the protons in benzene (δ 7.3).

A more exact analysis avoids the point dipole approximation considering that the density of the π-electrons is greatest where the carbon $2p_z$ orbitals overlap most strongly. This leads to two current loops, one above and one below the plane of the σ-bonds. For protons within the perimeter of the benzene ring an increased shielding results.

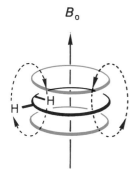

The magnitude of the induced field B' and the change in the shielding constant, $\Delta\sigma$, have been calculated and tabulated for the benzene nucleus. The Appendix contains a graphical representation of these results that can be used to calculate the contribution of a phenyl group to the chemical shift of a proton in a compound of interest. Moreover, by analogy with the simple ring current model, equation (4.6) can also be used for more distant protons if a value of $\Delta\chi = -630 \times 10^{-36}$ m³/ molecule is used for the diamagnetic anisotropy of benzene, which is assumed to originate from the centre of the ring.

Because of the inverse proportionality of $\Delta\sigma$ to R^3, the ring current model allows a qualitative interpretation of the spectra of *polynuclear* aromatic compounds if the observed shift is considered the sum of the contributions of the individual rings. Thus, the α-protons in naphthalene absorb at lower field than the β-protons because the contributions of the two rings are more important at the α-position since the latter is closer to both rings. The order of the proton resonance frequencies in anthracene can be explained in the same fashion. Here one finds $\delta_\gamma > \delta_\alpha > \delta_\beta$ (Figure 4.11). The calculation of $\Delta\sigma$ values in these systems can be made with the help of equation (4.13a).

$$\Delta\sigma = -\frac{\mu_0}{4\pi} \frac{e^2 r^2}{2m_e} \sum_i R_i^{-3} \qquad (4.13a)$$

or

$$\Delta\sigma[\text{ppm}] = -0.0276 \sum_i R_i^{-3} \qquad (4.13b)$$

if R_i is given in nm. Thereby, the total efect is obtained as the sum of the contributions of individual benzene units; R_i is the distance of the proton of interest from the centre of the ith ring.

Exercise 4.3 Calculate the ring current effect $\Delta\sigma$ (relative to benzene) for H-4 and H-9 in phenanthrene using equation (4.6).

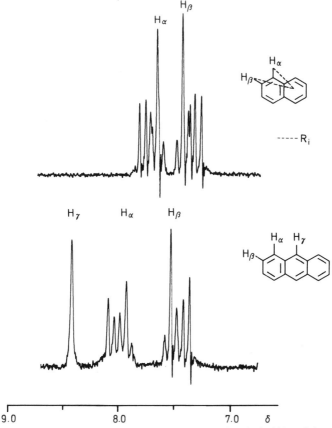

Figure 4.11   Correlation between the relative chemical shifts of the proton resonances in naphthalene and anthracene and the distance R_i of the proton from the centre of a specific benzene ring

In *non-alternating* hydrocarbons such as azulene the position of the individual proton resonances can be determined satisfactorily only if, in addition to the ring current effect, the different charge densities at the individual carbons are taken into consideration.

The results discussed above for benzene are fully supported by the proton resonance frequencies of the *annulenes*, a series of cyclic conjugated π-systems with more than six π-electrons. The three compounds 1,6-methano[10]annulene (**24**), *trans*-15,16-dimethyl-15,16-dihydropyrene (**25**) and [18]annulene (**26**) are presented here as examples.

Together with benzene, these molecules belong to the group of annulenes with $4n + 2$ π-electrons ($n = 0, 1, 2, ...$) that, following the well known *Hückel rule* possess aromatic character. The cyclic delocalization of the π-electrons in the ground state of these systems can therefore be demonstrated by means of n.m.r. spectroscopy as a ring current effect. According to a proposal of Elvidge and Jackman the presence of such a ring current effect is taken as a qualitative criterion for aromatic character. Such molecules are called *diatropic*.

δ (Ring) 7.27; 6.95
δ (CH₂) -0.51

δ (Ring) 8.14 - 8.64
δ (CH₃) -4.25

δ (H outer)   9.28
δ (H inner) - 2.99

24   **25**   **26**

How, on the other hand, do the annulenes with $4n$ π-electrons behave in a magnetic field? For these compounds quantum mechanical calculations predict a paramagnetic ring current effect with just the opposite consequences for the proton resonance frequencies as its diamagnetic counterpart discussed above. Protons within the perimeter are now deshielded, whereas those outside and in the plane of the ring are shielded.

This different behaviour of $4n+2$ and $4n$ π-electron systems can be rationalized with the help of a simple quantum mechanical model. Let us consider the movement of an electron along a circular path with a circumference L so that its wavelength λ can take only certain values. As a consequence of this, the electron can exist only in certain states, the so-called *eigenstates*. This phenomenon is similar to the situation postulated in Chapter 1 for the energy of a proton in an external magnetic field. In order for the electron to 'fit' the circle, the condition $L = q\lambda$ with $q = 0, \pm 1, \pm 2, \ldots$, obviously must be met. This is the *quantum condition* for our problem and q is the quantum number that characterizes each eigenstate:

$q = 0$ $q = \pm 1$ $q = \pm 2$

Following de Broglie, the momentum, p, of an electron is given by the relation $p = h/\lambda$, and the kinetic energy is then

$$E = mv^2/2 = p^2/2m = h^2/2m\lambda^2 = h^2q^2/2mL^2 \qquad (4.14)$$

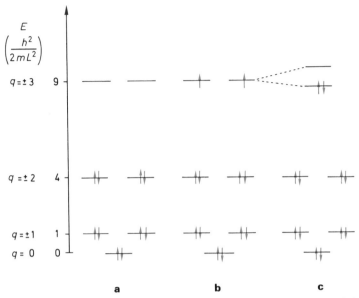

Figure 4.12 Energy level diagram for the model of an 'electron on a circle'

Accordingly, for each quantum number there is a corresponding energy value, the *eigenvalue*, and our model leads to the energy level diagram represented in Figure 4.12a.

If we choose the model of the 'electron on a circle' for the description of the π-electrons in cyclic conjugated systems, the energy level diagram must be filled with electrons according to the *aufbau principle*, that is, with regard to the Pauli exclusion principle and Hund's rule. Consequently, in a $4n+2$ π system a closed shell results (Figure 4.12a), and the occupied eigenstates or *orbitals* produce a diamagnetic contribution to the magnetic susceptibility. In contrast, in the $4n$ π systems the highest occupied orbitals contain only one electron each, the spins of which are unpaired (Figure 4.12b), and these compounds should be paramagnetic. Actually, neither cyclooctatetraene nor other [4n]annulenes exhibit molecular paramagnetism. As a theorem formulated by Jahn and Teller states, the degeneracy of the highest occupied orbital can be destroyed by a slight perturbation of the molecular symmetry, perhaps through alternating bond lengths, and this allows both electrons to occupy a single lower lying energy level. The resulting energy level diagram (Figure 4.12c) shows that accordingly there is only a small energy gap between the highest occupied and the lowest unoccupied eigenvalue. This difference is much smaller than the corresponding energy difference in the case of the $4n+2$ π systems. An interaction with the magnetic field B_0 will lead to a mixing of these electronic states and, in the sense of our analysis beginning on p. 17, produce a paramagnetic contribution to the shielding constant σ. This is larger in magnitude than the diamagnetic contribution of the lower orbitals so that the net result is a paramagnetic effect. Thus, in the case of [$4n$]annulenes we speak illustratively of a 'paramagnetic ring current' in analogy with the classical ring current model discussed at the outset. Molecules that show this behaviour are called *paratropic*, whereas those with no ring current at all are termed *atropic*.

A series experimental observations confirms the theoretical prediction by the detection of paramagnetic ring current effects. Two particularly impressive examples are presented here. By reduction with metallic potassium, the compound **25** is converted into the doubly charged anion **27**, which has 16 π-electrons. In this compound the methyl protons resonate at δ 21.0 and the ring protons at δ -3.2 to -4.0 ppm. The dramatic difference between the spectra of the neutral $4n+2$ π system and its charged $4n$ π counterpart is illustrated in Figure 4.13. The different charge density affects to a first approximation only the ring protons which, due to this factor, are shielded by 1.5 ppm (equation (4.2)).

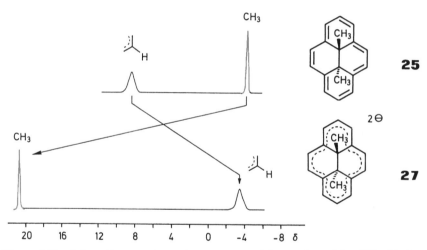

Figure 4.13 Schematic comparison of n.m.r. spectra of the annulenes **25** and **27** with 14 and 16 π-electrons, respectively

---

Exercise 4.4 For the dilithium salt of naphthalene dianion, which is formed by the reduction of the hydrocarbon with lithium metal in THF, ^{1}H resonance frequencies of δ 1.27 and 3.09 ppm are found. A Hückel MO calculation yields π-charge densities of 1.361 and 1.138 for the 1- and 2-position, respectively. Assign the ^{1}H n.m.r. signals and calculate with the data of the hydrocarbon (Table 12.1, p. 000) the experimental high field shift $\Delta\delta(1-H)$ and $\Delta\delta(2-H)$. Use equation (4.2) to estimate the contribution of the paramagnetic ring current sustained in the 12π system of the dianion to these values.

The two tricycloazines **28** and **29** represent a pair of compounds that clearly demonstrate the different properties of neutral [4n + 2]- and [4n]annulenes. The nitrogen atom here functions as a clamp that does not affect the resonance frequencies significantly. They are shifted to lower field in the case of the 10 π-system and to higher field in the case of the 12 π system.

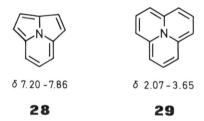

δ 7.20 - 7.86 δ 2.07 - 3.65

28   **29**

Non-planar cyclic π systems with pronounced bond alternation do not show a ring current effect since the delocalization of the π-electrons is diminished or completely quenched. Thus, the protons of cyclooctatetraene, which exists in the tub conformation **30**, have a resonance frequency of δ 5.80 that is practically identical with the resonance frequency of the protons of 1,3-cyclohexadiene (δ 5.85). In yet another example, a comparison of the proton chemical shifts of 1,6; 8,13-*anti*-bis-methano[14]annulene (**31a**) with the data for 1,6-methano[10]annulene or the *syn*-compound **31s** leads to the conclusion that the ring current expected for **31a** on the basis of the number of π-electrons does not exist. The spectra of the two differently bridged [14] annulenes (Figure 4.14) show distinct differences in the resonance frequencies of the bridge as well as the perimeter protons. The resonance frequencies of the CH_2 groups of the *syn*-system **31s** exhibit the expected diamagnetic shift while the resonances of the methylene protons of **31a** are recorded as two AB systems in the region characteristic of allylic methylene groups as that in 1,3,5-cycloheptatriene. Moreover, the deshielding of the perimeter protons observed for **31s** is absent in **31a**. As examinations of models show, an extensive twisting of the carbon–carbon bonds between the centres 6, 7, 8 and 13, 14, 1 obviously hinders the effective overlap of the carbon $2p_z$ orbitals so that here, for the first time, a compound which has the correct number of π-electrons to satisfy the Hückel rule and thus should be aromatic, exhibits olefinic characteristics. We shall return to this interesting molecule later.

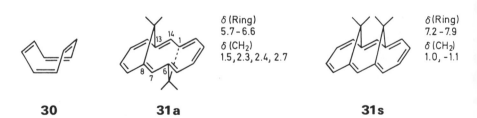

δ (Ring)
5.7 - 6.6
δ (CH_2)
1.5, 2.3, 2.4, 2.7

δ (Ring)
7.2 - 7.9
δ (CH_2)
1.0, -1.1

30   **31a**   **31s**

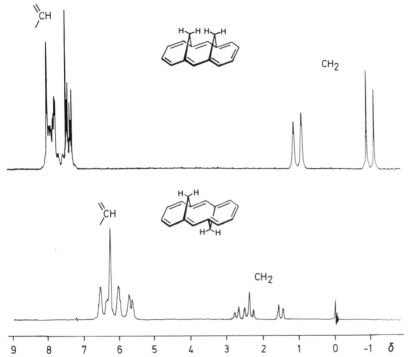

Figure 4.14 ¹H n.m.r. spectra of 1,6;8,13-*syn*- (above) and 1,6;8,13-*anti*-bismethano-[14]annulene (**31s, 31a**) (after Ref. 4)

The dependence of the ring current intensity, I, on the alternation of the bond lengths in cyclic π systems as obtained by means of quantum mechanical calculations is illustrated in Figure 4.15. It can be seen that paramagnetism decreases more rapidly than diamagnetism with increased bond alternation; conversely however, the ring current intensity is substantially higher for delocalized systems in [$4n$] π compounds. Figure 4.13 obviously confirms these predictions since, if compared to chemical shifts in non-delocalized model compounds, the chemical shifts in the dianion are considerably larger than in the hydrocarbon. If, in the hypothetical model with localized carbon–carbon bonds, δ values of about 6 and 1 ppm are taken for the resonances of the ring and methyl protons, respectively, then in the case of **25** shifts of 2–3 and 4–5 ppm are observed while for **27** the shifts are 9–10 and 20 ppm.

1.5   THE MAGNETIC ANISOTROPY OF THE CYCLOPROPANE RING

Finally, the cyclopropane ring should be considered because it also possesses a diamagnetic anisotropy perpendicular to the plane of the ring. However, as a

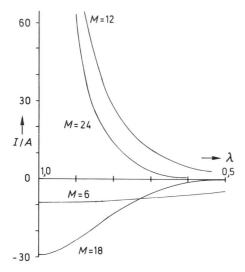

Figure 4.15 Ring current intensity, I, per unit area, A, in annulenes as a function of the alternance parameter, λ, which is a measure of the ratio of the resonance integrals, β, at adjacent carbon–carbon bonds. For completely equivalent bonds $\lambda = 1.0$. Negative signs signify diamagnetism; M indicates the number of π electrons present (after Ref. 5)

consequence of the different orientation of the C–H bonds in comparison with those in benzene, a shielding of the ring protons of cyclopropane results, so that the resonance frequency of δ 0.22 is considerably lower than for other saturated cyclic hydrocarbons. The two pairs of compounds **32** and **33** and **34** and **35** illustrate the shielding effect of the three-membered ring. Further, the temperature dependence of the resonance frequency of the proton H_b of vinyl cyclopropane is represented in Figure 4.16. In this compound there exists a rapid and reversible equilibrium between an *s-trans* and two *gauche* conformers (**36, 37**) in which the position of the vinyl group relative to the cyclopropane ring differs considerably.

The strong diamagnetic shielding of H_b with decreasing temperature indicates that the conformation of lower energy, that is more highly populated, is the *s-trans* form in which the proton H_b is in the shielding region of the cyclopropane ring.

1.6   THE ELECTRIC FIELD EFFECT OF POLAR GROUPS AND THE VAN DER WAALS EFFECT

Beside the previously discussed electronic and magnetic contributions to the chemical shift of proton resonances, two other effects that are in certain cases of substantial importance should be considered for the sake of completeness.

$\delta = 5.58$ $\delta = 5.43$

$\delta = 7.42$ $\delta = 6.91$

32   **33**   **34**   **35**

θ $\theta = 180°$

36   **37**

In molecules with highly polar groups it must be realized that the electric dipole moment may lead to a change of the charge density at particular protons because the charge cloud of the corresponding C–H bond can be distorted by electrostatic forces. Depending on the direction of the C–H bond relative to the field vector, the bonding electrons are shifted towards or away from the hydrogen atom with the result that the proton is either shielded or deshielded. As can be realized from Figure 4.17, the dipole moments in pyridine and nitrobenzene, which have been localized at the nitrogen and at the centre of the C–N bond of the nitro group, respectively, effect a

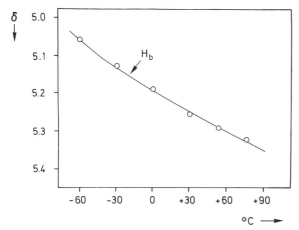

Figure 4.16 Temperature dependence of the resonance frequency of proton H_b in vinylcyclopropane (after Ref. 6)

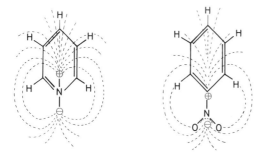

Figure 4.17   Electric field effect in pyridine and nitrobenzene

deshielding of the protons because the electrons are shifted along the lines of force toward the positive end of the dipole. According to the theory of Buckingham, this effect can be quantitatively described by the relation

$$\Delta\sigma = -AE_z - BE^2 \tag{4.15}$$

where E_z is the component of the electric field in the direction of the C–H bond and E^2 is the square of the field strength at the proton. Both terms are calculated from the known relation for the field of an electric dipole μ and A and B are constants with values of 2.0×10^{-12} and 10^{-18}, respectively, if μ is measured in debyes. Using equation (4.15), shielding contributions of -0.70, -0.19, and -0.14 ppm are calculated for the *ortho*-, *meta*-, and *para*-protons, respectively, in nitrobenzene, in qualitative agreement with the experimental results (-0.97, -0.30, and -0.45 in the order given above). A refinement of the model considers the polarization of medium by the polar solute that leads to a so-called reaction field, which also effects the shielding of the protons. We shall not consider it further here, however.

It can also be shown that the empirical relation (4.2) discussed on p. 73 follows from equation (4.15) for the electric field effect. The polarization of a C–H bond resulting from the localization of a positive or negative charge in the molecule and the subsequent change in the shielding, $\Delta\sigma$, for the affected proton can be quantitatively determined by the equation

$$\Delta\sigma = 0.125 \sum_i \frac{\Delta\rho_i}{R_i^2} \cos\theta_i - 0.170 \left(\sum_i \frac{\Delta\rho_i}{R_i^2} \right)^2 \tag{4.16}$$

where $\Delta\rho_i$ is the excess elementary charge at atom i, R_i is the distance (in nm) between atom i and the proton H, and θ is the angle between the C–H bond and the distance vector i, H.

Exercise 4.5   Determine by means of equation (4.16) the expected change in the shielding of the protons in the pyridinium ion on the basis of the indicated charge distribution. Consider the molecule to be a regular hexagon with bond lengths of 0.140 nm (C–C) and 0.110 nm (C–H).

The so-called *quadratic field effect* cited in equation (4.15), BE^2, is closely related to the *van-der-Waals effect* that arises when a strong steric interaction exists between a proton and a neighbouring group (possibly another proton). In this case we predict that the electron cloud around the proton becomes deformed. The diminished spherical symmetry of the electron distribution causes a paramagnetic contribution to the shielding constant (cf. p. 17), which results in a shift to lower field. Of known examples the deshielding of the indicated protons in compounds **38–40** might be substantially attributed to the van der Waals effect.

1.7   CHEMICAL SHIFTS THROUGH HYDROGEN BONDING

As already discussed on p. 23, no distinct region on the $\overset{o}{\delta}$ scale can be assigned to the resonances of exchangeable protons since the position of these resonance signals is strongly dependent upon the medium and temperature. In general, the formation of hydrogen bonds leads to significant shifts to low field although formally the electron density and with it the shielding at the proton should be increased through interaction with the free electron pair of the acceptor atom. The electrical dipole field of the hydrogen bond, that is formulated as a pure electrostatic attractive bond, however, appears to effect a deshielding. As is shown in Figure 4.18 for chloroform, there is a linear relationship between the deshielding of the chloroform proton and the dipole moment of the non-bonding orbitals of various acceptor atoms in different classes of compounds.

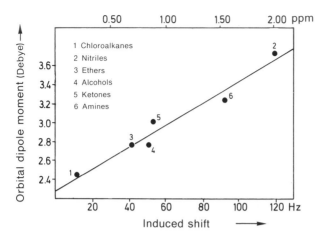

Figure 4.18. Correlation between the induced shift of the chloroform proton resonance frequency and the orbital dipole moment for different proton-acceptor atoms (after Ref. 6)

In the case of nitriles the chloroform proton resides within the shielding region of the triple bond (cf. p. 82) as shown in **41**. Accordingly, the data were corrected for the additional diamagnetic shift due to this arrangement. A similar effect exists in benzene, which acts as a π-electron donor. The chloroform proton in the benzene–chloroform complex (**42**) is therefore strongly shielded.

Intra- and intermolecular hydrogen bonds can easily be distinguished by means of nuclear magnetic resonance spectroscopy for only in the second case is the resonance frequency of the hydroxyl or amino proton strongly concentration dependent. As an illustration of this, the spectra of salicylaldehyde and ethanol at different concentrations are compared in Figure 4.19.

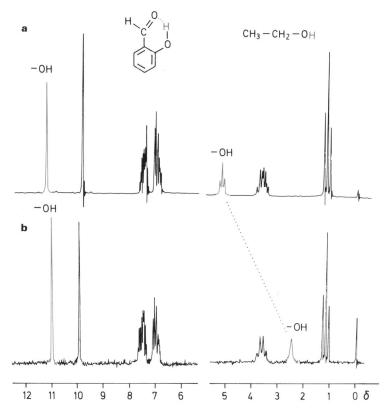

Figure 4.19 Concentration dependence of the proton resonance frequency of the hydroxyl protons of salicylaldehyde and ethanol: (a) neat; (b) 5% by volume in CCl_4

1.8 CHEMICAL SHIFTS OF PROTONS IN ORGANOMETALLIC COMPOUNDS

In Section 1.2 we discussed the shielding of protons in carbanions as a result of the increased electron density at the neighbouring carbon atom. In this section the shielding effect of metal carbonyl groups as illustrated by the spectra of metal carbonyl π-complexes of olefins and aromatic compounds will be discussed with reference to a few examples. As is shown in Figure 4.20, complex formation leads to shielding of the protons at the coordinated carbon atom of about 2–3 ppm. The cause of this shielding is still not fully understood. Certainly the presence of the metal plays a substantial role but the effect of the anisotropy of the metal carbonyl group can also be involved. Additional examples are compiled in Table 4.2. The corresponding proton resonances of the free ligands are collected in Table 12.1 in the Appendix (pp. 516 ff.).

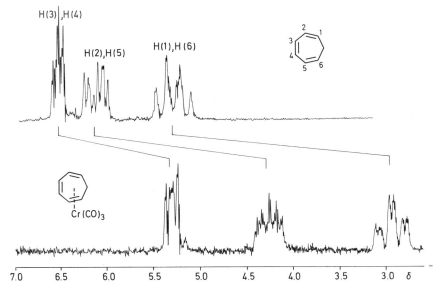

Figure 4.20 Nuclear magnetic resonance spectra of the olefinic protons in cycloheptatriene and in cycloheptatriene chromium tricarbonyl

Especially strong shieldings are observed for protons that are *directly* bonded to metals. Thus the resonance frequencies of the protons in transition metal hydride complexes are found in the region $\delta < 0$ ppm and in a few cases even at values up to -30 ppm. If in the complex the metal is positively charged the shielding is reduced, as expected. In protonated transition metal carbonyl complexes the formation of a metal–hydrogen bond, in contrast, again leads to an increased shielding of the proton. Table 4.2 also shows a few examples of these cases.

To a large extent the asymmetrical charge distribution in the valence orbitals of the metals are responsible for the strong diamagnetic shift of the proton resonance in these compounds. The resulting paramagnetic moment deshields the metal nucleus but shields the proton as indicated below:

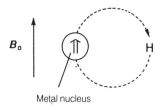

Table 4.2 Proton resonances in metal carbonyl π-complexes and in metal hybrides

1.9 SOLVENT EFFECTS

On p. 70 the influence of the solvent as a factor contributing to the magnitude of the proton screening constant was mentioned, and we want to include a brief discussion of the significance of solvent effects here for the sake of completeness. In general, it can be assumed that all of the effects that we have discussed up to now on an *intra*molecular basis also play a role on the *inter*molecular level. It has been observed, for example, that the resonance signals of substances dissolved in aromatic solvents appear at higher field than when dissolved in aliphatic solvents. This effect has been ascribed to the diamagnetic ring current of benzene and its derivatives. A similar influence of neighbouring molecules, however, associated with both shielding and deshielding can be expected from the effect of the magnetic anisotropy of multiple bonds or the electrical field effect of molecules with large dipole moments. Solvent effects become particularly significant when intermolecular interactions in the solvent lead to the formation of weak complexes. On the basis of dipole–dipole or van der Waals interactions, certain steric orientations become favoured with respect to others and, as a result, specific changes can be observed in the resonance frequencies of individual protons in the solute. This in turn can be used to obtain an insight into the structure of such complexes. Thus, n.m.r. spectroscopy has proven to be an important method for the study of intermolecular interactions. The changes in resonance frequency effected by the solvent in the case of proton resonances are usually smaller than 1 ppm. We have already considered special applications and the consequences for the resonance frequency of the reference substance in Chapter 3. If one wishes to avoid complications caused by solvent effects, the use of 'inert' solvents such as carbon tetrachloride and cyclohexane is recommended. On the other hand, concentration effects can be eliminated if several measurements at different sample concentrations are made and the data are extrapolated to infinite dilution. Only with the development of the Fourier transform method have measurements become feasible for compounds with high vapour pressures in the gas phase where intermolecular interactions are minimized.

1.10   EMPIRICAL SUBSTITUENT CONSTANTS

The observation that the influence of substituents on the resonance frequency is to a first approximation *additive* is of great importance for the interpretation of proton resonance spectra. On this basis it has been possible to derive empirical substituent constants $S(\delta)$ or *increments* that in general allow good predictions of resonance frequencies. Exceptions are, of course, to be expected when, because of strong electronic or steric interactions between the substituents, the condition for the additivity of the $S(\delta)$ values, namely their independence from the remainder of the molecule, is violated.

1.10.1   Aliphatic Compounds

The *Shoolery rule* provides information concerning the effect the stepwise substitution of the hydrogen atoms in methane by other groups has on the proton resonance frequency. According to this rule, the δ value of a proton in the compound in question is given by equation (4.17), in which the $S(\delta)$ values given in Table 4.3 are to be used.

$$\delta = 0.23 + \Sigma S(\delta) \qquad (4.17)$$

As an example, for the proton resonances in methyl chloride, methylene chloride and chloroform, equation (4.17) yields values of δ 2.76, 5.29, and 7.82, respectively, which can be compared with the experimental data of δ 3.1, 5.3, and 7.27. Similarly, the resonance frequency of the tertiary proton in triphenylmethane is calculated to be δ 4.22 and the experimental value is δ 4.70.

Table 4.3 Substituent constants, $S(\delta)$, for proton resonances in substituted methanes

Substituent	$S(\delta)$ (ppm)
Cl	2.53
Br	2.33
I	1.82
NRR'	1.57
OR	2.36
SR	1.64
CR−O	1.70
CR=CR'R'	1.32
C≡CH	1.44
C≡N	1.70
CH_3	0.47
Phenyl	1.85
OH	2.56
OCOR	3.13
COOR	1.55
CF_3	1.14

Table 4.4 Increments for dierent skeletal substitution (after Ref. 8)

Structure	$S(\delta)$[ppm]
$-C^2$	0.248
$-C^2-C^3$	0.244
$-C^2\begin{smallmatrix} \nearrow C^3 \\ \searrow C^3 \end{smallmatrix}$	0.147
$-C^2\!\!-\!\!C^3\begin{smallmatrix} \nearrow C^3 \\ \searrow C^3 \end{smallmatrix}$	0.006

A larger number of structural differences can be accommodated by means of the $S(\delta)$ increments collected in Tables 4.4 and 4.5.

First the carbon skeleton of the compound under consideration is numbered with the carbon α to the proton of interest numbered as position 1. Then, according to the structure, the β carbon atoms have the $S(\delta)$ values indicated in Table 4.4. Additional substituents are considered by means the $S(\delta)$ values in Table 4.5. The contributions of functional groups such as an ester are obtained by adding the increments for the

Table 4.5 Increments for individual substituents (after Ref. 8)

Substituent	Position	$S(\delta$ (ppm)
$-CR_3$	3	-0.038
Double bond	1	3.802
	2	0.583
	3	0.203
Triple bond	1	1.032
	2	0.694
$=O$	2	1.021
	3	0.004
$-OCH_3$	2	-0.374
$-OCH_2CR_3$	2	-0.237
	3	0.210
$-OH$	1	2.467
	2	0.048
	3	0.235
$-O-CO-CR_3$	1	2.931
	2	0.041
	3	-0.086
$-F$	2	0.089
	3	0.131
$-Cl$	1	2.170
	2	0.254
	3	0.177
$-Br$	1	1.995
	2	0.363
	3	0.023
$-I$	1	1.846
	2	0.388
$-NH_2$	2	0.094

$=O$ and the $-OR$ substituents. Finally, the δ value being calculated is determined according to the equation

$$\delta = 0.933 + \Sigma S(\delta) \tag{4.18}$$

Two examples serve to illustrate the procedure.

1. $CH_3-CO-CO-CH_3$

The resonance frequency of the methyl protons is to be determined. The numbering of the carbon skeleton leads to

$$\overset{1}{C}H_3-\overset{2}{C}O-\overset{3}{C}O-CH_3$$

Table 4.6 $S(\delta)$ values for substituted benzenes (after Ref. 9)

Substituent	$S(\delta)$ (ppm)		
	Ortho	Meta	Para
NO_2	0.95	0.17	0.33
CHO	0.58	0.21	0.27
COCl	0.83	0.16	0.3
COOH	0.8	0.14	0.2
$COOCH_3$	0.74	0.07	0.20
$COCH_3$	0.64	0.09	0.3
CN	0.27	0.11	0.3
C_6H_5	0.18	0.00	0.08
CCl_3	0.8	0.2	0.2
$CHCl_2$	0.1	0.06	0.1
CH_2Cl	−0.0	0.01	0.0
CH_3	−0.17	−0.09	−0.18
CH_2CH_3	−0.15	−0.06	−0.18
$CH(CH_3)_2$	−0.14	−0.09	−0.18
$C(CH_3)_3$	0.01	−0.10	−0.24
CH_2OH	−0.1	−0.1	−0.1
CH_2NH_2	−0.0	−0.0	−0.0
F	−0.30	−0.02	−0.22
Cl	0.02	−0.06	−0.04
Br	0.22	−0.13	−0.03
I	0.40	−0.26	−0.03
OCH_3	−0.43	−0.09	−0.37
$OCOCH_3$	−0.21	−0.02	—
OH	−0.50	−0.14	−0.4
$p-CH_3C_6H_4SO_3$	−0.26	−0.05	—
NH_2	−0.75	−0.24	−0.63
SCH_3	−0.03	−0.0	—
$N(CH_3)_2$	−0.60	−0.10	−0.62

so that $\Sigma S(\delta)$ is given by

$$
\begin{array}{ll}
C^2 - C^3 & +0.244 \\
= O \text{ at } C^2 & +1.021 \\
= O \text{ at } C^3 & +0.004 \\
-CR_3 \text{ at } C^3 & -0.038 \\
\hline
& +1.231
\end{array}
$$

and $\delta(CH_3)$ according to equation (4.18) is

$$\delta(CH_3) = 0.933 + 1.231 = 2.164$$

which compares well with the experimental value of δ 2.23.

2.

$$H_3\overset{2'}{C}-\overset{1}{C}H_2 \diagdown \underset{2}{C} \diagup \overset{3}{C}OOC_2H_5$$

$$\underset{\overset{3}{C}_2H_5}{} \diagup \overset{C}{} \diagdown \overset{3}{C}OOC_2H_5$$

The resonance frequency of the methylene protons is to be determined. According to the indicated numbering the following tabulation results:

$$-C^2\diagup^{C^3}_{\diagdown C^3}C^3$$

	+0.006
$-C^2$	+0.248
two $=O$ in 3-position	+0.008
two $-OC_2H_5$ in 3-position	+0.420
$-CR_3$ in 3-position	−0.038
	+0.644

This leads to a value of $\delta(CH_2) = 0.933 + 0.644 = 1.577$ ppm, while the value found experimentally is δ 1.84.

1.10.2 Substituted Benzenes

The systematic investigation of the resonance frequencies in substituted benzenes reveals that here too the influences of substituents to a good approximation are additive. From the data given in Table 4.6, predictions can be made concerning the $\delta(^1H)$ values in substituted benzenes relative to $\delta(^1H)$ in benzene. Thus, the increments for 1,4-dibromobenzene predict a deshielding of the proton resonances of 0.09 ppm while the experimental value is 0.06 ppm. For 3-bromonitrobenzene the following calculation can be made.

$$\delta(H_a) = 7.27 + 0.22 + 0.95 = 8.44$$
$$\delta(H_b) = 7.27 + 0.22 + 0.33 = 7.82$$
$$\delta(H_c) = 7.27 - 0.13 + 0.17 = 7.31$$
$$\delta(H_d) = 7.27 - 0.03 + 0.95 = 8.19$$

These results are in excellent agreement with the experimental data (δ8.38, 7.80, 7.36, and 8.16 in the same order).

   Deviations from the additivity principle are indeed observed when two or more substituents are adjacent to one another in a benzene nucleus. For example, the chemical shift of the *ortho* protons in dimethyl phthalate is calculated according to the data in Table 4.6 to be δ 8.08, while the experimental value is δ 7.66. The approximation that the substituents act independently obviously does not hold in these cases.

1.10.3   Olefinic Systems

Finally, in Table 4.7 $S(\delta)$ values are given for substituted ethylenes and these likewise are based on extensive experimental data. The resonance frequency of interest is calculated according to the equation

$$\delta = 5.28 + \Sigma S(\delta) \tag{4.19}$$

Table 4.7 $S(\delta)$ values for substituted ethylenes (after Ref. 10)

Substituent R*	S(δ)(ppm)			Substituent R*	S(δ)(ppm)		
	gem	cis	trans		gem	cis	trans
—H	0	0	0	H			
—Alkyl	0.44	−0.26	−0.29	—C=O	1.03	0.97	1.21
—Alkyl ring	0.71	−0.33	0.30	NR₂			
—CH₂O, —CH₂I	0.67	−0.02	−0.07	—C=O	1.37	0.93	0.35
—CH₂S	0.53	−0.15	−0.15	Cl			
—CH₂Cl, —CH₂Br	0.72	0.12	0.07	—C=O	1.10	1.41	0.99
—CH₂N	0.66	−0.05	−0.23	—OR (R aliph.)	1.18	−1.06	−1.28
—C≡C	0.50	0.35	0.10	—OR (R conj.)	1.14	−0.65	−1.05
—C≡N	0.23	0.78	0.58	—OCOR	2.09	−0.40	−0.67
—C=C (isol.)	0.98	−0.04	−0.21	—Aromatic	1.35	0.37	−0.10
—C=C (conj.)	1.26	0.08	−0.01	—Cl	1.00	0.19	0.03
—C=C (isol.)	1.10	1.13	0.81	—Br	1.04	0.40	0.55
—C=O (conj.)	1.06	1.01	0.95	—N（R aliph.)（＼R ／R)	0.69	−1.19	−1.31
—COOH (isol.)	1.00	1.35	0.74				
—COOH (conj.)	0.69	0.97	0.39	—N (R conj.)	2.30	−0.73	−0.81
—COOR (isol.)	0.84	1.15	0.56	—SR	1.00	−0.24	−0.04
—COOR (conj.)	0.68	1.02	0.33	—SO₂	1.58	1.15	0.95

* isol. = isolated; conj. = conjugated; aliph. = aliphatic. The increments for 'R conj.' are used instead of those for 'R isol.' when the substituent R or the double bond in question is conjugated with additional substituents. The increments for 'Alkyl ring' are used when the substitutent under consideration and the double bond form a ring.

1.11   TABLES OF PROTON RESONANCES IN ORGANIC MOLECULES

Proton resonance frequencies for different classes of organic compounds are tabulated in Table 12.1 in the Appendix (pp. 516 ff.). This general survey can be used as a data source and as an aid in becoming familiar with the resonance frequencies to be expected in various situations.

2. Spin–spin coupling and chemical structure

This section is devoted to the discussion of correlations between H–H coupling constants and chemical structure. A general survey of this subject has already been given at the end of Chapter 2 in Table 2.2. For the following discussion, which deals in detail with the several types of spin–spin interactions, we shall use a classification indicating the number of bonds between the coupled nuclei. Thus we differentiate between *geminal, vicinal*, and *long-range* coupling depending on whether the coupling occurs over *two, three*, or *more* bonds. The number of bonds, n, is used as a superscript in front of the symbol J. In unsaturated systems the position of the coupled nuclei relative to the double bond (*cis* or *trans*) can be indicated simultaneously by means of a subscript. In principle, there is no limit to the number of bonds over which coupling occurs, although coupling is seldom effective over more than five bonds and, in general, decreases in magnitude as n increases.

It should also be noted that the signs of all ^3J and most ^5J values are positive while for ^2J and ^4J values they can be either positive or negative. For a systematic discussion of the dependence of the coupling constants on structure it is essential to consider their signs. Their magnitude is measured in hertz (Hz).

2.1   THE GEMINAL COUPLING CONSTANT (^2J)

Following the spin–spin coupling over one bond in the hydrogen molecule, which amounts to 276 Hz and is of theoretical interest, the geminal coupling constants with values between -23 and $+42$ Hz form the group with the largest spin–spin interaction between protons. Many factors are responsible for the magnitude and the sign of ^2J.

2.1.1   *Dependence on the Hybridization of the Carbon*

In going from an sp^3 hybridized methylene group, as it exists in methane, to sp^2 hybridization in ethylene the geminal coupling constant changes from -12.4 to $+2.5$ Hz. Cyclopropane, because of its special bonding situation, has an intermediate value. Other strained ring systems exhibit ^2J values of up to -5.0 Hz while the geminal coupling in cyclobutane is not much different from that in methane (Table 4.9).

Table 4.8 Classification of spin–spin coupling

Type of coupling	Classification	n	Symbol
	Geminal	2	^2J
	Vicinal	3	^3J
	Vicinal	3	$^3J_{cis}$
	Vicinal	3	$^3J_{trans}$
Long-range coupling:			
H–C–C=C–H'	Allylic	4	^4J
H–C–C=C–C–H'	Homoallylic	5	^5J

2.1.2 The Effect of Substituents

The ^2J values are subject to the influence of both α and β substituents. In Table 4.10 characteristic data are presented. It can be seen that the substitution of an electronegative group to the methylene group in question leads to a positive change in the coupling constant. For sp^3 hybridized methylene groups the absolute value of the coupling constant therefore decreases. In polysubstituted methanes the effect of substituents, to a first approximation, is additive. The influence of oxygen in ethylene oxide results in a positive coupling constant for the methylene hydrogen

Table 4.9 The dependence of the geminal H–H coupling on carbon hybridization

– 12.4	– 4.3	+ 2.5

– 11...–15	– 5.4

Table 4.10 The influence of substituents on geminal-coupling constants

1. α-Substitution

CH_4	– 12.4	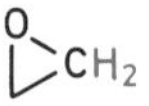 (O-CH_2 epoxide) CH_2	+ 5.5
CH_3Cl	– 10.8	$RN=CH_2$	+16.5
CH_3Cl_2	– 7.5	$O=CH_2$	+42.2
(HN-CH_2 aziridine) CH_2	+ 2.0	(six-membered dioxane) CH_2	– 6
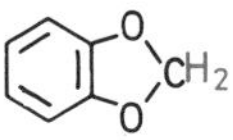 (benzodioxole) CH_2	± 1.5	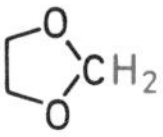 (dioxolane) CH_2	0

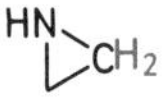

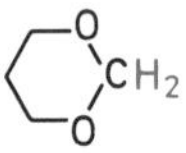

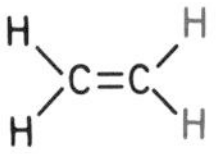

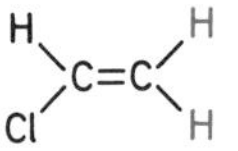

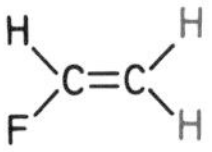

2. β-Substitution

$H_2C=CH_2$	+ 2.5	$ClHC=CH_2$	– 1.4
$FHC=CH_2$	– 3.2	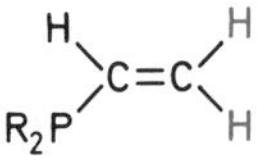 $R_2PHC=CH_2$	+ 2.0
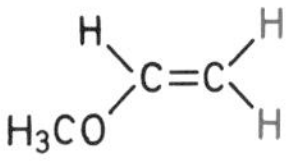 $H_3COHC=CH_2$	– 2.0	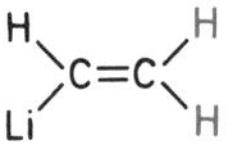 $LiHC=CH_2$	+ 7.1

3. Adjacent π bonds

CH_3CN	– 16.9	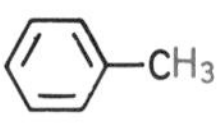 (toluene) $-CH_3$	– 14.5
$CN-CH_2-CN$	– 20.4		

atoms. The especially large positive value in formaldehyde is due to the additional influence of the non-bonded electron pair on the oxygen atom. The steric orientation of *non-bonding electron pairs* relative to the orientation of the C–H bonds under consideration is also of significance in the case of sp^3 hybridized methylene groups. The comparison of the ^2J values in 1,3-dioxane with those in the conformationally more rigid formaldehyde dioxolane illustrates this effect.

In contrast to the situation in the case of α substituents, an electronegative β substituent leads to a negative change in the coupling constant. The ^2J values in substituted ethylenes clearly indicate this. Conversely, an electropositive substituent such as lithium leads to a positive change in the coupling constant, which is found to

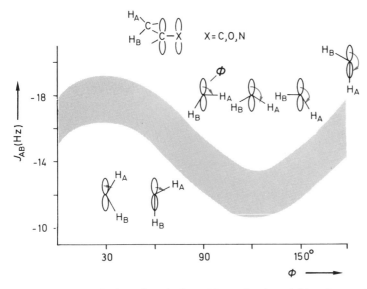

Figure 4.21 Perturbation of geminal H – H coupling by neighbouring π-orbitals (after Ref. 11)

be + 7.1 in vinyllithium. *Neighbouring π-bonds* also have a considerable influence on the magnitude of geminal coupling constants. They cause a negative change, that is, the absolute value of the constants increases. Thus, the magnitude of the geminal coupling constant changes from 12.4 Hz in methane to 20.4 Hz in malononitrile. It is important to note that the effect of neighbouring π-bonds on geminal coupling is a function of the stereochemistry of the system under consideration, just as in the case of the effect of non-bonding electron pairs mentioned above. Theoretical considerations and experimental data show that the effect of a neighbouring π-bond on the geminal coupling constant is a function of the angle ϕ between the π-orbital and the C–H bond. This dependence is clearly illustrated in Figure 4.21. The largest effect is observed when the neighbouring orbital and the plane of the methylene group are parallel.

The ^2J values in cyclopentedione (**43**) and fluorene (**44**) confirm this prediction as does the relative magnitude of the two constants found for compound **45**.

The dependence of ^2J on the hybridization of the carbon atom mentioned earlier leads one to expect that a characteristic correlation also exists involving the H–C–H bond angle that might be used to obtain information on the corresponding C–C–C bond angle as well. However, as a result of the variety of substituent effects to which the geminal coupling constants are subject, such a relation cannot be formulated with the accuracy necessary for reliable predictions.

43   **44**   **45**

2.1.3   A Molecular Orbital Model for the Interpretation of Substituent Effects on ^2J

In this section we consider a molecular orbital (MO) model allowing the rationalization of substituent effects for geminal H–H coupling constants discussed above. It is based on the theoretical result that the coupling constant, $J_{h,h'}$, between two protons is proportional to the so-called *mutual atom–atom polarizability*, $\pi_{h,h'}$, of the two hydrogen 1s orbitals:

$$J_{H,H'} \propto \pi_{h,h'} \tag{4.20}$$

This polarizability is defined in MO theory through the relation

$$\pi_{h,h'} = -4 \sum_{i}^{occ.} \sum_{j}^{unocc.} \frac{c_{ih} c_{ih'} c_{jh} c_{jh'}}{E_j - E_i} \tag{4.21}$$

where c_{ih}, $c_{ih'}$, etc., are the coefficients of the 1s orbitals h and h' in the molecular orbitals Ψ_i and Ψ_j which, as is well known, are formed by a linear combination of atomic orbitals:

$$\Psi_i = c_{iA} \phi_A + c_{iB} \phi_B + \cdots + c_{iN} \phi_N \tag{4.22}$$

E_j and E_i are the orbital energies and the summation includes all occupied and unoccupied molecular orbitals. Within the MO theory, atom–atom polarizabilities π_{ij} can be used, for example, to determine the effect of a perturbation of the coulomb integral at the centre i on the electron density at atom j.

For a CH_2 group we can construct molecular orbitals from four atomic orbitals if we use the two hydrogen 1s orbitals and two carbon hybrid orbitals which can be either sp^2 or sp^3. The following MO energy level diagram is obtained in which only the different signs of the coefficients c_{ih}, $c_{ih'}$, etc. symbolized by means of different colour, are given:

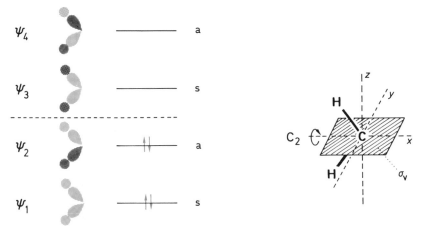

The energy sequence of the molecular orbitals Ψ_1–Ψ_4 follows from the number of bonding interactions between atomic orbital functions of the same sign. Relative to the symmetry plane σ_v of the CH_2 group the molecular orbitals can further be classified as either symmetric (s, reflection of Ψ_i yields Ψ_i) or antisymmetric (a, reflection of Ψ_i yields $-\Psi_i$).

The CH_2 moiety possesses four bonding electrons. In the ground state Ψ_1 and Ψ_2 are doubly occupied. Decisive in determining the magnitude of the coupling constants according to the proportionality (4.20), are the four possible electronic transitions A–D between the orbitals i and j. According to the signs of the coefficients $c_{ih'}$, $c_{ih'}$, c_{jh}, and $c_{jh'}$, they lead to contributions to the coupling with the following signs:

A: transition from Ψ_1 to Ψ_3: −
B: transition from Ψ_1 to Ψ_4: +
C: transition from Ψ_2 to Ψ_3: +
D: transition from Ψ_2 to Ψ_4: −

For the derivation of the substituent effects on $^2J_{H,H'}$ it is only necessary to investigate how the coefficients c_{ih} and $c_{ih'}$ are influenced by substitution. In so doing it is important to note that the interaction between an orbital of the substituent and the molecular orbitals of the CH_2 group is governed by symmetry. It is allowed only when both have the same symmetry relative to σ_v. Inductive effects through σ orbitals will therefore lead to changes in the coefficients in Ψ_1 (and Ψ_3, respectively) while a hyperconjugative interaction with a p_z orbital is restricted to Ψ_2 (or Ψ_4, respectively).

For substituents with a $-I$ effect we therefore expect a charge transfer out of Ψ_1 that will result in a decrease in c_{1h} and $c_{1h'}$. Since the sum of all the atomic orbitals employed in the formation of the molecular orbitals must remain constant, this means that at the same time the coefficients c_{3h} and $c_{3h'}$ in the symmetrical antibonding orbital Ψ_3 must become larger. According to equations (4.21) and (4.20) the contribution B to the geminal coupling decreases while that of C increases. A and D remain, to a first approximation, unchanged. Because of the smaller energy difference E_3-E_2 the increase in C predominates and ^2J must become more positive. This is observed experimentally (Table 4.10). Obviously the opposite predictions apply for the $+I$ effect.

Following similar reasoning it can be predicted that in case of a *hyperconjugative* interaction *electron withdrawal* will decrease the contribution C and increase the contribution B while, to a first approximation, A and D are again unaffected. Because of the smaller energy difference for C the change in this contribution assumes greater importance, and the coupling becomes more negative. These predictions have also been confirmed experimentally (Table 4.10; Figure 4.21). The striking increase in the magnitude of the geminal coupling in formaldehyde is an especially impressive example of the applicability of this simple MO model. Here the $-I$ effect of the oxygen atom and the hyperconjugative charge transfer of the CH_2 group by the non-bonding electron pairs on oxygen augment one another. Similarly, hyperconjugation in cyclic ethers leads to a positive change in ^2J. Finally, the conformational dependence of the effect of π-bonds and free electron pairs on ^2J mentioned above can also be understood because of the fact that the electronic interaction of these groups with Ψ_2 obeys a $\cos\phi$ relation in which ϕ is the angle between the z axis and the axis of the substituent orbital.

2.2   THE VICINAL COUPLING CONSTANT (^3J)

There are extensive data on vicinal coupling constants and their relation to chemical structure. In agreement with the results of theoretical calculations it has been shown that the magnitude of ^3J, the sign of which was earlier found to be always positive, depends in essence upon four factors:

1. The *dihedral angle*, ϕ, between the C–H bonds under consideration (a);
2. The C,C *bond length*, $R_{\mu\nu}$ (b);
3. The *H–C–C valence angles*, θ and θ' (c); and
4. The *electronegativity* of the substituent R on the H–C–C–H moiety.

a   b   c   d

2.2.1 Dependence on the Dihedral Angle

The dependence of vicinal coupling constants on the dihedral angle, ϕ, first theoretically predicted by Karplus and Conroy, is represented in Figure 4.22. The curve shown—now known as *Karplus curve*—is described by the relation

$$^3J = A + B \cos \phi + C \cos 2\phi \qquad (4.23)$$

where A, B, and C are constants with the values 4.22, -0.5, and 4.5, respectively. The experimental findings are in good qualitative agreement with the calculations that were worked out for a H–C–C–H fragment. Experience has shown, however, that the ^3J values for $\phi = 0°$ and $180°$ in general are about 2–4 Hz larger than the calculated values, although the prediction that $^3J_{180} > {}^3J_0$ is always confirmed. Therefore, the empirical constants A = 7, B = -1, and C = 5 have been proposed for equation (4.23).

A series of important regularities is explained by the *Karplus curve* (Table 4.11). For example, in olefinic systems the coupling of *trans* protons is always greater than that between *cis* protons. A clear distinction between *cis–trans isomers* can therefore be made. In 1,2-disubstituted ethane the corresponding sequence, $J_{gauche} < J_{trans}$ applies. Consequently, in the chair conformation of cyclohexane the coupling between two axial protons is larger than that between two equatorial protons or between an equatorial and an axial proton ($J_{aa} > J_{ea} \approx J_{ee}$). This is an important criterion in the conformational analysis of cyclohexane derivatives and carbohydrates. Thus, in the β form of glucose the anomeric proton possesses, in addition to the higher shielding mentioned earlier (p. 80), a larger vicinal coupling constant than in the α form.

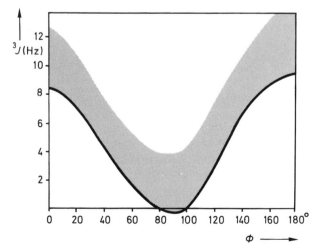

Figure 4.22 The Karplus curve for the dependence of vicinal H – H coupling on the dihedral angle ϕ: line, theoretical curve; shaded area, range of empirical results

Table 4.11   The dependence of vicinal H–H coupling on the dihedral angle, ϕ

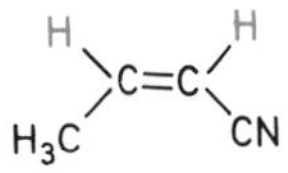

11.0 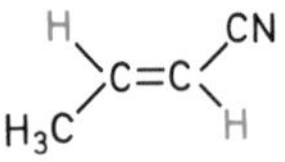   16.0 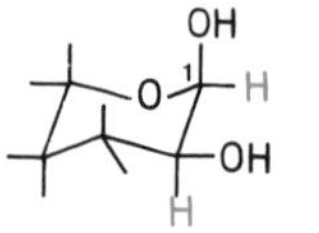   3.0 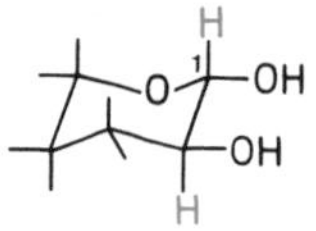   7.4

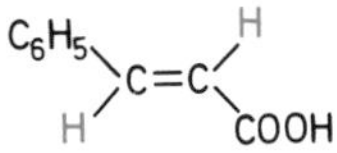

15.8 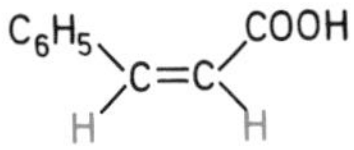   12.3    3.9    9.0    9.3

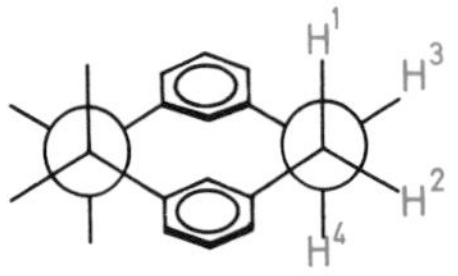

meta–Cyclophan

J_{12}	(^2J)	-12.0
J_{23}	$(^3J_{gauche})$	3.2
$J_{24} = J_{13}$	$(^3J_{gauche})$	4.0
J_{14}	$(^3J_{trans})$	12.3

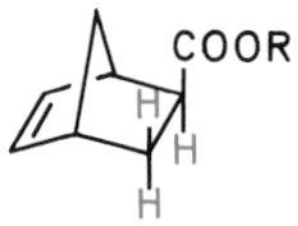

J_{cis} 9.0 J_{cis} 9.4

J_{trans} 4.4 J_{trans} 4.2

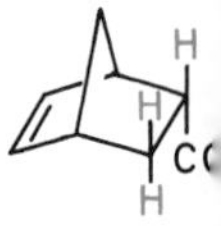

3.8    8.4 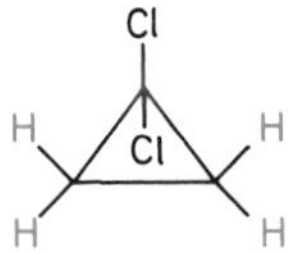 J_{cis} 11.2

J_{trans} 8.0

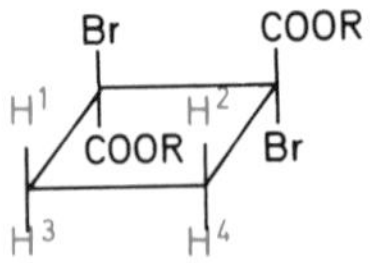

$J_{12} = J_{34}$ $(^3J_{cis})$	9.0	
J_{14} $(^3J_{trans})$	10.0	
J_{23} $(^3J_{trans})$	3.3	

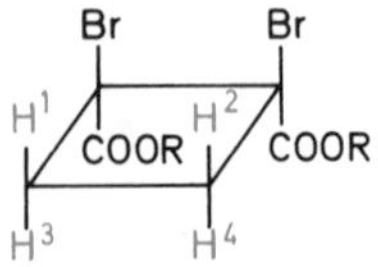

J_{12} $(^3J_{cis})$	9.8	
J_{34} $(^3J_{cis})$	8.6	
$J_{14} = J_{23}$ $(^3J_{trans})$	6.7	

Special circumstances exist in the case of the three-membered ring. Here the dihedral angle for *cis* protons is 0° and that for *trans* protons is ~130°. According to Figure 4.22 we would expect that $^3J_{cis} > {}^3J_{trans}$, and this is always found experimentally for a pair of *cis–trans* isomers of a substituted cyclopropane.

Analogous relations apply with oxirane and aziridine. For cyclobutane and cyclopentane derivatives, because of their greater flexibility, the dihedral angles are less well defined and an unequivocal assignment of the configuration on the basis of the ^3J values is, in general, not possible.

2.2.2 Dependence upon the Bond Length, $R_{\mu\nu}$

In Figure 4.23 the vicinal coupling constants in a few unsaturated six-membered rings are plotted against the bond lengths, $R_{\mu\nu}$ (in nm), which were determined by X-ray structural analyses.

The dihedral angle in the compounds under consideration can be assumed to be 0°, and since for hydrocarbons no substituent effects are expected, the linear relation

$$^3J = -351.0R_{\mu\nu} + 56.65 \tag{4.24}$$

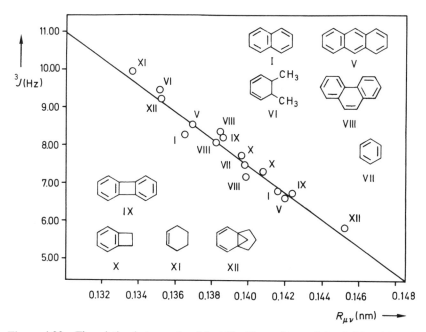

Figure 4.23   The relation between the vicinal H–H coupling and the C,C bond length, $R_{\mu\nu}$ in unsaturated six-membered rings (after Ref. 12): I, naphthalene; V, anthracene; VI, *cis*-5,6-dimethylcyclohex-1,3-diene; VII, benzene; VIII, phenanthrene; IX, biphenylene; X, benzocyclobutene; XI, cyclohexene; XII, tricyclo[4.3.1.0.^{1,6}]-2.4-decadiene

can be rationalized as the result of changes in the C–C bond length. The ^3J values are therefore *very sensitive* to small differences in the C–C bond length and, if other factors are considered to be constant, can give information concerning the degree of bond alternation in cyclic π systems. Since the π bond order, $P_{\mu v}$, of MO theory correlates linearly with the bond length $R_{\mu v}$, there also exists a linear relation between $P_{\mu v}$ and the ^3J values. For benzenoid aromatic compounds equation (4.25) holds:

$$^3J = 12.47P_{\mu v} - 0.71 \tag{4.25}$$

Similar relations, but with different constants due to HCC valence angle changes (see below), have been derived from planar five- and seven-membered rings:

Five-membered rings :
$$^3J = 7.12P_{\mu v} - 1.18 \tag{4.26}$$
$$^3J = -322.6R_{\mu v} + 48.45 \tag{4.27}$$

Seven-membered rings :
$$^3J = 21.91P_{\mu v} - 3.85 \tag{4.28}$$
$$^3J = -367.4R_{\mu v} + 60.68 \tag{4.29}$$

2.2.3 Dependence on HCC Valence Angles

The importance of this factor for the magnitude of ^3J is best demonstrated with the vicinal *cis* coupling constants across the double bond in cyclic monoolefins with different ring sizes. Here, a constant dihedral angle of $0°$ and the absence of substituent effects may be assumed. As is shown in Table 4.12 there is a steady increase in $^3J_{cis}$ in passing from cyclopropene to larger rings. Values as large as those found in acyclic olefins are observed in the eight-membered rings. We therefore conclude that a decrease in the HCC valence angles θ and θ' leads to an increase in ^3J. This observation is also supported by data for aromatic compounds.

Table 4.12 The dependence of vicinal coupling on the angles θ and θ'

2.2.4 Substituent Effects

In both saturated and unsaturated systems a decrease in the vicinal coupling is observed when an electronegative substituent is introduced at the H–C–C–H moiety. For substituted ethanes the relation between the electronegativity change, $\Delta E = E(X) - E(H)$, caused by the replacement of a hydrogen atom with a group X, and the coupling constant is given by

$$^3J = 9.41 - 0.80\Delta E \tag{4.30}$$

For substituted ethylenes similar relations result:

$$^3J_{trans} = 19.0 - 3.3\Delta E \tag{4.31}$$

$$^3J_{cis} = 11.7 - 4.7\Delta E \tag{4.32}$$

The following data serve to illustrate these effects:

	^3J		$^3J_{cis}$	$^3J_{trans}$
H_3C-CH_2-Li	8.9			
$-SiR_3$	8.0	$-Li$	19.8	23.9
$-CN$	7.6	$-SiR_3$	14.7	20.4
$-Cl$	7.2	$-CH_3$	10.0	16.8
$-OCH_2CH_3$	7.0	$-Cl$	7.3	14.4
$-\overset{\oplus}{OR_2}$	4.7	$-F$	4.7	12.8

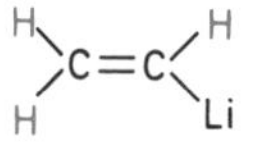

As the different constants in equations (4.31) and (4.32) intimate, the steric orientation of the substituent X in the H–C–C–H moiety is also of significance. An example is found with 4-phenyl-1,3-dioxane (46). For this molecule conformation 46a can be assumed because of the bulk of the phenyl substituent. The arrangement about the C5–C6 bond is represented by the Newman projection 46b. In spite of the possible flattening of the ring the dihedral angles ϕ_{cd} and ϕ_{ab} are equal. Nevertheless, J_{cd} is different from J_{ab}, as the experimental values given in formula 46b show. Accordingly, the electronegativity effect of the oxygen seems to predominantly affect the H_d, H_c proton pair. Other findings, such as the coupling

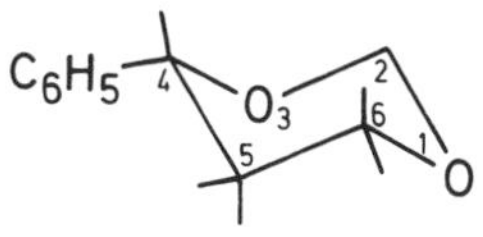

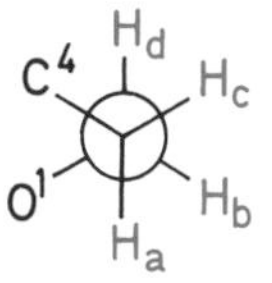

$J_{ab} = 5.1$ Hz
$J_{cd} = 2.9$ Hz

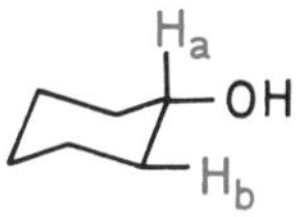

$J_{ab} = 4.2$ Hz

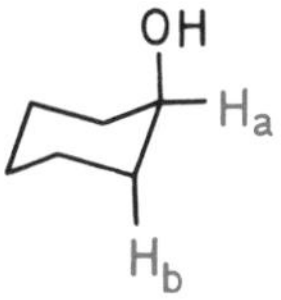

$J_{ab} = 2.7$ Hz

46a   **46b**   **47**   **48**

constants in the isomeric cyclohexanols **47** and **48**, are consistent with the observation that the maximum effect of a substituent on the vicinal coupling constant results when the substituent is *trans* to one of the protons at the neighbouring CH_2 group (**49**).

 49

The torsional angle dependence of the substituent effect on vicinal coupling constants in substituted ethanes derived from these data is confirmed by MO calculations. The Karplus curves for ethane and fluoroethane shown in Figure 4.24 indicate that the introduction of an electronegative substituent shifts the curve so that an increase in the ^3J value results in some conformations (for example, $\phi = 240°$, $\theta = 120°$). There are also experimental indications of this. It can be shown further that the substituent effect integrated over a complete rotation must lead to a diminution of the coupling constant that is obtained as the average of the couplings in all possible conformations. This conforms to the statement of the empirical relation (4.30).

The *mechanism* by which a substituent influences the magnitude of ^3J has been related in recent years to theoretical findings. Advanced calculations for both saturated and unsaturated systems indicated a widespread alternation of polarity in dipolar molecules, contrary to the earlier belief, where an attenuation, but not a charge alternation, was implied for the inductive substituent effect along a carbon chain. The alternation of sign for ΔJ, the change of the coupling constant, can thus

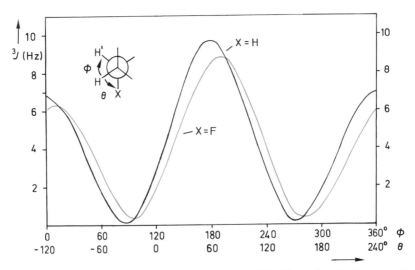

Figure 4.24   The effect of substituents on the vicinal H – H coupling in ethane as a function of torsional angle as derived from MO calculations (after Ref. 13)

be rationalized. These relations have been especially thoroughly investigated for monosubstituted benzenes. The dependence of the J_{23} coupling on the electronegativity of the substituent at C1 brought to light by this work is shown in Figure 4.25. Similarly, a comparison of the ^3J values in benzene and pyridine is most illustrative:

While the influence of a substituent on the value of ^3J in a HC–CH–C_αX unit is clearly detectable, the effect of substituents three bonds away from the HC–CH fragment as in the unit HC–CH–C_α–C_βX can be demonstrated only through very precise measurements. Exact analysis of the spectra of monosubstituted benzenes leads to the relation

$$^3J = 7.63 + 0.51\Delta E_\alpha - 0.10\Delta E_\beta \qquad (4.33)$$

which once again ensures a sign change for the substituent effect, ΔJ.

A special effect of substituents on vicinal coupling constants is observed in transition metal complexes of olefins and arenes. Here the $^3J_{cis}$ values decrease relative to those in uncomplexed double bonds by about 2–3 Hz and the decrease in iron carbonyl complexes is larger than in complexes of either chromium or molybdenum. The data for a few of these compounds, and for the free ligands, are presented in Table 4.13. In the iron carbonyl complexes of cyclobutadiene derivatives, the influence of the metal carbonyl groups has the result that the small vicinal coupling constant of the ring protons becomes zero in the complex.

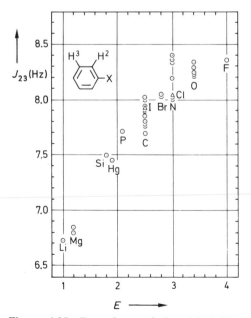

Figure 4.25 Dependence of the vicinal H – H coupling on the electronegativity of substituents in benzene derivatives (after Ref. 14)

2.3   LONG-RANGE COUPLING CONSTANTS (^4J, ^5J)

While the geminal and vicinal coupling constants usually have values of between 5 and 20 Hz and lead to easily recognizable line splitting in the spectra, the majority of long-range couplings over four, five and more bonds produce only small splittings of a few hertz or less. Therefore, these couplings were discovered only after the resolution of the n.m.r. spectrometers was greatly improved. Today, splittings of 0.2 Hz or even smaller are detectable without major difficulties and a wealth of structural and conformational information comes from long-range coupling constants. In general, this group of spin–spin interactions falls within the range

Table 4.13   Vicinal H,H coupling constants in transition metal complexes with organic ligands

0.1–3.0 Hz. Larger values are not unusual but they are observed less frequently than for ^2J or ^3J coupling constants.

For the interpretation of long-range coupling constants a consideration of σ and π mechanisms has proved useful. The spin–spin interaction is then approximated as the sum of two quantities, $J(\sigma)$ and $J(\pi)$ that are transmitted via the σ and π electrons, respectively.

The spin–spin interaction via σ-electrons has already been represented schematically in Figure 2.11. For the π mechanism an analogous diagram can be prepared. Let us consider a CH group with an sp^2 hybridized carbon atom (Figure 4.26a). To a first approximation, no interaction between the proton and π-electron in the carbon $2p_z$ orbital is possible since the proton lies exactly in the nodal plane of that orbital. From electron spin resonance (e.s.r.) spectroscopy it is known, however, that this deduction is not correct because the hyperfine splitting of e.s.r. spectral lines in radical ions of π systems comes about directly through a coupling of the proton with the unpaired electron residing in the carbon $2p_z$ orbital. To explain this one assumes that the unpaired electron polarizes the two electrons in the CH σ-bond in such a way that the one with the parallel spin prefers to remain at the carbon atom (Figure 4.26b). Thus, at the proton, the opposite magnetic polarization, which can be oriented either parallel or antiparallel to the nuclear magnetic moment, predominates. The two possibilities are energetically different and as a result there is a splitting of the Zeeman levels of the electron and consequently a splitting of the e.s.r. spectral lines.

We assume the same conclusions for the π mechanism of spin–spin coupling in nuclear magnetic resonance spectroscopy with the difference that here the spin polarization originates at one proton and is detected at another. Even in the case of the simple double bond we can discuss a σ and a π contribution to the vicinal coupling. This is represented schematically in Figure 4.26c. According to results of *valence bond* theory, the π contribution to the vicinal coupling, $^3J(\pi)$, is proportional to the product of the e.s.r. hyperfine coupling constants $\alpha(C–H)$, which are characteristic of the magnetic interaction between electron and nuclear spins in the $=C–H$ group. A detailed calculation shows that $J(\pi)$ in the case of vicinal coupling

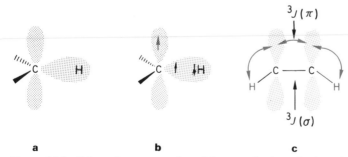

a   **b**   **c**

Figure 4.26 Schematic representation of the π-mechanism of spin–spin interaction between protons

constitutes about 10% of the total effect. Since the spin–spin interaction via the σ-electrons decreases rapidly with an increasing number of intervening bonds, the contribution of π-electrons to long-range coupling assumes a much greater importance. This is clearly shown by the results found in the case of unsaturated compounds. In the following section we first discuss the situation that exists in saturated compounds and then consider long-range coupling in unsaturated systems with special emphasis on the π contribution to it.

2.3.1   Saturated Systems

^4J and ^5J couplings are observed in saturated compounds, in particular when the C–H and C–C bonds exist in the zig-zag arrangement of the form

In the case of ^4J one speaks of the M or a W arrangement. Thus, in α-bromocyclohexanone, which exists in the chair conformation shown (50), spin–spin interactions of 1.1 Hz were found between the protons H_a, H_b, and H_c. In the bicyclic systems 51 and 52 the bonding arrangement of the coupled protons also meets the M criterion. For the assignment of stereochemistry in isomeric endo- and exo-bicyclo-heptane derivatives the magnitude of ^4J is of importance since only the endo proton couples with the anti bridge proton. Especially large ^4J values are found in strained systems such as bicyclo[2.1.1]hexane (53) and bicyclo[1.1.1]pentane (54). This is not surprising if one considers that in these compounds two or three routes are available for coupling between the protons compared with only one in the examples cited above.

^5J coupling is less frequently observed in saturated systems. As examples, compounds 55 and 56 are mentioned here. If the coplanar arrangement of the bonds is lost, the magnitude of ^4J and ^5J rapidly decreases. Nevertheless, in steroids the axial and equatorial orientation of angular methyl groups can still be distinguished on the basis of the different line width, despite methyl rotation (57, 58).

2.3.2   Unsaturated Systems

In compounds that contain π-bonds, both the σ and π contributions to the H–H coupling must be considered. For the latter, the valence bond calculations mentioned above lead to the following proportionalities between $J(\pi)$ and the e.s.r. hyperfine

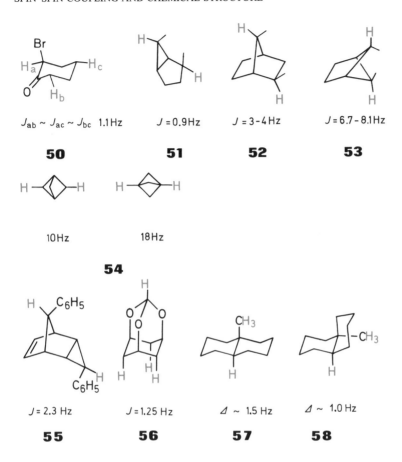

$J_{ab} \sim J_{ac} \sim J_{bc}$ 1.1 Hz $J = 0.9$ Hz $J = 3-4$ Hz $J = 6.7 - 8.1$ Hz

50   **51**   **52**   **53**

10 Hz   18 Hz

54

$J = 2.3$ Hz $J = 1.25$ Hz $\Delta \sim 1.5$ Hz $\Delta \sim 1.0$ Hz

55   **56**   **57**   **58**

coupling constants a that allow one to predict the sign of $J(\pi)$ and to estimate its magnitude:

$$HC = CH : \quad ^3J(\pi) \propto a(\dot{C} - H) \times a(C - H) \tag{4.34}$$

$$HC = C - CH : \quad ^4J(\pi) \propto a(\dot{C} - H) \times a(C - C - H) \tag{4.35}$$

$$HC = C - CH : \quad ^5J(\pi) \propto a(C - C - H) \times a(\dot{C} - C - H) \tag{4.36}$$

With $a(C-H) = -65 \times 10^6$ Hz and $a(C-C-H) = 150 \cos^2 \phi \times 10^6$ Hz, a *negative* sign results for $^4J(\pi)$ in an *allyl group* of the type HC=C–C–H and, in addition, a dependence on the torsional angle, ϕ, is expected.

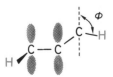

A large π-contribution to the coupling is consequently found in conformations with $\phi = 0°$ and $\phi = 180°$ while, because of the $\cos^2$ term, the π-contribution for $\phi = 90°$ and $\phi = 270°$ disappears.

In the propenes **59–61**, ^4J values have been observed which experimentally confirm the relations discussed above. Thus, the introduction of a large alkyl residue leads to a preference of the conformations shown for which the angle ϕ is approximately $270°$. Therefore, the π contribution to the allylic coupling decreases in the series **59–61** and the magnitude of ^4J becomes smaller. These results also confirm that $^4J(\pi)$ has a negative sign. In *cyclic* systems such as the lactone **62**, very large ^4J values are often found since here the favoured conformation is defined with $\phi \approx 0°$ or $\phi = 180°$. On the other hand, for the arrangement with $\phi = 90°$ we expect, according to these findings and the explanation given in Section 2.3.1, that $^4J(\sigma)$ should dominate and that the coupling constant should have a positive sign. Indeed, ^4J values of $+0.5$ to $+1.0$ Hz are found in cyclohexadienes (**63**). Of the same order and also with positive sign is the coupling constant between *meta* protons in benzene derivatives (**64**).

^4J_c	-1.33	-1.17	-0.10
^4J_t	-1.75	-1.43	-0.63
	59	**60**	**61**

$\mid ^4J \mid = 4.1$ Hz	$^4J = 0.5 - 1.1$ Hz	$^4J = 1 - 3$ Hz	$^5J = 5.5 - 11$ Hz
62	**63**	**64**	**65**

Based on a series of experimental data, the dependence of $^4J_{trans}$ in a $HC = C-C-H$ fragment on the torsional angle, ϕ, can be represented as the sum of $^4J(\sigma)$ and $^4J(\pi)$, as indicated in Figure 4.27. For $^4J_{cis}$ a similar result applies, but the large positive σ contribution is absent for the conformation with $\phi = 90°$.

For the *homoallylic* ^5J coupling observed in fragments of the type $HC-C = C-CH$ the conformational dependence is completely analogous to that discussed above for ^4J. Since, however, $^5J(\pi)$ has a positive sign (cf. equation 4.36), σ and π contributions to ^5J augment one another. In favourable cases very large ^5J values can be observed, as in 1,4-dihydrobenzenes (**65**), in which two routes are available for

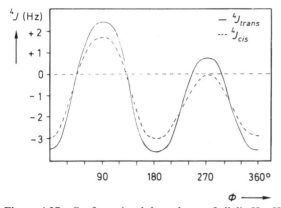

Figure 4.27 Conformational dependence of allylic H – H coupling (after Ref. 11)

the transmission of spin information. In contrast, in benzene itself the *para* coupling is only 0.69 Hz.

In a series of unsaturated systems couplings over five or more bonds are found which are probably transmitted largely over the σ-bonds that in these compounds assume the favourable zig-zag arrangement mentioned in Section 2.3.1. These can include the couplings between H-4 and H-8 and H-2 and H-7, respectively, in

$J \sim 0.8$ Hz

66

$J \sim 0.2$ Hz

$J \sim 0.5 - 1.0$

X = O, S, NH

67

$^5J_{tt} = 1.30$ Hz

68

$^5J_{tc} = 0.60$ Hz

$J = 0.4$ Hz

69

naphthalene (**66**), and also similar interactions in heterocyclic systems such as **67**. Also, the larger value of $^5J_{tt}$ in 1,3-butadiene compared to the value of $^5J_{tc}$ can be attributed to the additional σ contribution (**68**). Similarly, in the case of benzaldehyde it was shown that only the *meta* protons couple with the aldehyde proton (**69**), again an indication of the importance of $^5J(\sigma)$. On the other hand, for *polyacetylenes, allenes,* and *cumulenes* long-range coupling arises almost exclusively through the π mechanism. For allene and butatriene the following interaction diagrams can be formulated:

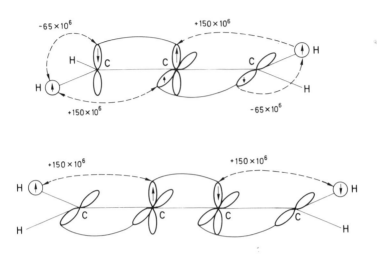

As the examples collected in Table 4.14 show, spin transmission through the π system is very effective. With an increasing number of bonds only relatively small decreases of the coupling constants occur. Thus, even over nine bonds an interaction can be observed. Furthermore, the substitution of a terminal hydrogen in the $CH_3-(C\equiv C)_x-H$ unit by a methyl group, formally a transformation of an allyl-type coupling into one of the homoallyl type, leads only to a sign change in the coupling constant while its magnitude remains unaffected. Actually, the sign changes for the examples given in Table 4.14 have not yet been verified experimentally, but they can be considered to be correct on the grounds of results obtained in other systems as well as on the basis of theoretical calculations.

Table 4.14 Long-range coupling in selected polyacetylenes and allenes

$H_3C-C\equiv C-H$	- 2.93	$(H_3C)_2C=C=CH_2$	3.0
$H_3C-C\equiv C-CH_3$	+ 2.7	$H_3C-CH=C=CHCl$	- 5.8
$H-C\equiv C-C\equiv C-H$	+ 2.2	$H_3C-CH=C=CHCl$	+ 2.4
$H_3C-C\equiv C-C\equiv C-H$	- 1.27		
$H_3C-C\equiv C-C\equiv C-CH_3$	+ 1.3		
$H_3C-C\equiv C-C\equiv C-C\equiv C-CH_2OH$	+ 0.4		4.58

2.4   DIRECT SPIN–SPIN INTERACTION AND THROUGH-SPACE COUPLING

Finally, we discuss here two mechanisms of spin–spin interaction that play only a limited or no role at all in the line splitting in high-resolution n.m.r. spectroscopy.

The first involves the direct magnetic interaction of nuclear moments through space mentioned earlier (p. 8), which can also be termed a dipole–dipole or simply a *dipolar coupling*. As a qualitative consideration shows, this interaction results in a splitting of the resonance signal by the amount

$$\Delta B = 3\mu(3\cos^2\theta - 1)r^{-3}(\mu_0/4\pi) \qquad (4.37)$$

where μ is the magnetic moment of the proton, r the distance between the two nuclei and θ the angle between the line joining the nuclei and the direction of the external field:

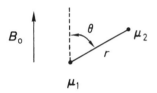

Equation (4.37) applies to a spin system of two isochronous protons, such as those in methylene chloride. In solution or in a pure liquid, the angle θ varies with time because of the rapid rotational motion of the molecule and the expression $3\cos^2\theta - 1$ vanishes. Consequently $\Delta B = 0$ and a line splitting as a result of dipolar coupling is not observed. In contrast, in a solid where the molecules assume a *fixed* orientation with respect to the direction of the external field B_0 and with respect to each other, splitting will occur. As a result, the *n.m.r. spectra of solids* have an entirely different appearance to those of liquids. In particular, the width of the resonance lines is different by several orders of magnitude (Figure 4.28). Since, as equation (4.37) indicates, distances between nuclei can be determined from these spectra, the n.m.r. of solids (also known as *wide-line spectroscopy*) forms a branch of solid-state physics.

The other mechanism we want to discuss here is in principle only a variation of the spin–spin coupling transmitted by electrons treated in detail earlier. It has been detected in a few cases when, as the result of steric compression, an extensive non-bonding or van der Waals interaction of orbitals occurs. Transmission of magnetic information then results through a "short circuit" where no formal bonds are present. Thus, a coupling of 1.1 Hz is observed between the protons H_a and H_b in the case of compound **70**. Since the two nuclei are separated by six σ-bonds with an unfavourable geometry for conventional coupling, a direct spin–spin interaction between the two hydrogen 1s orbitals is very probable. This mechanism—known as

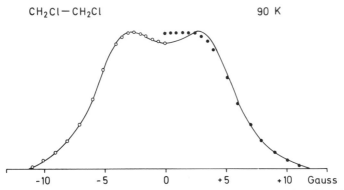

CH_2Cl-CH_2Cl 90 K

-10   -5   0   +5   +10   Gauss

Figure 4.28 The proton magnetic resonance spectrum of solid 1,2-dichloroethane at 29.5 MHz. Here the dipolar coupling of the protons in a methylene group dominates and the measured doublet splitting of 37.5 kHz leads to a calculated proton separation of 0.170 nm. The theoretical points (open circles) calculated with this value agree better with the experimental curve than those for which the calculation was based on a separation of 0.172 nm (solid circles) (after Ref. 15)

through-space coupling—has greater significance for spin–spin coupling between a proton and a fluorine nucleus as well as between two fluorine nuclei (cf. Chapter 10).

70

2.5   TABLES OF SPIN–SPIN COUPLING CONSTANTS IN ORGANIC MOLECULES

Just as for the chemical shift, characteristic data for spin–spin coupling in different classes of organic compounds are collected in Table 12.2 in the Appendix (p. 519 ff.).

Exercise 4.6 The proton resonance spectrum of 2,3-benzoxepine is given in Figure 4.29. Develop, on the basis of the integration and the splitting pattern, an assignment of the resonance signals and determine the coupling constants in the olefinic portion of the molecule. Check to see if your assignment of the olefinic resonances is unequivocal. How could the correct solution be determined?

Exercise 4.7 A 6-chloro-1-trimethylsilylhex-1-ene produces a proton resonance spectrum the olefinic region of which is shown in Figure 4.30. Determine the coupling constants and determine whether the spectrum is that of the *cis* compound, the *trans* compound, or a mixture of the two. Estimate, if necessary, the *cis/trans* ratio.

Exercise 4.8 Cycloheptatrienes that are mono-substituted in the 7-position can exist in either conformation **71a** or **71b**. Experimentally it was found that the allylic coupling constant J_{27} has a negative sign in 7-phenylcycloheptatriene. Which of the two conformations is consistent with this? In which conformation can a measurable homoallylic coupling J_{37} be expected?

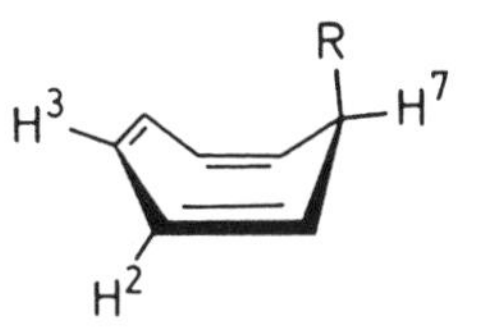

71a 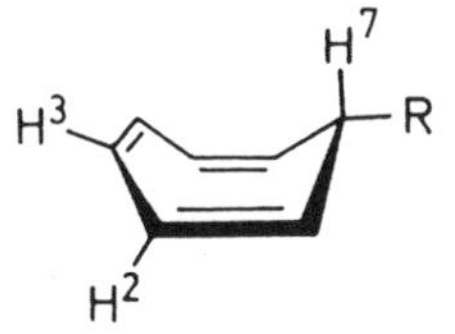   **71b**

Exercise 4.9 The conformations **72** and **73** of the A ring of an acetylated steroid are to be dierentiated. The signal for the proton H_a appears as a doublet of doublets with coupling constants of 13.1 and 6.6 Hz. Which conformation is consistent with this finding?

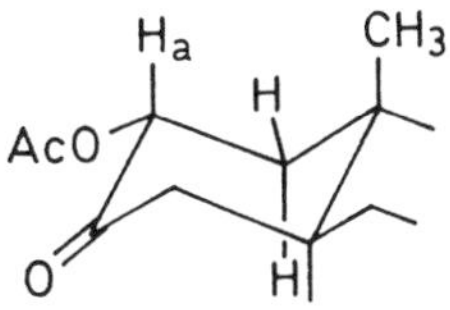

72 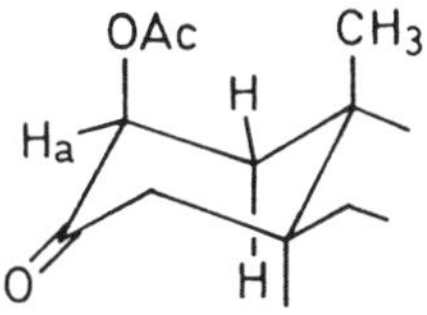   **73** 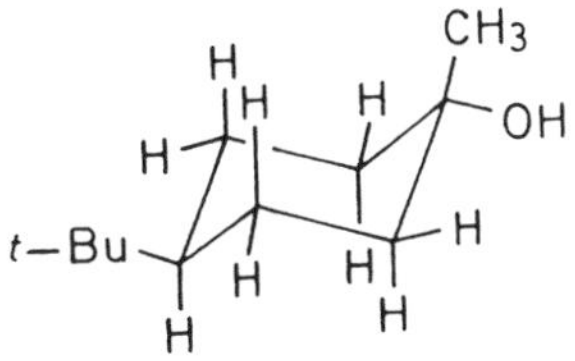   **74**

Exercise 4.10 The line widths of the methyl proton resonances in the isomeric 1-methyl-4-t-butylcyclohexanols **74** and **75** are 1.0–1.3 and 0.6–0.7 Hz, respectively. Explain this result and which assignment must be made?

Exercise 4.11 The two structures **76** and **77** are proposed for an unknown compound. The vicinal coupling constant of the olefinic protons is 2.8 Hz. Which structure is correct?

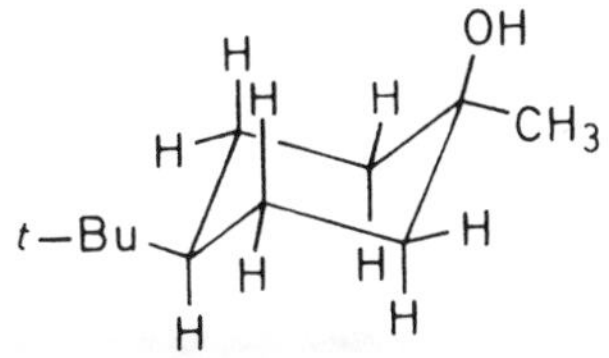

75 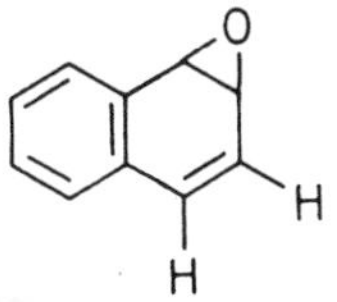   **76**

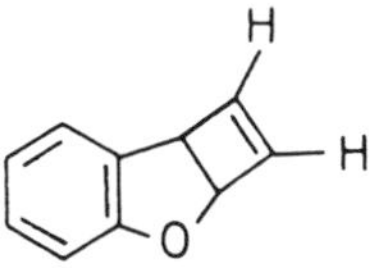

77

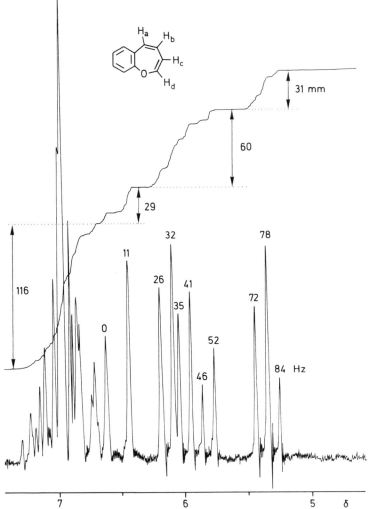

Figure 4.29   Proton magnetic resonance spectrum of 2,3-benzoxepine at 60 MHz

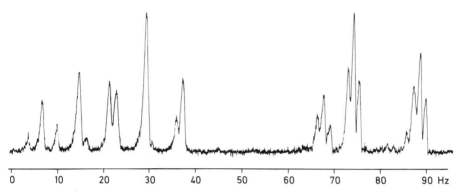

Figure 4.30 ¹H n.m.r. spectrum of the olefinic protons of a 6-chloro-1-trimethylsilylhexene at 60 MHz

3. References

1. N. S. Ham and T. Mole, in F5, **4**, 91 (1969).
2. W. Bremser, PhD thesis, University of Cologne, 1968.
3. J. R. Knowles, *Aldrichimica Acta* **22**, 59 (1989).
4. E. Vogel, *Pure Appl. Chem.* **54**, 1015 (1982); E. Vogel, U. Haberland, H. Günther, *Angew. Chem.* **82**, 510 (1970); *Angew. Chem., Int. Ed. Engl.,* **9**, 513 (1970).
5. J. A. Pople and K. G. Untch, *J. Am. Chem. Soc.,* **88**, 4811 (1966).
6. G. R. DeMare and J. S. Martin, *J. Am. Chem. Soc.,* **88**, 5033 (1966).
7. H. F. Friedrich, *Z. Naturforsch.,* **20b**, 1021 (1965).
8. H. Strehlow, *Magnetische Resonanz und Chemische Struktur,* 2nd ed., Steinkopff Verlag, Darmstadt 1968.
9. L. M. Jackman and S. Sternhell, *Application of Nuclear Magnetic Resonance Spectroscopy in Organic Chemistry,* Pergamon Press, Oxford, 1969.
10. C. Pascual, J. Meier and W. Simon, *Helv. Chim. Acta,* **48**, 164 (1969).
11. S. Sternhell, *Quart. Rev.,* **23**, 236 (1969).
12. J. B. Pawliczek and H. Günther, *Tetrahedron,* **26**, 1755 (1970).
13. K. G. R. Pachler, *Tetrahedron,* **27**, 187 (1971).
14. S. Castellano and C. Sun, *J. Am. Chem. Soc.,* **88**, 4741 (1966).
15. H. S. Gutowsky, G. B. Kistiakowski, G. E. Pake and E. M. Purcell, *J. Chem Phys.,* **17**, 972 (1949).

Review Articles

(a) C. W. Haigh and R. B. Mallion, Ring Current Theories in Nuclear Magnetic Resonance, in F5, **13**, 303 (1966).
(b) S. Sternhell, Correlation of Interproton Spin–Spin Coupling Constants with Structure, *Quart. Rev.,* **23**, 236 (1969).
(c) A. A. Bothner-By, Geminal and Vicinal Proton–Proton Coupling Constants in Organic Compounds, in F1, **1**, 195 (1965).
(d) R. C. Cookson, T. A. Crabb, J. J. Frankel, and J. Hudec, Geminal Coupling Constants in Methylene Groups, *Tetrahedron,* Suppl. No. 7, 355 (1966).

(e) S. Sternhell, Long-Range ^{1}H–^{1}H Spin–Spin Coupling in Nuclear Magnetic Resonance Spectroscopy *Rev. Pure Appl. Chem.,* **14**, 15 (1964).
(f) M. Barfield and B. Chakrabarti, Long-range Proton Spin–Spin Coupling, *Chem. Rev.,* **69**, 757 (1969).
(g) J. Hilton and L. H. Sutcliffe, The Through-space Mechanism in Spin–Spin Coupling, in F5, **10**, 27 (1975).
(h) J. Kowalewski, Calculations of Nuclear Spin–Spin Coupling Constants, in F5, **11**, 1 (1977).
(i) H. Günther and G. Jikeli, ^{1}H-NMR Spectra of Cyclic Monoenes: Hydrocarbons, Ketones, Heterocycles, and Benzo Derivatives, *Chem. Rev.,* **77**, 599 (1977).
(j) P. Laszlo, Solvent Effects and Nuclear Magnetic Resonance, in F5, **3**, 231 (1967).
(k) J. Ronayne and D. H. Williams, Solvent Effects in Proton Magnetic Resonance Spectroscopy, in F2, **2**, 83 (1969).

5 THE ANALYSIS OF HIGH-RESOLUTION NUCLEAR MAGNETIC RESONANCE SPECTRA

In Chapter 2 we introduced simple rules for the direct determination of chemical shifts and spin–spin coupling constants from the splitting patterns observed in a nuclear magnetic resonance spectrum. As already mentioned there, however, the application of these rules is limited to first order spectra and the more complicated spin systems found for many molecules can only be analysed by quantum chemical methods. The increased use of high magnetic fields provided by superconducting magnets has improved the situation to some extent and first order spectra which can be analysed by inspection are met today quite frequently. On the other side, spectra

of higher order are now observed for structures which could not be analysed by
n.m.r. before due to insufficient spectral resolution. In other words, many of the spin
systems illustrated here with 60 or 100 MHz spectra are met today at 300 or
500 MHz measuring frequency for molecules like steroids, peptides or carbohydrates. Furthermore, as a consequence their particular spin hamiltonian, quite a
number of spin systems, as, for example, the AA'XX' system, will never transform
into a first order spectrum even at the highest magnetic field available. The problem
of accurate spin analysis is thus a permanent challenge for the n.m.r. spectroscopist.
The following chapter is, therefore, devoted to the quantum chemical basis of n.m.r.
spectral analysis. First we shall attempt to present the essential principles, treating
later individual types of spectra, and finally a number of important generalizations.
We will limit ourselves, however, to the consideration of the more frequently
encountered spin systems since a comprehensive treatment of the subject is beyond
the scope of this introductory text.

The question that is of primary interest in this chapter can be formulated as
follows: 'How can the parameters—chemical shifts and coupling constants—of the
spin system under consideration be derived from the nuclear magnetic resonance
spectrum?' In order to answer this question we must familiarize ourselves with the
principles of the calculation of high-resolution n.m.r. spectra. Therefore, the
converse question: 'How do the line frequencies and the line intensities in a
spectrum follow from a known set of chemical shifts and coupling constants?' will
be investigated first. That is, before we consider the *analysis* of a spectrum we want
to understand its *synthesis*.

1. Notation for spin systems

Let us first introduce a nomenclature for the classification of different spin systems.
The notion 'spin system' is used for a group of n nuclei of spin quantum number
$\mathbf{I} = \frac{1}{2}$ that is characterized by no more than n resonance frequencies, v_i, and $n(n-1)/2$
2 coupling constants, J_{ij}. This group does not interact magnetically with any other
group of nuclei. Nuclei of equal chemical shift are labelled with the same capital
letter and the number of such nuclei in the system is indicated by a subscript. Thus,
the protons of a methyl group form an A_3 system while those of an ethyl group
constitute an A_3B_2 system. The relative chemical shifts of different nuclei in a spin
system are indicated by the position in the alphabet of the labelling letter. For a
CH_3CF_2-group the designation A_3X_2 is used to indicate the large difference between
the chemical shifts of the protons and the fluorine nuclei. Magnetically non-
equivalent nuclei such as the two protons and the two fluorine nuclei in 1,1-
difluoroethylene (cf. p. 43) and the two pairs of protons in 1,2-dichlorobenzene are
distinguished by using primed letters. These two spin systems are an AA'XX'
system and an AA'BB' system, respectively. Furthermore, whenever possible, the
nuclei are labelled in such a fashion that the sequence of the letters in the alphabet
matches the sequence of the resonance frequencies in order of increasing field
strength.

Exercise 5.1 Classify by spectral type the protons in the compounds (a)–(m).

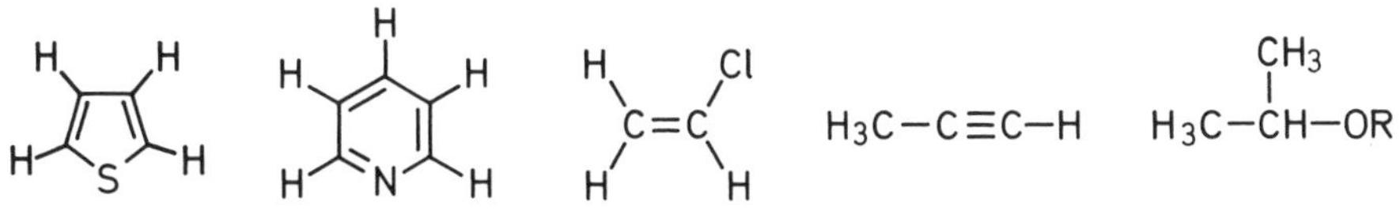

a   b   c   d   e

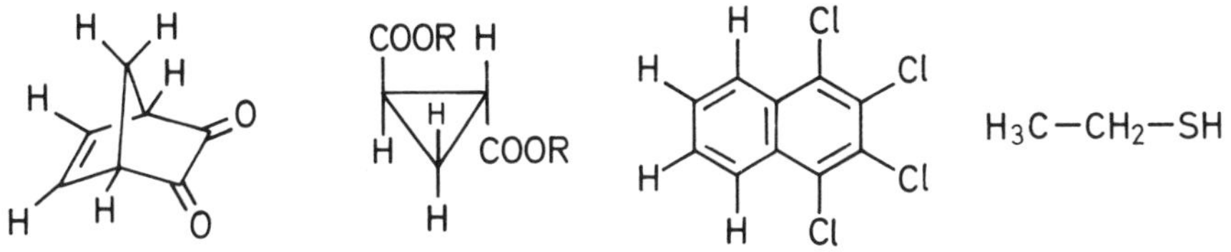

f   g   h   i

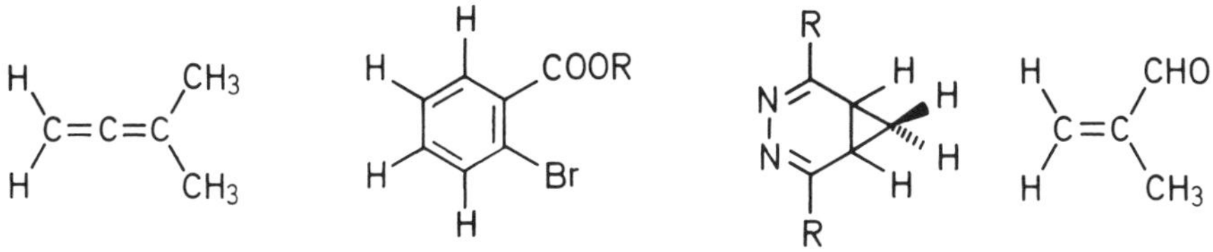

j   k   l   m

2. Quantum mechanical formalism

Since the nuclear spin is a non-classical property of atomic nuclei, we can solve the problem posed by the calculation of nuclear magnetic resonance spectra only with the aid of quantum mechanics. In the scope of this text we must introduce the necessary quantum mechanical principles and methods axiomatically since, on the one hand, we cannot presume a detailed knowledge of the theory by the reader and, on the other hand, we want to present as exact a derivation as possible. Our approach, however, has a certain rationale since quantum mechanics rests on postulates that can be neither proved nor rigorously derived mathematically. Instead they are based exclusively on experimental observations. Those readers who are familiar with the Hückel molecular orbital theory will soon discover that the mathematical formalism used there to calculate eigenvalues and wave functions of the electrons is the same that applies to the present problem. A number of interesting parallels exist and a comparison of the two theories is very illuminating.

The fact that we can observe a nuclear magnetic resonance spectrum with distinct spectral lines demonstrates that the energy of a spin system in a magnetic field is *quantized*. Just like the individual nuclei, the spin system as a whole can exist only in certain states, the *stationary states* or *eigen-states*. The energies of these eigenstates, the *eigenvalues*, are determined by the interaction between the nuclei and the external magnetic field, B_0, as well as by the spin–spin interaction of the nuclei with one another. Each eigenstate is characterized by a *wave function* or *eigenfunction*, ψ.

The frequencies, f_{pq}, of the n.m.r. signals correspond to the energy differences between the stationary states of the spin system. Their calculation therefore presupposes a knowledge of the eigenvalues E_p and E_q:

$$f_{pq} = \frac{1}{h}(E_q - E_{p)} \tag{5.1}$$

For the derivation of the relative intensities of the signals we must also know the eigenfunctions, Ψ, as will be shown later.

2.1 THE SCHRÖDINGER EQUATION

We postulate that at the atomic level the relation between the energy, E, of a particle and its wave function, Ψ, can be described by the *Schrödinger equation*. In its simplest time-independent form, this equation can be written as a so-called *eigenvalue equation*:

$$\mathscr{H}\Psi = E\Psi \tag{5.2}$$

where $\mathscr{H}$ is the Hamiltonian operator. Its application on the eigenfunction, Ψ, yields the product of eigenvalue and eigenfunction. Equation (5.2) enables us to calculate the stationary energy states of one or more particles, perhaps the electrons in a molecule or the magnetic nuclei of a spin system, as long as we know the Hamiltonian operator and the eigenfunctions. Since the energy of the system under consideration can be derived experimentally, E is called the *observable* appropriate to the particular Hamiltonian.

The notion of the *operator* that we shall use in what follows requires a short explanation. Operators are, first of all, nothing more than instructions. The symbol $\mathscr{H}\Psi$ signifies only that an operation prescribed by the Hamiltonian operator (and not yet explained in detail) will be executed on the wave function, Ψ. In this sense the square root and the integral signs are also operators since they prescribe definite operations to be carried out on a certain function. Further, if a function is to be differentiated with respect to x the differential operator reads d/dx. From these examples, it is apparent that operators are always written at the left of the function to which they are to be applied. The exchange $\mathscr{H}\Psi \to \Psi\mathscr{H}$ therefore is not allowed.

The central problem of the theory of chemical bonding is the motion of the electron in the potential field of the atomic nuclei and the other electrons in the molecule. In this case, the familiar Hamilton operator given in equation (5.3) is employed. Here the *Laplace operator* stands for the kinetic energy and the operator $\hat{V}$ stands for the potential energy:

$$\mathcal{H} = \frac{-h^2}{8\pi^2 m}\left(\frac{\partial^2}{\partial x^2} + \frac{\partial^2}{\partial y^2} + \frac{\partial^2}{\partial z^2}\right) + \hat{V} \tag{5.3}$$

The difficulties that prevent a simple treatment of the bonding problem according to equation (5.2) result principally from the complex form of the Hamiltonian operator according to which the energy is a function of the coordinates of all of the electrons. For the model of an electron on a circle that we used in Chapter 4 these difficulties are drastically reduced. Therefore, it seems worthwhile to demonstrate the application of equation (5.2) with this example.

Since the motion of the electron shall be confined to a circular path of radius r we can use the angle ϕ formed by the radius vector with an arbitrarily assumed starting point as the only variable to characterize the position of the particle:

Further, the potential energy shall be zero so that the Hamiltonian operator, written in polar coordinates (see Appendix), assumes the following form:

$$\mathcal{H} = \frac{-h^2}{8\pi^2 mr^2}\cdot\frac{\partial^2}{\partial\phi^2} \tag{5.4}$$

The wave nature of the electron suggests that a sine or a cosine function should be used as the eigenfunction. Let us try $\Psi = N\sin q\phi$, so that equation (5.2) with the substitution of equation (5.4) leads to

$$\frac{-h^2}{8\pi^2 mr^2}\cdot\frac{\partial^2}{\partial\phi^2}\cdot N\sin q\phi = E\cdot N\sin q\phi \tag{5.5}$$

which, on carrying out the operation, yields

$$\frac{h^2}{8\pi^2 mr^2}\cdot N\sin q\phi\cdot q^2 = E\cdot N\sin q\phi \tag{5.6}$$

From this it follows that

$$E = \frac{h^2}{2mL^2}\cdot q^2 \tag{5.7}$$

if we substitute L for the circumference of the circle, $2\pi r$. The quantum condition $q = 0, \pm 1, \pm 2, \ldots, \pm n$ results from the requirement that the eigenfunction for all values of ϕ must be unambiguous. Specifically on the circle the condition $\Psi(\phi) = \Psi(2\pi + \phi)$ must hold. The eigenvalues E_0, E_1, E_2, etc., accordingly correspond to the energies derived in Chapter 4. Let us consider further that the square of the eigenfunction Ψ_q is equal to the probability that the electron occupies a particular position on the circle so that

$$\int_0^{2\pi} N^2 \, \sin^2 q\phi \, d\phi = 1 \tag{5.8}$$

For N, the so-called *normalization constant*, $1/\sqrt{\pi}$ is obtained since

$$\int_0^{2\pi} \sin^2 q\phi \, d\phi = \pi \tag{5.9a}$$

The eigenfunction is then

$$\Psi_q = \sqrt{\frac{1}{\pi}} \sin q\phi \tag{5.9b}$$

with $q = 0, \pm 1, \pm 2, \pm 3, \ldots, \pm n$.

Exercise 5.2   Using equation (5.2), the Hamiltonian operator derived from equation (5.3),

$$\hat{\mathscr{H}} = \frac{-h^2}{8 \, \pi^2 m} \frac{\partial^2}{\partial x^2}$$

and the trial function $\psi = N \sin ax$, calculate the eigenvalues and the eigenfunctions of an electron that can move in a one-dimensional box of length L at the potential $V = 0$.

---

3. The Hamiltonian operator for high resolution nuclear magnetic resonance spectroscopy

We now make use of the Schrödinger equation for the solution of the problem of interest to us—the determination of the energy levels of a spin system in a magnetic field. The phenomenologically formulated Hamiltonian operator that applies here has the form

$$\mathscr{H} = \mathscr{H}^{(0)} + \mathscr{H}^{(1)} = \sum_i v_i \, \hat{I}_z^{(i)} + \sum_i \sum_{i<j} J_{ij} \, \hat{\mathbf{I}}(i) \, \hat{\mathbf{I}}(j) \tag{5.10}$$

In which the first term, $\mathscr{H}^{(0)}$, relates to the interaction of the nuclei with the external field, B_0, and the second term, $\mathscr{H}^{(1)}$, relates to the spin–spin coupling energy.

Correspondingly, $\mathscr{H}^{(0)}$ contains the resonance frequencies v_i and $\mathscr{H}^{(1)}$ contains the coupling constants J_{ij}. $\mathscr{H}^{(1)}$ thus corresponds exactly to equation (2.10) introduced earlier for the energy of spin–spin interaction, while the form of $\mathscr{H}^{(0)}$ can be understood on the grounds of the ideas developed in Chapter 1. There the relation $E = -\mu_z B_0$ was introduced for the potential energy of a nucleus in a magnetic field. With equation (1.5) and the resonance condition (1.10), it follows that $E = v_i \hat{I}_z(i)^*$. When divided by Planck's constant, h, the energy—as in the case of spin–spin interaction—is obtained in hertz. Thus the factor $1/h$ in equation (5.1) disappears.

Turning now to the wave functions, we shall use the functions α and β that were introduced earlier to characterize the antiparallel and parallel orientations of the nuclear magnetic moment with respect to the external magnetic field. Important properties of these functions will be introduced later.

The Hamiltonian operator (5.10) contains additional operators, namely the nuclear spin operators $\hat{I}_z$ and $\hat{I}$. Their properties are defined as postulates that tell us the results of their application on the wave functions α and β:

$$\hat{I}_x\alpha = \tfrac{1}{2}\beta \qquad \hat{I}_y\alpha = i\tfrac{1}{2}\beta \qquad \hat{I}_z\alpha = \tfrac{1}{2}\alpha$$
$$\hat{I}_x\beta = \tfrac{1}{2}\alpha \qquad \hat{I}_y\beta = -i\tfrac{1}{2}\alpha \qquad \hat{I}_z\beta = -\tfrac{1}{2}\beta \tag{5.11a}$$

In addition, the complex combinations

$$\hat{I}^+ = \hat{I}_x + i\hat{I}_y \tag{5.11b}$$

and

$$\hat{I}^- = \hat{I}_x - i\hat{I}_y \tag{5.11c}$$

are introduced, known as *raising* and *lowering operator*, respectively. Their application to a wave function leads to the wave function of the eigenstate with the next higher or lower quantum number, respectively. The reader can verify this statement by a straightforward application of $\hat{I}^+$ and $\hat{I}^-$ to the spin functions α and β. In Chapter 8 we will meet these operators again.

The vector $\hat{\mathbf{I}}$ is defined by its components $\hat{I}_x$, $\hat{I}_y$, and $\hat{I}_z$. We note that the relations for the operator $\hat{I}_z$ can also be interpreted as eigenvalue equations: α and β are then eigenfunctions of $\hat{I}_z$ having eigenvalues $+\tfrac{1}{2}$ and $-\tfrac{1}{2}$, respectively; that is, the magnetic quantum numbers m_I introduced in Chapter 1 are the eigenvalues of the $\hat{I}_z$ operator.

Relative to the wave functions we want to stipulate further that they are *orthogonal* and *normalized*. Then

$$\int \alpha\alpha d\upsilon = \int \beta\beta \, d\upsilon = 1 \tag{5.12a}$$

---

* B_0 is assumed to be directed along the negative z axis of the coordinate system so that E is positive.

and

$$\int \alpha\beta \; dv = \int \beta\alpha \; dv = 0 \qquad (5.12b)$$

where the integral is over all space.

The obvious significance of these conditions is that an individual nucleus can exist in either the α or the β state and that the probability of its existence in one of the two states, when integrated over all space, is exactly unity.

4. The calculation of individual spin systems

In principle, we are now in the position to calculate the eigenvalues of any spin system by using equation (5.2) in conjunction with the rules formulated in (5.11) and the properties of the wave functions α and β defined by (5.12). It is important, however, to point out that only relative energies for the eigenstates of a spin system can be determined by means of the formalism we have developed. We have practically eliminated the question of the absolute energy values since we have introduced the resonance frequencies, v_i, and the coupling constants, J_{ij}, as phenomenological parameters. This procedure obviated the much more complicated absolute determination of these quantities in which one encounters the same difficulties as in the exact treatment of chemical bonding since the Schrödinger equation must be solved for the unperturbed molecule before the shielding constants of the nuclei in the external magnetic field and the magnetic spin–spin interactions can be studied. However, a knowledge of the relative energies of the eigenstates of a system is all that is necessary in spectroscopy since the spectral frequencies depend only on the energy difference of the eigenvalues.

Below we shall carry out the calculations for a few simple spin systems using the background developed in the preceding sections and, in so doing, we shall introduce additional important rules from time to time.

4.1 THE STATIONARY STATES OF A SINGLE NUCLEUS A

This nearly trivial case will be considered at the outset for completeness. Here $\mathscr{H} = \mathscr{H}^{(0)}$ and it follows according to equations (5.2) and (5.11a) that the energy of the stationary state characterized by the wave function α is given by

$$\mathscr{H}^{(0)}\alpha(A) = E_{+1/2}\alpha(A)$$
$$v_A \hat{I}_z(A)\alpha(A) = E_{+1/2}\alpha(A)$$
$$v_A \tfrac{1}{2}\alpha(A) = E_{+1/2}\alpha(A)$$
$$E_{+1/2} = \tfrac{1}{2}v_A$$

Analogously, $E_{-1/2} = -\tfrac{1}{2}v_A$ and the frequency for the transition of a nucleus from $E_{-1/2}$ to $E_{1/2}$ is v_A.

4.2 TWO NUCLEI WITHOUT SPIN–SPIN INTERACTION ($J_{ij} = 0$); SELECTION RULES

The system under consideration consists of two nuclei, A and B, that are individually characterized by the wave functions α and β. Four product functions identified by the sums, m_T, of the magnetic quantum numbers $m_I(A)$ and $m_I(B)$ — that is, the *total spin* introduced on p. 32 — serve to describe the stationary states:

$$(1)\ \phi_1 = \alpha(A)\alpha(B) \qquad m_T = +1$$
$$(2)\ \phi_2 = \alpha(rmA)\beta(B) \qquad m_T = 0$$
$$(3)\ \phi_3 = \beta(A)\alpha(B) \qquad m_T = 0$$
$$(4)\ \phi_4 = \beta(A)\beta(B) \qquad m_T = -1$$

The use of product functions should become clear by means of the following consideration. The Hamiltonian operators for the individual nuclei would be $\mathscr{H}_A$ and $\mathscr{H}_B$, respectively, so that for the system of two nuclei $\mathscr{H} = \mathscr{H}_A + \mathscr{H}_B$ and correspondingly the energy of the system would be $E = E_A + E_B$. Let us use the product function ϕ_1 so that the left-hand side of the Schrödinger equation would be $\mathscr{H}\alpha(A)\alpha(B) = \mathscr{H}_A\alpha(A)\alpha(B) + \mathscr{H}_B\alpha(A)\alpha(B)$. As long as no interaction exists between the nuclei we can consider the wave function $\alpha(B)$ as a constant with respect to the operation $\mathscr{H}_A\alpha(A)\alpha(B)$. Proceeding similarly with $\mathscr{H}_B\alpha(A)\alpha(B)$, we find that $\mathscr{H}_A\alpha(A) = E_A\alpha(A)$ and $\mathscr{H}_B\alpha(B) = E_B\alpha(B)$:

$$\mathscr{H}\alpha(A)\alpha(B) = E_A\alpha(A)\alpha(B) + E_B\alpha(A)\alpha(B)$$
$$= (E_A + E_B)\alpha(A)\alpha(B)$$
$$= E\alpha(A)\alpha(B)$$

That is, the product function ϕ_1 satisfies the Schrödinger equation $\mathscr{H}\phi_1 = E\phi_1$.

We can now calculate the energies of the four spin states of the two spin system according to equation (5.2) using the product functions ϕ_1, ϕ_2, ϕ_3, and ϕ_4. Since there is no coupling, $\mathscr{H} = \mathscr{H}^{(0)}$ and it follows that

$$[v_A\hat{I}_z(A) + v_B\hat{I}_z(B)]\alpha(A)\alpha(B) = E_1\alpha(A)\alpha(B)$$
$$(\tfrac{1}{2}v_A + \tfrac{1}{2}v_B)\alpha(A)\alpha(B) = E_1\alpha(A)\alpha(B)$$
$$E_1 = \tfrac{1}{2}(v_A + v_B)$$

In the calculation it should be noticed that the operator $\hat{I}_z(A)$ is applied only to the wave function of the nucleus A. Correspondingly, $\hat{I}_z(B)$ is effective only on $\alpha(B)$.

In an analogous fashion we obtain

$$E_2 = \tfrac{1}{2}(v_A - v_B)$$
$$E_3 = -\tfrac{1}{2}(v_A - v_B)$$
$$E_4 = -\tfrac{1}{2}(v_A + v_B)$$

Consequently, the energy level diagram for the two-spin system without spin–spin coupling has the structure shown in Figure 5.1. The frequencies of the lines correspond to the differences between the eigenvalues and the spectrum consists of two lines at ν_A and ν_B.

The transitions $(3) \rightarrow (2)$ or $(4) \rightarrow (1)$ are not considered because, according to the applicable *selection rules*, only those transitions which result in a unit change of the total spin of the wave function ($\Delta m_T = \pm 1$) are allowed. This corresponds to the plausible requirement that one quantum of energy can effect the reorientation of *only one* nucleus. In Table 5.1(A) (p. 155) our results for the AB system without spin–spin coupling are tabulated. We shall treat the derivation of the relative intensities of the individual transitions in Section 4.3.2.

Exercise 5.3   Form the product functions for a spin system of three nuclei and calculate the eigenvalues assuming that there is no spin–spin coupling.

For the calculation of the eigenvalues in the examples considered so far, we were able to use the Schrödinger equation (5.2) directly. This was possible because the eigenfunctions of the corresponding systems were already available to us in the functions α and β or $\alpha\alpha$, $\alpha\beta$, $\beta\alpha$, and $\beta\beta$. This will not always be the case. On the contrary, in the future we shall usually have the problem of calculating the energy of a spin system by means of a trial function that is different from the true eigenfunction. It is here that a further postulate of quantum mechanics has its

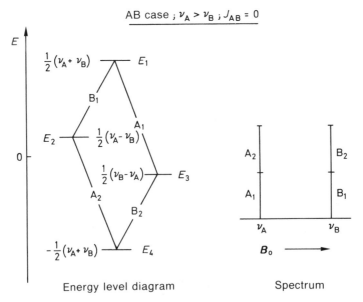

Figure 5.1   Energy level diagram and the spectrum of the AB system in which $J_{AB} = 0$

application. The energy of a spin system can be calculated using trial or approximate functions by means of the relation

$$E = \frac{\int \Psi \mathcal{H} \Psi \, d\upsilon}{\int \Psi^2 \, d\upsilon} = \frac{<\Psi \mid \mathcal{H} \mid \Psi>}{<\Psi \mid \Psi>} \tag{5.13}$$

Equation (5.13) follows from equation (5.2) by multiplication by Ψ in the usual manner followed by integration. The shorthand notation introduced here for the integral is due to Dirac.

4.3   TWO NUCLEI WITH SPIN–SPIN INTERACTION ($J_{ij} \neq 0$)

4.3.1   The A_2 case and the variational method

We shall now introduce spin–spin coupling between the nuclei as an additional magnetic interaction so that the complete Hamiltonian operator (5.10) must be used for the calculation of the eigenvalues. From now on we must first determine whether the simple product functions ϕ_1 to ϕ_4 are suitable for the description of the stationary states, i.e. are they eigenfunctions?

Let us consider a two-spin system where the nuclei have the same resonance frequency ($v_A = v_B$) and that is therefore to be classified as an A_2 system. Here we obviously can no longer differentiate between the nuclei A(1) and A(2) and the product functions $\alpha(1)\beta(2)$ or $\beta(1)\alpha(2)$ can no longer be assigned unequivocally to the discrete states (2) and (3). In this context it is said that the states (2) and (3) mix with one another. It is therefore necessary to look for new wave functions for these states. On the other hand, however, the functions ϕ_1 and ϕ_4 are applicable for the states (1) and (4) since $\alpha(1)\alpha(2)$ and $\alpha(2)\alpha(1)$ as well as $\beta(1)\beta(2)$ and $\beta(2)\beta(1)$ obviously are identical.

What wave functions should now be chosen for the eigenstates (2) and (3)? The variational method of quantum mechanics is used in cases of this kind. The wave function for the corresponding eigenstate is first approximated by a linear combination. The states have certain characteristics of the product functions ϕ_2 and ϕ_3 and thus can be described by the expression*

$$\Psi_{2,3} = c_2(\alpha\beta) + c_3(\beta\alpha) \tag{5.14}$$

This trial function requires that we now calculate the energy according to equation (5.13) rather than according to equation (5.2) as was done earlier. The variational theorem states that the energy value, ϵ, so obtained can never be less than the actual value and will equal the actual value only when the trial function and the true wave function are identical. The best solution is thus obtained when the calculated energy of the system is minimized. Since we have not yet defined the coefficients c_2 and c_3

---

* Here and in the future the indices of the wave functions α and β will be dispensed with for the sake of clarity. According to convention, the sequence of nuclei is always (1) (2) (3) ... (n).

in our trial function we can conveniently establish as the condition for obtaining the best possible solution the requirement that

$$\frac{\partial \epsilon}{\partial c_2} = \frac{\partial \epsilon}{\partial c_3} = 0$$

In other words, the best solution is obtained when a variation of the coefficients no longer has the result of reducing the energy. Substituting equation (5.14) into (5.13) and performing the indicated operations leads to

$$\epsilon = \frac{\langle [c_2(\alpha\beta) + c_3(\beta\alpha)] \mid \mathscr{H} \mid [c_2(\alpha\beta) + c_3(\beta\alpha)] \rangle}{\langle [c_2(\alpha\beta) + c_3(\beta\alpha)] \mid [c_2(\alpha\beta) + c_3(\beta\alpha)] \rangle}$$

$$= \frac{c_2^2 \langle \alpha\beta \mid \mathscr{H} \mid \alpha\beta \rangle + c_2 c_3 \langle \alpha\beta \mid \mathscr{H} \mid \beta\alpha \rangle + c_3 c_2 \langle \beta\alpha \mid \mathscr{H} \mid \alpha\beta \rangle + c_3^2 \langle \beta\alpha \mid \mathscr{H} \mid \beta\alpha \rangle}{c_2^2 \langle \alpha\beta \mid \alpha\beta \rangle + c_2 c_3 \langle \alpha\beta \mid \beta\alpha \rangle + c_3 c_2 \langle \beta\alpha \mid \alpha\beta \rangle + c_3^2 \langle \beta\alpha \mid \beta\alpha \rangle}$$

In order to improve the clarity of the expression we use the following abbreviations:

$$H_{22} = \langle \alpha\beta \mid \mathscr{H} \mid \alpha\beta \rangle; \qquad H_{23} = \langle \alpha\beta \mid \mathscr{H} \mid \beta\alpha \rangle$$
$$H_{33} = \langle \beta\alpha \mid \mathscr{H} \mid \beta\alpha \rangle; \qquad H_{32} = \langle \beta\alpha \mid \mathscr{H} \mid \alpha\beta \rangle$$

Imposing the implications of equation (5.12) and the identity $H_{32} = H_{23}$, we finally obtain

$$\epsilon = (c_2^2 H_{22} + 2c_2 c_3 H_{23} + c_3^2 H_{33})/(c_2^2 + c_3^2) = u/v$$

In the sense of the above-defined criterion for the best solution, ϵ must be partially differentiated with respect to c_2 and c_3. The rule for quotients leads to

$$\frac{\partial \epsilon}{\partial c_2} = \frac{1}{v}\left(\frac{\partial u}{\partial c_2} - \frac{u}{v}\cdot\frac{\partial v}{\partial c_2}\right)$$

Since the quotient u/v is equal to ϵ:

$$\frac{\partial \epsilon}{\partial c_2} = \frac{1}{v}\left(\frac{\partial u}{\partial c_2} - \epsilon\frac{\partial v}{\partial c_2}\right)$$

$$= \frac{1}{c_2^2 + c_3^2}(2c_2 H_{22} + 2c_3 H_{23} - \epsilon 2c_2)$$

To minimize ϵ with respect to c_2, we equate $\partial \epsilon/\partial c_2$ with zero. This can be satisfied only when the quantity in the parentheses is zero and we find that

$$c_2(H_{22} - \epsilon) + c_3 H_{23} = 0 \qquad\qquad (5.15)$$

Analogously, for $\partial \epsilon/\partial c_3 = 0$ (with $H_{23} = H_{32}$) we have

$$c_2 H_{32} + c_3(H_{33} - \epsilon) = 0 \qquad\qquad (5.16)$$

The equations (5.15) and (5.16) obtained in this fashion are called *homogeneous linear equations* with the coefficients c_2 and c_3 as unknowns. They are also called *secular equations*. According to a theorem of algebra, a system of equations of this type possesses non-trivial, i.e. non-zero, solutions for the coefficients only if the determinant of the system, known as *secular determinant*, is zero. In our case this requires that

$$\begin{vmatrix} H_{22} - \epsilon & H_{23} \\ H_{32} & H_{33} - \epsilon \end{vmatrix} = 0 \tag{5.17}$$

Through the solution of this second-order determinant a quadratic equation is obtained from which the energy ϵ can be calculated.

In order to go through this calculation we must first introduce the elements H_{23}, H_{32}, H_{22}, and H_{33} explicitly in equation (5.17). With the aid of the Hamiltonian operator (5.10) we obtain

$$H_{22} = \langle \alpha\beta \mid \mathscr{H} \mid \alpha\beta \rangle = \langle \alpha\beta \mid \mathscr{H}^{(0)} \mid \alpha\beta \rangle + \langle \alpha\beta \mid \mathscr{H}^{(1)} \mid \alpha\beta \rangle$$

We treat the individual terms separately and find that

$$\begin{aligned} \langle \alpha\beta \mid \mathscr{H}^{(0)} \mid \alpha\beta \rangle &= \langle \alpha\beta \mid \nu_A \hat{I}_z(1) + \nu_A \hat{I}_z(2) \mid \alpha\beta \rangle \\ &= \langle \alpha\beta \mid (\tfrac{1}{2}\nu_A - \tfrac{1}{2}\nu_A)\alpha\beta \rangle \\ &= 0 \end{aligned}$$

$$\begin{aligned} \langle \alpha\beta \mid \mathscr{H}^{(1)} \mid \alpha\beta \rangle &= \langle \alpha\beta \mid J\hat{\mathbf{I}}(1)\hat{\mathbf{I}}(2) \mid \alpha\beta \rangle \\ &= J\langle \alpha\beta \mid \hat{I}_x \hat{I}_x + \hat{I}_y \hat{I}_y + \hat{I}_z \hat{I}_z \mid \alpha\beta \rangle^* \\ &= J(\langle \alpha\beta \mid \hat{I}_x \hat{I}_x \mid \alpha\beta \rangle + \langle \alpha\beta \mid \hat{I}_y \hat{I}_y \mid \alpha\beta \rangle + \langle \alpha\beta \mid \hat{I}_z \hat{I}_z \alpha\beta \rangle) \\ &= J(\tfrac{1}{4}\langle \alpha\beta \mid \beta\alpha \rangle + \tfrac{1}{4}\langle \alpha\beta \mid \beta\alpha \rangle - \tfrac{1}{4}\langle \alpha\beta \mid \alpha\beta \rangle) \\ &= -\tfrac{1}{4}J \end{aligned}$$

Thus $H_{22} = -\tfrac{1}{4}J$, and analogously we obtain $H_{33} = -\tfrac{1}{4}J$. The calculations for the off-diagonal elements proceeds as follows:

$$H_{23} = \langle \alpha\beta \mid \mathscr{H} \mid \beta\alpha \rangle = \langle \alpha\beta \mid \mathscr{H}^{(0)} \mid \beta\alpha \rangle + \langle \alpha\beta \mid \mathscr{H}^{(1)} \mid \beta\alpha \rangle$$

$$\begin{aligned} \langle \alpha\beta \mid \mathscr{H}^{(0)} \mid \beta\alpha \rangle &= \langle \alpha\beta \mid \nu_A \hat{I}_z(1) + \nu_A \hat{I}_z(2) \mid \beta\alpha \rangle \\ &= \langle \alpha\beta \mid (-\tfrac{1}{2}\nu_A + \tfrac{1}{2}\nu_A) \mid \beta\alpha \rangle \\ &= 0 \end{aligned}$$

$$\begin{aligned} \langle \alpha\beta \mid \mathscr{H}^{(1)} \mid \beta\alpha \rangle &= \langle \alpha\beta \mid J\hat{\mathbf{I}}(1)\hat{\mathbf{I}}(2) \mid \beta\alpha \rangle \\ &= J(\langle \alpha\beta \mid \hat{I}_x\hat{I}_x \mid \beta\alpha \rangle + \langle \alpha\beta \mid \hat{I}_y\hat{I}_y \mid \beta\alpha \rangle + \langle \alpha\beta \mid \hat{I}_z\hat{I}_z \mid \beta\alpha \rangle) \\ &= J(\tfrac{1}{4}\langle \alpha\beta \mid \alpha\beta \rangle + \tfrac{1}{4}\langle \alpha\beta \mid \alpha\beta \rangle - \tfrac{1}{4}\langle \alpha\beta \mid \beta\alpha \rangle) \\ &= \tfrac{1}{2}J \end{aligned}$$

H_{32} also equals $\tfrac{1}{2}J$ and the determinant reduces to

$$\begin{vmatrix} -\tfrac{1}{4}J - \epsilon & \tfrac{1}{2}J \\ \tfrac{1}{2}J & -\tfrac{1}{4}J - \epsilon \end{vmatrix} = 0$$

* Here the scalar product $\hat{\mathbf{I}}\hat{\mathbf{I}}$ is resolved into $\hat{I}_x\hat{I}_x + \hat{I}_y\hat{I}_y + \hat{I}_z\hat{I}_z$ and the indices of the operators have been omitted for the sake of clarity.

This leads to the quadratic equation $(-\frac{1}{4}J - \epsilon)^2 - \frac{1}{4}J^2 = 0$ that has the solutions $\epsilon_2 = +\frac{1}{4}J$ and $\epsilon_3 = -\frac{3}{4}J$.

The variational method with the appropriate trial function $\Psi_{2,3}$ thus leads us to two energy values, one of which corresponds to a destabilization and the other to a stabilization of the system. The fact that two nuclei of equal resonance frequencies interact with one another through spin–spin coupling thus leads to a splitting of the energy values ϵ_2 and ϵ_3 which in the case $J = 0$ and $\nu_A = \nu_B$ were degenerate (cf. Section 4.2). We can state further without proof that the approximation of the variational method is sufficiently exact in the present case so that the energies ϵ_2 and ϵ_3 are the new eigenvalues E_2 and E_3. Consequently, the energy level diagram for the A_2 case has the form shown in Figure 5.2. The eigenvalues E_1 and E_4, $\nu_A + \frac{1}{4}J$ and $-\nu_A + \frac{1}{4}J$, respectively, result from the substitution of the corresponding product functions $\alpha\alpha$ and $\beta\beta$ into equation (5.2) since the latter are always true eigenfunctions.

We now complete our consideration of the A_2 case with the calculation of the coefficients c_2 and c_3 in our linear combination (5.14). Substitution of the solution of E_2 in equations (5.15) and (5.16) results in

$$c_2(-\tfrac{1}{2}J) + c_3 \tfrac{1}{2}J = 0$$
$$c_2 \tfrac{1}{2}J + c_3(-\tfrac{1}{2}J) = 0$$

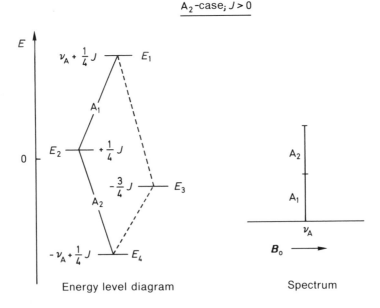

Figure 5.2 Energy level diagram and the spectrum of the A_2 system

From this it follows that $c_2 = c_3$. As an additional equation for the unknown coefficient we employ the normalization condition (5.12a), which also must hold for our linear combination. It requires that $\langle \Psi | \Psi \rangle = 1$ and this leads to the result that $c_2^2 + c_3^2 = 1$. Thus $c_2 = c_3 = 1/\sqrt{2}$ and the correct wave function has the form $\Psi_2 = (1/\sqrt{2})(\alpha\beta + \beta\alpha)$. Through substitution of E_3 we find that $c_2 = -c_3$ and in a manner analogous to that above it results that

$$\Psi_3 = \frac{1}{\sqrt{2}}(\alpha\beta - \beta\alpha).$$

Thus, the two eigenstates with total spin $m_T = 0$ are characterized by different wave functions. In summary we can conclude that the variational principle leads us in our quest for the wave functions of the states (2) and (3) first to the energies of those states and from those energies to the coefficients c_2 and c_3 in the linear combination (5.14).

4.3.2   The calculation of the relative intensities

We have previously calculated transition energies by determining the differences in the eigenvalues of the corresponding spin system on the basis of the selection rule $\Delta m_T = \pm 1$. However, in so doing we did not concern ourselves with the relative intensities of the lines, that is, with the relative probabilities of the transitions. For the A_2 case we want to proceed differently and stipulate first that, in general, the relative intensity of a line is proportional to the square of the so-called *transition moment*, M, between the eigenstates under consideration. The transition moment between two stationary states Ψ_m and Ψ_n is defined by equation (5.18), in which the operator $\hat{I}_x$ is involved:

$$M = \langle \Psi_m \mid \sum_i \hat{I}_x(i) \mid \Psi_n \rangle \tag{5.18}$$

Applying equation (5.18) to the A_2 case we obtain the following relative intensities:
For the transition $\Psi_2 \to \Psi_1$:

$$M^2 = \left\langle \frac{1}{\sqrt{2}}(\alpha\beta + \beta\alpha) \mid \hat{I}_x(1) + \hat{I}_x(2) \mid \alpha\alpha \right\rangle^2$$
$$= \tfrac{1}{2}(\tfrac{1}{2}\langle \alpha\beta \mid \beta\alpha \rangle + \tfrac{1}{2}\langle \alpha\beta \mid \alpha\beta \rangle + \tfrac{1}{2}\langle \beta\alpha \mid \beta\alpha \rangle + \tfrac{1}{2}\langle \beta\alpha \mid \alpha\beta \rangle)^2$$
$$= \tfrac{1}{2}$$

For the transition $\Psi_3 \to \Psi_1$

$$M^2 = \left\langle \frac{1}{\sqrt{2}}(\alpha\beta - \beta\alpha) \mid \hat{I}_x(1) + \hat{I}_x(2) \mid \alpha\alpha \right\rangle^2$$
$$= \tfrac{1}{2}(\tfrac{1}{2}\langle \alpha\beta \mid \beta\alpha \rangle + \tfrac{1}{2}\langle \alpha\beta \mid \alpha\beta \rangle - \tfrac{1}{2}\langle \beta\alpha \mid \beta\alpha \rangle - \tfrac{1}{2}\langle \beta\alpha \mid \alpha\beta \rangle)^2$$
$$= 0$$

In the same fashion the calculations for the transitions $\Psi_4 \to \Psi_2$ and $\Psi_4 \to \Psi_3$ result in relative intensities of 1/2 and 0, respectively. Transitions that involve the eigenvalue E_3 thus have an intensity of zero so that only two lines of the same frequency appear in the spectrum, that is, only a single line at ν_A is observed. This is a confirmation of the earlier postulate that spin–spin coupling between magnetically equivalent nuclei does not affect the experimental spectrum.

Exercise 5.4   Using the wave functions of Table 5.1(A) and equation (5.18), calculate the relative intensities of the transitions for the AB case assuming that there is no spin–spin interaction.

4.3.3   Symmetric and antisymmetric wave functions

A consideration of the A_2 case proceeding on the basis of its symmetry leads to the same result and we will go through it because in so doing we will become acquainted with some important properties of operators.

Neglecting the axis of rotation coincident with the internuclear axis, the A_2 group possesses a plane of reflection, σ, and a two-fold axis of rotation, C_2, as symmetry elements:

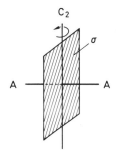

Relative to these symmetry elements the eigenfunctions Ψ_2 and Ψ_3 are different. While Ψ_2 remains unchanged as a result of the symmetry operations, that is by an exchange of the two nuclei, Ψ_3 changes its sign. Thus, Ψ_2 and Ψ_3 are designated as symmetric and antisymmetric wave functions, respectively:

$$\frac{1}{\sqrt{2}}(\alpha\beta + \beta\alpha) \xrightarrow{\sigma \text{ or } C_2} \frac{1}{\sqrt{2}}(\beta\alpha + \alpha\beta) \equiv \frac{1}{\sqrt{2}}(\alpha\beta + \beta\alpha)$$

$$\frac{1}{\sqrt{2}}(\alpha\beta - \beta\alpha) \xrightarrow{\sigma \text{ or } C_2} \frac{1}{\sqrt{2}}(\beta\alpha - \alpha\beta) \equiv -\frac{1}{\sqrt{2}}(\alpha\beta - \beta\alpha)$$

To the corresponding symmetry operations, reflection in σ or rotation around C_2, an operator, $\hat{S}$, can be assigned that obviously must have the eigenvalues $s = +1$ and $s = -1$ since the eigenvalue equations $\hat{S}\Psi_2 = (+1)\Psi_2$ and $\hat{S}\Psi_3 = (-1)\Psi_3$ apply Now, a theorem of quantum mechanics states that for two commuting operators, $\hat{Q}$

and $\hat{R}$ (that is, $\hat{Q}\hat{R}\Psi = \hat{R}\hat{Q}\Psi$) expressions of the type $\langle \Psi_n \mid \hat{R} \mid \Psi_m \rangle$ vanish if Ψ_n and $\Psi_{\dot{m}}$ are eigenfunctions of the operator $\hat{Q}$ belonging to different eigenvalues q_n and q_m, that is if the condition $q_n \neq q_m$ holds.

For the specific case of the transition probability in the A_2 system this statement means that the transitions $\Psi_3 \to \Psi_1$ and $\Psi_4 \to \Psi_3$ are 'forbidden'. The proof hereof appears in the Appendix. Since the line intensity obviously must be independent of whether the symmetry operation is executed or not, the operators $\hat{S}$ and $\hat{I}_x$ must commute. Now, the wave functions Ψ_1, Ψ_2, and Ψ_4 are eigenfunctions of $\hat{S}$ for the eigenvalue $s = +1$ while Ψ_3 is an eigenfunction for the eigenvalue $s = -1$. Thus the expressions

$$< \Psi_3 \mid \hat{I}_x \mid \Psi_1 > \quad \text{and} \quad < \Psi_4 \mid \hat{I}_x \mid \Psi_3 >$$

must vanish and (according to equation (5.18)) the intensity of the corresponding transitions must be zero.

The general conclusion to be drawn from this is that *transitions between wave functions of different symmetry are forbidden*. In this connection the eigenvalues of the operator $\hat{S}$ can also be considered as 'good quantum numbers' that are not changed in the n.m.r. experiment. As a further selection rule for allowed transitions, it follows that $\Delta s = 0$.

The results we have obtained for the A_2 system are tabulated in Tables 5.1(B) and 5.1(C) (p. 155). The wave functions are labelled by their total spin, m_T, and their symmetry properties. As can be seen, the introduction of spin–spin coupling causes a destabilization of $\frac{1}{4}J$ for the symmetric states while in the antisymmetric state a stabilization of $\frac{3}{4}J$ results. This is in agreement with the tenets of valence theory concerning the coupling of electron spins in chemical bonds. The three symmetric spin functions describe the state of two particles that formally possess parallel spin orientations and consequently the spin quantum number of $I = 1$ with the magnetic quantum numbers $+1$, 0, and -1. These three functions represent a so-called *triplet state**. The *singlet state* with $I = 0$, on the other hand, is characterized by the antisymmetric function a_0, the stabilization of which justifies the well known statement of the Pauli principle that the bonding state of two electrons is characterized by an antiparallel arrangement of their spins. The selection rules for the transitions in an A_2 system discussed above can also be expressed in this terminology. According to a general law of quantum mechanics, transitions between term systems of different multiplicity, that is between singlet and triplet states in this context, are forbidden. On this basis *ortho*-hydrogen, for example, is metastable near 0 K for several months.

---

* The multiplicity of a state (singlet, doublet, triplet, etc.) is determined by the spin quantum number, I, according to the formula $2I + 1$ (cf. the quantum condition 1.2). It indicates the number of magnetic quantum numbers a state has and thus the number of possible orientations with respect to the direction of an external magnetic field.

4.4   THE AB SYSTEM

The variational principle and the relation (5.18) for the determination of the transition moment also enable us to treat spin systems that are not simplified by restrictive conditions. Let us now turn to the AB system in which both parameters $v_0\delta$ and J are of comparable magnitude.

We follow the treatment of the A_2 case completely up to the derivation of the determinant (5.17). Here the simplifying condition that $v_A = v_B$ does not hold and for the matrix elements H_{22}, H_{23} and H_{33} the following expressions are obtained:

$$H_{22} = \tfrac{1}{2}v_0\delta - \tfrac{1}{4}J$$
$$H_{23} = H_{32} = \tfrac{1}{2}J$$
$$H_{33} = -\tfrac{1}{2}v_0\delta - \tfrac{1}{4}J$$

The determinant is then

$$\begin{vmatrix} (\tfrac{1}{2}v_0\delta - \tfrac{1}{4}J) - E & \tfrac{1}{2}J \\ \tfrac{1}{2}J & (-\tfrac{1}{2}v_0\delta - \tfrac{1}{4}J) - E \end{vmatrix} = 0 \qquad (5.19)$$

Expanding the determinant leads to the quadratic equation

$$E^2 + \tfrac{1}{2}JE - \tfrac{1}{4}(v_0\delta)^2 - \tfrac{3}{16}J^2 = 0$$

which has solutions

$$E_{2,3} = -\tfrac{1}{4}J \pm \tfrac{1}{2}\sqrt{J^2 + v_0\delta^2}$$

The eigenvalues E_1 and E_4 once more result from equation 5.2 and the product functions $\alpha\alpha$ and $\beta\beta$, respectively:

$$E_1 = \tfrac{1}{2}(v_A + v_B) + \tfrac{1}{4}J$$
$$E_4 = -\tfrac{1}{2}(v_A + v_B) + \tfrac{1}{4}$$

The next step consists in the calculation of the coefficients in our trial function (5.14). For this we first substitute the solution E_2 in equation (5.15). For the sake of brevity we let

$$C = \tfrac{1}{2}\sqrt{J^2 + v_0\delta^2}$$

and obtain

$$c_2(\tfrac{1}{2}v_0\delta - C) + c_3\tfrac{1}{2}J = 0$$

The calculations now necessary can be elegantly simplified if an angle 2θ defined by the following relations is introduced:

$$v_0\delta/2C = \cos 2\theta \quad \text{and} \quad J/2C = \sin 2\theta$$

It follows that

$$c_2(1 - \cos 2\theta) - c_3 \sin 2\theta = 0$$

and

$$c_2 = c_3 \sin 2\theta / (1 - \cos 2\theta)$$

The application of the identities $\sin 2\theta = 2 \cos \theta \sin \theta$ and $\cos 2\theta = \cos^2 \theta - \sin^2 \theta$ leads to the result $c_2 = c_3 \cos \theta / \sin \theta$ and, with the aid of the normalization condition $c_2^2 + c_3^2 = 1$, the values $c_2 = \cos \theta$ and $c_3 = \sin \theta$ are obtained for the coefficients.

Substitution of E_3 in an analogous fashion results in the coefficients $c_2 = -\sin \theta$ and $c_3 = \cos \theta$. Thus, the correct wave functions for the states (2) and (3) are

$$\Psi_2 = \quad \cos \theta (\alpha\beta) + \sin \theta (\beta\alpha)$$
$$\Psi_3 = - \sin \theta (\alpha\beta) + \cos \theta (\beta\alpha)$$

We now turn to the calculation of the relative intensities where we again use relation (5.18). For the $\Psi_2 \to \Psi_1$ transition, for example, this yields

$$M^2 = \langle [\cos \theta (\alpha\beta) + \sin \theta (\beta\alpha)] \mid \hat{I}_x(A) + \hat{I}_x(B) \mid \alpha\alpha \rangle^2$$
$$= \tfrac{1}{4} (1 + \sin 2\theta)$$

The relative intensities of the other transitions are obtained in an analogous fashion.

The complete results of our consideration of the AB system are given in Table 5.1D. The spectrum consists of four lines that are arranged symmetrically about the centre $(\nu_A + \nu_B)/2$. The outer lines of the AB quartet are of lower intensity than the central lines, a result that we introduced empirically on p. 44 as the 'roof effect'. The energy level diagram is different from that shown in Figure 5.1 only in that the eigenvalues are differently stabilized or destabilized and that the lines $A_1(f_1)$ and $A_2(f_2)$ and $B_1(f_3)$, and $B_2(f_4)$, respectively, are no longer degenerate.

In order to illustrate our conclusions derived for the AB system with a practical example, the spectrum calculated with the assumed parameters $\nu_0\delta = 15$ Hz and $J = 12$ Hz is shown in Figure 5.3.

Exercise 5.5   In the preceding sections the expressions for the eigenvalues, wave functions and transition probabilities were not explicitly derived in all cases. Confirm the results given by working through the necessary derivations.

Exercise 5.6   Using the relations in Table 5.1(D) calculate the line frequencies and intensities for an AB system with $\nu_0\delta = 20$ Hz and $J = 15$ Hz.

Exercise 5.7   Write the secular determinant according to the variational method for the linear combination

$$\Psi = c_2(\alpha\alpha\beta) + c_3(\alpha\beta\alpha) + c_4(\beta\alpha\alpha)$$

Exercise 5.8   For an AB system it is observed that $f_1 - f_2 = f_2 - f_3 = f_3 - f_4$.
   (a) What is the ratio $\nu_0\delta/J$?
   (b) Calculate the relative intensities of the lines.
   (c) How can one determine that the signals do not comprise a first-order quartet?

Table 5.1 Eigenfunctions, eigenvalues, transition energies, and transition probabilities for the two-spin systems of the A_2, AB, and AX types

Eigenfunctions			Eigenvalues	Transitions	Transition energy	Relative intensity
(A) The AB case; $J_{AB}=0$						
m_T						
(1) +1	$\alpha\alpha$		$\frac{1}{2}(\nu_A + \nu_B)$	$(3) \rightarrow (1)(A)$	ν_A	1
(2) 0	$\alpha\beta$		$\frac{1}{2}(\nu_A - \nu_B)$	$(4) \rightarrow (2)\,(A)$	ν_A	1
(3) 0	$\beta\alpha$		$-\frac{1}{2}(\nu_A - \nu_B)$	$(2) \rightarrow (1)\,(B)$	ν_B	1
(4) −1	$\beta\beta$		$-\frac{1}{2}(\nu_A + \nu_B)$	$(4) \rightarrow (3)\,(B)$	ν_B	1
(B) The A_2 case; $J = 0$						
m_T						
(1) +1	$\alpha\alpha$		ν_A	$(3) \rightarrow (1)\,(A)$	ν_A	1
(2) 0	$\alpha\beta$		0	$(4) \rightarrow (2)\,(A)$	ν_A	1
(3) 0	$\beta\alpha$		0	$(2) \rightarrow (1)\,(A)$	ν_A	1
(4) −1	$\beta\beta$		$-\nu_A$	$(4) \rightarrow (3)\,(A)$	ν_A	1
(C) The A_2 case; $J > 0$						
m_T						
(1) +1	$\alpha\alpha$	s_{+1}	$\nu_A + \frac{1}{4}J$	$a_0 \rightarrow s_{+1}(A)$	$\nu_A + J$	0
(2) 0	$(\alpha\beta + \beta\alpha)/\sqrt{2}$	s_0	$+\frac{1}{4}J$	$s_{-1} \rightarrow s_0\,(A)$	ν_A	2
(3) 0	$(\alpha\beta - \beta\alpha)/\sqrt{2}$	a_0	$-\frac{3}{4}J$	$s_0 \rightarrow s_{+1}(A)$	ν_A	2
(4) −1	$\beta\beta$	s_{-1}	$-\nu_A + \frac{1}{4}J$	$s_{-1} \rightarrow a_0\,(A)$	$\nu_A - J$	0

(D) The AB case: $J_{AB} > 0$

m_T					
(1)	+1	$\alpha\alpha$	$\frac{1}{2}(\nu_A + \nu_B) + \frac{1}{4}J_{AB}$		
(2)	0	$\cos\theta(\alpha\beta) + \sin\theta(\beta\alpha)$	$-\frac{1}{4}J_{AB} + C$		
(3)	0	$-\sin\theta(\alpha\beta) + \cos\theta(\beta\alpha)$	$-\frac{1}{4}J_{AB} - C$		
(4)	-1	$\beta\beta$	$-\frac{1}{2}(\nu_A + \nu_B) + \frac{1}{4}J_{AB}$		
(3) $\rightarrow$ (1) (A) f_1				$\frac{1}{2}(\nu_A + \nu_B) + \frac{1}{2}J_{AB} + C$	$1 - \sin 2\theta$
(4) $\rightarrow$ (2) (A) f_2				$\frac{1}{2}(\nu_A + \nu_B) - \frac{1}{2}J_{AB} + C$	$1 + \sin 2\theta$
(2) $\rightarrow$ (1) (B) f_3				$\frac{1}{2}(\nu_A + \nu_B) + \frac{1}{2}J_{AB} - C$	$1 + \sin 2\theta$
(4) $\rightarrow$ (3) (B) f_4				$\frac{1}{2}(\nu_A + \nu_B) - \frac{1}{2}J_{AB} - C$	$1 - \sin 2\theta$

(E) The AX case; $J_{AX} > 0$

m_T					
(1)	+1	$\alpha\alpha$	$\frac{1}{2}(\nu_A + \nu_X) + \frac{1}{4}J_{AX}$		
(2)	0	$\alpha\beta$	$\frac{1}{2}(\nu_A - \nu_X) - \frac{1}{4}J_{AX}$		
(3)	0	$\beta\alpha$	$-\frac{1}{2}(\nu_A - \nu_X) - \frac{1}{4}J_{AX}$		
(4)	-1	$\beta\beta$	$-\frac{1}{2}(\nu_A + \nu_X) + \frac{1}{4}J_{AX}$		
(3) $\rightarrow$ (1) (A)				$\nu_A + \frac{1}{2}J_{AX}$	1
(4) $\rightarrow$ (2) (A)				$\nu_A - \frac{1}{2}J_{AX}$	1
(2) $\rightarrow$ (1) (X)				$\nu_X + \frac{1}{2}J_{AX}$	1
(4) $\rightarrow$ (3) (X)				$\nu_X - \frac{1}{2}J_{AX}$	1

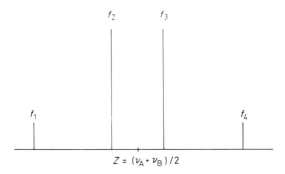

Figure 5.3   Spectrum of an AB system with $|J_{AB}| > 0$

4.5   THE AX SYSTEM AND THE FIRST-ORDER APPROXIMATION

Having determined the eigenvalues and the eigenfunctions of the AB system, it would be of interest at this juncture to investigate the dependence of the line frequencies and intensities on the ratio of the parameters $v_0\delta$ and J.

Let us first discuss the case in which the *relative chemical shift $v_0\delta$ is very large compared with the coupling constant*. As a result of this the parameter C approaches $\frac{1}{2}(v_0\delta)$ and the expression $\sin 2\theta$ approaches zero. However, since $\sin 2\theta = 2 \sin \theta \cos \theta$, either $\sin \theta$ or $\cos \theta$ must be zero. Further, since $\sin^2 \theta + \cos^2 \theta = 1$, when $\sin \theta = 0$ it follows that $\cos \theta = 1$. The eigenfunctions and eigenvalues for this limiting case, classified as an AX case, are thus

(1) $\alpha\alpha$ $\qquad \frac{1}{2}(v_A + v_X) + \frac{1}{4}J$

(2) $\alpha\beta$ $\qquad \frac{1}{2}(v_A - v_X) - \frac{1}{4}J$

(3) $\beta\alpha$ $\qquad -\frac{1}{2}(v_A - v_X) - \frac{1}{4}$

(4) $\beta\beta$ $\qquad -\frac{1}{2}(v_A + v_X) + \frac{1}{4}J$

and the transition energies and the intensities for this system given in Table 5.1(E) are obtained in the usual manner.

We now go back a step further to the determinant (5.19) and see what the eigenvalues and the eigenfunctions of the states (2) and (3) would be if the off-diagonal elements H_{23} and H_{32} were neglected and set to zero:

$$\begin{vmatrix} H_{22} - E & 0 \\ 0 & H_{33} - E \end{vmatrix} = \begin{vmatrix} (\frac{1}{2}v_0\delta - \frac{1}{4}J) - E & 0 \\ 0 & (-\frac{1}{2}v_0\delta - \frac{1}{4}J) - E \end{vmatrix} = 0$$

Since a determinant becomes zero when the elements of one of its columns or rows are exactly zero, we immediately obtain

$$E_2 = H_{22} = \frac{1}{2}v_0\delta - \frac{1}{4}J$$

and

$$E_3 = H_{33} = -\tfrac{1}{2} v_0 \delta - \tfrac{1}{4} J$$

Substitution of these solutions in the secular equations (5.15) and (5.16) in conjunction with the normalization condition, $c_2^2 + c_3^2 = 1$, leads to values for the coefficients c_2 and c_3 of 1 and 0 for Ψ_2 and 0 and 1 for Ψ_3, respectively. The wave functions are therefore $\Psi_2 = \alpha\beta$ and $\Psi_3 = \beta\alpha$.

Thus, the neglect of the off-diagonal elements leads directly from the general AB case to the special AX case. Clearly, this simplification is justified only when these elements are substantially smaller than the diagonal elements. The appropriate situation is obtained when $v_0 \delta \gg J$, a criterion we introduced on p. 43 for the application of the first-order rules. As can now be clearly understood, these rules are a special case resulting from the general derivation and strictly applying for very large chemical shifts only.

We encounter a second limiting case when the *relative chemical shift*, $v_0 \delta$, *becomes very small compared with the coupling constant, J*. The parameter C then approaches J and $\sin 2\theta$ approaches 1. According to Table 5.1(D) the intensities of the lines f_1 and f_4 then decrease to the extent that the spectrum degenerates to that of an A_2 system, since in addition the transitions f_2 and f_3 coincide at $(v_A + v_B)/2$.

In practice, AB systems that closely approach the A_2 case are very often met. On the one hand, the centre lines lie so close to one another that the spectral resolution is not sufficient to separate them. On the other hand, the intensity of the outer lines is so small that the sensitivity of the spectrometer does not suffice for their detection. Such spectra are termed 'deceptively simple' spectra. The criteria for such spectra in the AB case are given by the relations

$$\frac{v_0 \delta^2}{2J} < \Delta \quad \text{and} \quad \frac{v_0 \delta^2}{2J^2} < i$$

where Δ is the natural width of the spectral line and i is the lower limit of the detectable intensity. We shall encounter this phenomenon again later in our discussion of other spin systems.

4.6   GENERAL RULES FOR THE TREATMENT OF MORE COMPLEX SPIN SYSTEMS

The preceding sections have shown how the eigenvalues and the eigenfunctions for stationary states with the same total spin can be obtained by means of the variational method. The same formalism can be used for more complex spin systems since the simple product functions of the type $\alpha\alpha \ldots \beta$ always serve as the basis for the linear combinations. The method is thus very easily generalized.

First the Pascal triangle (p. 34) gives a systematic survey of the number of eigenstates and product functions, grouped according to their total spin, that are to be expected for a particular spin system with n nuclei of spin $I = \tfrac{1}{2}$. In general there are 2^n eigenstates for a system with n nuclei, so it is apparent that the number increases rapidly for more complex spin systems.

Thus, in the *three-spin* case the three basis functions $\alpha\alpha\beta$, $\alpha\beta\alpha$, and $\beta\alpha\alpha$ correspond to the m_T value $+\frac{1}{2}$. Then the expression

$$\Psi = c_2(\alpha\alpha\beta) + c_3(\alpha\beta\alpha) + c_4(\beta\alpha\alpha)$$

serves as our linear combination. More generally in vector notation we have

$$\Psi = c_i \times \phi_i$$

where c_i is a row vector and ϕ_i is column vector:

$$\Psi = (c_1, c_2, \ldots, c_n) \times \begin{pmatrix} \phi_1 \\ \phi_2 \\ \vdots \\ \phi_n \end{pmatrix}$$

Since the coefficients c_i convert the corresponding set of basis functions, ϕ_i, to eigenfunctions of the spin system they are termed *eigenvectors*. For the AB case there results for the states (2) and (3):

$$(\cos\theta,\ \sin\theta) \times \begin{pmatrix} \alpha\beta \\ \beta\alpha \end{pmatrix} \text{ and } (-\sin\theta,\ \cos\theta) \times \begin{pmatrix} \alpha\beta \\ \beta\alpha \end{pmatrix}.$$

Finally, from the Pascal triangle we also can obtain the theoretically possible number of lines for a spin system if the selection rule $\Delta m_T = \pm 1$ is observed. Of course, this number also includes the so-called *combination lines* for which the spin orientation of several nuclei is changed simultaneously and that are therefore forbidden (e.g. $\alpha\beta\beta \to \beta\alpha\alpha$). To be more correct, the selection rule must be reformulated with respect to the magnetic quantum number, m_I, of the individual nuclei:

$$\sum_{i=1}^{n} \Delta m_I(i) = +1 \text{ with } \Delta m_I(i) = 0, +1 \tag{5.20}$$

If we now apply the Hamiltonian operator (5.10) to the basis functions of a spin system of interest, we obtain the quantities $H_{11}, \ldots, H_{kk}$ and $H_{12}, \ldots, H_{kl}$, which most clearly can be arranged in the *Hamilton matrix*, $\hat{H}$. For a two-spin system this matrix has the form

$$\hat{H} = \begin{vmatrix} H_{11} & 0 & 0 & 0 \\ 0 & H_{22} & H_{23} & 0 \\ 0 & H_{32} & H_{33} & 0 \\ 0 & 0 & 0 & H_{44} \end{vmatrix} \tag{5.21}$$

The following points deserve attention:

1. The *Hamilton matrix* is a square matrix and because of the identity $H_{23} = H_{32}$ (in general $H_{kl} = H_{lk}$) symmetric with respect to the principal diagonal.

2. The *off-diagonal matrix elements* between eigenstates with different total spins are zero. The matrix can therefore be factorized into submatrices:

$$\hat{\mathbf{H}} = |\ H_{11}\ | \times \begin{vmatrix} H_{22} & H_{23} \\ H_{32} & H_{33} \end{vmatrix} \times |\ H_{44}\ |$$

This result is likewise a consequence of the theorem introduced earlier (p. 150 f.) concerning commuting operators. Here the Hamiltonian operator and the operator $\hat{F}_z$ commute (cf. Appendix) and the matrix elements $\langle \Psi_n | H | \Psi_m \rangle$ for eigenfunctions that belong to different eigenvalues n and m of F_z vanish.

3. If we subtract the energy E from the diagonal matrix elements H_{kk} and after factorizing set the individual factors equal to zero, secular determinants of the general form

$$|\ H_{kl} - \delta_{kl}E\ | = 0 \quad \text{with} \quad \delta_{kl} = 1 \text{ for } k = l$$

$$\text{and} \quad \delta_{kl} = 0 \text{ for } k \neq l$$

result. Because of paragraph (2) their number is equal to the number of different m_T values of the corresponding spin system. Their dimensions follow clearly from the number of basis functions belonging to the respective total spin. They can also be determined directly by reference to the Pascal triangle. Solving the secular determinants yields the eigenvalues of the corresponding spin system and via the secular equations the eigenvectors as coefficients of the eigenfunctions.

4. Independent of the size of the spin system, the diagonal elements H_{11} and H_{kk} are always correct eigenvalues for the states with total spin $+n/2$ or $-n/2$ and the basis functions $\alpha\alpha \ldots \alpha$ or $\beta\beta \ldots \beta$ are correct eigenfunctions of these spin states.

The Hamilton matrix can be set up using the basis function ϕ_k for any spin system by the application of simple rules. For the diagonal elements:

$$\langle \phi_k | \mathscr{H} | \phi_l \rangle = H_{kk} \sum_{i=1}^{n} v_i\, m_I(i) + \frac{1}{4} \sum_{i<j} \sum J_{ij}\, T_{ij}$$

with $T_{ij} = +1$ if the nuclei i and j have parallel spin in the corresponding basis function
and $T_{ij} = -1$ if the nuclei i and j have antiparallel spin in the corresponding basis functions.
This formula corresponds to equation (2.32) (p. 9).

The off-diagonal elements between two basis functions ϕ_k and ϕ_l are given by

$$\langle \phi_k | \mathscr{H} | \phi_l \rangle = H_{kl} = \tfrac{1}{2} J_{ij} U \text{ for } i \neq j$$

with $U = 1$ if ϕ_k and ϕ_l differ only by the exchange of the spin functions of the nuclei i and j (e.g. $\alpha\beta\alpha\beta$ and $\alpha\beta\beta\alpha$), and $U = 0$ in all other cases (e.g. $\alpha\beta\alpha\beta$ and $\beta\alpha\beta\alpha$).

Let us consider as an illustration of these rules the spin system of three nuclei A, B, and C. Here the complete set of the basis functions is:

$m_T = \frac{3}{2}$	$m_T = \frac{1}{2}$	$m_T = -\frac{1}{2}$	$m_T = -\frac{3}{2}$
(1) $\alpha\alpha\alpha$	(2) $\alpha\alpha\beta$	(5) $\alpha\beta\beta$	(8) $\beta\beta\beta$
	(3) $\alpha\beta\alpha$	(6) $\beta\alpha\beta$	
	(4) $\beta\alpha\alpha$	(7) $\beta\beta\alpha$	

Then, for the diagonal elements it follows

$$H_{11} = \tfrac{1}{2}(v_A + v_B + v_C) + \tfrac{1}{4}(J_{AB} + J_{AC} + J_{BC})$$
$$H_{22} = \tfrac{1}{2}(v_A + v_B - v_C) + \tfrac{1}{4}(J_{AB} - J_{AC} - J_{BC})$$
$$\vdots$$
$$H_{88} = -\tfrac{1}{2}(v_A + v_B + v_C) + \tfrac{1}{4}(J_{AB} + J_{AC} + J_{CB})$$

and for the off-diagonal elements

$$H_{12} = H_{13} = \cdots H_{18} = 0$$
$$H_{23} = \tfrac{1}{2}J_{BC}$$
$$H_{24} = \tfrac{1}{2}J_{AC}$$
$$H_{34} = \tfrac{1}{2}J_{AB}$$

A simple example with the parameters

$$v_A = -v, \quad v_B = 0, \quad v_C = v, \quad J_{AB} = J_{AC} = J_{BC} + J$$

then produces the Hamilton matrix

$$
\begin{bmatrix}
\frac{3}{4}J & 0 & 0 & 0 & 0 & 0 & 0 & 0 \\
0 & -v - \frac{1}{4}J & \frac{1}{2}J & \frac{1}{2}J & 0 & 0 & 0 & 0 \\
0 & \frac{1}{2}J & -\frac{1}{4}J & \frac{1}{2}J & 0 & 0 & 0 & 0 \\
0 & \frac{1}{2}J & \frac{1}{2}J & v - \frac{1}{4}J & 0 & 0 & 0 & 0 \\
0 & 0 & 0 & 0 & -v - \frac{1}{4}J & \frac{1}{2}J & \frac{1}{2}J & 0 \\
0 & 0 & 0 & 0 & \frac{1}{2}J & -\frac{1}{4}J & \frac{1}{2}J & 0 \\
0 & 0 & 0 & 0 & \frac{1}{2}J & \frac{1}{2}J & v - \frac{1}{4}J & 0 \\
0 & 0 & 0 & 0 & 0 & 0 & 0 & \frac{3}{4}J
\end{bmatrix}
$$

Without going into the details of the mathematical treatment of such matrices here, we want to indicate a way that represents an alternative to the procedure of factoring the matrix into secular determinants as discussed above. It forms the basis for a series of computer programs for the treatment of quantum mechanical problems. The area of concern here is treated in mathematical textbooks under the rubric of the *eigenvalue problem*.

It can be shown that a quadratic matrix such as the Hamilton matrix $\hat{\mathbf{H}}$ is linked through the matrix equation.

$$\hat{\mathbf{H}}\hat{\mathbf{U}} = \hat{\mathbf{U}}\hat{\mathbf{D}} \tag{5.22}$$

to the diagonal matrix $\hat{\mathbf{D}}$ of the eigenvalues, that is, to a matrix the elements of which are $D_{kk} = E_k$ and $D_{kl} = D_{lk} = 0$.

$\hat{\mathbf{U}}$ is a transformation matrix of a special kind that transforms, in the sense of equation (5.23), the matrix $\hat{\mathbf{H}}$ to the diagonal form. Its special feature is that it is exactly the matrix of the eigenvectors so that it contains the coefficients $c_1, \ldots, c_k$ of the linear combinations of the type (5.14). $\hat{\mathbf{U}}$ is called either *orthogonal* or *unitary* and equation (5.23) is a *unitary transformation*.

$$\hat{\mathbf{U}}^{-1}\hat{\mathbf{H}}\hat{\mathbf{U}} = \hat{\mathbf{D}} \tag{5.23}$$

For the eigenstates (2) and (3) of the A_2 system, for example, it follows according to equation (5.22) that

$$\begin{bmatrix} -\dfrac{1}{4}J & \dfrac{1}{2}J \\[2mm] \dfrac{1}{2}J & -\dfrac{1}{4}J \end{bmatrix} \times \begin{bmatrix} \dfrac{1}{\sqrt{2}} & \dfrac{1}{\sqrt{2}} \\[2mm] \dfrac{1}{\sqrt{2}} & -\dfrac{1}{\sqrt{2}} \end{bmatrix} = \begin{bmatrix} \dfrac{1}{\sqrt{2}} & \dfrac{1}{\sqrt{2}} \\[2mm] \dfrac{1}{\sqrt{2}} & -\dfrac{1}{\sqrt{2}} \end{bmatrix} \times \begin{bmatrix} \dfrac{1}{4}J & 0 \\[2mm] 0 & -\dfrac{3}{4}J \end{bmatrix}$$

This can easily be verified by performing the indicated operations.

Standard mathematical procedures allowing to diagonalize the Hamiltonian matrix $\hat{\mathbf{H}}$ are available. Their application not only yields the diagonal matrix $\hat{\mathbf{D}}$ of the eigenvalues, but also the matrix $\hat{\mathbf{U}}$ of the coefficients. With that the eigenvalue problem is solved and the frequencies and intensities of the spectral lines can be calculated according to equation (5.1) or (5.18).

Since the mathematical formalism mentioned here can easily be programmed, spectra for various spin systems are most conveniently calculated using a digital computer. We shall return later to the programs that can be employed to calculate a theoretical spectrum from a set of resonance frequencies and coupling constants. Nevertheless, we shall already use the results of such calculations in the following sections in order to check the parameters we obtain in the analyses of spin systems.

5. The calculation of the parameters v_i and J_{ij} from the experimental spectrum

We now want to consider the question posed in the introduction of to this chapter: 'How can the parameters of interest of the spin system under consideration, the chemical shifts and the coupling constants, be calculated from observed line frequencies and intensities?' That is, how is an experimental spectrum analysed?

Naively considered, this problem should be solved simply by a 'reversal' of the mathematical derivations presented in the foregoing sections. This is possible in practice, however, only for a two-spin system, since only here we obtain second-order equations. In general all more complex spin systems yield equations of higher order that cannot be solved explicitly. A direct analysis in the sense indicated is thus impossible. There are, however, certain strategies that allow us to simplify the problem and second-order equations can be derived even for four-spin systems. How this actually can be done shall be illustrated in the following sections with a number of examples from the more common spin systems.

5.1   THE DIRECT ANALYSIS OF THE AB SYSTEM

Spectra of the AB type, an example of which is shown in Figure 5.4 for the aromatic protons in 1-amino-3,6-dimethyl-2-nitrobenzene, are encountered in a large number of organic compounds.

2-Bromo-5-chlorothiophene (**78**), 1-bromo-1-chloroethylene (**79**), 2.5-dibromo-1,6-methano[10]annulene (**80**) and acetaldehyde dibenzylacetal (**81**) represent additional examples. As Figure 5.5 illustrates, the appearance of the spectrum is determined by the ratio $J/v_0\delta$.

78   **79**   **80**   **81**

For the analysis we infer the trivial conclusion from Table 5.1 that the coupling constant, J, is equal to the difference $f_1 - f_2$ or $f_3 - f_4$. Moreover, $f_1 - f_3 = f_2 - f_4 = 2C$ and, since $C = \sqrt{J^2 + v_0\delta^2}/2$, the relative chemical shift in an AB system is given by

$$v_0\delta = \sqrt{4C^2 - J^2}$$
$$= \sqrt{(2C - J)(2C + J)}$$
$$= \sqrt{(f_2 - f_3)(f_1 - f_4)}$$

If Z is defined as the centre of the multiplet, that is, the mid-point between f_1 and f_4 or f_2 and f_3, then

$$v_A = Z - \tfrac{1}{2}v_0\delta \quad \text{and} \quad v_B = Z + \tfrac{1}{2}v_0\delta$$

Further, as can be derived easily with reference to the expressions in Table 5.1 (pp. 154 and 155), for the ratio of the intensities we have $I_2/I_1 = I_3/I_4 = (f_1 - f_4)/(f_2 - f_3)$.

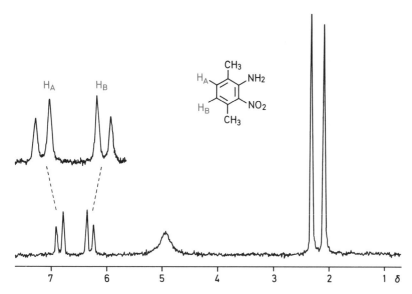

Figure 5.4 ^{1}H n.m.r. spectrum of 1-amino-3,6-dimethyl-2-nitrobenzene at 60 MHz

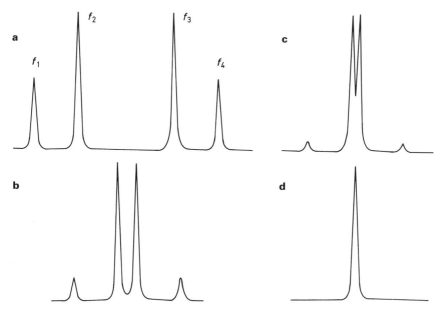

Figure 5.5 Dependence of the AB system on the ratio $J/v_0\delta$; spectra illustrated are for values of $J/v_0\delta$ of (a) 1:3, (b) 1:1, (c) 5:3, and (d) 5:1.

164   ANALYSIS OF HIGH-RESOLUTION NMR SPECTRA

For the AB system there is also a geometric solution. A circle with a radius $2C = f_2 - f_4 = \sqrt{J^2 + v_0\delta^2}$ is drawn about the point P_1 and there is thus obtained the right triangle $P_1P < S_2P_3$. Since P_1P_2 is J, $P_2P_3 = v_0\delta$. The angle $\not< P_1P_3P_2$ is the angle 2θ introduced earlier (cf. p. 152).

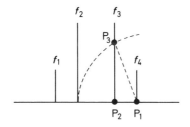

Exercise 5.9 Determine the resonance frequencies v_A and v_B for an AB system with the lines $f_1 = 38.5$, $f_2 = 28.0$, $f_3 = 20.5$, and $f_4 = 10.0$ Hz. Also calculate the relative intensities of the lines.

5.2 SPIN SYSTEMS WITH THREE NUCLEI

The product functions tabulated on p. 160 according to their total spin serve as the basis for a general three-spin system. The variational method must be employed for the functions with total spin $m_T = +\frac{1}{2}$ and $m_T = -\frac{1}{2}$ in order to determine the correct eigenfunctions and eigenvalues. Only the basis functions $\alpha\alpha\alpha$ and $\beta\beta\beta$ are already eigenfunctions and the appropriate eigenvalues can be calculated by direct substitution in equation (5.2).

In the following we shall treat spin systems in which, because of special properties resulting from symmetry considerations or the existence of large relative chemical shifts between individual nuclei, we encounter simplifications that enable us to derive additional eigenfunctions directly from the basis functions without going through the complete variational calculation. The eigenvalues are then obtained by means of equation (5.2). In this manner the calculations for the three-spin system can be limited to quadratic equations so that explicit solutions for the spectral parameters can be obtained.

5.2.1 The AB$_2$ (A$_2$B) system

AB$_2$ spectra are observed for compounds that possess a two-fold axis of symmetry such as 2,6-dimethylpyridine (**82**) and 1,2,3-trichlorobenzene (**83**). Other examples are found in trisubstituted cyclopropanes with C_s symmetry (**84, 85**) and in benzylmalonic esters (**86**).

A typical AB$_2$ system is shown in Figure 5.6. In general, seven or eight (at the most, nine) transitions are observed, of which four (f_1 to f_4) are in the A portion. In the B_2 portion the lines f_7 and f_8 are well separated whereas f_5 and f_6 are often not resolved.

As we explained at the outset, the general case of a three-spin system is characterized by three resonance frequencies and three coupling constants*. In the process of the analysis of such a system two 3×3 secular determinants that arise through mixing of states with m_T values of $+\frac{1}{2}$ and $-\frac{1}{2}$ must be solved. This leads to third-order equations that cannot be explicitly solved for the parameters v_i and J_{ij}.

In the case of the AB_2 system this difficulty can be obviated if we make use of the results obtained for the A_2 case. The wave functions derived there can now be used for the magnetically equivalent nuclei of the B_2 group. Since α or β are the only possibilities for the wave function of the A nucleus, eight product functions are obtained as basis functions of the AB_2 system by simple multiplication. These are classified in Table 5.2a according to their symmetry and their total spin, m_T. If we

82   **83**   **84**   **85**

86

recall that transitions between states of different symmetry are forbidden, it follows that a line in the A portion of the spectrum must correspond to the antisymmetric transition $(8)\rightarrow(7)$. Since the basis functions ϕ_7 and ϕ_8 are already eigenfunctions we obtain the eigenvalues E_7 and E_8 by direct substitution of ϕ_7 and ϕ_8 into equation (5.2). The result is $E_7 = \frac{1}{2}v_A - \frac{3}{4}J_{BB}$ and $E_8 = -\frac{1}{2}v_A - \frac{3}{4}J_{BB}$, so that the frequency of the transition is v_A. Each AB_2 spectrum thus contains a line in the A portion that is independent of the coupling constant J_{AB} and that directly yields the resonance frequency of the A nucleus. Further, the following transitions are allowed:

A lines†	B lines†	Combination line
$(3) \rightarrow (1)$	$(2) \rightarrow (1)$	$(4) \rightarrow (3)$
$(6) \rightarrow (4)$	$(4) \rightarrow (2)$	
$(5) \rightarrow (2)$	$(6) \rightarrow (5)$	
	$(5) \rightarrow (3)$	

---

* Five parameters are sufficient since the appearance of the spectrum depends only on the relative chemical shifts.
† This classification holds rigorously only for the AX_2 limiting case.

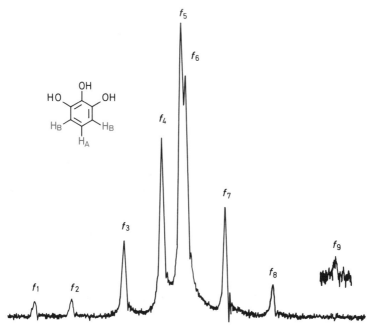

Figure 5.6 The AB_2 system of the aromatic protons of pyrogallol (in chloroform) at 60 MHz

Table 5.2a Basis functions of the AB_2 system

Symmetric functions:

	A	B_2	m_T
(1) ϕ_1	α	$\alpha\alpha$	$+\frac{3}{3}$
(2) ϕ_2	α	$(\alpha\beta + \beta\alpha)/\sqrt{2}$	$+\frac{1}{2}$
(3) ϕ_3	β	$\alpha\alpha$	$+\frac{1}{2}$
(4) ϕ_4	α	$\beta\beta$	$-\frac{1}{2}$
(5) ϕ_5	β	$(\alpha\beta + \beta\alpha)/\sqrt{2}$	$-\frac{1}{2}$
(6) ϕ_6	β	$\beta\beta$	$-\frac{3}{2}$

Antisymmetric functions:

	A	B_2	m_T
(7) ϕ_7	α	$(\alpha\beta - \beta\alpha)/\sqrt{2}$	$+\frac{1}{2}$
(8) ϕ_8	β	$(\alpha\beta - \beta\alpha)/\sqrt{2}$	$-\frac{1}{2}$

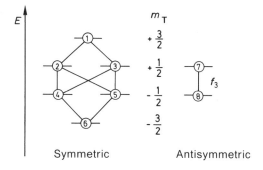

Figure 5.7   Energy level diagram of the AB_2 system

The resulting energy level diagram is shown in Figure 5.7.

Of the symmetric basis functions, ϕ_1 and ϕ_6 are already eigenfunctions of the Hamiltonian operator and the corresponding eigenvalues can be calculated according to equation (5.2). The correct eigenfunctions Ψ_2 to Ψ_5 are determined by means of the variational method. Since only functions of the same total spin mix with one another, we obtain the following secular determinants:

$$\begin{vmatrix} H_{22} - E & H_{23} \\ H_{32} & H_{33} - E \end{vmatrix} \quad \text{and} \quad \begin{vmatrix} H_{44} - E & H_{45} \\ H_{54} & H_{55} - E \end{vmatrix}$$

The elements of these determinants $(H_{22} = \langle \phi_2 | \mathscr{H} | \phi_2 \rangle, \; H_{23} = \langle \phi_2 | \mathscr{H} | \phi_3 \rangle$, etc.) can be obtained explicitly so that the eigenvalues and the eigenvectors of the states (2) to (5) can be derived. We shall forego a complete treatment here that yields expressions for all of the frequencies and relative intensities of the AB_2 system; this material appears in the Appendix.

From the results of such a direct analysis of the AB_2 system the following important rules are obtained:

$$v_A = f_3$$
$$v_B = (f_5 + f_7)/2$$
$$J_{AB} = [(f_1 - f_4) + (f_6 - f_8)]/3$$

Since the coupling between the magnetically equivalent B nuclei does not influence the spectrum, the appearance of the AB_2 spectrum is dependent only on the ratio $J_{AB}/v_0\delta$ and thus the line frequencies and their intensities in such spectra can be

Exercise 5.10   Analyse the AB_2 spectrum below and determine the parameters v_A, v_B, and J_{AB}.

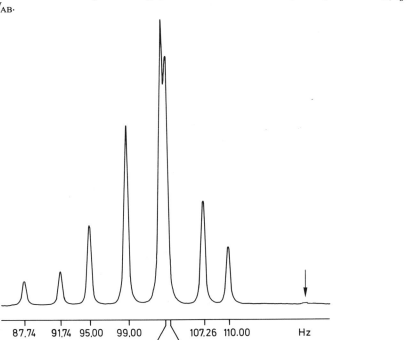

| 87.74 | 91.74 | 95.00 | 99.00 | | 107.26 | 110.00 | Hz |

102.74 103.26

tabulated on the basis of this ratio (cf. Appendix). In Figure 5.8 a few theoretical spectra are reproduced that illustrate the transition of an AB_2 spectrum, via the degenerate A_3 case, to an A_2B system. As we have already mentioned (p. 58), these spectral changes can be observed experimentally using benzyl alcohol.

The line f_9 deserves special attention. This is one of the combination lines mentioned previously that corresponds to the forbidden transition $\alpha\beta\beta \rightarrow \beta\alpha\alpha$. Its intensity is therefore generally very low (cf. Figure 5.6 and Exercise 5.10).[†]

5.2.2   The Particle Spin

We used the AB_2 system in order to illustrate the simplification that applies in the analysis when the symmetry present with magnetically equivalent groups is considered. A further important method for the treatment of equivalent nuclei that leads to the same result can also be illustrated using the AB_2 system. This involves the so-called *particle spin*.

[†] With the exception of transitions forbidden by symmetry the selection rules, as well as other statements of quantum mechanics, possess only probability character.

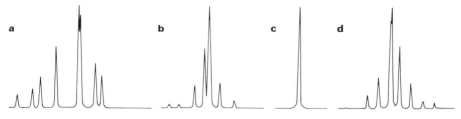

Figure 5.8 Transition from the AB_2 to the A_2B system: (a) $v_0\delta = 14.0$ Hz; (b) $v_0\delta = 6.0$ Hz; (c) $v_0\delta = 1.0$ Hz; (d) $v_0\delta = -8.0$ Hz; $J = 6.0$ Hz in each instance

If we consider the B_2 group as a simple particle its total spin I^* obviously must have the values 0 (for antiparallel spin orientations) or 1 (for parallel spin orientations); that is, the B_2 group exists either in a singlet (S) or a triplet (T) state. The A nucleus, because of its spin of $\frac{1}{2}$, is in a doublet (D) state and accordingly is at one time bound to a hypothetical nucleus with spin quantum number 0 and at another time to one of spin quantum number 1. I^* is a good quantum number that is not changed by means of the n.m.r. experiment. As a further selection rule for allowed transitions, it follows that $\Delta I^* = 0$. The spectrum can thus be considered to arise from two *subspectra* that are characterized by the notations $A_{1/2}B_0$ and $A_{1/2}B_1$ and that are completely independent of one another. The two subspectra are also known as the *irreducible representations* of the AB_2 system.

The eigenfunctions for the AB_2 system can now again be represented as product functions arranged according to total spin m_T (Table 5.2b).

In the $A_{1/2}B_0$ subsystem the B particle is not magnetic so that only one line, namely the transition $A(\beta) \rightarrow A(\alpha)$ or $\Psi_8 \rightarrow \Psi_7$ can be observed. As can be seen immediately, the principle of particle spin leads to a considerable simplification in the treatment of spin systems with groups of n magnetically equivalent nuclei. If n is an even number there always exists a non-magnetic state for the group, while for uneven n the particle spin $I^* = \frac{1}{2}$ results. For AB_n systems a subspectrum of the A type is observed in the first case while in the second case an AB subspectrum results.

Table 5.2b Basis functions of the AB_2 system

m_T	Basis function	m_T	Basis function
	$A_{1/2}B_1$		$A_{1/2}B_0$
(1) $+\frac{3}{2}$	$D^{+1/2}$ T^{+1}	(7) $+\frac{1}{2}$	$D^{+1/2}$ S^0
(2) $+\frac{1}{2}$	$D^{+1/2}$ T^0	(8) $-\frac{1}{2}$	$D^{-1/2}$ S^0
(3) $+\frac{1}{2}$	$D^{-1/2}$ T^{+1}		
(4) $-\frac{1}{2}$	$D^{+1/2}$ T^{-1}		
(5) $-\frac{1}{2}$	$D^{-1/2}$ T^0		
(6) $-\frac{3}{3}$	$D^{-1/2}$ T^{-1}		

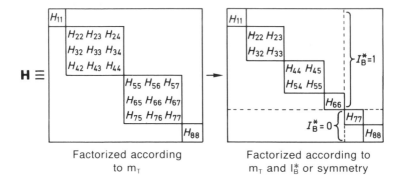

Factorized according   Factorized according to
to m_T m_T and I_B^* or symmetry

The latter contains all the information characteristic of the spin system (ν_A, ν_B, and J_{AB}).

If we now turn to the Hamiltonian matrix for the general three-spin case the simplifications arising from the use of symmetry or the principle of the particle spin are evident. It should be noted that the indices of the matrix elements refer to the basis functions of p. 160 in the first case and to the functions in Table 5.2a or 5.2b in the second case.

Instead of two third-order submatrices, the Hamiltonian matrix now possesses only two second-order submatrices. We mention here that the particle spin approach belongs to the methods that are generally called 'subspectral analysis'. In the following, subspectra will be characterized by small letters.

5.2.3 The ABX System

As another three-spin system we want to investigate the ABX system. As the notation indicates, this is a system in which two nuclei, A and B, having similar chemical shifts are coupled with a third nucleus the resonance frequency of which is very different from ν_A and ν_B. The X nucleus in such systems is said to be weakly coupled and the A and B nuclei are said to be strongly coupled. Examples of such systems that are characterized by three resonance frequencies (ν_A, ν_B, and ν_X) and three coupling constants (J_{AB}, J_{AX}, and J_{BX}) are found in 1,2,4-trichlorobenzene (**87**), 2-fluoro-4,6-dichlorophenol (**88**), 2-iodothiophene (**89**), and styrene oxide (**90**).

87   **88**   **89**   **90**

Another important principle is used in the analysis of these systems. It is known as the *X approximation* and is based on the fact that those off-diagonal elements of the Hamilton matrix for the three-spin case that occur between states with different magnetic quantum numbers $m(X)$ of the X nucleus are negligibly small compared with the diagonal elements. Omitting these off-diagonal elements leads to a Hamilton matrix that is considerably simplified, as schematically illustrated below, where the indices of the elements refer to the basis functions in Table 5.3:

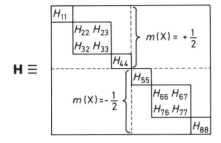

In this table the basis functions are arranged separately according to the eigenvalues $m_I(X)$ of the operator $\hat{I}_z(X)$. As a result we get two sets of four functions each. These sets contain product functions for the AB part of the spin system that are identical with those already introduced as basis functions for the isolated AB system.

The AB portion of the ABX spectrum thus consists of two ab subspectra, one for each of the two $m_I(X)$ values of $+\frac{1}{2}$ and $-\frac{1}{2}$. The eigenvalues E_1, E_4, E_5, and E_8 are immediately obtained through substitution of the corresponding product functions $\alpha\alpha\alpha$, $\beta\beta\alpha$, $\alpha\alpha\beta$, and $\beta\beta\beta$ in equation (5.2), while it is only necessary to solve two second-order determinants for the determination of E_2, E_3, E_6, and E_7.

Table 5.3 Product functions of the ABX system

	$m_I(X) = +\frac{1}{2}$				$m_I(X) = -\frac{1}{2}$		
	A	B	X		A	B	X
m_T			m_T				
(1) $+\frac{3}{2}$	α	α	α	(5) $+\frac{1}{2}$	α	α	β
(2) $+\frac{1}{2}$	α	β	α	(6) $-\frac{1}{2}$	α	β	β
(3) $+\frac{1}{2}$	β	α	α	(7) $-\frac{1}{2}$	β	α	β
(4) $-\frac{1}{2}$	β	β	α	(8) $-\frac{1}{2}$	β	β	β

$$\Downarrow \qquad\qquad\qquad\qquad \Downarrow$$

(1)	$\alpha\alpha$	α	(5)	$\alpha\alpha$		β
(2)	$\cos\theta(\alpha\beta) + \sin\theta(\beta\alpha)$	α	(6)	$\cos\theta(\alpha\beta) + \sin\theta(\beta\alpha)$		β
(3)	$-\sin\theta(\alpha\beta) + \cos\theta(\beta\alpha)$	α	(7)	$-\sin\theta(\alpha\beta) + \cos\theta(\beta\alpha)$		β
(4)	$\beta\beta$	α	(8)	$\beta\beta$		β

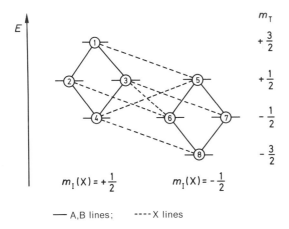

Figure 5.9   Energy level diagram of the ABX system

The advantage of the X approximation becomes abundantly clear in the energy level diagram of the ABX system as shown in Figure 5.9. For transitions within each subspectrum the spin orientation of the X nucleus is not changed.

The eigenvalues $m_I(X)$ of $\hat{I}_z(X)$ are to be considered good quantum numbers and as a special selection rule for allowed AB lines it follows that $\Delta m_I(X) = 0$. The spectrum is independent of the shift differences $\nu_A - \nu_X$ and $\nu_B - \nu_X$.

According to this analysis there are a total of eight AB and six X lines in the ABX system. Of these, the two arising from the transitions $\Psi_4 \rightarrow \Psi_5$ and $\Psi_6 \rightarrow \Psi_3$ are combination lines and in general are of low intensity. The X portion of the spectrum is symmetrical about ν_X (Figure 5.10).

The principle of the approximation discussed here states in its generalized form that for a group, X_n, of n magnetically equivalent nuclei that are weakly coupled with nuclei of another group, the eigenvalues $m_T(X)$ of the operator $\hat{F}_z(X)[=\sum_n \hat{I}_z(X)]$ are good quantum numbers. Therefore, for an ABX_2 system one would expect three ab subspectra and for an ABX_3 system one would expect four ab subspectra. The relative intensities of these subspectra can be obtained from the Pascal triangle (ABX, 1:1; ABX_2, 1:2:1; ABX_3, 1:3:3:1).

If the effect of the X nucleus is treated as a first-order perturbation, then the ab subspectra are characterized by the *effective Larmor frequencies, ν^**. Specifically, in the ABX case we have

$$\begin{aligned} \nu_a^* &= \nu_A + m_I(X)J_{AX} = \nu_A \pm \tfrac{1}{2}J_{AX} \\ \nu_b^* &= \nu_B + m_I(X)J_{BX} = \nu_B \pm \tfrac{1}{2}J_{BX} \end{aligned} \tag{5.24}$$

This approach is thus known as the *method of the effective Larmor frequencies*.

For the analysis of the ABX system it is advantageous to proceed from the AB portion. There is the problem of locating the two ab subspectra and analysing them according to the rules of Section 5.1. To illustrate the procedure we shall present an

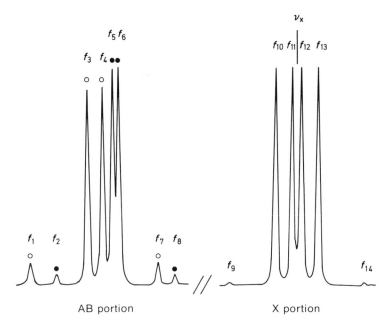

Figure 5.10   The ABX system with the parameters $v_0\delta$ (AB) = 5.0 Hz, J_{AB} = 8 Hz, J_{AX} = 4.2 Hz and J_{BX} = 1.8 Hz. The ab subspectra in the AB portion are identified by the open and closed circles. The parameters used are those of 2-chloro-3-aminopyridine (after Ref. 1)

example. Figure 5.11a shows the AB portion of the ABX system of protons H^1, H^2, and H^3 in 4-bromo-3-t-butylcyclopentan-2-one (**91**).

91

We observe that a line separation of 19.2 Hz occurs four times, i.e. between the pairs f_1–f_3, f_5–f_7, f_2–f_4, and f_6–f_8. By consideration of the relative intensities it is possible to make the following assignments:

ab subspectrum I :   lines $f_{1,}$, f_3, f_5, and f_7

ab subspectrum II :   lines f_2, f_4, f_6, and f_8

The analysis of these subspectra according to the rules derived for AB spectra (Section 5.1) leads to the following results:

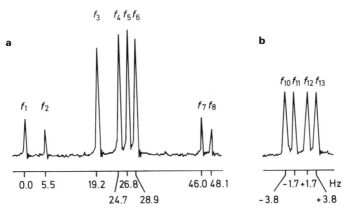

Figure 5.11   The ABX spectrum of 4-bromo-3-t-butylcyclopentene-2-one (**91**) (after Ref. 2): (a) AB portion, relative line frequencies in Hz; (b) X portion (in this case this lies at lower field); 60 MHz

$$\text{I.} \quad v_a^* = 13.6 \text{ Hz} \qquad \text{II.} \quad v_a^* = 20.1 \text{ Hz}$$
$$v_b^* = 32.4 \text{ Hz} \qquad\qquad v_b^* = 33.5 \text{ Hz}$$
$$J_{ab} = 19.2 \text{ Hz} \qquad\qquad J_{ab} = 19.2 \text{ Hz}$$

If we now apply the relations (5.24) for the determination of v_A, v_B, J_{AX}, and J_{BX}, two solutions result since the relative assignment of the effective Larmor frequencies in the subspectra can be exchanged:

$$\text{Solution 1:} \qquad v_A = 16.9 \text{ Hz} \qquad J_{AX} = 6.5 \text{ Hz}$$
$$v_B = 33.0 \text{ Hz} \qquad J_{BX} = 1.1 \text{ Hz}$$
$$|J_{AB}| = 19.2 \text{ Hz}$$

$$\text{Solution 2:} \qquad v_A = 23.6 \text{ Hz} \qquad J_{AX} = 19.9 \text{ Hz}$$
$$v_B = 26.3 \text{ Hz} \qquad J_{BX} = -12.3 \text{ Hz}$$
$$|J_{AB}| = 19.2 \text{ Hz}$$

This generally applies for each ABX system. We must therefore seek criteria by which the correct solution can be identified. The intensities of the lines in the X part of the spectrum enable us to do this. As we have already explained, the X portion of the spectrum consists of six lines symmetrically arranged about v_X. Two of these lines have relative intensities of 1 and the frequencies $v_X \pm \frac{1}{2}(J_{AX} + J_{BX})$. Their separation thus directly yields the sum $J_{AX} + J_{BX}$. These lines are assigned to the transitions $\Psi_5 \rightarrow \Psi_1$ and $\Psi_8 \rightarrow \Psi_4$ and they constitute X subspectra with the effective Larmor frequencies

$$v_X^* = v_X + \tfrac{1}{2}(J_{AX} + J_{BX}) \qquad \text{and} \qquad v_X^* = v_X - \tfrac{1}{2}(J_{AX} + J_{BX})$$

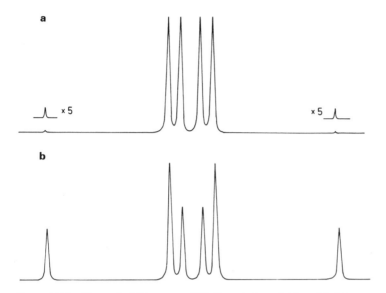

Figure 5.12   Theoretical X spectra of **91** for solutions 1 (a) and 2 (b).

In the spectrum of compound **91** we observed only four lines of approximately equal intensity (Figure 5.11b). Of those, the lines f_{10} and f_{13} have the expected separation of 7.6 Hz. A decision about the correct solution can now be made by means of a computed spectrum (cf. Sections 4.6 and 5.4). The theoretical spectra obtained using the parameters of solutions 1 and 2 are shown in Figure 5.12. As can be seen, the *frequencies* of the lines are indeed virtually identical but the *intensities* indicate distinct differences. Accordingly, solution 2 can be discarded.

One can also forego a complete calculation of the spectrum and only compare the relative intensities of lines f_{10} and f_{11} of the X portion with one another. If we set the intensity of the line $f_{10} = 1$, then the intensity of the line f_{11} is given by equation (5.25), which we present here without proof. Since the line f_9 is not observed in our example, equation (5.25) of course cannot be applied.

$$I_{11} = \left[f_9^2 - 0.25(J_{AX} - J_{BX})^2 \right] / (f_9^2 - f_{11}^2) \qquad (5.25)$$

Exercise 5.11   A calculation made for the ABX system of compound **91** yields a value of 25.2 Hz as the frequency of line f_9 for both solutions. Calculate the intensity of the line f_{11} for both solutions using equation (5.25).

With solution 1 we now have the spectral parameters of the H^1–H^3 protons. However, it is only with the aid of the empirical correlations discussed in Chapter 4 that we are able to assign the spectrum since our analysis gives us no indication of which of the individual parameters ν_A, ν_B, etc., must be attributed to which proton. Also, the relative sign of J_{AB} has not been determined by the analysis.

If we assume that the protons H^1 and H^2 should have similar resonance frequencies, H^3 is established as the X proton. Because $J_{AX} \gg J_{BX}$, it follows on the basis of the Karplus curve that $H^1 \neq H_A$ and $H^2 \neq H_B$ (p. 115). Further, J_{AB} must be negative (p. 108).

With the ABX spectrum we encounter for the first time a spectral type in which, by means of equation (5.24), the *relative signs* of two coupling constants can be determined. In the AB case the spectrum is, as inspection of Table 5.1 shows, independent of the sign of the spin–spin coupling. Indeed, if the sign of J were reversed a different assignment of transitions would have to be made but the appearance of the spectrum would not change. This is also true for the AB_2 system.

Exercise 5.12 Calculate the two solutions for the ABX system of 2-fluoro-4,6-dichlorophenol with reference to the AB portion shown in Figure 5.13.

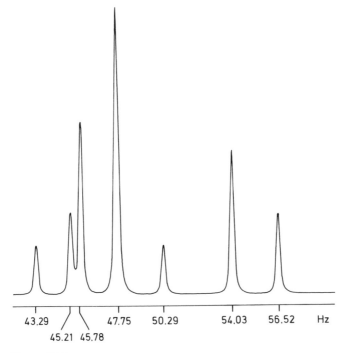

Figure 5.13

How the relative signs of J_{AX} and J_{BX} can influence the appearance of an ABX spectrum can be seen from the two calculated spectra in Figure 5.14. The parameters used for the two systems are identical except for the relative signs of the coupling constants J_{AX} and J_{BX}. One further notes that the X portion of the spectrum is also

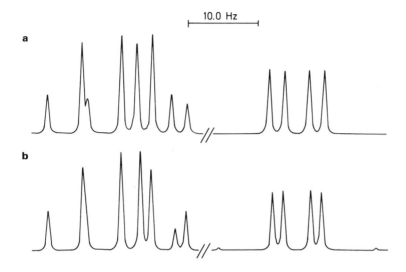

Figure 5.14 Theoretical ABX spectra with the following parameters:

(a) $v_0\delta(AB) = 10.0$ Hz	(b) $v_0\delta(AB) = 10.0$ Hz
J_{AB} = 5.0 Hz	J_{AB} = 5.0 Hz
J_{AX} = 6.0 Hz	J_{AX} = 6.0 Hz
J_{BX} = 2.0 Hz	J_{BX} = -2.0 Hz

sensitive to this difference. In contrast the appearance of the spectrum is insensitive to the sign of J_{AB}.

The dependence of the ABX spectrum on the relative signs of the coupling parameters J_{AX} and J_{BX} leads to the expectation that, in general, there should be three different types of AB portions for ABX spectra. These are represented in Figure 5.15. In case (a), both ab subspectra are clearly separated. The relative signs of J_{AX} and J_{BX} must then be the same. In case (b), one ab subspectrum is framed by the other and consequently for both solutions the relative signs must be different. Finally, in case (c), the one most frequently observed, the two subspectra overlap and both the same and different signs are possible for the coupling constants J_{AX} and J_{BX}.

Not always all fourteen lines of the ABX spectrum are observed. Sometimes one of the ab subspectra degenerates to a deceptively simple AB system, that is, to an A_2 system. An example of this is shown in Figure 5.16 for the 60-MHz spectrum of 1,2-dibromo-1-phenylethane. The AB portion exhibits only five lines.

An approximate analysis of such a spectrum is possible since J_{AB} can be determined and this separation can be subtracted from the signal f_2 in order to define the outer lines of the second ab subspectrum.

a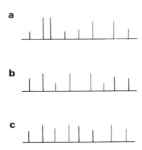

b

c

Figure 5.15

With reference to a series of theoretical spectra, Figure 5.17 illustrates how the appearance of an ABX spectrum is influenced by the shift difference $v_A - v_B$ between the nuclei A and B. As expected, not only the AB portion but also the X portion of the spectrum is sensitive to this parameter. The fact that for all of these spectra $J_{AX} = 0$ deserves special attention. The multiplicity of the X portion in these

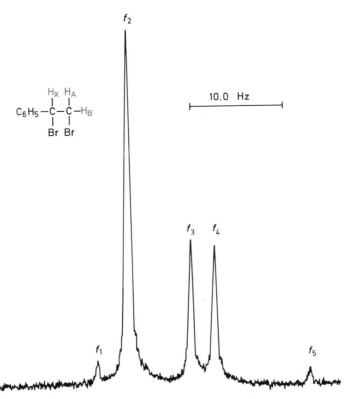

Figure 5.16 The AB portion of the ABX spectrum of the aliphatic protons of 1,2-dibromo-1-phenylethane; 60 MHz

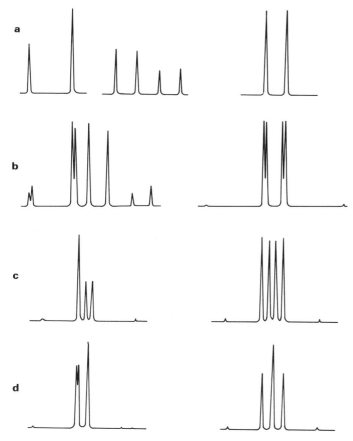

Figure 5.17 Dependence of the ABX system on the parameter $v_A - v_B$: left, the AB portion; right, the X portion. The following parameters apply in all examples: J_{AB} = 15.7 Hz; J_{AX} = 0 Hz; and J_{BX} = 7.7 Hz. The relative chemical shifts $v_0\delta$ (AB) amount to (a) 56.7 Hz, (b) 18.7 Hz, (c) 5.0 Hz, and (d) −0.6 Hz. Experimental data from 2-furfuryl-(2)-acrolein form the basis for the calculated spectra (after Ref. 3)

examples is thus not the result of direct spin–spin interaction with the nucleus A *and* the nucleus B, as the incorrect application of the first-order rules would suggest. Therefore, the conclusion that $J_{AX} = J_{BX}$ cannot be drawn from the "triplet splitting" in case (d). Also, the appearance of the A lines as doublets in case (b) is not caused by a finite coupling constant J_{AX}.

The phenomenon observed here has been called *virtual coupling* in order to indicate that the multiplicity of the X portion of an ABX system can be higher than a simple first order approach suggests. This observation is not limited to ABX systems but is also encountered in other cases if one nucleus of a set of strongly coupled nuclei is additionally coupled to a third nucleus with a very different resonance frequency. The introduction of a special notation here seems superfluous, however, and in addition is misleading since the circumstances described merely demonstrate that the first-order rules may not be applied to the ABX system, a statement that follows immediately from the Hamilton matrix (p. 171). By no means is virtual coupling a special variety of spin–spin coupling or a property of the spin system that requires special treatment for analysis.

Exercise 5.13   Analyse the ABX spectrum of the aliphatic proton of L-asparagine shown in Figure 5.18.

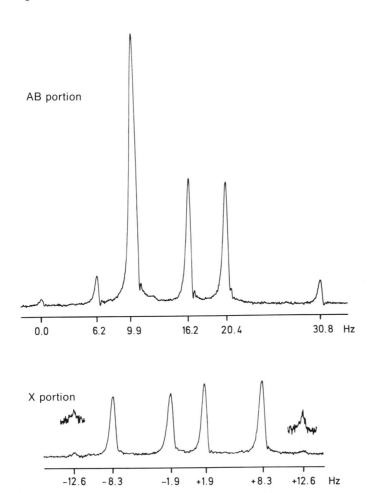

Figure 5.18   The ABX spectrum of the aliphatic protons in L-aspargine; the X portion in this case lies at lower field; 100 MHz

The analyses of ABC systems often observed for vinyl groups turns out to be more difficult. Here no simplifying conditions apply and the maximum number of fifteen theoretically possible transitions can be observed. Special procedures for the direct analysis of these spectra are known but their treatment is beyond the scope of this book.

5.3   SPIN SYSTEMS WITH FOUR NUCLEI

For a four-spin system we use sixteen product functions as a basis:

$m_T = +2$	$m_T = +1$	$m_T = 0$	$m_T = -1$	$m_T = -2$
(1) $\alpha\alpha\alpha\alpha$	(2) $\alpha\alpha\alpha\beta$	(6) $\alpha\alpha\beta\beta$	(12) $\alpha\beta\beta\beta$	(16) $\beta\beta\beta\beta$
	(3) $\alpha\alpha\beta\alpha$	(7) $\alpha\beta\alpha\beta$	(13) $\beta\alpha\beta\beta$	
	(4) $\alpha\beta\alpha\alpha$	(8) $\beta\alpha\alpha\beta$	(14) $\beta\beta\alpha\beta$	
	(5) $\beta\alpha\alpha\alpha$	(9) $\beta\alpha\beta\alpha$	(15) $\beta\beta\beta\alpha$	
		(10) $\beta\beta\alpha\alpha$		
		(11) $\alpha\beta\beta\alpha$		

In the general case the determination of the parameters of the system — four chemical shifts and six coupling constants — therefore requires the solving of one sixth-order and two fourth-order determinants. For the AA'XX' system that we want to treat here the analysis can, however, be simplified substantially by means of the principles discussed in the preceding sections.

We encounter AA'XX' systems in molecules such as *para*-disubstituted benzenes (**92**), furan (**2**), and 1,2-difluoroethylene (**93**), to name only a few examples. Owing to the *chemical* equivalence of the two A and the two X nuclei, respectively, and also to the molecular symmetry, they are characterized by only two resonance frequencies and four coupling constants, v_A, v_X, $J_{AA'}(=J_A)$, $J_{XX'}(=J_X)$, $J_{AX}(=J)$, and $J_{AX'}(=J')$.

92   **93**

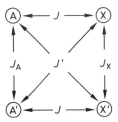

The appearance of the spectrum is, by definition, independent of the difference $v_A - v_X$. Well resolved spectra have twenty lines, ten each for the AA' and the XX' parts. They are symmetric about the centre and the AA' and the XX' portions are symmetric about v_A and v_X, respectively. Since $J \neq J'$, the nuclei of the A or the X groups are not *magnetically* equivalent. This means that in spite of the large relative chemical shift, $v_0\delta = v_A - v_X$, we cannot use first-order rules for the analysis of the spectrum. The latter apply only for the special case when $J = J'$, classified as an A_2X_2 system. Here two triplets with intensity distributions of 1:2:1 are observed.

For constructing the energy level diagram for the AA'XX' system we refer to the known functions of the A_2 case, s_{+1}, s_0, s_{-1} and a_0 (cf. Table 5.1), which, because of the symmetry in the AA'XX' system, can be used as basis functions for the AA' and the XX' groups. We then obtain suitable basis functions for the four-spin case by forming all possible products:

$$\phi_1 = s_{+1}(AA') \times s_{+1}(XX'); \qquad \phi_2 = s_0(AA') \times s_{+1}(XX'); \quad \text{etc.}$$

If we arrange these products according to their symmetry*, their total spin and the magnetic quantum number, $m_T(XX')$, of the XX' group, we obtain Scheme 5.1 for the AA' portion. A completely analogous scheme is obtained for the XX' portion if the classification is made according to $m_T(AA')$.

Consequently, the Hamiltonian matrix of the general four-spin case is drastically simplified. Instead of a 6×6 and two 4×4 submatrices, it now consists of only two 2×2 submatrices. The remaining elements are 1×1 matrices and already represent the correct eigenvalues:

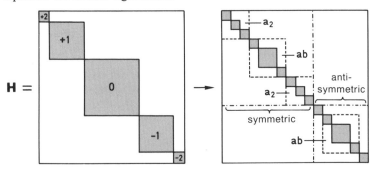

* The general rule that the products of symmetric functions are symmetric, the products of antisymmetric functions are likewise symmetric, and the products of symmetric and antisymmetric functions are antisymmetric applies.

183

Scheme 5.1

Symmetric product functions

m_T	$m_T(XX') = +1$	$m_T(XX') = 0$	$m_T(XX') = -1$
+2	(1) $s_{+1}(AA')\,s_{+1}(XX')$		
+1	(2) $s_0\,(AA')\,s_{+1}(XX')$	(4) $s_{+1}(AA')\,s_0(XX')$	(8) $s_{+1}(AA')\,s_{-1}(XX')$
0	(3) $s_{-1}(AA')\,s_{+1}(XX')$	(5) $s_0\,(AA')\,s_0(XX')$	(9) $s_0\,(AA')\,s_{-1}(XX')$
−1		(6) $a_0(AA')a_0(XX')$	(10) $s_{-1}(AA')\,s_{-1}(XX')$
−2		(7) $s_{-1}(AA')\,s_0(XX')$	
$m_T(XX')$	+1	0	−1
	a_2 subspectrum	ab subspectrum	a_2 subspectrum

Antisymmetric product functions

m_T	$m_T(XX') = +1$	$m_T(XX') = 0$	$m_T(XX') = -1$
+1	(11) $a_0(AA')s_{+1}(XX')$	(12) $s_{+1}(AA')\,a_0(XX')$	(16) $a_0(AA')\,s_{-1}(XX')$
0		(13) $a_0\,(AA')\,s_0(XX')$	
−1		(14) $s_0(AA')\,a_0(XX')$	
		(15) $s_{-1}(AA')\,a_0(XX')$	
$m_T(XX')$	+1	0	−1
		ab subspectrum	

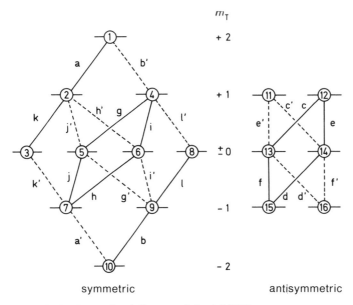

Figure 5.19   Energy level diagram of the AA'XX' system

A further inspection shows that the matrix contains only substructures that are known from the A_2 and AB cases and in the experimental spectrum for the AA' or the XX' portion we actually expect to find two subspectra of the a_2 type and two subspectra of the ab type. Thus there results a total of twelve A and twelve X transitions eight of which (four pairs) are degenerate so that the number of experimental lines is reduced to a total of 20 (10 for each half spectrum). The energy level diagram for the AA'XX' system thus has the structure shown in Figure 5.19.

Since, as we have already noted, the spectrum is symmetric about the centre $(v_A + v_X)/2$, we can limit our discussion to a consideration of one half spectrum. For this Figure 5.20 shows the AA' portion of a theoretical spectrum that was calculated with the parameters $J_A = 9$ Hz, $J_X = 4$ Hz, $J = 8$ Hz, and $J' = 2$ Hz. The labelling of the lines follows from the nomenclature introduced by Dischler [Ref. (a)] and corresponds to that in Figure 5.19. We may obviously assign the lines a, k and b, l to the a_2-systems because of their high intensity. The two ab systems are represented by the lines c, d, e and f and g, h, i, and j, respectively.

The relation between the spectral parameters of the subspectra (v_a, v_b, and J_{ab}) and those of the AA'XX' system (J_A, J_X, J, and J') must now be established. For this it is necessary to calculate the eigenvalues to the functions introduced in Scheme 5.1. Except for the four product functions ϕ_5, ϕ_6, ϕ_{13}, and ϕ_{14}, all of the other functions are already eigenfunctions the Hamiltonian operator and the corresponding energies can be derived directly by the application of equation (5.2). It is to our advantage to

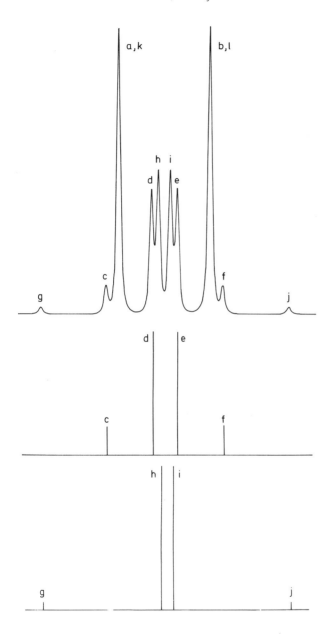

Figure 5.20   A half spectrum of an AA′XX′ system; for the sake of clarity the ab subspectra (antisymmetric c, d, e, f: symmetric g, h, i, j) are shown separately

make use of the following shorthand notation:

$$J_A + J_X = K$$
$$J_A - J_X = M$$
$$J + J' = N$$
$$J - J' = L$$

and we thus obtain the energies given in Table 5.4.

Table 5.4   Eigenvalues of the AA'XX' system

(1)	$v_A + v_X + \frac{1}{2}N + \frac{1}{4}K$	(9)	$-v_X + \frac{1}{4}K$
(2)	$v_X + \frac{1}{4}K$	(10)	$-v_A - v_X + \frac{1}{2} - N\frac{1}{4}K$
(3)	$-v_A + v_X - \frac{1}{2}N + \frac{1}{4}K$	(11)	$v_X - \frac{1}{4}K - \frac{1}{2}M$
(4)	$v_A + \frac{1}{4}K$	(12)	$v_A - \frac{1}{4}K + \frac{1}{2}M$
(7)	$v_X + \frac{1}{4}K$	(15)	$-v_A - \frac{1}{4}K + \frac{1}{2}M$
(8)	$v_A - v_X - \frac{1}{2}N + \frac{1}{4}K$	(16)	$-v_X - \frac{1}{4}K - \frac{1}{2}M$

Exercise 5.14   Verify the results presented in Table 5.4.

The states (5) and (6) as well as (13) and (14) mix with one another. Their eigenvalues must therefore be determined according to the variational method. Besides the diagonal elements $H_{kk} = \langle \phi_5 | \mathcal{H} | \phi_5 \rangle$ and $\langle \phi_6 | \mathcal{H} | \phi_6 \rangle$ and $\langle \phi_{13} | \mathcal{H} | \phi_{13} \rangle$ and $\langle \phi_{14} | \mathcal{H} | \phi_{14} \rangle$ the off-diagonal elements $H_{kl} = \langle \phi_5 | \mathcal{H} | \phi_6 \rangle$ and $\langle \phi_{13} | \mathcal{H} | \phi_{14} \rangle$ must also be calculated. Using the identity $H_{kl} = H_{lk}$ we obtain as determinants

$$\begin{vmatrix} \frac{1}{4}K - E & -\frac{1}{2}L \\ -\frac{1}{2}L & -\frac{3}{4}K - E \end{vmatrix} = 0$$

for the symmetric ab subspectrum and, after addition of $\frac{1}{4}(K-M)$ to the diagonal elements,

$$\begin{vmatrix} -\frac{3}{4}M - E & -\frac{1}{2}L \\ -\frac{1}{2}L & \frac{1}{4}M - E \end{vmatrix} = 0$$

for the antisymmetric case.

The energies of the states (5) and (6) or (13) and (14) are then

$$E_{5,6} = -\frac{1}{4}K \pm \frac{1}{2}\sqrt{K^2 + L^2}$$
$$E_{13,14} = -\frac{1}{4}M \pm \frac{1}{2}\sqrt{M^2 + L^2}$$

If these expressions are compared with the solutions for the eigenvalues E_2 and E_3 of the AB system, it is quite obvious that the parameters K and M represent the effective coupling constants of the symmetric and antisymmetric ab subspectra, and that both ab subspectra are characterized by the effective chemical shift difference $v_0\delta = L$.

Together with the data in Table 5.4 and after subtraction of $\frac{1}{4}(K - M)$ from the eigenvalue $E_{13,14}$, the transition energies in Table 5.5 are obtained for the lines of the AA'XX' system. In the direct analysis of the AA'XX' system, on the other hand, the identification of the subspectra is the central problem. If it is solved, the relations derived for the AB and the A_2 systems are applicable and the following equations hold:

$$N = a - b = k - l$$
$$M = c - d = e - f$$
$$K = g - h = i - j$$
$$L = \sqrt{(h - l)(g - j)} = \sqrt{(c - f)(d - e)}$$

Furthermore,

$$v_A \text{ or } v_X = \tfrac{1}{2}(a + b) = \tfrac{1}{2}(k + l)$$
$$J_A = \tfrac{1}{2}(K + M) \quad J = \tfrac{1}{2}(N + L)$$
$$J_X = \tfrac{1}{2}(K_M) \quad J' = \tfrac{1}{2}(N_L)$$

Exercise 5.15 Analyse the AA' portion of the AA'XX' system in Figure 5.21.
Exercise 5.16 Figure 5.22 shows the AA' portion of the AA'BB' system of the protons H^1–H^4 in 2-methylbenztriazole. Attempt an analysis of this system using the procedures we have derived for the AA'XX' system.

Table 5.5 Transition energies of the AA' nuclei of the AA'XX' system

Transition	Eigenvalues involved	Frequency relative to v_A
a	$(2) \rightarrow (1)$	$\frac{1}{2}N$
b	$(10) \rightarrow (9)$	$-\frac{1}{2}N$
c	$(13) \rightarrow (12)$	$\frac{1}{2}M - \frac{1}{2}\sqrt{M^2 + L^2}$
d	$(15) \rightarrow (14)$	$-\frac{1}{2}M - \frac{1}{2}\sqrt{M^2 + L^2}$
e	$(14) \rightarrow (12)$	$\frac{1}{2}M + \frac{1}{2}\sqrt{M^2 + L^2}$
f	$(15) \rightarrow (13)$	$-\frac{1}{2}M + \frac{1}{2}\sqrt{M^2 + L^2}$
g	$(5) \rightarrow (4)$	$\frac{1}{2}K - \frac{1}{2}\sqrt{M^2 + L^2}$
h	$(7) \rightarrow (6)$	$-\frac{1}{2}K - \frac{1}{2}\sqrt{M^2 + L^2}$
i	$(6) \rightarrow (4)$	$\frac{1}{2}K + \frac{1}{2}\sqrt{M^2 + L^2}$
j	$(7) \rightarrow (5)$	$-\frac{1}{2}K + \frac{1}{2}\sqrt{M^2 + L^2}$
k	$(3) \rightarrow (2)$	$\frac{1}{2}N$
l	$(9) \rightarrow (8)$	$-\frac{1}{2}N$

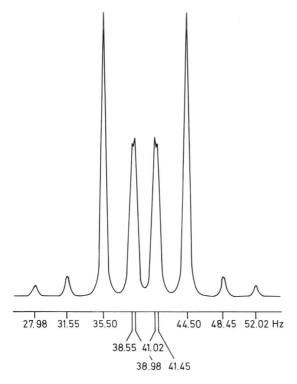

27.98 31.55 35.50   44.50 48.45 52.02 Hz

38.55 41.02

38.98 41.45

Figure 5.21

In connection with the treatment of the AA′XX′ system developed so far a few important points should be emphasized. Through the analysis of an AA′XX′ system a differentiation between the parameters N and L, but not between K and M, can be made. This follows from the fact that we cannot define which ab subspectrum has to be assigned to the symmetric and which to the antisymmetric transitions. This can be achieved, however, as we will see later, if a double resonance technique is used. Thus, in general, only the relative signs of J and J' can be determined in an AA′XX′ system. Further, the assignment of the parameters obtained by the analysis to the spin system under consideration deserves attention. Since the spectrum is not altered if we exchange ν_A and ν_X, neither the assignment of the resonance frequencies nor that of the coupling constants J_A and J_X and also J and J' is obvious. However, through comparison with known values obtained with similar compounds, this problem can be solved in most cases without difficulty.

With reference to the rules that we have established for AA′XX′ systems the *multiplicity* and the *intensity distributions* in some typical spectra will now be discussed. Since we are concerned here with proton spectra it must be emphasized that, as in the case of 2-methylbenztriazole, the criterion for the AA′XX′ case—a

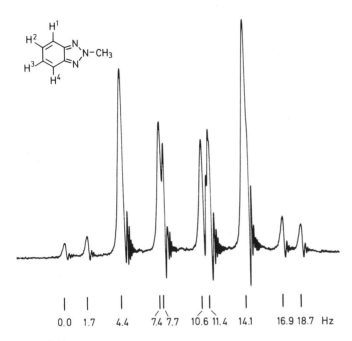

Figure 5.22

very large chemical shift between the A and the X nuclei—is not always rigorously met. Correctly these systems should be classified as AA'BB' spectra, the characteristics of which we will briefly consider on page 193. For the examples treated here, however, the AA'XX' formalism is a very good approximation and since the essential features of the AA'XX' system reappear in the AA'BB' system the following discussion is justified.

As the first example let us consider the spectra of *para*-disubstituted benzenes of the type **92**. Representative of this class of compounds is the AA' portion of the four-spin system for the aromatic protons in 4-bromoanisole shown in Figure 5.23.

The spectrum can be interpreted easily if we consider the coupling constants expected for this molecule. The parameter N here consists of the relatively large *ortho* coupling (J) and the smaller *para* coupling (J'). The two a_2 subspectra are consequently separated by about 7–9 Hz (*cf.* Table 2.2, p. 45) and can immediately be assigned to the intense lines 1 and 2. It follows that the parameter L should be about 5–7 Hz since it represents the difference of the *ortho* and *para* couplings. The effective coupling constants of the ab subspectra, the parameters K and M, differ considerably in the present case since the remaining two *meta* couplings (J_A and J_X) are similar in magnitude. Therefore, M is smaller than 1 Hz and a value of 4–6 Hz is expected for K. The symmetrical ab quartet (*g, h, i,* and *j*) can thus be assigned to the lines 3, 4, 5, and 6. In the antisymmetric ab subspectrum only the inner lines (*d* and

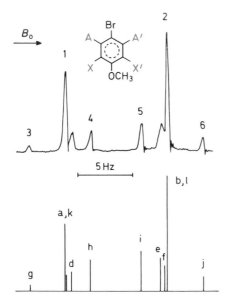

Figure 5.23 The AA′ portion of the AA′XX′ system of the aromatic protons of 4-bromoanisole at 60 MHz (after Ref. 4)

e) are discernable. The outer lines (*c* and *f*) coincide with the a_2 subspectra and because of this the parameter M can be determined only approximately. Of importance for the characteristic appearance of the spectral type discussed here is the relatively large shift difference between the two proton pairs in the positions *ortho* to the bromine atom and the methoxyl group. If the substituents X and Y become similar relative to their influence on the proton resonance, $\nu_0\delta$ decreases and consequently the X approximation is no longer applicable. The appearance of the spectrum becomes more complicated as it approaches the AA′BB′ case.

In the case of *substituted ethanes* of the type XCH_2CH_2Y, AA′XX′ spectra are also observed. As an example, we consider morpholine. At room temperature rapid ring inversion occurs (I $\rightleftharpoons$ II), effectively reducing the complicated four-spin system of the protons of the CH_2CH_2 groups to an AA'XX' system.

If we use Newman projections to represent these conformations we can show that:

1. The equilibrium I $\rightleftharpoons$ II has the effect of exchanging protons H^1 and H^2 for H^3 and H^4, respectively, and consequently $\nu_1 = \nu_3 = \nu_A$ and $\nu_2 = \nu_4 = \nu_X$

2. In addition to the two geminal coupling constants, $J_{13} = J_A$ and $J_{24} = J_X$, only two other time-averaged coupling constants are obtained; namely:

$$
\begin{aligned}
J &= \tfrac{1}{2}[J_{14}(\mathrm{I}) + J_{14}(\mathrm{II})] = \tfrac{1}{2}(J_{trans} + J_{gauche}) \\
&= \tfrac{1}{2}[J_{23}(\mathrm{I}) + J_{23}(\mathrm{II})] = \tfrac{1}{2}(J_{gauche} + J_{trans}) \\
J' &= \tfrac{1}{2}[J_{12}(\mathrm{I}) + J_{12}(\mathrm{II})] = \tfrac{1}{2}(J'_{gauche} + J''_{gauche}) \\
&= \tfrac{1}{2}[J_{34}(\mathrm{I}) + J_{34}(\mathrm{II})] = \tfrac{1}{2}(J''_{gauche} + J'_{gauche})
\end{aligned}
$$

Thus the criteria for an AA'XX' system are met. The appearance of the spectrum here is determined by the large geminal coupling constants of about -10 Hz. Consequently, only K of the parameters $K, L, M,$ and N becomes very large and M, as the difference of two large values, becomes relatively small. Since $J > J'$ it follows that $N > L$. Therefore, in the symmetric quartet of the ab subspectra only the inner lines, h, and i, that sometimes degenerate to a singlet, are easily assignable, and the antisymmetric quartet should approach an AX system in appearance (Figure 5.24).

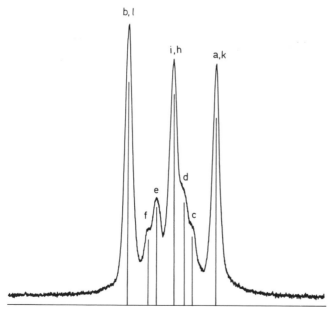

Figure 5.24 The XX' portion of the AA'XX' system of the methylene protons in morpholine at 100 MHz

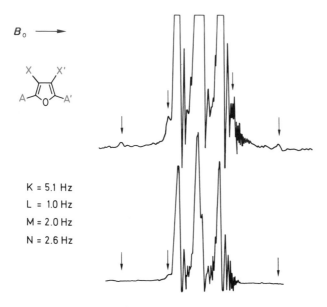

B_0 ⟶

K = 5.1 Hz
L = 1.0 Hz
M = 2.0 Hz
N = 2.6 Hz

Figure 5.25   AA' part ^{1}H n.m.r. spectrum of furan at 60 MHz (after Ref. 4)

In the event $M \approx 0$ it degenerates to a doublet. The parameter K cannot be determined, so that from the coupling constants J_A and J_X only the difference (M) is accessible.

The special case of deceptively simple AA'XX' system is found in furan (cf. Figure 2.19c). The erroneous interpretation of this spectrum as an A_2X_2 system leads to the conclusion that $J = J'$, i.e. that $L = 0$. This is the condition for the magnetic equivalence of the A and the X nuclei, respectively. In the case at hand this cannot be true since if it were, the intensity ratio in both triplets would be 1:2:1. The simple appearance of the spectrum arises mainly from two sources. Firstly, the inner lines of the ab subspectra (d, e, h, and i) are so close together that they cannot be resolved. Secondly, the intensities of the outer lines of the symmetric ab subspectrum lie below the limit of detection. In addition, the outer lines of the antisymmetric quartet are also of low intensity and very close to the intense lines a and k or b and l. Recording the spectrum with greater receiver gain makes these points clear (Figure 5.25).

Another type of simplified AA'XX' system is met if J_A or $J_X \approx 0$. In this case K becomes almost equal to M ($K \approx M$) and the ab subspectra degenerate to the extent that only six lines appear in the AA' and the XX' portion of the spectrum. The spectrum of the olefinic protons of the iron tricarbonyl complex of tricyclo[4.3.1.0$^{1.6}$]deca-2,4-diene provides an example of this (Figure 5.26).

If the relative chemical shift between the A and X nuclei decreases, the spectrum changes gradually from the AA'XX' to the AA'BB' type. It is then sensitive to the

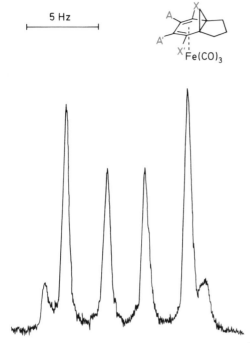

Figure 5.26 The half spectrum of the AA'XX' system of the olefinic protons of tricyclo[4.3.1.0^{1,6}]deca-2,4-dieneiron tricarbonyl (after Ref. 5)

chemical shift difference $v_0\delta(AB)$ and as a first indication of this new situation we note changes in the line intensities.

In the case of 4-bromoanisole (p. 189) these intensity changes are illustrated by the roof effect for the lines 1 and 2 which would not be expected for a true AA'XX' system as shown in Figure 5.20. The error that is made in this case by using the AA'XX' formalism is however, rather small. By further diminishing the relative chemical shift, the error increases rapidly so that we finally must treat the spin system correctly as an AA'BB' system. The eigenvalues $m_T(BB')$ are then no longer good quantum numbers and a glance at Scheme 5.1 shows that a fourth-order determinant must be solved in the course of the analysis. Therefore, a straightforward *direct analysis* of the AA'BB' case is *not possible*. It can be shown, however, that the four unknown eigenvalues of the states (3), (5), (6), and (8) can be eliminated. Without going into the details of this procedure, we mention that this allows the derivation of equations that relate the measured line frequencies directly to the parameters of the spectrum (cf. Appendix). Direct analysis is thus also possible for AA'BB' spectra.

Experimentally, the AA'BB' system can be recognized by the intensity distribution mentioned above that results in progressively heightened intensities of the lines flanking the centre $(v_A + v_B)/2$ at the expense of the intensity of the outer lines. Consequently, AA'BB' systems show the roof effect but are symmetric about

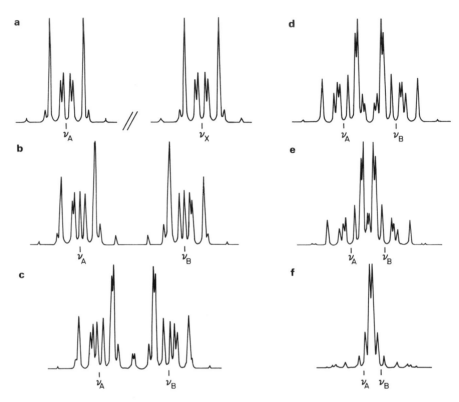

Figure 5.27 Dependence of the AA'BB' spectrum on the shift difference $v_0\delta$(AB): (a) AA'XX' limiting case; (b) $v_0\delta = 30.0$ Hz; (c) $v_0\delta = 20.0$ Hz; (d) $v_0\delta = 15.0$ Hz; (e) $v_0\delta = 10.0$ Hz; (f) $v_0\delta = 5$ Hz. The following coupling parameters apply in all of the examples: $J = 8.2$ Hz; $J' = 1.5$ Hz; $J_A = 7.5$ Hz; and $J_B = 3.0$ Hz

the centre. Furthermore, the transitions a and k or b and l are no longer degenerate so that each half of the spectrum consists of twelve lines. Figure 5.27 shows the transition from the AA'XX' limiting case to the AA'BB' case with reference to a series of calculated spectra in which the relative chemical shift was decreased while the original values of the coupling constants were maintained.

5.4   COMPUTER ANALYSIS

As mentioned before, computers play an important role in the analysis of complicated spectra that arise from spin systems without symmetry or with a large number of nuclei. In these cases the simplifications discussed in the previous sections do not apply and computer programs are used to solve the eigenvalue problem. In addition, the results obtained by a direct analysis of a spin system are always checked by comparing a calculated spectrum with the experimental

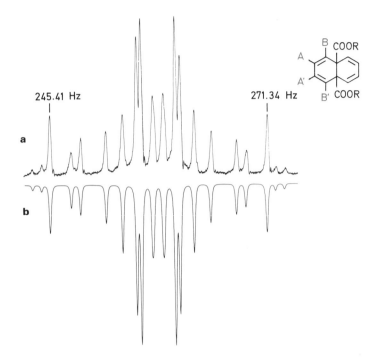

Figure 5.28   (a) Experimental and (b) calculated ^{1}H n.m.r. spectrum of the olefinic protons of 9,10-dicarbethoxy-9,10-dihydronaphthalene at 60 MHz (after Ref. 6)T144

spectrum. This comparison is a stringent test especially since the line shape of the n.m.r. signals can be simulated. Figure 5.28 produces such a comparison for the spectrum of the olefinic protons of 9,10-dicarboethoxy-9,10-dihydronaphthalene.

The solution of the eigenvalue problem formally presented in Section 4, that is the calculation of transition frequencies and intensities on the basis of a given set of chemical shifts and coupling constants, can be easily programmed. Since for the more complicated spectra in general no explicit equations can be derived for the parameters, the 'trial and error' procedure forms the basis of the computer analysis. From the consideration of known data for similar compounds and possibly with the aid of recognizable or familiar features in the experimental spectrum—recurring line separations, for example—a set of trial parameters is estimated and in turn used to calculate a trial spectrum. The comparison of the calculated with the experimental spectrum suggests changes for chemical shifts and coupling constants in the initial set of parameters that are believed to improve the agreement between the calculated and the experimental spectrum. Depending upon the degree of complexity of the spectrum and the extent of one's experience and sophistication, a set of parameters is finally reached that can be accepted as a solution of the problem since the spectrum calculated will be consistent with the experimental one relative to the line positions as well as to the line intensities.

The disadvantages of this method are obvious. First, it is very tedious and time consuming in its application, and secondly, it lacks any indication that a further change of one or more parameters will not allow a still better fit between theory and experiment and thus a more exact analysis. Definite progress was therefore made when programs were developed that enabled the computer to perform the comparison between the calculated and the experimental spectra and criteria established that guarantee a convergence to the correct solution. From several approaches the one developed by Castellano and Bothner-By, known as LAOCOON* (*least* *s*quares *a*djustment *o*f *c*alculated *o*n *o*bserved *n*.m.r. transitions), is generally used today and forms the basis of the computer software which is commercially available for spectral analysis. Without going into the details of the mathematics of this program, we shall attempt to describe the essential principle.

The analysis starts with the calculation of a trial spectrum with a set of estimated starting parameters $v_i^{(0)}$, $J_{ij}^{(0)}$. In the process of this calculation, i.e. by the diagonalization of the Hamilton matrix set up according to the rules of Section 4.4, one obtains in addition to the eigenvalues the unitary matrix $\hat{U}$ of the eigenvectors. Now comes the very important step of assigning the lines of the experimental spectrum to the lines of the trial spectrum, that is the investigator provides the computer with the information

$$E_p - E_q = f_{pq} \tag{5.26}$$

where E_p and E_q are the calculated eigenvalues and f_{pq} is the experimentally measured energy difference. The basic idea is then that the best set of parameters is the one that leads to the smallest sum of the squares of the errors (equation 5.27):

$$\sum_{i=1}^{k} (f_{\text{exp.}} - f_{\text{calc.}})_i^2 \tag{5.27}$$

where k is the number of the measured lines and $f_{\text{exp}} - f_{\text{calc.}}$ is the frequency difference between the observed and calculated transitions for the ith line. For each parameters p_j, then, the condition

$$\partial \sum_{i=1}^{k} (f_{\text{exp.}} - f_{\text{calc.}})_i^2 / \partial p_j = 0 \tag{5.28}$$

or

$$-2 \sum_{i=1}^{k} (f_{\text{exp.}} - f_{\text{calc.}})_i \left(\frac{\partial f_{\text{calc.}}}{\partial p_j} \right)_i = 0 \tag{5.29}$$

should hold. If one assumes for small parameter changes a linear dependence of the frequencies one can write

---

* This acronym signifies the effort necessary to unravel complicated n.m.r. spectra.

$$\Delta f_i = \left(\frac{\partial f_i}{\partial p_j}\right) \Delta p_j \tag{5.30}$$

and for the best solution

$$\left(\frac{\partial f_i}{\partial p_j}\right) \Delta p_j = (f_{\text{exp.}} - f_{\text{calc.}})_i \tag{5.31}$$

That is, what is sought are those parameter changes Δp_j that make the experimental and the calculated frequencies equal.

The assignment of the experimental lines to the lines of the first calculated spectrum provides the computer with the information $f_{\text{exp.}} - f_{\text{calc.}}$. The partial derivatives are obtained from the eigenvalues of the trial spectrum in a manner we shall not discuss here. They are approximated just as the first parameter correction Δp_j. Various iteration cycles lead finally to convergence toward the correct solution (Scheme 5.2). As one can easily see, the process of assignment is of crucial

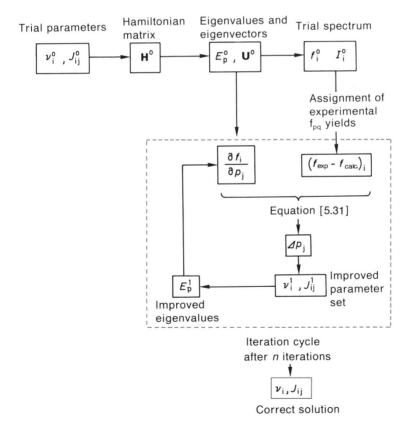

Scheme 5.2

significance to the analysis since too many 'incorrect' input data lead necessarily to erroneous results. In this connection it is interesting to ask whether the solution for a particular spin system is unique or whether perhaps several parameter sets that describe the experimental spectrum equally well within the limits of experimental error exist. Fortunately, only infrequently has the latter situation been found to be the case. The agreement between experiment and theory, not only with respect to frequencies but also with respect to intensities, therefore applies as the criterion for the correct solution. For the comparison it is thus advantageous to make use of the previously mentioned possibilities to simulate the natural line shapes of the n.m.r. signals. Especially with strongly overlapping lines an examination of the results of the spectral analysis is almost impossible without recourse to this aid.

6. References

1. F. A. Bovey, *Nuclear Magnetic Resonance Spectroscopy*, Academic Press, New York, 1969.
2. E. W. Garbisch, Jr., *J. Chem. Educ.*, **45**, 410 (1968).
3. T. Schaefer, *Can. J. Chem.*, **40**, 1678 (1962).
4. D. M. Grant, R. C. Hirst, and H. S. Gutowsky, *J. Chem. Phys.*, **38**, 470 (1963).
5. W. E. Bleck, PhD thesis, University of Cologne 1969.
6. G. Günther and H.-H. Hinrichs, *Justus Liebigs Ann. Chem.*, **706**, 1 1967).

RECOMMENDED READING

R. J. Abraham, *Analysis of High Resolution NMR Spectra*, Elsevier, Amsterdam, 1971, 324 S.

P. L. Corio, *Structure of High-Resolution NMR Spectra*, Academic Press, New York, 1966, 548 S.

Review Articles

(a) B. Dischler, Systematics and Analysis of NMR Spectra, *Angew. Chem.*, **78**, 653 (1966); *Angew. Chem. Int. Ed.*, **5**, 623 (1968).
(b) P. L. Corio and R. C. Hirst, Magnetic Resonance Spectra of Multispin Systems, *J. Chem. Educ.*, **46**, 345 (1969).
(c) A. Ault, Classification of Spin Systems in NMR Spectroscopy, *J. Chem. Educ.*, **47**, 812 (1970).
(d) E. W. Garbisch, Jr., Analysis of Complex NMR Spectra for the Organic Chemist, *J. Chem. Educ.*, **45**, 311, 402, and 480 (1968).
(e) P. Diehl, R. K. Harris, and R. G. Jones, Sub-Spectral Analysis, in F5, **3**, 1 (1967).
(f) R. G. Jones, The Use of Symmetry in Nuclear Magnetic Resonance, in F3, **1**, 97 (1969).
(g) E. O. Bishop, The Interpretation of High-Resolution Nuclear Magnetic Resonance Spectra, in F2, **1**, 91 (1968).
(h) R. A. Homann, S. Forsén, and B. Gestblom, Analysis of NMR Spectra, in F3, **5**, 1 (1971).
(i) P. Diehl, H. Kellerhals, and E. Lustig, Computer Assistance in the Analysis of High Resolution NMR Spectra, in F3, **6**, 1 (1972).
(j) H. Günther, ^{1}H-NMR Spectra of the AA'XX' and AA'BB' type—Analysis and Systematics, *Angew. Chem.*, **84**, 907 (1972); *Angew. Chem. Int. Ed.*, **11**, 861 (1972).

6 THE INFLUENCE OF MOLECULAR SYMMETRY AND CHIRALITY ON PROTON MAGNETIC RESONANCE SPECTRA

The success of n.m.r. spectroscopy in chemistry is due primarily to the fact that the information obtained from n.m.r. spectra corresponds closely to the model-like thinking of chemists. The association of definite spectral regions with certain types of differently bonded protons such as 'aromatic' and 'olefinic', and the multiplicity of the signals, provide information that can be transformed more easily into conceptions about structure and stereochemistry than can the absorption bands in infrared or ultraviolet spectra. Of special significance is the fact that the symmetry of a molecule, because of the sensitivity of the n.m.r. parameters to the molecular environment of the nuclei, is also reflected in the spectrum.

1. Spectral types and structural isomerism

As in infrared spectroscopy, a *highly symmetric* compound can be recognized from its n.m.r. spectrum by the small number of signals it presents. Of the isomeric compounds **94–97** with the molecular formula C_5H_6O compound **97** that has two planes of symmetry and a two-fold axis of rotation produces only two signals with

94   **95**   **96**   **97**

an intensity ratio of 2:1 and thereby can be differentiated from all of the other structures. Also in the case of ketene dimer the proposed structures **99–101** can be immediately eliminated from consideration since only two signals of equal intensity, due to the olefinic and the aliphatic methylene protons, are observed in the spectrum of the compound.

98   **99**   **100**   **101**

In the case of the bismethylene adducts of the symmetrical hexahydroanthracene the possible *syn*- and *anti*-isomers (**102** and **103**, respectively) can be easily differentiated since on the one hand an AB system is expected for the central methylene groups while an A_2 system is expected on the other. Analogously, a differentiation of the cyclopropyl methylene group in the adducts **104** and **105** is possible. Moreover, the number of the olefinic protons and the environment of the methylene groups in the six-membered ring are also different here.

102   **103**   **104**   **105**

If, from the point of view of symmetry, we consider the spectra of disubstituted benzenes in which the two substituents are identical, an unambiguous assignment of structure can be made without a detailed analysis through the mere classification of the spectra. Thus for the *ortho*-derivative (**106**) a spectrum of the AA'BB' type is observed while for the *meta* and *para* derivatives (**107** and **108**) AB$_2$M and A$_4$ spectra, respectively, are recorded.

106   **107**   **108**

A similar situation results in the case of disubstituted cyclopropanes with identical substituents as is shown, for example, with the three isomeric dichlorocyclopropanes (**109–111**) that have spectra of the A_4, ABC_2 and $AA'BB'$ type, respectively. For trisubstituted cyclopropanes with identical substituents the spectra are likewise clearly distinguished and the two isomers, **112** and **113** have A_3- and AB_2-type spectra, respectively. Fewer favourable relationships exist in the case of cyclobutanes. The isomeric tetrachlorocyclobutanes (**114–118**) all have spectra of the A_4 type and individual structures cannot, *a priori*, be differentiated even though the resonance frequency of the protons in all five compounds is different. In the case of the isomeric dichlorocyclobutanes (**119–123**) the spectra are again of different symmetry, but they are so complicated that they cannot be interpreted without analysis.

109   **110**   **111**   **112**   **113**

114   **115**   **116**

117   **118**

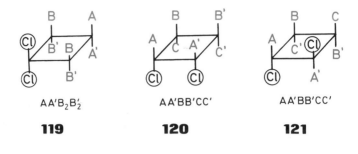

AA'B$_2$B$_2'$

119

AA'BB'CC'

120

AA'BB'CC'

121

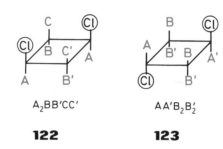

A$_2$BB'CC'

122

AA'B$_2$B$_2'$

123

In the dibromonaphthalenes (**124** and **125**) the assignment indicated in Figure 6.1 is straightforward since for **124**, if one neglects H–H coupling over more than four bonds, an ABC system is expected, whereas the spectrum of **125** should exhibit an AA'BB' and an A$_2$ system.

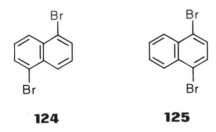

124   **125**

Exercise 6.1   Which spectral types do the protons in the following compounds exhibit?
a. 1,3-Dioxane
b. *cis*-Cyclopropane-1,2-dicarboxylic acid
c. *trans*-Cyclopropane-1,2-dicarboxylic acid
d. 1,4-Dichlorobenzene
e. 2-Chlorophenol
f. 4-Chlorophenol
Exercise 6.2   Sketch the proton resonance spectra of the three isomeric acetylpyridines.

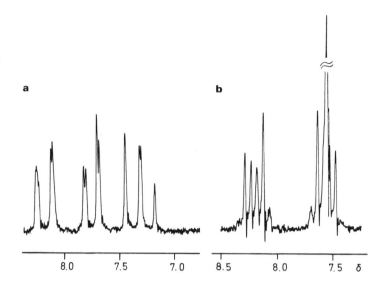

Figure 6.1 60-MHz proton resonance spectrum of (a) 1,5- and (b) 1,4-dibromonaphthalene

Exercise 6.3 In 1,6;8,13-bis-[oxido]-[14]annulene only *one* AA'BB' system is observed for the perimeter protons. On the basis of this observation should the compound be assigned the structure (**a**) or (**b**)?

Exercise 6.4 1,4-Disila-octamethyl(6)radialene (**c**) can exist in the chair conformation **d** which has C_{2h} symmetry or in the twist conformation **e** with D_2 symmetry. Assign spectra 1 and 2 to **d** and **e**, respectively.

From the above considerations it can be concluded that compounds that have little or no symmetry will have more complex spectra than compounds that have a number of symmetry elements. Nevertheless, relatively simple spectra are often observed for

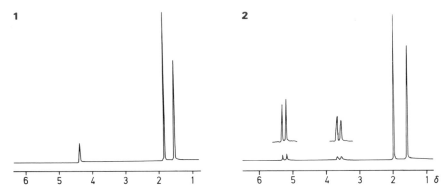

Figure 6.2   80 MHz ^{1}H n.m.r. spectra of the two conformations of the disila(6)radialene **c** (after Ref. 1)

high molecular weight natural products that are considered to be complicated in the chemical sense. In these cases groups of protons form isolated spin systems of A$_2$, AB, AB$_2$, or ABX types that can be easily analysed and assigned and only the symmetry of the group under consideration, and not the symmetry of the entire molecule, determines the spectral type. The prerequisite for the appearance of subspectra of this type is that the spin–spin interactions between the individual proton groups lie below the experimental limits of detection. The assignment of the signals explained in the following example serves to illustrate this fact.

In the spectrum of flemingin B (Figure 6.3), an African drug, an AB$_2$ and two AB systems are recognized. The different coupling constants of 15.5 and 10.0 Hz allow an unambiguous assignment of the AB systems to the protons of the *trans* double bond and the C3–C4 double bond, respectively, while only one phenyl ring, namely (a), possesses the symmetry necessary for the AB$_2$ system. A singlet at δ7.34 arises from the isolated proton at C7 while the broadened triplet at δ5.12 comes from the olefinic proton of the side-chain, the resonance of which is split by the protons of the neighbouring methylene group. The allylic spin–spin coupling with the methyl protons here leads merely to a broadening of the signals. This is reflected in the methyl region for the signals at δ1.63 and δ1.57, that are assigned to the protons of the geminal methyl groups in the side-chain. The signals of the remaining methylene protons at about δ2.0 cannot be clearly distinguished since the protons of the incompletely deuterated solvent absorb here. Noteworthy is the low-field position of one of the OH resonances. This is the absorption of the C5 hydroxyl group that can form a hydrogen-bond with the neighbouring ketone function.

In contrast to the case discussed above, Figure 6.4 shows the complicated spectra of two 'chemically simple' molecules. While for acrylonitrile (cyanoethylene) a strongly coupled ABC system is found, an AA$'$BB$'$ spectrum is observed for 1,2-dichlorobenzene. In both cases the n.m.r. parameters can be determined only through a spectral analysis based on quantum mechanical principles.

205

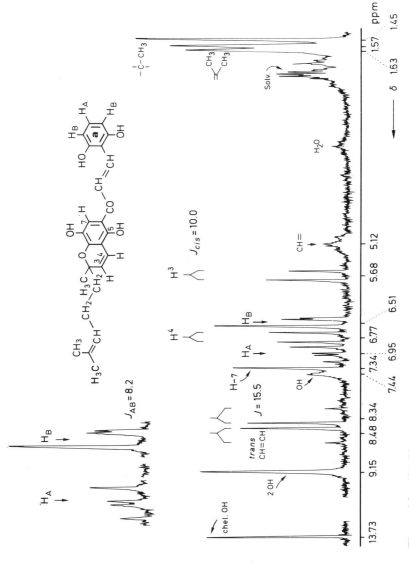

Figure 6.3 60-MHz proton resonance spectrum of a natural product (see text); in acetone-d_6 (after Ref. 2)

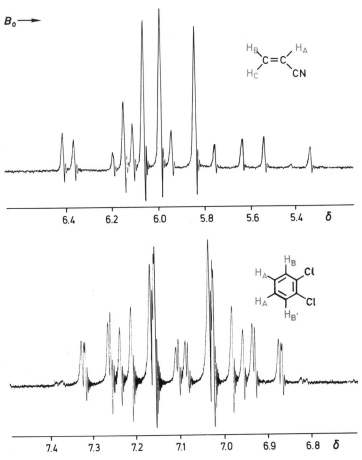

Figure 6.4   60-MHz proton resonance spectrum of (a) acrylonitrile and (b) o-dichlorobenzene

Beside the simplification of a spectrum due to molecular symmetry and the absence of spin–spin coupling, a diminution of the number of resonance signals is observed in several cases because of an accidental superposition of two or more resonance frequencies. Such a case occurs with the spectrum of 3,6-dihydrobenzocyclobutene (Figure 6.5), where the proton resonances of all four methylene groups are isochronous.

2. The influence of chirality on the NMR spectrum

Turning our attention to compounds with asymmetric centres, we find that, just as with the other physical properties of optical antipodes—except for their interaction

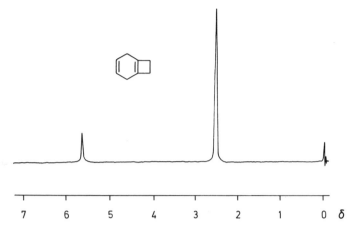

Figure 6.5 60-MHz proton resonance spectrum of 3,6-dihydrobenzocyclobutene (after Ref. 3)

with polarized light—the D- and the L-forms cannot be distinguished. Their n.m.r. spectra are therefore superimposible and also correspond exactly to that of the racemate. However, it is possible through the use of *optically active solvents* to engender in a racemate a chemical shift between the signals of the D- and L-forms. Thus, with a solution of phenylisopropylcarbinol (**126**) in D-1-naphthylethylamine (**127**) two equally intense doublets are observed for the resonance of the tertiary proton, H_t. Their separation amounts to 1.6 and 2.5 Hz at 60 and 100 MHz, respectively. Accordingly, the splitting must be the result of a difference in resonance frequencies since spin–spin coupling is independent of the field strength.

126

127

This finding does not contradict the introductory statement that enantiomers have the same n.m.r. spectra since in optically active solvents, (+)SOL or (−)SOL, diastereomeric complexes D-X/(+)SOL and L-X/(+)SOL or D-X/(−)SOL and L-X/(−)SOL can form through intermolecular interaction between the solvent (SOL) and the dissolved substance, D,L-X, and lead to different spectra. The magnitude of the splitting depends upon the asymmetry or chirality of the solvent and also upon the degree of association between substrate and solvent and therefore upon the temperature. Thus, with 1-cocaine (128) the difference of the resonance frequencies for the proton, H_a, is 0.14 ppm when the spectra are measured at 20 °C in 30% (v/v) solutions of (+)- and (−)-1-phenylethanol in carbon disulphide. At −40 °C, on the other hand, a difference of 0.47 ppm is observed. The example also shows that even when the optically active solvent is diluted with the optically inactive carbon disulphide the diastereomeric solvation effect is still observed.

128

As one can see immediately, integration of the n.m.r. absorptions of the affected signals provides a means for determining the *optical purity* of incompletely resolved racemates. Another method consists of reacting the enantiomeric mixture under consideration with an optically active compound to form a mixture of diastereomeric products that can be investigated by integration of its n.m.r. spectrum. Figure 6.6 shows the spectrum of the product (4-methyl-benzenesulphonic acid 3,3-dimethylbutyl-2-ate) of such a reaction that was obtained by treating racemic *p*-toluenesulphonyl chloride with pinacoline alcohol that was nearly completely (97%) in the *S*-configuration. After reaction the product consisted of a mixture of two components (+)A(+)B and (−)A(+)B in which A and B represent the *p*-toluenesulphonyl and the pinacoline moieties, respectively. As the spectrum demonstrates, all of the resonances of the aliphatic protons, with the exception of that of the methyl group on the benzene ring, are duplicated. The two singlets of the t-butyl group and the doublets of the methyl group can be recognized especially clearly. The quartets of the tertiary proton overlap to form a quintet. Using this method it is even possible to determine the optical purity of compounds that owe their chirality merely to the substitution of a proton by a deuterium.

Let us now investigate the intramolecular influence of optically active centres. If a molecule contains an *asymmetric* carbon atom the magnetic equivalence of neighbouring protons or groups of protons can be destroyed by its presence. Choosing a typical example, one thus finds two doublets at $\delta 0.90$ and $\delta 0.83$ for the CH_3 protons of the isopropyl group in L-valine (129). In another case, instead of the expected quartet and triplet, the ethoxyl group in the methylene cyclobutene derivative (130) gives rise to a complex splitting pattern (Figure 6.7) that must be classified as an ABX_3 spectrum. Other chiral centres such as the sulphite group in diethyl sulphite (131) also have this effect.

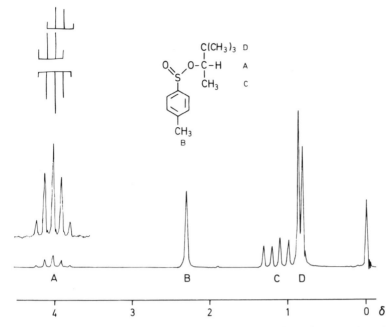

Figure 6.6 0-MHz proton resonance spectrum of the product of the reaction of racemic *p*-toluenesulphinyl chloride with (*S*)-3,3-dimethyl-2-butanol (after Ref. 4)

129

130

131

To explain these findings, we consider a molecule of the general structure

$$b-\overset{\overset{\displaystyle a}{|}}{C}-\overset{\overset{\displaystyle H}{|}}{C}-R$$
$$\overset{\displaystyle |}{c}\quad\overset{\displaystyle |}{H}$$

and its conformations I–III that are represented as Newman projections. As can be seen, H^1 and H^2 are always located in different chemical environments since even if the populations of the three rotamers I, II, and III are equal, which is generally not the case, the non-equivalence between H^1 and H^2 remains since, because of the presence of the group R, the resonance frequencies $v_1(I) \neq v_2(III)$, $v_2(I) \neq v_1(II)$, and $v_2(II) \neq v_1(III)$. If the substituent R is replaced by a hydrogen atom, this difference is

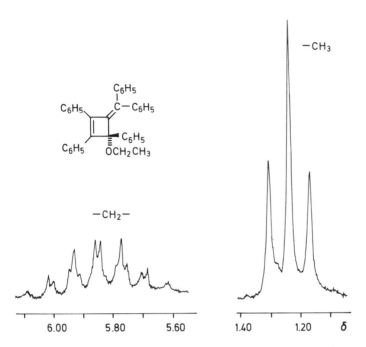

Figure 6.7   60-MHz proton resonance spectrum of the ethyl group in compound **130**

eliminated and on the time average the protons of a methyl group have the same resonance frequency (see also pp. 39 ff.). A compound for which the inherent asymmetric structural contribution to magnetic non-equivalence is independent of

the populations of the individual conformations is the derivative of the 'propeller' molecule 4-methyl-2,6,7-trithiabicyclo[2.2.2]octane-2,6,7-trioxide **(132)**. The 2-hydroxyisopropyl group at the 1-position shows a shift difference of 0.04 ppm (in pyridine as solvent) for the resonances of the methyl protons. The field dependence of the splitting (60 MHz, 2.3 Hz; 90 MHz, 3.6 Hz) proves that it is indeed a matter of a difference in the resonance frequencies.

132

As the last example suggests, the phenomenon discussed here is not limited to molecules with optically active carbon atoms. Thus, in the general case treated above, the residue (a) can be substituted by another CH_2R group, leading to a prochiral arrangement, as for example in acetaldehyde diethylacetal, the spectrum of which is shown in Figure 6.8. Always then, if a substituent of the general structure CX_2R is in the neighbourhood of a prochiral $CR^1R^2R^3$ group or a chiral centre, the environments of the substituents X become non-equivalent, or *diastereotopic*. In contrast, groups whose environments are mirror images are designated as *enantiotopic*. The X–C–X angle between enantiotopic groups X is therefore bisected by a mirror plane, σ (**133a**). The groups X are also equivalent if the particular molecule has C_2 symmetry and the C_2 axis passes through the carbon atom of the CX_2 moiety and is perpendicular to the line joining the X groups (**133b**). The A_2 system of the methylene protons in 2,7-dibromo-1-6-methano[10]annulene (**134**) provides an example of this.

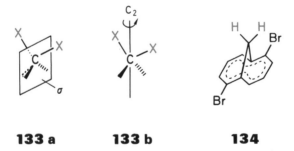

133 a   **133 b**   **134**

A CX_2R group can also serve as a probe for the demonstration of the chirality of a larger ring system. Thus one observes the expected AB system for the methylene protons of the benzyloxy group in 2-benzyloxy-1,6-methano[10]-annulene (**135**). Likewise, the non-equivalence of the methyl groups in 2-isopropyl-1,6-oxido[10]an-nulene shows that this compound is not planar. Moreover, a rapid inversion of the oxygen bridge through the carbon perimeter can be excluded since this would result, on the time average, in an effective plane of symmetry and a consequent loss of chirality.

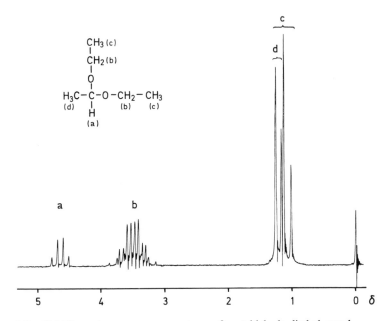

Figure 6.8   60-MHz proton resonance spectrum of acetaldehyde diethyl acetal

135   **136**

Exercise 6.5   Decide whether the indicated protons in the compounds (a)–(j) are enantiotopic or diastereoetopic.

a   b

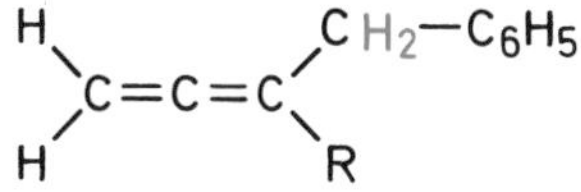

H
$CH_2-C_6H_5$
$C=C=C$
H R

c

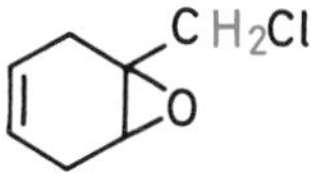

CH_2Cl
O

d

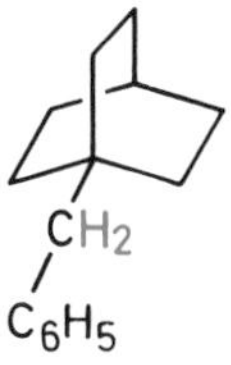

CH_2
C_6H_5

e

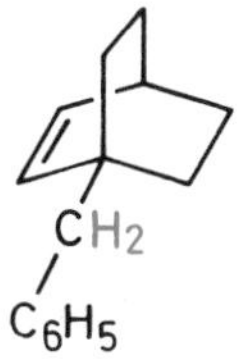

CH_2
C_6H_5

f

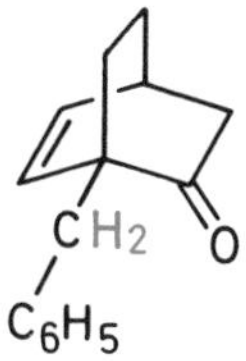

CH_2 O
C_6H_5

g

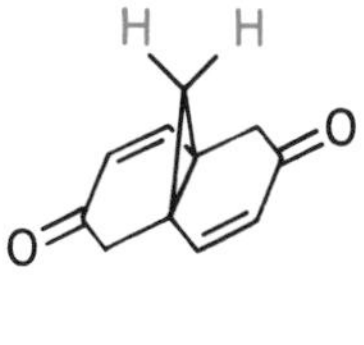

H H
O
O

h

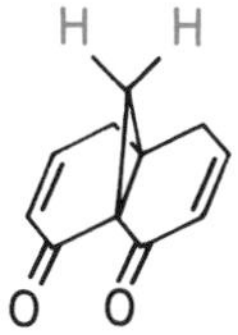

H H
O O

i

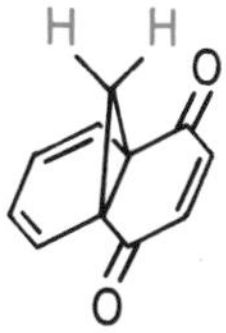

H H O
O

j

The sensitivity of the chemical shift to the symmetry of the molecular environment has also led to progress in *polymer* spectroscopy. This is revealed by observations made for poly(methyl methacrylate) polymer chains with different stereochemistry. Let us first single out a sequence of six carbon atoms for which the conformations (a), (b), and (c) are conceivable.

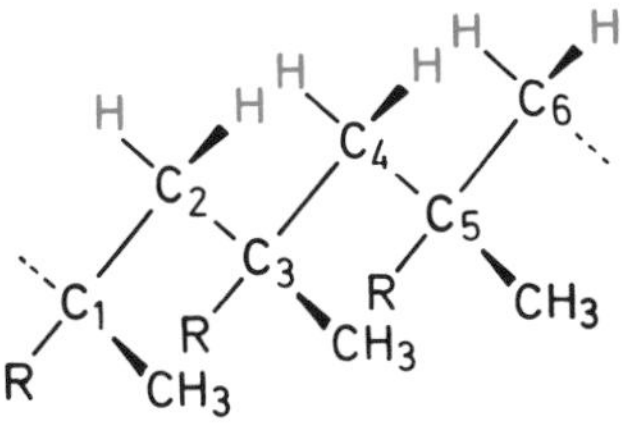

a

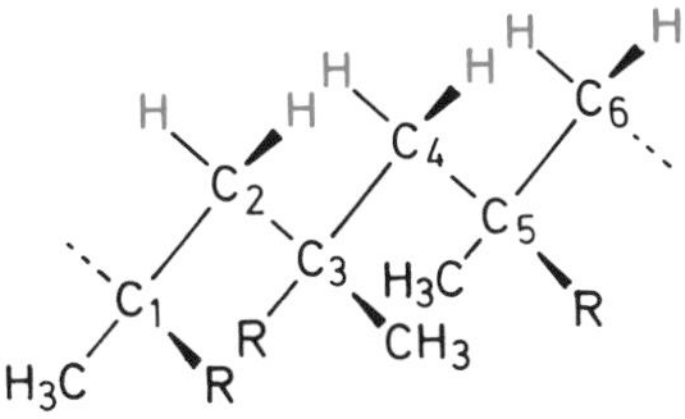

b

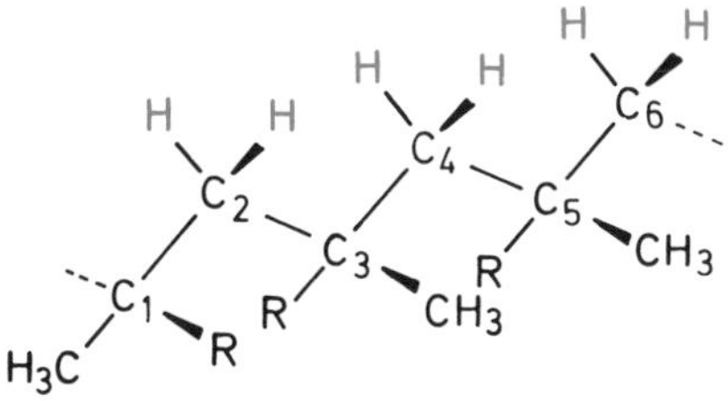

c

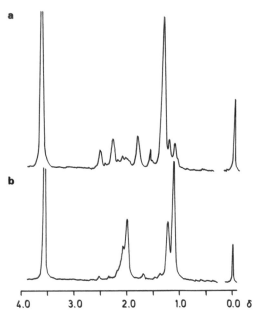

Figure 6.9 60-MHz proton resonance spectra of (a) isotactic and (b) syndiotactic poly(methyl methacrylate) (after Ref. 5)

These sequences are called 'triads' since they are formed from three monomers and the linkage of the monomers is designated as isotactic (a), syndiotactic (b), and heterotactic (c). If we limit ourselves to the first two systems it is easy to see that the methylene protons in (a) should absorb as an AB system while in contrast those in (b) should absorb as an A_2 system, since in the latter case the $C_1–C_2–C_3$ segment possesses C_2 symmetry. As Figure 6.9 demonstrates, this prediction is confirmed by experiment.

3. The analysis of degenerate spin systems by means of ^{13}C satellites and H/D substitution

If the resonance signal of chloroform is recorded at high gain one observes a low-intensity singlet on both sides of the principal absorption separated *ca.* 104 Hz from the main signal. The position of these weak signals is not dependent upon the spinning rate of the sample cell (Figure 6.10) and therefore they are not spinning side bands. Instead, these signals are the so-called ^{13}C *satellites* of the chloroform signal.

Each organic compound contains 1.1% of the stable isotope carbon-13 in natural abundance. Thus, of 1000 chloroform molecules 989 are ^{12}CHCl$_3$ and 11 are ^{13}CHCl$_3$. Since carbon-13 has a nuclear spin of $I = \frac{1}{2}$ these molecules show a spin–spin interaction between ^{13}C and the proton that leads to the doublet splitting in the

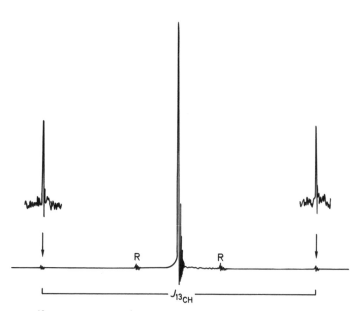

Figure 6.10 ^{13}C satellites in the ^{1}H n.m.r. spectrum of chloroform; spinning side bands are labelled R

proton n.m.r. spectrum. The same splitting is found in the ^{13}C n.m.r. spectrum of ^{13}CHCl$_3$.

The phenomenon of carbon-13 satellites can be used to advantage in order to measure coupling constants between magnetically equivalent nuclei that are normally not available from the spectra of ^{12}C molecules, as was shown before (see Chapter 5). This will be demonstrated with the example of *trans*-1,2-dichloroethylene (see below). As a consequence of molecular symmetry the protons in this compound form an A$_2$ system and the coupling constant $^3J_{trans}$ is not accessible. However, if we consider the molecules that have one ^{13}C nucleus, an AMX spin system with ^{13}C as the X-part is expected since $J(^{13}$C–H$_A) \gg J(^{13}$C–H$_M)$. The effective Larmor frequencies of the protons are then

$$\left. \begin{array}{l} \nu'_A = \nu_0 + \tfrac{1}{2}J(^{13}\mathrm{CH_A}) \\ \nu'_M = \nu_0 + \tfrac{1}{2}J(^{13}\mathrm{CH_M}) \end{array} \right\} m_I(^{13}\mathrm{C}) = +\tfrac{1}{2}$$

$$\left. \begin{array}{l} \nu''_A = \nu_0 - \tfrac{1}{2}J(^{13}\mathrm{CH_A}) \\ \nu''_M = \nu_0 - \tfrac{1}{2}J(^{13}\mathrm{CH_M}) \end{array} \right\} m_I(^{13}\mathrm{C}) = -\tfrac{1}{2}$$

and in the proton spectrum we observe two AM subspectra for the ^{13}C satellites that are represented in Figure 6.11 with the assumption that $J(^{13}$C–H$_A)$ and $J(^{13}$C–H$_M)$ have the same sign. These subspectra now clearly contain the information $J($A,M$)$,

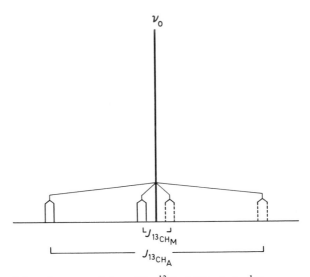

Figure 6.11   Schematic representation of the ^{13}C satellites in the ^{1}H n.m.r. spectrum of *trans*-1,2-dichloroethylene

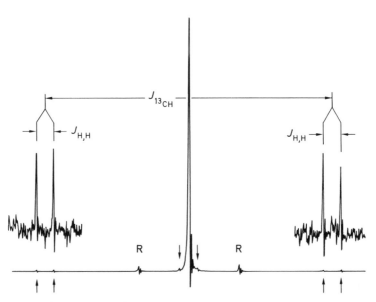

Figure 6.12   Experimental ^{1}H n.m.r. spectrum of *trans*-1,2-dichloroethylene with ^{13}C satellites; spinning side bands are labelled R

which is the coupling of interest here. The signal of the ^{12}C molecules appears undisturbed at ν_0*.

The experimental spectrum of the compound (Figure 6.12) confirms the above analysis. Indeed, $J(^{13}C,H_M)$ is approximately zero so that the two doublets in the neighbourhood of the main signal are superimposed. $J(^{13}C,H_A)$ and $J_{H,H}$ are determined to be 199 Hz and 12.5 Hz, respectively. For *cis*-1,2-dichloroethylene one finds in an analogous experiment that $J_{H,H} = 5.3$ Hz. The differentiation of the two isomer on the basis of their vicinal coupling is therefore easily accomplished through the observation of the ^{13}C satellites.

The analysis of ^{13}C satellite spectra also yields valuable information about spin–spin coupling in more complicated cases. Thus, in the proton spectrum of 1,4-dioxane one measures an AA′XX′ system for the ^{13}C satellites and in benzene an ABB′CC′X system is observed. Through the analyses of these spectra the desired parameters can be obtained.

Exercise 6.6 Construct and classify the spectra of the ^{13}C satellites of the following molecules:

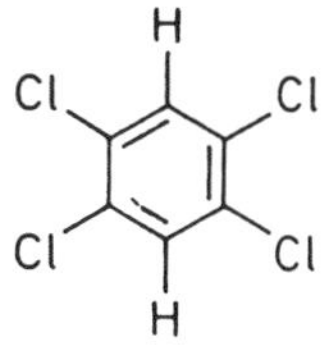

a

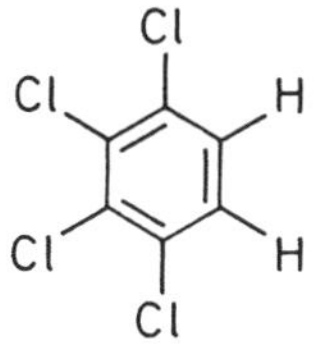

b

$^{13}CHCl_2-^{12}CH_2Cl$

c

$^{13}CH_2Cl-^{12}CH_2Cl$

d

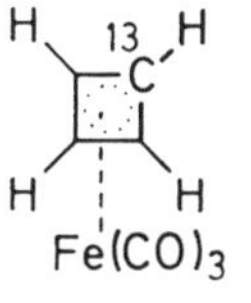

e

$H_3C-^{13}CH_2-CH_3$

f

If the protons of interest are bound to the same carbon atom no information can be extracted from the ^{13}C satellites since both nuclei have the same ^{13}C, ^1H coupling

---

* A possible isotope effect on the resonance frequency is neglected in this consideration.

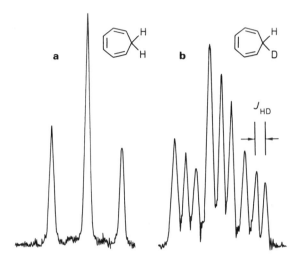

Figure 6.13 ^{1}H n.m.r. absorption of the methylene protons in 1,3,5-cycloheptatriene and 7-deutero-1,3,5-cycloheptatriene; $|{}^2J_{H,D}| = 2.0$ Hz and with equation (5.1) $|{}^2J_{H,H}| = 13.0$ Hz

constant and the ^{13}C satellite spectra are of the A_2X or A_3X type and independent of the spin–spin interaction within the A_2 or A_3 group. In the case of methylene and methyl groups this problem can be obviated by the substitution of a proton by a deuterium. The H–D coupling constant can then be measured and converted to $J_{H,H}$ by equation (6.1):

$$J_{H,H} = 6.5144 \, J_{H,D} \tag{6.1}$$

The use of this strategy is illustrated in Figure 6.13 for the methylene protons of 1,3,5-cycloheptatriene that are equivalent at room temperature as a consequence of fast ring inversion.

In conclusion, it should be mentioned that other magnetic nuclei also give rise to satellite spectra (cf. p. 36 ff.). Thus, the ^{1}H resonance line of tetramethylsilane is always accompanied by the ^{29}Si satellites that arise through geminal ^{1}H,^{29}Si coupling. The magnitude of the coupling constants here is 6.8 Hz. We shall consider another example on p. 386.

4. References

1. A. Maercker, F. Brauers, W. Brieden, and B. Engelen, *J. Organomet. Chem.*, **377**, C45–C51 (1989); W. Brieden, PhD thesis, University of Siegen, 1990.
2. G. Cardillo, L. Merlinie, and R. Mondelli, *Tetrahedron*, **24**, 497 (1968).
3. W. Grimme and K. Redecker, personal communication.
4. M. Raban and K. Mislow, Ref. (b) below.
5. F. A. Bovey, *Nuclear Magnetic Resonance Spectroscopy*, Academic Press, New York, 1969.

RECOMMENDED READING

Review Articles

(a) J. Jacobus and M. Raban, An NMR Determination of Optical Purity, *J. Chem. Educ.,* **46,** 351 (1969).
(b) M. Raban and K. Mislow, Modern Methods of the Determination of Optical Purity, in N. L. Allinger and E. L. Eliel (Editors), *Topics in Stereochemistry,* Vol. 2, Interscience, New York, 1969.
(c) M. V. Gorkom and G. E. Hall, Equivalence of Nuclei in High-Resolution Nuclear Magnetic Resonance Spectroscopy, *Quart. Rev.,* **22,** 14 (1968).
(d) W. B. Jennings, Chemical Shift Nonequivalence in Prochiral Groups, *Chem. Rev.,* **75,** 307 (1975).
(e) D. Parker, NMR Determination of Enantiomeric Purity, *Chem. Rev.,* **91,** 1441 (1991).

7 THE PHYSICAL BASIS OF THE NUCLEAR MAGNETIC RESONANCE EXPERIMENT. PART II: FOURIER TRANSFORM AND PULSE NUCLEAR MAGNETIC RESONANCE

In the first chapter the physical background of the nuclear magnetic resonance experiment was described in terms of quantum theory. Equally useful, however, is the classical description, even if the quantization of angular momentum cannot be explained on a purely classical basis. The physical concepts behind the n.m.r. experiment, the construction of the n.m.r. spectrometer, and a large number of other aspects can be demonstrated most clearly, however, by using a classical approach. Particularly in recent years the growing importance of pulse spectroscopy, also in the area of high-resolution n.m.r., where it forms the basis of the Fourier transform technique, has emphasized the need to understand n.m.r. experiments in terms of classical interactions between magnetic moments and magnetic fields. In fact, nuclear magnetism is not in the domain of either quantum mechanics or classical physics; rather, it forms an exercise for the combination of both concepts.

1. Resonance for the isolated nucleus

The energy diagram shown in Figure 1.2 (p. 3) for the two spin states of nuclei with $I = \frac{1}{2}$ has its classical equivalent in the parallel (ground state) and antiparallel (excited state) orientation of the z-component of the nuclear magnetic moment μ relative to the external field B_0. In this model absorption of energy via interaction of the electromagnetic radiation with the nuclear moment leads to inversion of the magnetic vector μ.

The magnetic dipole in a homogeneous magnetic field B_0 experiences a torsional moment that attempts to align it with the direction of the field. The angular momentum of the nucleus therefore causes a precessional motion of μ around the z-axis that can be easily understood according to the principles of gyration theory (Figure 7.1a)*. The angular velocity of this precessional motion, known as Larmor precession, is given by $\omega_0 = -\gamma B_0$, since the vector $\boldsymbol{\omega}_0$ points into the negative z-direction. The Larmor frequency is thus $\omega_0 = \gamma B_0$.

For the resonance process it is important to note that a magnetic field B_1 can affect the inversion of the magnetic moment μ mentioned above. In order to achieve this, B_1 must be directed at right angles to the x,y-component of μ and rotate in the x,y-plane with an angular velocity equal in sign and magnitude to the Larmor frequency. At this point it proves advantageous to introduce, in addition to the fixed coordinate system C (x, y, z), known as *laboratory frame*, a rotating coordinate system C' (x', y', z) (Figure 7.1c). In this *rotating frame*, as C' is called, the magnetic moment no longer feels the effect of the static magnetic field B_0 but rather that of a magnetic field

$$B' = B_0 + \omega/\gamma \tag{7.1}$$

where ω is the angular velocity of C' and ω/γ is a fictitious field B_f that exists only as a result of the relative motion of the coordinate systems C and C'. For $\omega = 0$ B_f vanishes while for $\omega = -\gamma B_0$, B' becomes zero. This obviously corresponds to the statement that the vector assumes a fixed position in the rotating frame if ω is equal

---

* Here and in the following discussion B_0 will be directed in the positive z-direction.

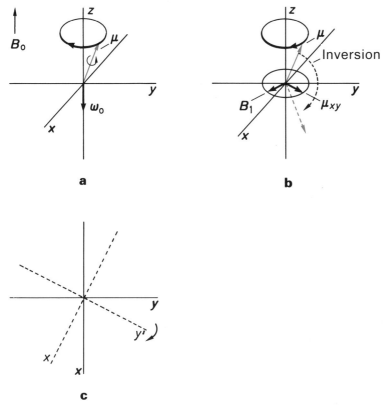

Figure 7.1 (a) Precession of the nuclear moment μ in the fixed laboratory system C; (b) effect of the rotating field vector B_1 on the nuclear moment μ; (c) rotating and fixed coordinate systems C' and C

both in sign and magnitude to the Larmor frequency. The angular velocity and sign of rotation of C' then coincides with the precessional motion.

If we now turn on the magnetic field B_1 that is assumed stationary in the rotating frame and directed along the x'-axis perpendicular to B_0, the effective field according to equation (7.1) is given by

$$
\begin{aligned}
B_{\text{eff}} &= B' + B_1 \\
&= B_0 + \omega/\gamma + B_1 \\
&= B_0(1 - \omega/\omega_0) + B_1
\end{aligned}
\tag{7.2}
$$

The angle θ (Figure 7.2a) formed by B_{eff} with the z-axis is then defined by

$$
\tan \theta = \frac{B_1}{B_0\left(1 - \dfrac{\omega}{\omega_0}\right)}
\tag{7.3}
$$

With the condition $B_0 \gg B_1$ for the magnitudes of the individual fields variation of B_0 and thus the Larmor frequency ω_0 leads to the following situation:

(1) If the magnitude of ω_0 and β are very different, the effective field is aligned *parallel* to the z-axis, because according to equation (7.3), tan θ becomes approximately equal to zero, i.e. $\theta \approx 0°$ or $180°$ for $\omega_0 < \omega$ or $\omega_0 > \omega$, respectively (note the condition $B_0 \gg B_1$ introduced above).

(2) On the other hand, if $\omega_0 \approx \omega$, tan θ approaches infinity and $\theta = 90°$; B_e is then equal to B_1 and the vector μ precesses with frequency ω_1 around the direction of B_1, that is, around the x'-axis (Figure 7.2b). Thus, μ passes from the ground to the excited state. Because $B_0 \gg B_1$ the situation described under (2) represents a typical resonance phenomenon, since a small periodic perturbation of the system leads to a large variation. The system is affected by the perturbing field, however, only when the Larmor frequency and the frequency ω are identical.

In practice, the rotating field B_1 is generated by an oscillator along the x-axis of the fixed coordinate system C. A magnetic field B_x linearly polarized in the x-direction with frequency ω and amplitude $2B_1$ can be represented by two rotating magnetic vectors $B_1(l)$ and $B_1(r)$, one of which, $B_1(r)$, has the desired rotational sense. The other vector has practically no effect on the experiment (Figure 7.3).

2. Resonance for a macroscopic sample and the CW NMR signal

The foregoing model was based upon an isolated nucleus. We shall now attempt to extend our analysis to a macroscopic sample and thus to a large number of nuclei.

After turning on the magnetic field B_0, the nuclei approach an equilibrium distribution between energy levels α and β. This process, which occurs within a

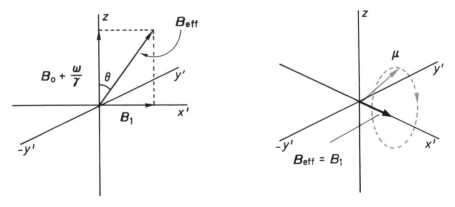

Figure 7.2 (a) The effective magnetic field, B_{eff}, in the rotating frame. (b) Precession of the nuclear moment, μ, around B_1

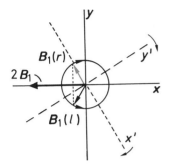

Figure 7.3 The resolution of a linearly polarized field with an amplitude of $2B_1$ into two rotating components $B_1(l)$ and $B_1(r)$

certain time interval, yields $N_\beta > N_\alpha$ according to the Boltzmann distribution law. The result of this process is the build-up of macroscopic equilibrium magnetization M of magnitude M_0, which is the resultant of individual magnetic moments of those nuclei that form the excess population of the ground state (Figure 7.4). Since the nuclear moments do not rotate in phase but are statistically distributed over a conical envelope, no component of the macroscopic magnetization in the x,y plane exists. By means of a transmitter on the x-axis a linearly polarized electromagnetic field B_1 of frequency ω and amplitude $2B_1$ stationary in the rotating frame is now generated. At resonance ($\omega_0 = \omega$) an interaction between the individual nuclear moments and the field B_1 occurs, which deflects M from its equilibrium position along the z-axis. This in turn creates a finite transverse magnetization $M_{y'}$ in the y'-direction (Figure 7.5a). In contrast to the case considered above for individual nuclear magnetic moments, here the vector M is not inverted, because when the amplitude of B_1 is small, not all nuclear moments μ can absorb energy. Consequently, in the fixed coordinate system M executes a precessional motion around the z-axis (Figure 7.5b).

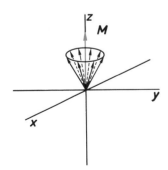

Figure 7.4 Longitudinal macroscopic magnetization M as the resultant of the individual nuclear moments μ; only the moments of the excess ground-state nuclei are shown

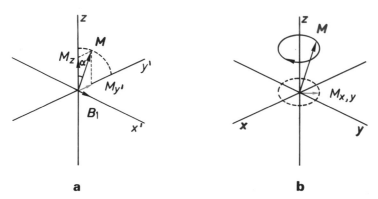

Figure 7.5 (a) Generation of the transverse magnetization, $M_{y'}$. (b) Precession of the vector M in the fixed coordinate system, C

As a result the transverse magnetization produced also rotates in the coordinate system C and can be detected by means of a receiver coil on the y-axis. The deviation of M is proportional to the energy take-up of the spin system from the r.f. field B_1 and the CW n.m.r. signal corresponds to the stationary state between nuclear excitation and relaxation.

Since in practice the ideal case, in which all nuclei of a macroscopic sample have the same Larmor frequency, is not encountered, a transverse magnetization is induced both before and after attainment of the exact resonance condition. When ω_0 is varied sufficiently slowly, the vector M traces a circle in the rotating frame (Figure 7.6). If one plots its components $M_{x'}$ and $M_{y'}$ (symbolized by u and v, respectively) as a function of the frequency difference $\Delta\omega = \omega_0 - \omega$, one obtains a dispersion curve for $M_{x'}$ and an absorption curve for $M_{y'}$ (Figure 7.7). These components of the transverse magnetization differ in phase by 90 °, but both can be measured, since according to Faraday's law the induced electric current in the fixed coordinate system

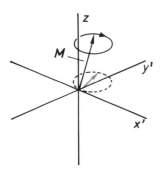

Figure 7.6 Motion of the vector M in the rotating coordinate system C' during slow passage through the resonance condition

C is proportional to the periodic variation dM_x/dt or dM_y/dt. For obvious reasons the receiver coil is mounted on the y-axis.

The quantitative mathematical basis for the phenomenological treatment we have given here was developed by Bloch. It culminates in the famous *Bloch equations*, which are derived in the Appendix (p. 527f.), since a detailed treatment would be contrary to the concept of our non-mathematical approach. For the present, we cite as a result of the full theory merely the following quantitative expression for the transverse magnetization $M_{y'}$ which corresponds to the absorption signal:

$$M_{y'} = \frac{-M_0 \gamma B_1 T_2}{1 + T_2^2(\omega_0 - \omega)^2 + \gamma^2 B_1^2 T_1 T_2} \tag{7.4}$$

Similar relationships are obtained for $M_{x'}$ and, correspondingly, for M_y and M_x in the laboratory frame. The Bloch equations thus allow the calculation of the components of the transverse magnetization as a function of the frequency difference $\omega_0 - \omega$, the amplitude of the B_1 field, the equilibrium magnetization M_0, and the two relaxation times T_1 and T_2 (with which we shall become acquainted in Section 3). In other words, on the basis of the Bloch equations, the line shape of the resonance signal, as it is represented in Figure 7.7, can be calculated.

The experimental arrangement for detection of nuclear magnetic resonance signals is embodied in the Bloch induction spectrometer, which has separate transmitter and receiver coils. This principle was already illustrated in Figure 1.5 (p. 7).

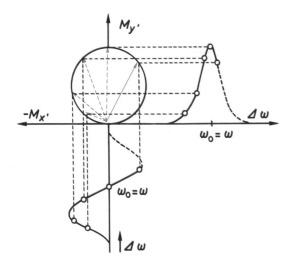

Figure 7.7 Transverse magnetization $M_{x',y'}$ in the rotating frame as a function of the difference $\Delta\omega$ and the resolution of the vector $M_{x',y'}$ into its components $M_{y'}$ and $M_{x'}$.

3. Relaxation effects

As has been shown in the preceding section, two macroscopic magnetizations are distinguished in a nuclear magnetic resonance experiment, the longitudinal magnetization along the z-axis and the transverse magnetization in the x,y plane. Both are subject to relaxation phenomena, that is, their magnitudes are time-dependent.

3.1 LONGITUDINAL RELAXATION

Immediately after exposing the spins of a sample to the external magnetic field B_0, they are in a non-equilibrium state because all spin states are equally populated and $M_0 = 0$. The build-up of the equilibrium magnetization M_0 then requires a time T_1 and the variation of the z component of the macroscopic magnetization obeys a first-order differential equation:

$$dM_z/dt = (M_0 - M_z)/T_1 \qquad (7.5)$$

$1/T_1$ is thus the rate constant for the transition of the perturbed system to the equilibrium state. During T_1, energy is transferred from the spins to the environment, the so-called *lattice*. This process, characterized by equation (7.5), is called *longitudinal relaxation*. Accordingly, T_1 is known as the *longitudinal* or *spin-lattice relaxation time*.

Relaxation plays an important role in the observation of the resonance phenomenon described in Section 2 of this chapter. As we have already explained, in contrast to the μ_z components of the individual nuclear moments, the macroscopic magnetization is not rotated by 180° in the negative z-direction by the application of B_1 fields with small amplitudes, but is only deflected from the z-axis by a small angle α. Consequently, even during resonance, z-magnetization is maintained since the system tries by relaxation to restore the normal Boltzmann distribution (equation (1.11)). A portion of the absorbed energy of the radiofrequency field is therefore finally transferred to the environment. The magnitude of the new equilibrium magnetization M_z is obviously a function of the longitudinal relaxation time and the amplitude of the B_1 field. As can be derived from equation (7.4) for strong B_1 fields, the maximum intensity I of the CW signal at $\omega = \omega_0$ is given by

$$I(\omega_0) = \text{constant}/B_1 T_1 \qquad (7.6)$$

Long relaxation times T_1 and high amplitudes of the oscillating field therefore reduce the signal intensity, i.e. saturate the resonance line. By analogy with Figure 7.7, Figure 7.8 shows the variation of the transverse magnetization in the case of the application of a stronger B_1 field. Instead of a circle the vector M traces an ellipse, the short axis of which is coincident with the y'-axis. The diagram also indicates that for the dispersion curve recorded for $dM_{x'}/dt$ no saturation effect is to be expected, but the linewidth will increase enormously.

Shorter relaxation times, on the other hand, broaden the resonance lines. This

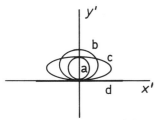

Figure 7.8 Movement of the transverse magnetization in the rotating frame as a function of the amplitude of the B_1 field; at d one has complete saturation

arises because the lifetime of nuclei in the excited state is decreased, which causes an uncertainty in the determination of the energy difference. According to the uncertainty principle $\Delta E \Delta t \approx h$ and with $\Delta E = h\Delta v$ this leads to $\Delta v \Delta t \approx 1/2\pi$ or $\Delta v = 1/2\pi\Delta t$ for the uncertainty in the determination of the resonance frequency. The line width therefore contains the quantity $1/\Delta t$ or $1/T_1$. In organic liquids T_1 for protons is generally in the order of a few seconds or less so that spin–lattice relaxation contributes not more than 0.1 Hz to the line width.

Now, by which mechanism is energy exchanged between the lattice and the nuclear spin system? In liquids magnetic dipole–dipole interaction is mainly responsible and thus is significant for high-resolution nuclear magnetic resonance spectroscopy even if, as mentioned on p. 9 it does not lead to a line splitting. Rotational and translational motions of a molecule in a liquid occasion a fluctuating, i.e. time-dependent magnetic field, which can be described simply as magnetic noise. This fluctuating field possesses components $B_{x'}$ and $B_{y'}$ with frequency ω_0 which satisfies the resonance condition and can stimulate transitions between the stationary states of the nuclear spin system, in other words, they act as built-in r.f. transmitters. The magnetic energy received by the lattice is then transformed into thermal energy. The longitudinal relaxation process is especially effective if paramagnetic substances are present in the solution. This is because the relaxation time T_1 theoretically is inversely proportional to the square of the magnetic moment that gives rise to the above-mentioned fluctuating field. The magnetic moment of an unpaired electron is larger than the nuclear magnetic moment by a factor of about 10^3. T_1 therefore becomes smaller than 10^{-1} s and the resonance lines become very much broadened. Even the presence of trace amounts of oxygen, a paramagnetic molecule, shows this effect (Figure 7.9), which is illustrated also by the longitudinal relaxation times measured for the protons of benzene under different experimental

Table 7.1 Longitudinal proton relaxation times T_1 for benzene (s)

Benzene	(degassed) at 20 °C	19.9
Benzene	in CS_2 (11 vol.-%, degassed)	60.0
Benzene	in the presence of air	2.7

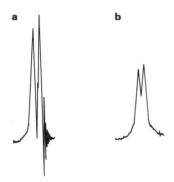

Figure 7.9   Influence of atmospheric oxygen on the shape of the resonance signal: shown are the lines from the spectrum of *o*-dichlorobenzene previously presented in Figure 3.6 (d). (a) A degassed sample; (b) an air-containing sample. Both signals were recorded with the same spectrometer adjustments—only the field homogeneity was optimized

conditions (Table 7.1). Although an especially long relaxation time T_1 is observed in carbon disulphide, a molecule with only a few magnetic nuclei (1.1% ^{13}C), the presence of atmospheric oxygen accelerates the spin-lattice relaxation process.

Because protons normally occupy positions at the molecular surface, their relaxation is strongly determined by *inter*molecular interactions. For nuclei more shielded against their environment, as for instance ^{13}C nuclei, relaxation processes based on *intra*molecular interactions dominate. For this situation theory yields for the dipolar relaxation rate, R_1^{DD}, of a ^{13}C,^{1}H AX spin system the relation

$$R_1^{DD} = 1/T_1^{DD} = (\mu_0/2\pi)\hbar^2\gamma_{^{13}C}^2\gamma_{^1H}^2 r_{CH}^{-6}\tau_c \tag{7.7}$$

which includes, besides well known constants, the ^{13}C,^{1}H distance, r_{CH}, and the so-called correlation time τ_c; τ_c characterizes the dynamic behaviour of molecules and is regarded as the average time which a molecule in solution needs for rotation around one radiant.

For heavy nuclei with $I = \frac{1}{2}$ further contributions to longitudinal relaxation exist. For example, in many cases the anisotropy of the chemical shift, intramolecular dynamic processes like the rotation of methyl groups (spin rotation), or scalar coupling to fast relaxing quadrupolar nuclei (e.g. ^{35}Cl, ^{79}Br) may be responsible for further relaxation mechanisms.

Nuclei with spin quantum numbers $I > \frac{1}{2}$, as already mentioned on p. 10, possess a charge distribution which is not spherically symmetric. These nuclei therefore have an electric quadrupole moment Q. This can interact with the electric field gradient at the nucleus and thus contribute to the relaxation. For the halogen nuclei chlorine, bromine and iodine, for example, this mechanism is so effective that these nuclei, although they have large magnetic moments, are practically nonmagnetic for the purpose of high-resolution nuclear magnetic resonance spectroscopy. In the cases of nitrogen-14 (^{14}N) and deuterium (^{2}H) quadrupolar relaxation is less effective so that their resonance lines can be observed more easily. Those of ^{14}N ($Q = 2 \times 10^{-2}$)*, however, are strongly broadened (half-width up to several hundred hertz), whereas those for ^{2}H are broadened less because of the smaller quadrupolar moment ($Q = 2.77 \times 10^{-3}$). An experimental method for measuring T_1 values will be discussed in Section 4.1 (p. 235).

3.2 TRANSVERSE RELAXATION

In the classical description of the n.m.r. experiment we have learned that, in addition to the z-magnetization there exists a second magnetization in the x,y plane, usually termed transverse or x,y-magnetization ($M_{x,y}$). It seems therefore reasonable to introduce a second relaxation time T_2, the so-called transverse relaxation time, especially since it turns out that the time dependence of $M_{x,y}$ usually differs from that observed for M_z. T_2 is also known as the spin–spin relaxation time after the mechanism responsible for transverse relaxation (energy transfer between individual spins) which will be discussed below.

Another justification for the introduction of T_2 comes from consideration of the line width of the n.m.r. transitions. As was mentioned above, longitudinal relaxation usually contributes less than 0.1 Hz. Nevertheless, observed line widths are larger and may amount to several kilohertz in the case of solids. It is therefore convenient to define another characteristic time T_2, shorter than T_1, to deal with this situation.

In the simplest case $T_2 = T_1$ for liquids since, after resonance, the x,y-component of the magnetization vanishes at the same rate as the longitudinal magnetization attains its previous value M_0 along the z-axis. On the other hand, the transverse magnetization can be reduced without the simultaneous increase in the z-component ($T_2 < T_1$). As in the case of spin–lattice relaxation, fluctuating fields can interact with the transverse component $M_{x,y}$, thereby reducing its magnitude. Whereas time dependent fields $B_{x'}(t)$ and $B_{y'}(t)$, stationary in the rotating frame, interact with M_z, $M_{x,y}$ can interact not only with $B_{x'}(t)$ and $B_{y'}(t)$ but also with B_z. The component B_z, however, is static in the laboratory frame; thus transverse relaxation can also originate from the presence of static dipolar fields.

An important mechanism for transverse relaxation is based on an energy transfer within the spin system. Any transition of a nucleus between its spin states changes the local field at nearby nuclei at the correct frequency to stimulate a transition in the opposite direction. The lifetime of the spin states will be shortened by this process and it therefore contributes to the n.m.r. line width in a manner similar to the spin-lattice relaxation process. The total energy of the spin system does not change, however, and transverse relaxation of this kind can be regarded as an entropy process. Spin-lattice relaxation, on the other hand, is classified as an enthalpy process.

In solids transverse relaxation is strongly affected by the static dipolar fields present. In the absence of motion, each spin experiences a slightly different local

---

* In units of $e \times 10^{-28}$ m^2, where e is the charge of the proton.

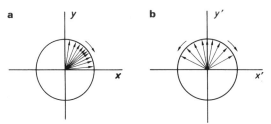

Figure 7.10   Transverse relaxation: (a) in the laboratory and (b) in the rotating frame

field as a result of dipolar interactions with its neighbours. If we now remember that the transverse magnetization $M_{x,y}$ is a macroscopic quantity that can arise only if the individual magnetic moments in the sample have the same Larmor frequency and the same phase angle with respect to an arbitrary origin, it is clear that the spread in Larmor frequencies resulting from the different local fields will destroy $M_{x,y}$. Graphically this process can be described as a fanning out of the vector $M_{x,y}$ (Figure 7.10). Since M_z is in principle not affected by this process, the total energy of the spin system in the external field B_0 does not change (see above).

In liquids, the inhomogeneity, ΔB_0, of the magnetic field B_0 is by far the most important factor for the time dependence of $M_{x,y}$. Exposure of the individual nuclear spins to different external fields $B_0 \pm \Delta B_0$ will result in a spread of their Larmor frequencies and in a fanning out process for $M_{x,y}$ which is completely analogous to that shown in Figure 7.10. In order to avoid the resulting line broadening, each determination of an n.m.r. spectrum should be preceded by optimization of field homogeneity through adjustment of the field gradients (cf. Chapter 3).

According to the quantitative classical treatment of the resonance process (equation (7.4)), for small amplitudes of the B_1 field, i.e. for $\gamma^2 B_1^2 T_1 T_2 \ll 1$, the resonance signal is described by

$$I(\omega) = \frac{\text{constant} \times B_1 T_2}{1 + (\omega_0 - \omega)^2 T_2^2} \qquad (7.8)$$

The signal intensity at the point of resonance ($\omega = \omega_0$) is then directly proportional to the transverse relaxation time:

$$I(\omega_0) = \text{constant} \times B_1 T_2 \qquad (7.9)$$

Including B_1 in the constant, it follows for the intensity at half the signal height that

$$I_{1/2} = \text{constant} \times \tfrac{1}{2} T_2 \qquad (7.10)$$

As this value for $I_{1/2}$ must also satisfy equation (7.8), there results

$$\frac{T_2}{2} = \frac{T_2}{1 + (\omega_0 - \omega_{1/2})^2 T_2^2} \qquad (7.11)$$

and one obtains

$$\omega_0 - \omega_{1/2} = \frac{1}{T_2} \qquad \text{or} \qquad \Delta = \frac{2}{T_2} \qquad (7.12)$$

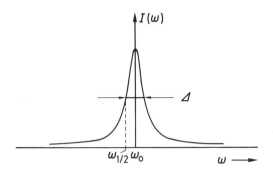

Figure 7.11 The n.m.r. absorption signal (Lorentz curve)

where Δ is the line width of the resonance signal at half-height (Figure 7.11). Since the decay of $M_{x,y}$ is caused by field inhomogeneity and natural spin–spin relaxation as well, one usually writes

$$\Delta = \frac{2}{T_2^*} \tag{7.13}$$

with

$$\frac{1}{T_2^*} = \frac{\gamma \Delta B_0}{2} + \frac{1}{T_2} \tag{7.14}$$

where the first term is the inhomogeneity contribution to the line width. In hertz one has $\Delta = 1/\pi T_2^*$ if T_2^* is measured in seconds. Equation (7.8) describes a *Lorentz curve* and the signal is said to have a *Lorentzian line shape*.

A very effective reduction of the transverse relaxation time also occurs when the nuclei under consideration periodically change their Larmor frequency. This is of great importance for chemistry since in the case of intra- and intermolecular dynamic processes such as proton transfers, conformational equilibria, or valence tautomerism, rapid and reversible variations of the resonance frequencies can occur for particular protons, as we shall see in the following chapter. In cases where this is the dominant mechanism for transverse relaxation, reaction rates can be derived from the temperature-dependent T_2 values which are themselves related to the line widths (equation 7.9)). Chemical kinetics can thus be studied with the help of nuclear magnetic resonance spectroscopy and the method plays an important part in research on rapid reversible reactions.

4. Pulse spectroscopy

Our description of the n.m.r. experiment so far has dealt with situations where relatively weak B_1 fields (of the order of a few tenths of a micro tesla (μT) or less than 1 W) are used for excitation. In this section we shall discuss experiments which

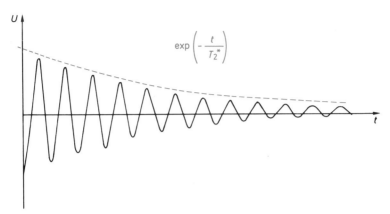

Figure 7.12 Time dependence of the voltage, U, induced in the receiver coil by the rotating component $M_{x,y}$

use strong B_1 fields in the range of 0.01–0.4 T or several hundred kilowatts. To avoid complete saturation of the spin system, these strong fields can be applied only for a short period. Fields that fulfil these conditions are known as *radio-frequency impulses* or simply as *r.f. pulses*.

Applying an r.f. pulse to the spin system causes a deflection of the vector M from the z-axis and induces x,y-magnetization as shown in Figure 7.5a. Because Larmor precession around the z-axis starts, the transverse magnetization in the y-direction of the laboratory frame is time dependent and as a result, an alternating voltage is detected in the receiver coil on the x-axis which exponentially decays to zero with time constant T_2^*. The receiver signal is known as the *free induction decay* (FID) and is shown in Figure 7.12. In contrast to the CW signal, the time signal detected here is an emission signal, because the r.f. field B_1 is turned off during signal detection. The experiment yields practically a linear polarized r.f. field in the y-direction. This is nothing else than an oscillator or transmitter with the Larmor frequency of the particular nucleus.

The *pulse angle* or *flip angle*, α, for M (Figure 7.13a) is given by the relation

$$\alpha = \gamma B_1 t_p \qquad (7.15)$$

where γ/B_1 is the *amplitude* or *power* of the pulse and t_p its *length* or *width*. Both can be varied in order to obtain certain flip angles of interest. One of these is $\alpha = 90°$, where the total magnetization is in the x,y-plane along the y-axis of the rotating frame and the signal has its maximum intensity. Another is $\alpha = 180°$, where M is inverted and points into the negative z-direction. This corresponds to an inverse polarization of the spin system and formally to a negative spin temperature if we remember the Boltzmann distribution law (p. 6). R.f. fields that yield these angles are called 90° $(\pi/2)$- and 180° (π)-pulses. The usefulness of pulses will now be demonstrated with two experiments designed to measure the relaxation times T_1 and T_2, respectively.

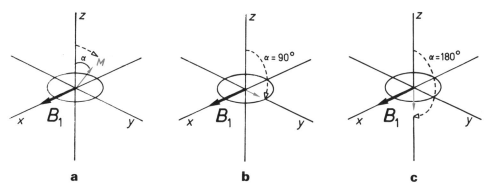

Figure 7.13 Pulse angle α and position of the vector M in the rotating coordinate system after applying a B_1 r.f. field in x-direction; (b) 90°_x pulse; (c) 180°_x pulse

4.1 T_1 MEASUREMENTS

T_1 values for individual nuclei have been recognized as significant parameters related to the dynamic properties of molecules. Longitudinal relaxation rates R_1 ($=1/T_1$) for carbon-13 are now measured frequently. From the various methods available for T_1 determinations we shall describe only the most often used *inversion recovery* experiment.

Let us first consider the macroscopic magnetization M in the rotating coordinate system (Figure 7.14a).* A 180° pulse at the beginning of the experiment brings the vector M into the negative z-direction (b). As a result of spin-lattice relaxation the value of M decreases (c), passes through zero (d), begins to increase in the positive z-direction (e), and finally reaches its initial value. If we characterize the situations (c), (d), and (e) by the time τ_1, τ_0, and τ_2 after the 180° pulse, the magnetization can be detected by 90° pulses at τ_1 and τ_2 which align M along the negative or positive y-

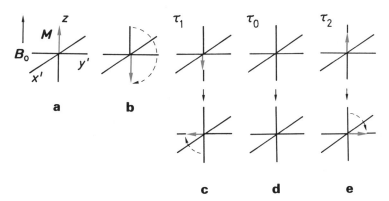

Figure 7.14 The principle of the inversion-recovery experiment for T_1 measurements

---

* From now on x and y refer to the rotating frame of reference if not stated otherwise.

direction, respectively. The two signals differ in phase by 180° and thus lead to an emission and absorption line, respectively. At time τ_0 no signal can be observed since here the sample is not magnetized. For this situation the relation

$$\tau_0 = T_1 \ln 2 = 0.693 \ T_1 \qquad (7.16)$$

holds and from it the relaxation time T_1 can be determined. Alternatively, T_1 can be obtained more accurately from a semilogarithmic plot of the intensity changes $M_0 - M_z$ against τ, since from equation (7.5) one derives

$$\ln(M_0 - M_z) = \ln 2 M_0 - \tau/T_1 \qquad (7.17)$$

by integration.

An application of such experiments is shown in Figure 7.15 for the ^{1}H n.m.r. spectrum of toluene. Each spectral trace is the result of the pulse sequence 180°, τ, 90° applied to the sample, and the delay time was varied from 0.01 to 50 s. From the plot shown a τ_0 of 2.25 s can be estimated for the aromatic proton signal that leads, according to equation (7.12), to a value for T_1 of about 3 s. For the methyl protons T_1 is slightly shorter.

4.2 THE SPIN ECHO EXPERIMENT

The physics of this experiment can best be understood by reference to the diagrams shown in Figure 7.16 (p. 238). Case (a) shows the macroscopic magnetization vector M along the z-axis of the laboratory system. A pulse in the x-direction of the rotating frame leads to a deflection of the vector as discussed above. If one chooses a 90° pulse, M ends up along the y-axis (b). As a result of the inhomogeneity of the B_0 field the individual nuclear spins begin to fan out and the magnitude of the transverse magnetization decreases (c). After a certain time τ a 180° pulse is applied so that all vectors are turned around into the negative y-direction (d). Now, however, their relative motion follows a course such that after a time 2τ they become focused in the negative y-direction. The resultant transverse magnetization can now be detected in the receiver coil as a signal, the so-called *spin echo*. The experiment is conveniently formulated as a *pulse sequence*:

$$90^\circ_x \cdots \tau \cdots 180^\circ_x \cdots \tau \cdots FID$$

From the above analysis it becomes clear that the intensity of the spin echo should depend only on the transverse relaxation rate, i.e. the irreversible loss of transverse magnetization during the period 2τ, since contributions of the field inhomogeneity to the fanning out process for the elementary spins have been eliminated by the refocusing process. If this were true, the echo amplitude should be proportional to $\exp(-2\tau/T_2)$. In practice, diffusion processes complicate the situation by changing the positions of the spins in the magnetic field and thereby increasing the spread of the Larmor frequencies. However, this complicating factor can be eliminated in an

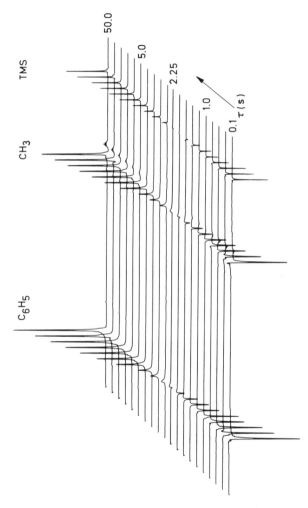

Figure 7.15 Inversion-recovery experiment for the ^{1}H n.m.r. spectrum of toluene

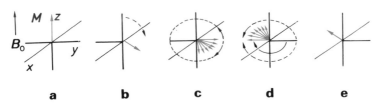

Figure 7.16   The spin echo experiment

elegant manner if, instead of using a single 180° pulse at time τ, one uses a whole sequence of such pulses at τ, 3τ, 5τ, etc. (Carr–Purcell pulse train). The decrease in the amplitude of the spin echo that in turn is recorded at 2τ, 4τ, 6τ, etc., is now proportional to $\exp(-t/T_2)$ and the effect of diffusion becomes negligible if the interval between the pulses is small.

4.3   PULSE FOURIER TRANSFORM (FT) SPECTROSCOPY

The foregoing description of two pulse experiments was introduced in order to show how strong r.f. fields can be used to move the magnetization vector M in certain directions of the coordinate system and to follow its relaxation behaviour. Many other experiments of this type exist, especially such using certain pulse sequences on both liquid and solid samples. These form the basis of the branch of n.m.r. spectroscopy known as *pulse spectroscopy*. The most important application of pulse spectroscopy developed after it was recognized by Ernst and Anderson that r.f. pulses also can be used to excite normal high-resolution n.m.r. spectra and after means were found to analyse the signals detected after such an excitation.

4.3.1   Pulse excitation

If a strong r.f. field is applied to the spin system repeatedly for short periods, a situation results where nuclei with Larmor frequencies ν_i within a certain range $\Delta\nu$ can be excited simultaneously. The reason for this is that a pulse or a pulse-modulated r.f. field with a carrier frequency ν_0, that is, a train of pulses with frequency ν_0 and small width t_p, produces side bands within a range $\pm 1/t_p$ and separated by a frequency difference $1/t_r$, where t_r is the repetition time for the individual pulses. This is illustrated most clearly in Figure 7.17 where the pulse train shown on a time scale in Figure 7.17a has the frequency spectrum shown in Figure 7.17b.

Two points require special attention. On the one hand the time function of Figure 7.17a, that is the pulse train with the repetition time t_r and the pulse width t_p has an equivalent frequency function which is shown in Figure 7.17b on the frequency scale. On the other hand, such an experiment obviously is equivalent to application of a large number of B_1 fields of different frequency ν_i. Thus, the experiment corresponds in principle to having a multichannel spectrometer with numerous

a

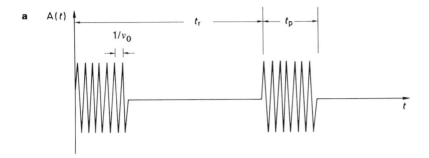

b

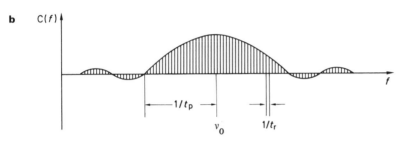

Figure 7.17 (a) Sequence of r.f. pulses of frequency v_0, with width t_p and repetition time t_r. (b) The corresponding frequency components

transmitters distributed equally over the spectral range of interest and available for simultaneous excitation of all resonance lines. That strong r.f. fields indeed can be used for this purpose follows from equation (7.2), which gives the magnitude of the effective field B_{eff} as

$$|B_{eff}| = (2\pi/\gamma)[v - v_i)^2 + (\gamma B_1)^2]^{1/2} \tag{7.18}$$

For large B_1 fields satisfying the condition $\gamma B_1 \gg \Delta v$, the term $(v - v_i)$ can be neglected and the approximate relationship

$$B_{eff} \approx B_1 \tag{7.19}$$

holds. For nuclei having resonance frequencies within the range Δv the magnetization precesses about B_1. For a 90° pulse ($\gamma B_1 t_p = \pi/2$), the above condition for γB_1 is met when

$$t_p \ll 1/4\pi\Delta v \tag{7.20}$$

Accordingly, small pulse widths, typically a few microseconds, are necessary.

The same conclusions can be drawn from another inspection of Figure 7.17b, where the relationship between the frequency range and the number of the side bands on the one hand, and the parameters t_p and t_r on the other, is illustrated. Pulse excitation in high-resolution n.m.r thus requires small t_p and large t_r values. In the limit, if t_p increases Δv will reduce to zero and the side bands disappear. We then have a situation similar to that in the traditional experiment with continuously applied B_1 field. On the other hand, if t_r decreases, the frequency difference between the individual side bands will increase until finally again we have continuous-wave (CW) conditions.

We now ask, what is the advantage of this excitation technique compared to the older CW spectroscopy? This is explained by a simple consideration. Suppose an n.m.r. spectrum of width 500 Hz shows 10 lines of 0.5 Hz half-width. In order to record this spectrum we typically choose a sweep of 250 or 500 s. Only 2% of this sweep time, however, is used to record the information of interest, since this is the time we need to measure the resonance signals. The remaining time is actually wasted recording the noise. With the single transmitter of the traditional CW spectrometer, however, we have no alternative means of recording an unknown spectrum other than by sweeping slowly through the spectral range, checking point by point if absorption occurs or not. Only the pulse technique gives us a method that allows us to reduce considerably the time necessary for this part of the experiment. In practice, our r.f. field has become polychromatic.

4.3.2   The receiver signal and its analysis

Next, we must consider what signal we detect after such an excitation and how we can analyse it. If we remember the comparison made in Chapter 1 between n.m.r. and optical spectroscopy, we conclude that, if polychromatic excitation is used, we obviously need a device equivalent to the prism in optical spectroscopy.

In order to understand this part of the experiment as well, let us first discuss the receiver signal more closely. For a single line we record the free induction decay shown in Figure 7.12. We know already that the envelope of this curve is determined by T_2^*, but the time interval between the maxima of this decaying sine function also has a definite meaning: it corresponds to the reciprocal of the frequency difference Δv_i between the pulse frequency v_0 and the Larmor frequency v_i of the resonance line which is excited. This curve, recorded on a time scale, hence contains all of the information needed to characterize the n.m.r. signal also on the frequency scale, because Δv_i gives us the position of the line (with respect to v_0) and T_2^* determines the line shape. Recording the time dependence of the decaying x,y-magnetization is thus fully equivalent to recording the spectrum in the traditional way on the frequency scale (Figure 7.18), but requires less than 1 s.

It is important to realize that the two forms of the n.m.r. spectrum shown in Figure 7.18 are two representations of the same data set. One representation is on time scale or in the *time domain* and the spectral trace is a function of time, f(t). The other representation is on a frequency scale or in the *frequency domain* and the spectral trace is a function of frequency, F(v). A transformation from one domain into the other is possible through a well known mathematical operation: the *Fourier transformation*. In this sense, the free induction decay and the CW spectrum form a Fourier transform pair.

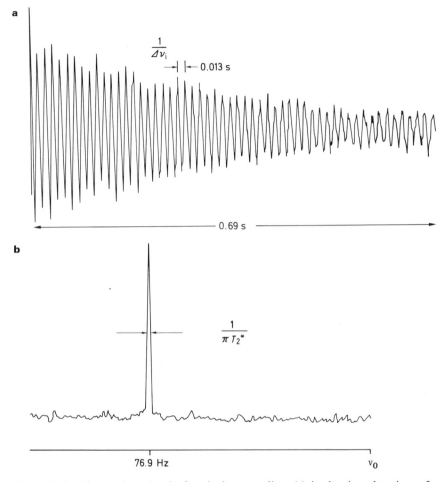

Figure 7.18 The receiver signal of a single n.m.r. line: (a) in the time domain as free induction decay (FID); (b) in the frequency domain as Lorentz curve

In mathematical terms, the two representations of the receiver signal are given by the expressions

$$f(t) = \int_{-\infty}^{+\infty} F(v) \exp(-i2\pi vt) \, dv \qquad (7.21)$$

for the time domain function $f(t)$ and

$$F(v) = \int_{-\infty}^{+\infty} f(t) \exp(+i2\pi vt) \, dt \qquad (7.22)$$

for the frequency domain function F(v), where $i = \sqrt{-1}$. The transformation of (t) → F(v) is achieved point by point following the relation

$$F_j = \frac{1}{N} \sum_{k=0}^{N-1} T_k \exp\left(\frac{-2\pi i j k}{N}\right) \qquad (7.23)$$

where F_j is the jth point in the frequency domain, T_k is the kth point in the time domain and N is the total number of points. Expression (7.23) can be executed by a small computer using a standard algorithm derived by Cooley and Tukey, details of which are beyond the scope of this discussion. For the following considerations it is adequate to note that the mathematical operation indicated, and viewed much as a 'black box', fulfils the requirements of the prism and allows us to analyse the receiver signal and pick out the frequencies responsible for the free induction decay.

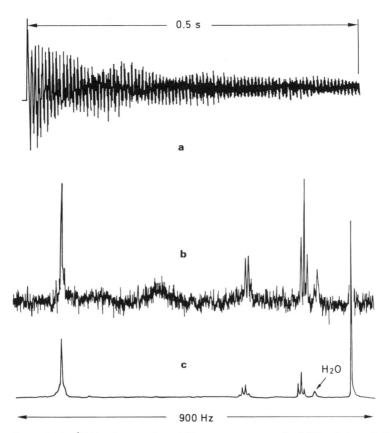

Figure 7.19 ¹H Fourier transform (FT) n.m.r. spectrum of a 0.1% solution of ethylbenzene: (a) free induction decay; (b) conventional CW spectrum, sweep time 1000 s; (c) FT spectrum of 1000 transients of 1 s each (after Ref. 1)

In practice, the free induction decay of an n.m.r. spectrum is much more complicated than the function shown in Figure 7.18 since it results from a superposition of the FIDs of the individual resonance lines, including the noise. An example is given in Figure 7.19a. Nevertheless, it may be stored in digital form in the memory of the computer of the FT spectrometer and later transformed into the frequency domain spectrum.

Since data collection is now so fast, requiring about 1 s or less, it is of course advantageous to accumulate data before carrying out the Fourier transformation. Recording several hundred transients is only a matter of minutes and hence signal-to-noise ratios can be increased considerably (Figure 7.19). This makes the FT technique the method of choice for the n.m.r. spectroscopy, in particular for less sensitive nuclei such as carbon-13 or nitrogen-15, and these fields have made substantial progress after this technique became available. What, at an early stage, may have seemed to be a special experimental aid for selected applications soon developed into the most powerful tool of present day n.m.r. spectroscopy and has opened up new areas of application. Pulse spectroscopy also forms the basis for *two-dimensional n.m.r. spectroscopy* which will be treated in Chapter 8. In the remaining sections of the present chapter, the experimental aspects and requirements of pulse FT experiments are discussed in more detail. Because a complete account is beyond the scope of the present introduction, our discussion must be limited to the most important aspects.

5. Experimental aspects of pulse Fourier transform spectroscopy

5.1   THE FT NMR SPECTROMETER—BASIC PRINCIPLES AND OPERATION

The basic principles of an FT n.m.r. spectrometer are best explained with the aid of a simplified block diagram as given in Figure 7.20. Its different parts will be discussed in the following sections.

5.1.1   The Computer and the Analog-Digital Converter

The most prominent feature of the FT n.m.r. spectrometer is the *digital computer* that plays a central role in the experimental set-up. It controls both the transmitter and the receiver, stores and processes the incoming data and transfers the results to display units such as the oscilloscope or the recorder. The software provides the necessary basis for the commands given to the computer by the operator through an input device, a keyboard, a mouse or a light pen. Practically all functions of the spectrometer—from the transmitter to the recorder pen—are included in this program and are thus under computer control.

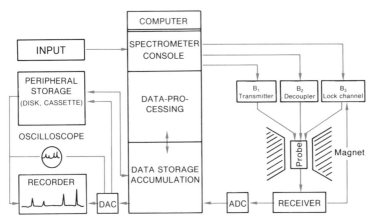

Figure 7.20   Block diagram of an FT n.m.r. spectrometer

The computer itself is characterized by two important parameters that define its storage capacity: the *number of memory locations* (x-axis) and the *word length* (y-axis). Memory locations are counted in multiples of K, which stands for $2^{10} = 1024$. Computers with a memory of 12K can be regarded as the absolute minimum requirement for an FT spectrometer; 4K are reserved for the software, including the Fourier transform routine, and 8K are available for the actual data. Since the Fourier transformation yields the real and imaginary part of the frequency domain function, the two have to be separated and for the final spectrum that corresponds to the real part 4K data points can be used. Modern n.m.r. instruments are equipped with computers of at least 32 or 64K, mostly considerable more memory blocks. Hard disks or tape units lead to external storage capacities in the megabyte range.

The *wordlength* determines the amount of data or its magnitude that can be stored in each memory location. Its unit is the *bit* (*binary digit*), and the number of bits defines the word length. The information, usually a number as a result of a measurement, is stored in binary form, where each decimal number is expressed as the sum of powers of two; for example, $7 = 2^0 + 2^1 + 2^2$. Each bit is a dual unit that can take the value 1 or 0, indicating whether the particular power of two is necessary to represent the decimal number in question or not. With a wordlength of 4 bits, for example, the largest decimal number that can be stored is 15, which in binary representation takes the form 1111 ($2^0 + 2^1 + 2^2 + 2^3$). In general, for *n* bits the largest possible decimal number that can be represented is $2^n - 1$. Furthermore, one bit must be provided for sign information.

From the foregoing it is clear that the raw data somehow must be converted from analogue to digital form before they are acceptable to the computer. This is accomplished by means of an *analogue-to-digital converter* (ADC), also known as *digitizer*. This device samples the free induction decay at regular time intervals and converts each voltage measured to a binary number that can be stored in the corresponding memory location of the computer. Two important aspects have to be considered: one is the *resolution* of the ADC, measured in bits, the other is the *sampling rate*.

With a voltage range normally covering ± 10 V, a 12-bit resolution for the ADC means that the voltage is measured in steps of $10\ 000/(2^{12} - 1) = 2.44$ mV. The integers thus produced are then converted to binary numbers. Data points below one step, in our case 2.44 mV, are not detected. The wordlength of the ADC, in addition to that of the computer, is therefore important for the *dynamic range* available, that is, the capacity to detect weak signals in the presence of strong signals. For the present example of a 12-bit ADC, the limit is given by the intensity ratio $2^{12}:1 = 4096:1$; for an ADC with a 4-bit resolution this ratio is only 16:1. Full use of the dynamic range of the ADC is therefore advisable in order to characterize the free induction decay correctly. On the other hand, it also follows that for data accumulation the word length of the computer must exceed the resolution of the ADC, otherwise memory overflow will result with a consequent loss of information. This is a particular property of the FT experiment that follows from the fact that the frequency domain spectrum is the result of the transformation of the entire free-induction decay. Whereas memory overflow during accumulation of CW spectra as described in Chapter 3 affects only the particular region of the spectrum, for example, a large solvent peak, 'chopping off' a part of the free induction decay falsifies the time domain function and may destroy the frequency domain spectrum completely.

A further point of interest is the rate of data collection in the time domain. Here we remember that the free-induction decay contains frequency components Δv_i given by the difference between the pulse frequency v_0 and the frequency of the n.m.r. signal of interest, v_i. According to the *Nyquist theorem* of information theory, in order to characterize properly each Δv_i at least two data points per cycle must be measured. Therefore, the sampling rate is determined by the spectral width we choose to investigate. If a range of 5 kHz is of interest, data must be collected at a rate of 10 000 points s^{-1} or 10 kHz. If, to use another example, the sampling rate is only 5 kHz, the highest frequency that can be recorded is 2500 Hz.

In addition to sampling rate and dynamic range there are a number of other, not necessarily independent, parameters connected with certain aspects of data collection in FT n.m.r. One of these is the *dwell time*, t_{dw}, which is the time used to produce a particular data point. It is given by the reciprocal of the sampling rate. Thus, for a spectral width of 5 kHz $t_{dw} = 10^{-4}$ s or 100 μs. Given an available computer storage of 8K, the total time during which a free induction decay can be measured in this particular example is then 0.82 s. This time is known as *acquisition time*. It is important to realize that spectral width, sampling rate, dwell time, and acquisition time are all interrelated and, for a given experimental set-up, it is not possible to change one of these parameters without affecting the others. Sampling more slowly, for example, decreases the spectral width and an increase in acquisition time at a given computer storage is possible only if the dwell time is reduced through an increase in the sampling rate. The sometimes confusing interrelations of these

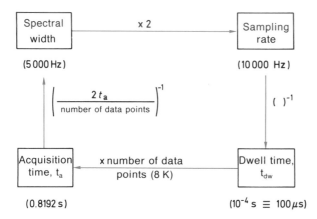

Figure 7.21 Interrelation of experimental parameters in FT n.m.r.

parameters just discussed are illustrated graphically with a numerical example in Figure 7.21.

After transformation we again store the results in the computer, and the available number of memory locations now becomes critical for the resolution of the frequency domain spectrum. Before this curve can be recorded, a *digital-to-analogue converter* (DAC) must be applied. In the example chosen above, the 4K data points available for the spectral range of 5 kHz yield a digital resolution of 1.22 Hz for the frequency spectrum, no matter how good the actual resolution due to magnet homogeneity was. Limited computer space, therefore, may severely affect the quality of the spectrum, especially as far as small line splittings are concerned.

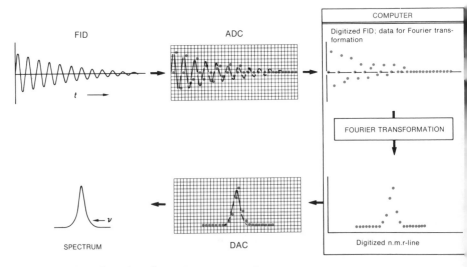

Figure 7.22 Data flow chart for an FT n.m.r. experiment

The data flow and transformation, beginning with the free induction decay, through the ADC, the computer, the DAC, and finally to the recorder, where the conventional n.m.r. signal is displayed, is represented in Figure 7.22.

5.1.2   The RF sources of an FT NMR spectrometer

A number of aspects concerned with the r.f. sources of the FT n.m.r. spectrometer require special attention. Data accumulation over large time intervals requires high field/frequency stability and an internal heterolock system that usually employs the ^{2}H resonance of a deuterated solvent ($CDCl_3$, C_6D_6, etc.) is therefore essential. The lock channel operates in the CW mode. Furthermore, in order to allow double resonance experiments of various types, a second r.f. source of variable frequency should be available. Finally, the pulse transmitter provides the r.f. power for the nucleus of interest. It usually has a fixed value for γB_1 and is characterized by the width t_p (in µs) necessary for a 90° pulse of a standard sample. Typical values for $t_p(90°)$ range from a few microseconds for protons to up to 100 µs for the less sensitive nuclei with small magneto-gyric ratios (cf. equation (7.15)). Whereas the maximum signal results from a 90° pulse, small flip angles are advisable for data accumulation in order to reduce the recovery time for the z-magnetization that is governed by spin–lattice relaxation. As can be seen from Figure 7.13a (p. 235) the induced transverse magnetization along the y-axis is equal to $M_0 \sin \alpha$. At the same time the z-magnetization is reduced to a value of $M_0 - M_0 \cos \alpha$. For small flip angles ($\alpha < 30°–50°$), we have $\sin \alpha > 1 - \cos \alpha$ and the detected signal is much larger than the loss of longitudinal magnetization. For the optimal pulse angle, known as *Ernst angle*, the equation

$$\cos \alpha = \exp\left(-\frac{t_r}{T_1}\right) \tag{7.24}$$

holds. The *delay time* between individual pulses — for a sequence of 90° pulses of the order of 5 times T_1 — and consequently the *repetition time* for the experimental sequence can thus be shortened considerably.

A further *delay time*, however much shorter (of the order of one or two times t_{dw}), is necessary to allow for the recovery of the receiver after application of the strong r.f. pulse, even if time sharing is used, a technique where transmitter and receiver are turned on and off alternately. Finally, as will be discussed below, the correct setting of the pulse frequency with respect to the spectral range is of crucial importance for the results. Figure 7.23 summarizes the sequence of an FT experiment in graphical form.

5.2   EXCITATION AND DETECTION

5.2.1   Transmitter and signal phase

Modern FT n.m.r. spectrometers are equipped with pulse transmitters which can then generate transverse magnetization in each direction of the x,y-plane of the laboratory

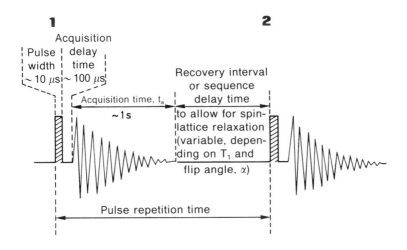

Figure 7.23   Time sequence in FT n.m.r. with data accumulation

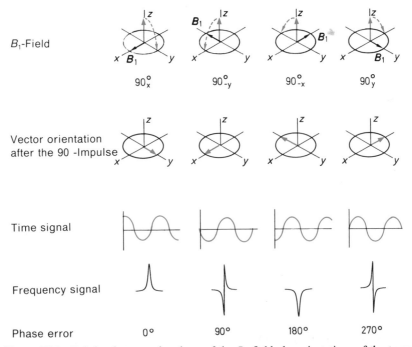

Figure 7.24   Relation between the phase of the B_1 field, the orientations of the transverse magnetization vector after a 90° pulse, and the phases of the respective time and frequency signals in a FT n.m.r. experiment

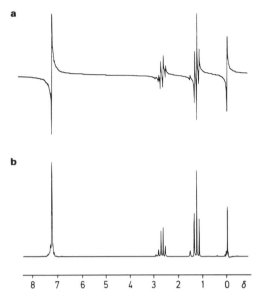

Figure 7.25 Phase correction for a FT n.m.r. spectrum: (a) ^{1}H n.m.r. spectrum of ethylbenzene after Fourier transformation with frequency dependent phase errors; (b) the same spectrum after phase correction

frame. The orientation of the respective vector, $M_{x,y}$, is known as its *phase*, which is measured relative to the y-axis of the rotating frame. The vector orientation at the time when data accumulation starts determines the phase of the time signal as well as that of the frequency signal after Fourier transformation. These relations are illustrated in Figure 7.24. Because n.m.r. spectrometers are equipped with phase-sensitive detectors the dispersion signals with phase errors of 90° or 270° can be suppressed.

A special feature of the FT n.m.r. experiment is the *frequency dependent phase error* that results after Fourier transformation of the time domain data. It originates from the delay time necessary for receiver recovery. During this time the individual cosine components of the free-induction decay progress by different phase angles, thus giving rise to phase shifts in the transformed signals. A straightforward correction of this effect can be made by multiplying the n.m.r. line shape by a frequency-dependent phase factor which is a standard routine in the FT program. An illustration of this aspect is given in Figure 7.25.

Phase errors can also be caused by insufficient power of the pulse transmitter. As shown in Figure 7.17, the intensity of the Fourier components decreases with their distance from the carrier frequency. For large spectral widths the condition $B_{eff} = B_1$ (equation (7.15), p. 239) is then not valid for all frequencies. Accordingly, for some frequencies B_{eff} makes an angle θ with the z-axis, which is smaller than 90°. Since

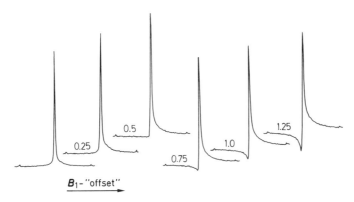

Figure 7.26   Phase error resulting from frequency offset in units of B_1

the vector M rotates around B_{eff}, a phase error arises which is dependent on the frequency offset (Figure 7.26).

5.2.2   Selective pulses in FT NMR spectroscopy

As discussed on p. 240, the basis of FT n.m.r. spectroscopy is polychromatic signal excitation. This means that the pulses used extend always over the whole frequency range. These pulses are called *hard pulses*. In a number of cases, however, it is desirable to use selective or *soft pulses* with a small frequency range, for instance if solvent signals are to be saturated or if certain resonances are to be excited selectively.

In case of heteronuclear experiments the decoupler channel provides the simplest means for the generation of selective $180\,^\circ$ ^{1}H-pulses. For example, ^{13}C resonances can be observed while at the same time the proton magnetization of one particular ^{13}C-satellite in the ^{1}H spectrum is inverted by a short application of the CW ^{1}H transmitter. This experiment is used for ^{13}C n.m.r. assignments and will be discussed in more detail in Chapter 10 (p. 408). For homonuclear cases, on the other hand, selectivity can be achieved in two ways: by the application of GAUSS pulses or by the DANTE pulse sequence.

GAUSS pulses have, as their name implies, instead of the common rectangular envelope the shape of a GAUSS curve and relatively long pulse times (60–100 ms). In order to achieve the GAUSS envelope, the pulse amplitude must be controlled during the pulse time. A narrow excitation range results and the pulse power readily falls off with the frequency offset. Special pulse transmitters have been developed for this purpose.

With the DANTE sequence selective excitation is achieved through a train of n rectangular pulses with small repetition times, t_r, and small pulse angles α. Such a pulse modulation of the carrier frequency corresponds to frequency spectrum with side bands separated by $1/t_p$, which can be regarded as selective pulse transmitters

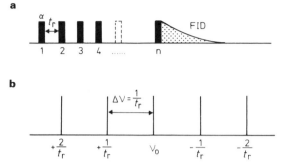

Figure 7.27 DANTE pulse sequence for selective excitation (a) and corresponding frequency spectrum (b); (DANTE = *d*elays *a*lternating with *n*utation for *t*ailored *e*xcitation)

(Figure 7.27). The carrier frequency of the pulse transmitter has to be arranged in such a way that signals, which are not to be excited, fall into the gaps between the side bands. For a selective 90° pulse one chooses $\alpha = 90/n$. Typical values are $t_r = 2$ ms, $n = 200$, $\alpha = 0.45°$. This technique finds its applications, for example, in the excitation of individual spin multiplets or in a number of one-dimensional measuring techniques which are based on selective pulses (cf. Chapter 10).

5.2.3 Pulse calibration

The success of modern n.m.r. experiments depends to a great extent on the precision with which certain pulse angles can be experimentally verified. Pulse angle calibration is, therefore, of great practical importance and pulse angles must be checked quite often, in 2D n.m.r. before each experiment. In principle, several methods are available for this purpose, but we will mention only the most widely used, the determination of the pulse time for the 360° pulse.

In a series of measurements, where the conditions for sensitivity enhancement and phase correction are kept constant, the singlet of a standard sample, for example benzene, is measured. Starting with small pulse angles the signal amplitude increases and becomes negative after going through zero at $t_p(180°)$ (Figure 7.28). A second zero transition is then found at $t_p(360°)$.

In order to avoid offset effects, the carrier frequency is positioned directly at or close to the frequency of the signal used for calibration. Since approximate data for the pulse lengths in the frequency range of interest are normally available, results are obtained most quickly with the $t_p(360°)$ search because it requires the shortest relaxation delays.

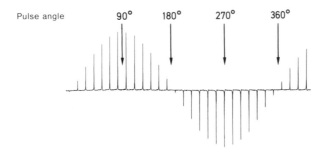

Figure 7.28 Dependence of signal amplitude on the pulse angle α

5.2.4 Composite pulses

Even with modern instruments and notwithstanding careful pulse calibration, small pulse errors are unavoidable. These errors result from the offset effects as well as from field inhomogeneity and limited r.f. power. Additional means to correct for these imperfections are, therefore, of general importance. The most effective is the method of *composite pulses*, where the desired pulse angle results after the application of a number of individual pulses. For example, a 180° pulse which inverts longitudinal magnetization can be replaced by the sequence 90°_x, 180°_y, 90°_x. This *pulse sandwich* reduces the error in the pulse time from up to $+20\%$ to $+1\%$ during an inversion-recovery experiment (Figure 7.29). In order to eliminate at the same time offset effects more extensive *pulse clusters* have been developed.

5.2.5 Single and quadrature detection

As already mentioned on p. 239, the Fourier components which arise during pulse excitation extend to frequency ranges above and below the carrier frequency.

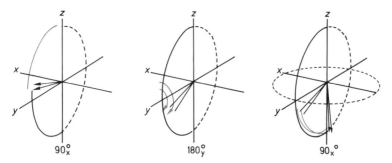

Figure 7.29 Illustration of the principle of *composite pulses* with the example of a 90°_x, 180°_y, 90°_x sandwich for the inversion of longitudinal magnetization M_z; errors in the pulse length are largely removed.

Because positive and negative frequency differences cannot be distinguished after Fourier transformation, it is necessary to position the pulse transmitter at the upper or lower limit of the sweep range of interest if only one detector is used. This experimental set-up is called *single detection*. It has the disadvantage that half of the transmitter power is lost and the sweep range which can be excited with uniform pulse power is limited. The addition of noise by frequency folding from the unused region of the pulse spectrum (cf. Section 5.3) must be prevented by the application of frequency filters.

The alternative method of *quadrature detection*, which is employed in all modern spectrometers, avoids these disadvantages. The carrier frequency is now positioned in the centre of the spectral window. Two phase sensitive detectors with a phase difference of 90° are employed and this arrangement allows one to determine the sign of the measured frequency relative to the frequency of the carrier. In order to understand the principle of this technique, imagine the projections of the rotating transverse magnetization on one or two axes of the rotating coordinate system, respectively. With single detection on one axis we observe only one-dimensional oscillations. With double detection on two axes in quadrature, the sense of rotation and, therefore, the relative sign of the frequency can be determined. Frequencies of different sign behave like sine or cosine functions (Figure 7.30). Their individual Fourier transformation yields frequency pairs, which, if added, eliminate one frequency. Consequently, in comparison to single detection the Nyquist frequency

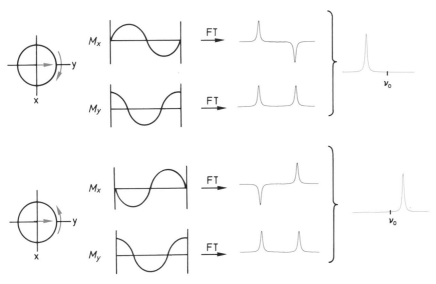

Figure 7.30 The principle of quadrature detection; (a) For a vector M which rotates clockwise, we detect on the x-axis a positive sine and on the y-axis the positive cosine; (b) if M rotates anti-clockwise, we detect on the x-axis a negative sine and on the y-axis again a positive cosine; (c) result of the Fourier transformation of the signals and addition of the frequency spectra (after Ref. 1)

during quadrature detection is given by half of the sweep width and the dwell time is $1/2SW$ instead of $1/SW$ (Figure 7.21).

Instrumental imperfections can lead to situation where signal selection by quadrature detection is not completely successful. So-called *quad-images* are then observed at $+v$ or $-v$ which can, however, be recognized by their different phase properties. Their intensity seldom exceeds 1% of the true line. Therefore, this problem is only important with intensive signals, for example those of solvents. In addition, hardware errors which lead to quad-images and a number of other artefacts can be removed or greatly attenuated by the method of *phase cycles*.

5.2.6   Phase cycles

As mentioned above, with modern n.m.r. spectrometers transmitter and detector phases can be varied and adjusted to the needs of the individual experiments. It is therefore possible, in particular by a certain choice for the receiver phase, to select signals of interest and suppress unwanted signals or artefacts. This is achieved by the use of *phase cycles*, the principle of which will be illustrated with two examples.

Many artefacts which result from inaccuracies of the experimental parameters can already be eliminated by a simple 180° transmitter *phase shift*. Such a variation allows us, for instance, to eliminate error signals which arise from an inaccurate pulse length. In an inversion recovery experiment, for example, an imperfect 180°_x pulse produces transverse magnetization along the $+y$-axis, which affects the result of T_1 measurement. If, however, in every second experiment a 180°_{-x} pulse is used, in other words the transmitter phase is shifted by 180°, $-y$-magnetization results and the disturbing transverse magnetization is eliminated by adding the individual experiments.

An elimination of the artefact of quad-images, mentioned above, results if the signals of the two receiver channels are interchanged in order to improve the balance. This is achieved by a 90 ° phase shift of the transmitter, which leads to a situation where the cosine signal of channel 1 and the sine signal of channel 2 are detected (cf. Figure 7.30). An additional 180 ° phase shift leads to further improvements. The complete phase cycle, known as CYCLOPS, is composed of the following four individual experiments:

	transmitter phase	code	receiver signal channel 1 M_x	channel 2 M_y	memory locations A	B
(1)	x	0	$\sin \omega t$	$\cos \omega t$	M_x	M_y
(2)	y	1	$-\cos \omega t$	$\sin \omega t$	M_y	$-M_x$
(3)	$-x$	2	$-\sin \omega t$	$-\cos \omega t$	$-M_x$	$-M_y$
(4)	$-y$	3	$\cos \omega t$	$-\sin \omega t$	$-M_y$	M_x
					$4 \sin \omega t$	$4 \cos \omega t$

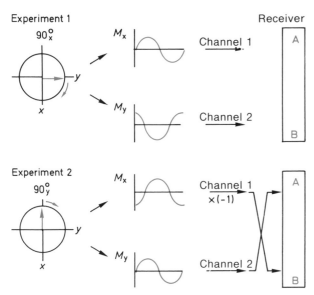

Figure 7.31 Transmitter and receiver phase during the CYCLOPS phase cycle for experiments (1) and (2) of the Table on page 254 opposite (after Ref. 1)

The phases of the receiver channels are fixed on the $+x$- and $+y$-axis, respectively. Because of the phase shift for the transmitter, the signals have to be stored in such a way that signal addition results. This necessitates multiplication by (-1) in four cases. As one sees, the resulting signal always contains two components which have passed through channel 1 and two which have passed through channel 2. Any existing imbalance of the two channels is thereby reduced (Figure 7.31).

5.3   COMPLICATIONS IN FT NMR SPECTROSCOPY

From the foregoing discussion it appears that a number of complications and perhaps disadvantages are typical of the FT method. One of the most common is known as *frequency folding*, which results from an improper choice of the spectral width with regard to the actual spectrum investigated. If there is a signal outside this range at a higher frequency $\Delta v + \delta v$, following the Nyquist theorem, it cannot be recognized. It can be shown, however, that the data points resulting from this frequency are treated by the computer as if they belonged to a frequency $\Delta v - \delta v$. Hence the term "folding", since transformation faithfully produces a signal at $\Delta v - \delta v$ in the frequency domain. This is illustrated in Figure 7.32.

With normal FT n.m.r., unlike the CW technique, it is therefore in general difficult, if not impossible, to investigate smaller spectral regions separately. In such cases the possibility of frequency folding has to be considered. In addition, and again contrary to the CW experiment, the elimination of strong solvent lines presents a

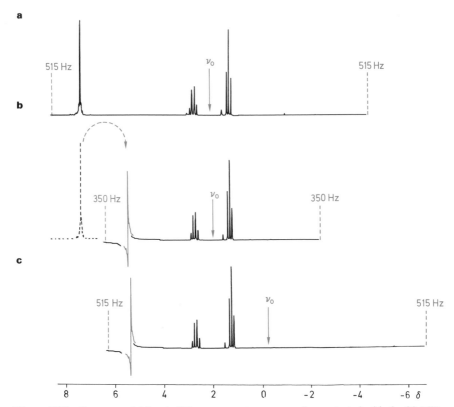

Figure 7.32   Frequency folding in FT n.m.r. spectroscopy as demonstrated with the 80 MHz ^{1}H n.m.r. spectrum of ethylbenzene observed with quadrature detection; (a) Nyquist frequency $+515$ Hz; (b) Nyquist frequency $+350$ Hz, spectral window too small; (c) Nyquist frequency as for (a), however, wrong choice of the transmitter position. As in the cases shown, the folded signals can be recognized mostly by the phase error which remains after phase correction. This criterion is, however, not always valid, especially in ^{13}C n.m.r. spectroscopy

problem. In particular, strong signals may cause a storage overflow during data accumulation. Special techniques are therefore necessary to deal with these shortcomings.

The phenomenon of frequency folding is also important with respect to the noise present at higher frequencies outside the spectral range. To prevent a fold-back of these data that would add to the noise in the spectral range, *filtering* devices that attenuate these signals are employed.

Another aspect related to the spectral width and the computer capacity arises with FT n.m.r. of nuclei covering a large chemical shift range (^{19}F, ^{31}P). In addition to the pulse power, the sampling rate must be high in these cases and the acquisition time may be severely limited by the available computer space. Distorted signals may then result from a truncated free-induction decay (see Figure 7.33, p. 259). Furthermore, the digital resolution in the frequency domain decreases.

In the case of a small Δv, on the other hand, provided that there is no frequency folding, the digital resolution improves. At the same time, the acquisition time increases and the conditions for data accumulation become less favourable. Thus in FT n.m.r. high sensitivity can be achieved only at the expense of resolution and *vice versa*.

A point of special interest that should be included here concerns *intensity measurements* in FT spectra. Errors can arise from various source, such as low pulse power or a pulse sequence with insufficient delay times. In the first case, the power distribution of the B_1 field over the spectral range varies and different flip angles for the individual resonance signals result. Since the magnitude of the induced transverse magnetization is a function of the flip angle, the intensities become distorted. If, on the other hand, the pulse sequence is too fast, nuclei with high T_1 values suffer from incomplete recovery of their z-magnetization and their intensities are systematically recorded too low. In order to avoid such shortcomings the experimental settings must be carefully checked. In general, then, the determination of correct integrals requires greater care in FT n.m.r. than with the older CW technique.

Problems with intensive solvent signals with respect to the dynamic range (see page 245) are met today in ^{1}H FT n.m.r. spectroscopy only if non-deuterated solvents have to be used. This is the case, for example, during measurements on biochemically or biologically interesting samples which have to be performed in H_2O instead of D_2O. From the various techniques for the suppression of the water signal, among them a number of pulse sequences which we cannot discuss in detail here, two relatively simple methods have been found useful in practice: presaturation and the selective inversion recovery experiment (page 235). In the first case the H_2O resonance is saturated just before the measurement of the spectrum of interest, and in the second case the remaining spectrum is excited when the H_2O signal just passes through zero intensity.

5.4 DATA IMPROVEMENT

FT n.m.r. spectroscopy provides several techniques which can be used to improve the experimental results in order to achieve better resolution or a more favourable signal-to-noise ratio. For this purpose a number of mathematical operations can be applied to the time domain data before Fourier transformation.

It turns out, for example, that after half of the acquisition time, which is usually of the order of 1 s, most of the signals have decayed to zero owing to transverse relaxation. The free-induction decay then primarily contains noise. This can be drastically reduced if each data point of the FID is multiplied by an exponential function $\exp(-jTC/N)$ where TC is an empirical time constant, j the number of the particular data point, and N the total number of data points. This corresponds to a multiplication of the time signal with the function

$$F_E = \exp\left(-\frac{t}{a}\right) \qquad (7.25)$$

As the exponent indicates, the higher data points are more strongly affected, thus reducing the noise. The effective transverse relaxation rate, $1/T_2^*$, which determines the time dependence of the FID signal and the half width of the n.m.r. signal in the frequency domain (see Figure 7.12 and equation (7.13), is then governed by the following relation:

$$(1/T_2^*)' = 1/T_2^* + 1/a \qquad (7.26)$$

Of course, the apparently shorter T_2 leads to an artificial line broadening, but this can usually be accepted for the benefit of a better signal-to-noise ratio. Figure 7.33 (p. 259) illustrates the result obtained by this procedure. The best compromise between reduction of signal-to-noise ratio and line broadening is generally obtained if we choose a $= T_2^*$. In this case the time constant in equation 7.25 is equal to the decay time of the FID.

In ^1H n.m.r. spectroscopy the procedure described is less important due to the high sensitivity available, especially since information about small spin–spin couplings may be lost through line broadening effects. In the field of n.m.r. spectroscopy of insensitive and rare nuclei, as for example ^{13}C, the method is, however, of great importance in order to detect signals of low intensity (Figure 7.33d).

In some cases, multiplication of the FID by an exponential function with positive exponent, $\exp(t/a)$, may be desirable in order to improve the resolution at the

---

Figure 7.33 Aspects of data treatment in FT n.m.r. spectroscopy; the time and frequency domain signals of the ^1H AMX spinsystem perylene dianion are used for demonstration in the examples (a)–(f); (a) time and frequency domain signal with 8K data points; (b) influence of exponential multiplication; line broadening 1 Hz. The time domain signal is artificially damped, the signal-to-noise ratio is improved, small signal splittings are lost; (c) Lorentz-Gauss transformation for resolution enhancement; (d) truncated time domain signal; the base line shows oscillations; (e) signal distortions in the frequency domain as a consequence of insufficient number of data points (2K data points, digital resolution 0.78 Hz/point); (f) improvement of spectrum (e) by zero filling to 16K data points (digital resolution 0.09 Hz/point); the density of data points can be observed at the monitor and the possibility to improve the digital resolution by zero filling may thus be checked; (g) line shape changes after extreme Lorentz–Gauss transformations using equation 7.27 broadened with the example of a singlet and a triplet: (1) without data treatment; (2) a = 0.318, b = 0.309; (3) a = 0.159, b = 1.41; (4) a = 0.106, b = 0.252; the a-values correspond to a decrease in line width of -1.0, -2.0 and -3.0 Hz (cf. equation 7.27); (h) detection of two broadened ^{13}C singlets of low intensity by exponential multiplication using equation (7.25): (1) without data treatment; (2) a = 0.0637; (3) a = 0.0318; the a-values correspond to line broadening effects of 5 and 10 Hz, respectively

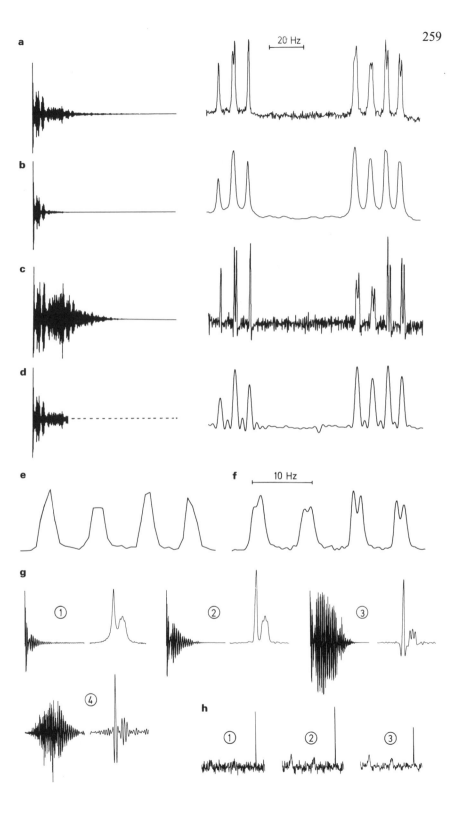

expense of the signal-to-noise ratio. The most effective function for resolution enhancement is the product function

$$F_G = \exp\left(\frac{t}{a}\right) \exp\left(-\frac{t^2}{b}\right) \qquad (7.27)$$

for which the parameters a and b are determined empirically. The first term corresponds to resolution enhancement by exponential multiplication as discussed above, the second term corresponds to a Lorentz–Gauss transformation of the signal. This leads to narrow lines which can show artefacts, however, at the basis. A disadvantage is that small signals may be lost in the noise and the peculiar signal shape prevents the correct integration of the spectrum (Figure 7.33c,g).

As a result of a data treatment by application of equation (7.27), certain regions of the time domain signals are enhanced or attenuated (Figure 7.33). The functions used for this purpose are, therefore, also called *window* or *weighting* functions. They also play an important rôle in two-dimensional n.m.r. spectroscopy (see Chapter 8).

Resolution enhancement may also be effected in the frequency domain if enough computer space is available to add several memory blocks of zeros to the free induction decay (zero-filling). This provides a larger number of data points in the frequency domain and allows for a better reproduction of the signals without affecting the acquisition time. This is especially important for accurate intensity measurements as well as for resolution enhancement with the Lorentz–Gauss transformation.

6. Double resonance experiments

We shall not conclude this chapter without describing an important group of experiments which are known as *double resonance experiments*. Such experiments may differ with respect to their specific application, however, a general and typical feature of all double resonance experiments is that besides the transmitter B_1 a second r.f. field B_2 is applied to the spin system. The results of such experiments can vary widely, depending upon the frequency and amplitude of B_2.

6.1 HOMONUCLEAR SPIN DECOUPLING

The most important area for the application of double resonance experiments is *spin decoupling* which was already mentioned shortly in Chapter 2. It leads to a simplification of n.m.r. spectra by removing the multiplet structure of individual resonances which is due to scalar spin–spin interactions. This facilitates spectral interpretations and yields at the same time, on the basis of the relations between the magnitude of homonuclear $^1H,^1H$ spin–spin coupling constants and chemical structure, as discussed in detail in Chapter 4, an assignment for individual resonances. One thus obtains information about neighbouring nuclei and bonding pathways in the particular molecule.

In a spin decoupling experiment the transmitter B_2 of frequency v_2 is positioned with sufficient power at the resonance of a certain nucleus while the spectrum is observed with the transmitter B_1. In a CW experiment B_2 is produced by modulation of the CW transmitter and the spectrum is observed with a B_1 frequency sweep. In the FT experiment the same result is produced by a time-sharing procedure. It takes advantage of the fact that for recording a data point of the free induction decay only a fraction of the dwell time is needed. The receiver and the decoupler channel can thus be synchronized in such a way that decoupling is applied between data collection. Because of the short dwell time interval, which is in the μs region, the B_2 field is practically pulse modulated (cf. p. 238). A centre band at v_2 and side bands in a distance $1/t_{dw}$ result because the relation $t_r = t_{dw}$ holds. For $t_{dw} \rightarrow 0$ the experiment is identical to continuous wave decoupling. In both cases the result is the same: all line splittings which are due to couplings with the irradiated nucleus are removed.

The basis for a theoretical analysis of double resonance spectra is the modified Hamilton operator, equation (7.28). In a coordinate system which rotates with frequency $\omega_2 = 2\pi v_2$, it has the form:

$$\hat{H} = \sum_i (v_i - v_2)\hat{I}_z(i) + \sum_{i<j}\sum J_{ij}\hat{I}(i)\hat{I}(j) - \sum_i \gamma_i B_2 \hat{I}_x(i) \tag{7.28}$$

In contrast to the familiar expression (5.10), the resonance frequencies of the nuclei i are fixed relative to the frequency of the field B_2. The sum $\Sigma \gamma_i B_2 I_x(i)$ describes the interaction of the B_2 field with the spin system which leads to a mixing of eigenstates with different total spin. It is of the order of magnitude of the spin–spin coupling and is therefore, in contrast to the effect of the weak field B_1, included in the Hamiltonian operator.

As in the case of the simple resonance spectrum the frequencies and the intensities of the transitions must now be determined with the help of equation (7.28) and the basis functions for the corresponding spin systems. We shall not go through such a calculation here but merely cite the most important result for the case of an AX system. If the frequency v_2 of the second r.f. field is equal to the Larmor frequency of the A nucleus and if for the strength of B_2 the condition $\gamma B_2 > 2J_{AX}$ holds, the X resonance is observed as a singlet. Figure 7.34 shows the experimental verification for this statement with a double resonance experiment for the AX system of the methylene protons of the iron carbonyl complex of o-xylylene [1,2-bis(methylene)-cyclohexadiene]. Here the X resonance is observed as a function of the parameter $v_A - v_2$. When $v_A - v_2 = 0$ a singlet is observed, i.e. the spin–spin coupling between the A and X nuclei vanishes. The significance of the double resonance technique thus becomes immediately obvious: with the aid of a second field spin–spin splittings in complicated spectra can be intentionally eliminated. This leads to a simplification of the spectrum that facilitates its analysis. Furthermore, the proton sequence in a molecule can be determined from double resonance experiments in conjunction with the empirical correlations between stereochemistry and spin–spin coupling (Chapter 4). In Chapter 8 we introduce the alternative two-dimensional experiment which serves the same purpose.

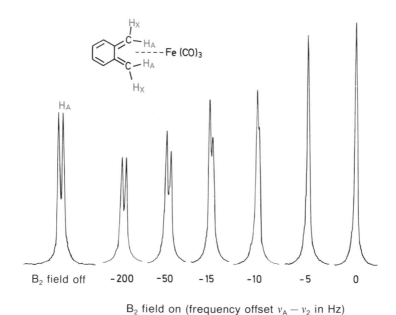

Figure 7.34 Dependence of the X resonance in an AX system on the frequency difference, $v_A - v_2$, between the A resonance and the B_2 field in a double resonance experiment

Without going into the details of a rigorous treatment, the results of a double resonance or spin-decoupling experiment can also be rationalized on the basis of the classical model for the n.m.r. experiment as was described above. For the case in which the frequency v_2 of the second field coincides with the Larmor frequency of the A nucleus $B_{eff} = B_2$, i.e. the vector μ_A precesses around B_2 and thus around the x'-axis. Consequently, μ_A is directed practically perpendicular to the vector μ_X. The nuclear spin vectors $I(X)$ and $I(A)$ are then quantized along the z- and the x'-axis, respectively. They are thus orthogonal and their scalar product, i.e. the scalar spin–spin coupling according to equation (2.8), vanishes.

One further example from a spectrum of a natural product may suffice to demonstrate the importance of spin decoupling experiments for structural investigations mentioned above. Figure 7.35 shows the 100 MHz ^{1}H n.m.r. spectrum of *helenaline*, a Mexican natural product from the family of *Tribus helenicae* with sesquiterpene lactone structure, where the resonances of the individual protons are nicely resolved. Aside from four groups of signals in the olefinic region (δ5.8–7.7), two absorptions of CH$-$OH protons (δ4.45 and 5.5), and two protons in allylic position (δ3.1, 3.6) can clearly be recognized (Figure 7.35a). In the first double resonance experiment (Figure 7.35b) irradiation is applied in the multiplet at 3.6 ppm (H^7) with the result that the couplings to the neighbouring CH$-$OH protons H^6 and H^8 disappear. Because H^8 must yield the multiplet with the stronger splitting

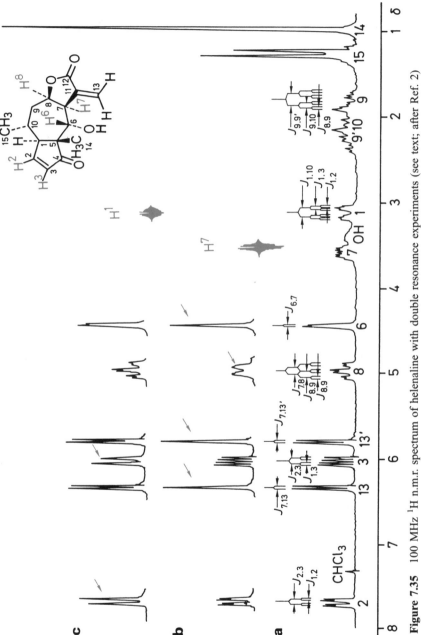

Figure 7.35 100 MHz ^{1}H n.m.r. spectrum of helenaline with double resonance experiments (see text; after Ref. 2)

(neighbourhood of a CH_2 group) and in addition shows the larger coupling at H^7, the β-lactone ring must be *cis*-fused ($J_{7,8} = 8.0$ Hz). On the other hand, $J_{6,7}$ amounts to only 1.7 Hz. The experiment also eliminates the allylic coupling of ca. 3.0 Hz to H^{13} and $H^{13'}$ which, because of the favourable conformation, is larger than the geminal coupling $J_{13,13'}$.

It is important to note in this discussion that such assignments of n.m.r. signals with the aid of double resonance experiments cannot be made in the absolute sense in most cases. With these experiments one can only determine which proton or which group of protons couples with the nucleus that is being irradiated. Only the consideration of the empirical relations between stereochemistry and the magnitude of the coupling constants will allow a definite decision as to which position the nucleus in question occupies in the proposed structure.

In addition to the application of double resonance experiments for the simplification of spectra and for the assignment of resonance frequencies, it is significant that by means of such experiments the resonance frequencies of nuclei can be determined even when their signals cannot be identified because of superposition with the absorption lines of other nuclei. For example, if one observes in the olefinic region the A portion of an AX system, the X resonance of which is masked by the multiplet absorption of a series of methylene protons, selective irradiation with B_2 in the CH_2 region allows one to determine the frequency difference $\nu_A - \nu_X$ and from that the chemical shift for H_X. We must point out, however, that in the case of a large B_2 amplitude and a small frequency difference this method may lead to erroneous results, since according to the rigorous theory for double resonance the relation

$$\nu'_A = \nu_A + \frac{\gamma^2 B_2^2}{8\pi^2(\nu_A - \nu_2)} \approx \nu_A + \frac{\gamma B_2^2}{8\pi^2(\nu_A - \nu_X)} \qquad (7.29)$$

holds for the A nucleus. The shift $\nu_A \rightarrow \nu'_A$ is called the *Bloch–Siegert shift*.

6.2 HETERONUCLEAR DOUBLE RESONANCE

The previously described variety of double resonance techniques is carried out on nuclei of a single type and therefore the term *homonuclear* double resonance is applied to them. The extension to different nuclei leads to *heteronuclear* double resonance, which differs from the homonuclear technique only in that the frequency difference $\nu_2 - \nu_1$ lies in the megahertz range. The second field B_2 is most advantageously produced by a separate transmitter such as a quartz-controlled frequency synthesizer. Such experiments are useful for the simplification of spectra that are complicated by heteronuclear spin–spin coupling such as $J(^1H,^{19}F)$ or $J(^1H,^{31}P)$. The line broadening caused by ^{14}N (cf. p. 387) can also be eliminated (Figure 7.36) so that the spectra can be analysed more easily. In the consideration of a specific pair of nuclei, e.g. ^1H and ^{19}F, a notation has been developed to indicate

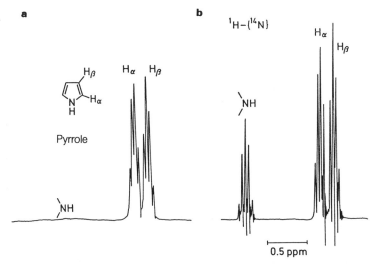

Figure 7.36 ^{1}H n.m.r. spectrum of pyrrole: (a) without and (b) with simultaneous irradiation of the ^{14}N nucleus (Joel-Kontron, Technical Bulletin)

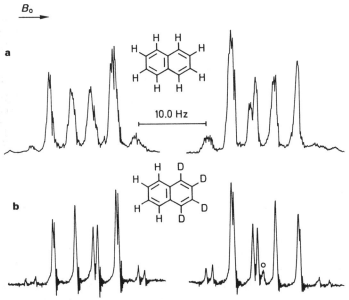

Figure 7.37 ^{1}H n.m.r. spectra of (a) naphthalene and (b) 1,2,3,4-tetradeuteronaphthalene with deuterium decoupling; 100 MHz (after Ref. 3)

which nucleus is observed and which is decoupled. The case in which fluorine is irradiated and the proton resonance is observed would be represented as $^1H\{^{19}F\}$.

Double resonance experiments of the type $^1H\{^2H\}$ (or $^1H\{D\}$ in connection with the synthesis of partially deuterated compounds are an important aid in the simplification of proton resonance spectra. For example, the eight-proton system of naphthalene can be transformed into an AA'BB' system if one synthesizes 1,2,3,4-tetradeuteronaphthalene and performs a $^1H\{D\}$ double resonance experiment (Figure 7.37).

6.3   BROADBAND DECOUPLING

6.3.1   Broadband decoupling by CW modulation

The technique of *broadband decoupling* represents an important development in the field of heteronuclear double resonance. The limits of the conventional heteronuclear decoupling technique are met very soon if it proves necessary to irradiate a larger spectral region. Since the amplitude of the B_2 field cannot be increased indefinitely, its application is confined to relatively small parts of the spectrum. In the case where the resonances of the heteronucleus extend over a range of many parts per million, as, for example, in the case of ^{19}F, complete decoupling would be impossible. The method of choice then becomes broadband decoupling. Here modulation techniques of various kinds are used to effectively produce a frequency band that extends over several kilohertz and covers the whole spectral area of the nucleus to be decoupled. Most widely known is *noise decoupling*, in which a noise generator is employed to produce the desired effect. Phase modulation techniques can be used just as effectively. The power of the method is demonstrated in Figure 7.38 for an $^{19}F\{^1H\}$ experiment. In comparison with the standard procedure, because of the width of the frequency band, this experiment is far less sensitive to the correct choice of ω_2. In order to achieve complete spin decoupling, however, it is important that the power of the B_2 field is large enough to encompass the complete spectral region of the particular nucleus.

Broadband decoupling now finds its application routinely in ^{13}C *n.m.r. spectroscopy* (see also page 465), where ^{13}C n.m.r. spectra are measured with broadband ^1H decoupling. An additional bonus of this experiment is an intensity enhancement for the ^{13}C resonances that has two sources. Firstly, the collapse of multiplet structures into singlets improves the signal-to-noise ratio in a straightforward manner. Secondly, the ^1H decoupling is accompanied by an intensity increase for the ^{13}C resonance that amounts to 200%. The reason for this is the so-called nuclear Overhauser effect which will be discussed in detail in Chapter 10. The effect of ^1H-broadband decoupling is demonstrated in Figure 7.39 with the example of the ^{13}C n.m.r. spectrum of αD-glucose.

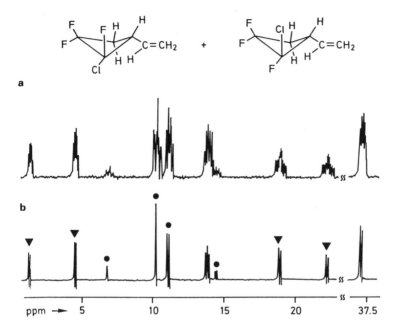

Figure 7.38 ^{19}F resonance spectra of a mixture of two isomeric cyclobutane derivatives: (a) normal spectrum; (b) with proton-noise decoupling (after Ref. 4). The 'AB systems' of the CF$_2$ groups in the two isomers are denoted with different symbols

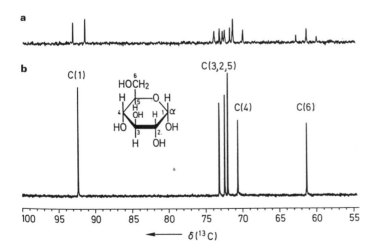

Figure 7.39 100 MHz ^{13}C n.m.r. spectrum of α-D-glucose in DMSO-d$_6$ without (a) and with (b) ^{1}H decoupling

6.3.2 Broad band decoupling by pulse methods

As an unwanted side effect, the application of strong r.f. fields for spin decoupling can lead to a considerable temperature rise in the sample, especially if high power is necessary in order to decouple a large frequency range. Sample cooling is a straightforward measure to prevent uncontrolled temperature changes, but alternative decoupler methods are available that are less critical in this respect. A breakthrough was possible in recent years by the introducing of spin decoupling techniques which are based on the spin echo experiment. As shown in Figure 7.40, the evolution of transverse X magnetization observed in the case of a heteronuclear scalar coupled AX spin system, for example ^{1}H,^{13}C, can be reversed by a 180°(A) pulse at time τ. Digital detection of the A-magnetization at times 2τ, 4τ, 6τ etc. yields data for a decoupled X-resonance. While the two doublet components still possess different Larmor frequencies $\pm J/2$, the transmitter does not recognize this difference since the magnetization is just measured when both vectors are superimposed.

The important practical aspect of this spin flip method, called MLEV after its inventor Malcolm Levitt, is the fact that the frequency range for the decoupled spin is three times as large as in experiments with conventional modulation techniques (for example noise modulation) (Figure 7.40b). Broadband decoupling of nuclei with large chemical shift ranges, as ^{19}F or ^{31}P, thus becomes possible. Further

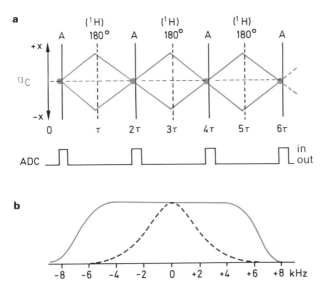

Figure 7.40 (a) Time dependence of transverse X-magnetization of an AX system with application of a train of 180°(A) pulses. The gating of the analogue to digital converter (ADC) shows the data acquisition; (b) comparison of the band width of the decoupler field during conventional modulation technique (– – – –) and pulse decoupling (————)

developments of this technique on the basis of composite pulses which greatly eliminate pulse errors have considerably improved the results. The methods used today are known under the acronyms WALTZ-16 and GARP.

6.3.3. Off-resonance decoupling

In addition to the gain in sensitivity discussed above, several techniques of heteronuclear spin–spin decoupling provide important experimental possibilities for the assignment of n.m.r. signals of various heteronuclei. One of these methods shall be discussed here, again with an example from carbon-13 n.m.r. spectroscopy. It is known as *off-resonance decoupling*. As the name indicates, it is really a partial decoupling technique that uses a strong r.f. field in the ^{1}H n.m.r. region with a frequency v_2 just "off" the resonance signal irradiated. The important aspect of this experiment becomes clear immediately if we realize that in such a partial decoupling experiment the line splittings are still visible. Of course, these splittings are smaller than the coupling constants, but the typical multiplet structure of certain resonances is retained. In the case of ^{13}C, it turns out that all of the smaller ^{13}C,^{1}H coupling constants (geminal, vicinal, long range) are eliminated, whereas the line splittings caused by the larger one-bond coupling constants are visible. As a consequence, ^{13}C resonances in an off-resonance ^{1}H-decoupling experiment show first-order multiplicity and can thus be distinguished. Quartets, triplets, doublets, and singlets are observed for primary (CH$_3$), secondary (CH$_2$), tertiary (CH) and quaternary carbon atoms, respectively. An example is shown in Figure 7.41 (p. 270).

From the theory of off-resonance decoupling an approximate equation that relates the residual splitting J_R to the frequency offset Δv and the unperturbed coupling constant J_0 can be derived:

$$\Delta v = \gamma B_2 J_R (J_0^2 - J_R^2)^{-1/2} \qquad (7.30)$$

This equation was used earlier to interrelate ^{1}H and ^{13}C spectra of a certain compound. For this purpose, the ^{1}H decoupler is swept through the proton spectrum and the J_R values measured for individual ^{13}C resonances are plotted against the decoupler position. Extrapolation to $J_R = 0$, that is complete decoupling, yields the resonance frequencies of the ^{13}C attached protons.

An additional decoupling technique, known as a gated decoupling and most frequently applied in ^{13}C n.m.r., will be treated in Chapter 11.

6.4 DOUBLE RESONANCE EXPERIMENTS FOR SPECTRAL ANALYSIS

For the experiments discussed so far B_2 fields of relatively high amplitude had to be used in order to achieve the desired effect of spin decoupling. Different results are obtained with *weak* B_2 fields with amplitudes in the order of the half width of the spectral line, $\gamma B_2 \approx \Delta$. Typical experiments of this type are the *spin tickling*

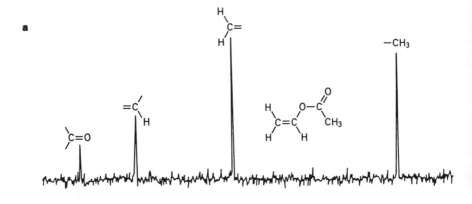

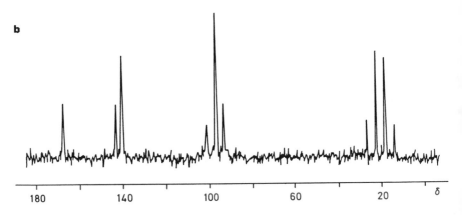

Figure 7.41 Off-resonance ^{1}H-decoupled ^{13}C-n.m.r. spectrum of vinyl acetate with resonances for the carbonyl-, methine-, methylene-, and methyl carbon at 167.2, 141.8, 96.8, and 20.2 ppm, respectively, relative to the ^{13}C resonance of TMS (not shown): (a) ^{1}H-decoupled; (b) off-resonance ^{1}H-decoupled; frequency offset 3 kHz

experiment, INDOR-spectroscopy (INDOR = internuclear double resonance) or the method of *Torrey oscillations*. They are used to test an energy level diagram of a spin system experimentally and to determine the relative signals of coupling constants in first-order spectra. In particular one can distinguish between progressively and regressively connected transitions. These differ in so far that the three eigenvalues characterizing connected transitions change their energy steadily (progressive connection) or that the central eigenvalue is larger or smaller than the eigenvalue of the initial and the final state (regressive connection):

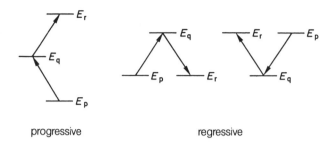

progressive   regressive

As many homonuclear experiments with selective excitation, these methods are typical CW-techniques which have lost their importance after the introduction of the pulse Fourier transform technique and the information they yield is today more easily obtained from two-dimensional experiments (see Chapter 8). We, therefore, forego a detailed discussion of these techniques and mention only that spin tickling experiments can also be performed with the pulse Fourier transform technique. The interpretation of the results obtained with this type of experiments is facilitated by the use of difference spectroscopy (cf. Chapter 10).

7. References

1. A. E. Derome, see below.
2. P. Joseph-Nathan, *Resonancia Magnetica Nuclear de Hidrogeno*, OEA, Monograph Nr. 9, Chemistry Series, Washington, 1973; P. Joseph-Nathan and E. Diaz, *Rev. Latinoamer. Quim.*, **2**, 54 (1971).
3. J. B. Pawliczek and H. Günther, *Tetrahedron*, **26**, 1755 (1970).
4. R. R. Ernst, *J. Chem. Phys.*, **45**, 3845 (1966).

RECOMMENDED READING

R. K. Harris, *Nuclear Magnetic Resonance Spectroscopy*, Longman, Harlow, 1986.

A. E. Derome, *Modern NMR Techniques for Chemistry Research*, Pergamon Press, Oxford, 1987.

D. Shaw, *Fourier Transform NMR Spectroscopy*, Elsevier, Amsterdam, 1984.

R. Freeman, *A Handbook of Nuclear Magnetic Resonance*, Longman, Harlow, 1987.

M. L. Martin, J.-J. Delpuech and G. J. Martin, *Practical NMR–Spectroscopy*, Heyden, London, 1980.

E. Fukushima and S. B. W. Roeder, *Experimental Pulse NMR: A Nut and Bolts Approach*, Addison-Wesley, London, 1981.

Review articles

(a) A. G. Marshall and M. B. Comisarow, Fourier Transform Methods in Spectroscopy, *J. Chem. Educ.*, **52**, 638 (1975).
(b) J. W. Cooper, The Computer in Fourier-Transform NMR, in F6, **2**, 391 (1976).
(c) L. J. Schwartz, Step-by-Step Picture of Pulsed (Time Domain) NMR, *J. Chem. Educ.*, **65**, 752 (1988).

(d) R. W. King and K. R. Williams, The Fourier Transform in Chemistry, Part 1, 2, *J. Chem. Educ.,* **66**, A213, A243 (1989); **67**, A93 (1990).
(e) G. H. Weiss and J. A. Ferretti, The Optimal Design of Relaxation Time Experiments, in F5, **20**, 317 (1988).
(f) H. Kessler, S. Mronga and G. Gemmecker, Multi-Dimensional NMR Experiments Using Selective Pulses, *Magn. Reson. Chem.,* **29**, 527 (1991).
(g) R. Freeman, Selective Excitation in High-Resolution NMR, *Chem. Rev.,* **91** (1991).
(h) M. H. Levitt, Composite Pulses, in F5, **18**, 61 (1986).
(i) J. C. Lindon and A. G. Ferrige, Digitisation and Data Processing in Fourier Transform NMR, in F5, **14**, 27 (1982).
(j) R. A. Hoffmann and S. Forsén, High Resolution Nuclear Magnetic Double and Multiple Resonance, in F5, **1**, 15 (1966).
(k) W. v. Philipsborn, Methods and Applications of Nuclear Magnetic Double Resonance, *Angew. Chem.,* **83**, 470 (1970); *Angew. Chem. Int. Ed.,* **10**, 472 (1971).
(l) M. H. Levitt, R. Freeman and T. Frenkiel, Broadband Decoupling in High-Resolution NMR Spectroscopy, in F1, **11**, 48 (1983).
(m) A. J. Shaka and J. Keeler, Broadband Spin Decoupling in Isotropic Liquids, in F5, **19**, 47 (1987).
(n) M. Guéron, P. Plateau and M. Decorps, Solvent Signal Suppression in NMR, in F5, **23**, 135 (1991).

8 TWO-DIMENSIONAL NUCLEAR MAGNETIC RESONANCE SPECTROSCOPY

As already mentioned in the Introduction, no other development has influenced magnetic resonance spectroscopy in the last twenty years so profoundly as the concept of two-dimensional (2D) n.m.r. spectroscopy. This technique, which emerged from an idea of the Belgian physicist Jeener, has been developed for applications primarily in the laboratories of R. R. Ernst at the ETH Zürich and R. Freeman at the University of Oxford. Today it forms the basis for a large number of experiments in all branches of nuclear magnetic resonance. It induced not only major improvements for the determination of the important spectral parameters, but also paved the way for the discovery of completely new facets in the properties of spin systems. Because a detailed discussion of all theoretical and practical aspects of two-dimensional n.m.r. spectroscopy would extend far beyond the limits of the present text, we restrict ourselves to an introduction into the basic principles and a discussion of the most popular experiments. Therefore, the present chapter describes the principles of 2D n.m.r. and two basic 2D experiments, J,δ- and COSY spectroscopy. In Section 5, the simplest formalism for the theoretical interpretation of 2D spectroscopy is introduced. Further 2D methods, like the INADEQUATE experiment, exchange (EXSY) and nuclear Overhauser (NOESY) spectroscopy as well as heteronuclear shift correlations will be treated in Section 6 and Chapters 9, 10, and 11.

1. The principles of two-dimensional NMR spectroscopy

The 2D n.m.r. experiment is a domain of Fourier transform and pulse spectroscopy. It is basically characterized by three time intervals: *preparation, evolution* and *detection* (Figure 8.1a). In a number of 2D experiments a further interval is added before detection, the so-called *mixing time* (Figure 8.1b).

During the preparation time the spin system of interest is prepared for the experiment, for example by application of decoupler experiments or simply by the

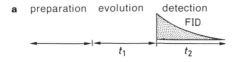

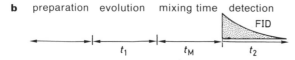

Figure 8.1 Time scales of 2D n.m.r. experiments

generation of transverse magnetization through a 90° pulse. In the evolution time t_1 it then develops under the influence of different factors, as for example Larmor precession or scalar spin–spin coupling, before a signal is detected during the detection time t_2.

The sequence described in Figure 8.1a does not yet constitute a 2D n.m.r. experiment. Only if in a series of experiments the sequence is repeated with a systematic variation of the evolution time t_1 by adding time increments Δt_1, and if after the first Fourier transformation of the resulting t_2 signals a number of one-dimensional spectra is obtained in the frequency domain F_2 which show a modulation in amplitude or phase, can a second Fourier transformation be applied. The data of these spectra are then transformed with respect to the time axis t_1. A frequency axis F_1 results which now contains the frequencies of those mechanisms which have been effective during the evolution time t_1 and which caused the observed modulation of signal amplitude or signal phase (Figure 8.2). If for instance spin–spin coupling was effective during t_1 and Larmor precession during t_2, F_1 contains coupling constants while the chemical shifts appear on the frequency axis F_2. Resonance frequencies and spin–spin coupling constants which are a priori indistinguishable in a conventional one-dimensional n.m.r. spectrum can thus be separated and presented on two distinct frequency axes.

Even if the ingenious idea of two-dimensional n.m.r. spectroscopy is basically simple, experience tells us that the principles are quite new and difficult to understand at the beginning. Therefore, the concise description given above will be supplemented by additional explanations.

Let us state again that a two-dimensional n.m.r. experiment is possible only after a series of n one-dimensional spectra has been measured. Thereby the evolution time

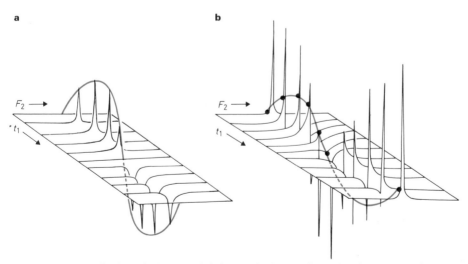

Figure 8.2 Amplitude and phase modulation (a, b) in one-dimensional spectrum after a systematic increase of the evolution time t_1

is systematically increased by adding time increments Δt_1: Normally, 32 1D experiments are used as a minimum, but 128 or 256 single 1D experiments are not uncommon. We shall see later what effect the number of t_1 increments on the final spectrum has.

The experimental data of such a series of 1D experiments are not individually Fourier-transformed but rather stored in the computer memory. They yield a data matrix which is characterized by two time axes: t_1 and t_2. The 2D spectrum is thus a function of two variables: $S(t_1, t_2)$. A first Fourier transformation with respect to t_2 yields $S(t_1, F_2)$. This function must be seen as a series of one dimensional spectra, the signals of which are modulated with respect to their amplitude or phase (Figure 8.2). There exists a periodical behaviour along t_1, characterized by a frequency which can be obtained through a second Fourier transformation with respect to the time axis t_1. The final 2D spectrum is thus a function of two frequencies variables: $S(F_1, F_2)$. These relations are again illustrated in Figure 8.3.

Only the detection time is a real time axis in a 2D experiment, that is, real FID signals are only detected in t_2. In contrast, the FID for the Fourier transformation along t_1 is constructed point by point. Therefore, the t_1 increments Δt_1 determine the Nyquist frequency in the F_1 domain. Thus a value of 1 ms for Δt_1 means, for example, that along t_1 frequencies of 1 kHz can be recognized if quadrature detection can be used in t_1. The t_1 increment Δt_1 thus corresponds to the dwell time of the t_1 dimension. Quadrature detection in t_1 can be achieved through certain phase cycles, which yield a phase shift of 90° for the receiver signal (see page 252). Two experiments must then be performed for each t_1 increment.

From our description of two-dimensional n.m.r. spectroscopy one can easily conclude that normally long measuring times are required. This is also due to the fact that nearly always 90° excitation pulses are used. Therefore, after each individual t_1 experiment a relaxation delay is necessary in order to provide uniform starting conditions. Furthermore, data accumulation has to be used also in 2D n.m.r. experiments in order to achieve sufficient sensitivity. Finally, the mass of data which has to be mathematically processed is much larger than in the one-dimensional experiment. Without modern computers, especially efficient data storage capacities, two-dimensional n.m.r. spectroscopy would not be practicable.

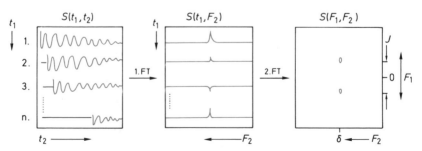

Figure 8.3 Data flow in a 2D n.m.r. experiment

1.1 GRAPHICAL PRESENTATION OF TWO-DIMENSIONAL NMR SPECTRA

For the presentation of the results of a two-dimensional n.m.r. experiment several choices are available. The total spectrum can be reproduced either in a sort of perspective diagram (3D or *stacked plot*) as shown in Figure 8.4a and already known from T_1 experiments (Figure 7.15, p. 237) or as *contour diagram* as shown in Figure 8.4b. The latter is better suited for spectral analysis.

Besides the two representations mentioned one can plot the rows or columns of the two-dimensional data matrix $S(F_1, F_2)$ as pseudo 1D spectra. This is especially advisable if signals of smaller intensity are to be detected, which often disappear in the noise of the contour diagram. Also projections of the 2D spectrum on the frequency axes F_1 or F_2 are possible. We shall come back to these aspects during the discussion of the individual experimental methods. We should mention already here, however, that the separation of the real and imaginary part of the spectrum after Fourier transformation, which is necessary to derive phase sensitive spectra, is by no means trivial in 2D n.m.r. spectroscopy. Quite often, therefore, 2D signals are

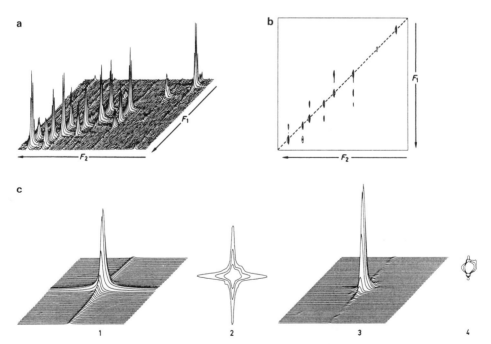

Figure 8.4 (a) 2D n.m.r. spectrum as stacked plot; (b) the same spectrum as contour plot; (c) stacked and contour diagram of the line shape of a 2D n.m.r. singlet in magnitude calculation (1, 2) and after data treatment with a sine function (3, 4)

displayed in the so-called *magnitude* or *absolute value* representation, which uses the square root of the sum of the squares of real and imaginary part:

$$M = \sqrt{Re^2 + Im^2} \qquad (8.1)$$

This representation has, however, the disadvantage that the n.m.r. signals are strongly broadened near the baseline in comparison to the Lorentzian signals which result from the real part of the Fourier transform data. Signals of low intensity which are close to intensive signals are thus often lost. It is therefore important that techniques which allow us to record *phase sensitive* 2D spectra have been introduced and phase sensitive representations of 2D n.m.r. spectra are today generally preferred.

A major practical aspect of 2D n.m.r. spectroscopy is data processing by the application of *filter* or *weighting functions* which improve the n.m.r. line shape (see Chapter 7, p. 257). Most widely used are sine functions of the form $\sin(\pi t/t_{max})$ or $\sin^2(\pi t/t_{max})$ (sine-bell, sine-bell squared) which have their maximum at $t = t_{max}/2$. In this way the dispersive parts of the 2D signals are suppressed (Figure 8.4c, p. 277). A wrong selection of the maxima of such functions can, however, lead to the situation that broad signals are eliminated. This problem can be overcome if functions which have their maximum shifted to smaller values of t_2 (shifted sine bell, shifted sine bell squared) are used. Filter functions are indispensable if magnitude spectra are processed.

2. The spin echo experiment in modern NMR spectroscopy

Before discussing different techniques of 2D n.m.r. spectroscopy in detail, some aspects of the spin echo experiment, introduced in Chapter 7, will be discussed. This experiment has emerged as one of the important building blocks of 2D pulse sequences. In connection with our explanations we shall also treat a number of relations which are of general importance for the understanding of multi-pulse experiments.

2.1 THE TIME-DEPENDENCE OF TRANSVERSE MAGNETIZATION

The transverse magnetization $M_{x,y}$ generated in an FT n.m.r. experiment is subject to changes caused by different mechanisms. Because all modern one- and two-dimensional n.m.r. experiments incorporate an evolution time and thus require delayed data detection (see Figure 8.1, p. 274), it is illustrative to study the time-dependence of $M_{x,y}$ in more detail. During the evolution time magnitude and direction of $M_{x,y}$ are subject to the influence of four different factors:

(1) transverse relaxation,
(2) inhomogeneity of the magnetic field B_0,
(3) Larmor precession, and
(4) scalar spin–spin coupling.

Mechanisms (1) and (2) determine the line widths of the n.m.r. signals through the effective transverse relaxation time T_2^*, which was introduced in Chapter 7 (p. 233). In this respect, true transverse relaxation leads to an unavoidable signal loss. In practice, however, the inhomogeneity contribution to the effective transverse relaxation rate is by far the most important factor. This effect can be largely eliminated through the spin echo experiment, as discussed on p. 237. Spin echo spectra are therefore distinguished by narrow lines, since the influence of field inhomogeneity has been eliminated.

2.2 CHEMICAL SHIFTS AND SPIN–SPIN COUPLING CONSTANTS AND THE SPIN ECHO EXPERIMENT

Transverse relaxation and field inhomogeneity, factors (1) and (2) of our list above, can now be excluded from the following discussion of individual pulse sequences because they are not related to the spin physics which governs these experiments, which depends instead on Larmor precession and scalar spin–spin coupling of the nuclei involved.

After excitation of the spins, the transverse magnetization components, which correspond to the individual signals of different Larmor frequency in the n.m.r. spectrum of interest (Figure 8.5, p. 280), are separated by Larmor precession (factor 3). Delayed data detection then causes the phase error already discussed (p. 249). If time intervals in the order of ms are involved, as in many 2D experiments, the spin echo experiment is used for phase correction. Generally, with this technique all effects which arise from different Larmor frequencies can be refocussed. Figure 8.5 illustrates this application of the spin echo experiment with a simple example. On the other hand, transverse magnetization is in many cases affected by scalar spin–spin coupling with neighbouring nuclei of the same or of a different nuclide (homo- or heteronuclear spin–spin coupling). After excitation a separation of transverse magnetization into magnetization vectors sets in, which correspond to the individual components of spin multiplets (doublets, triplets, etc.). In a coordinate system which rotates with the Larmor frequency of the nucleus in question the situations for a doublet or triplet are illustrated in Figure 8.6a and 8.6b, respectively. Within the X approximation, in the case of the doublet the rotation of the vector A1 is induced by the X-nucleus with α-spin, that of vector A2 by the X-nucleus with β-spin.

If we now turn our attention to the spin echo experiment for scalar coupled spin systems, in the simplest case an AX system, different situations can be envisaged (Figure 8.7, p. 281).

For a homonuclear AX spin system the 180° pulse after the interval τ is not selective, it affects the A- as well as the X-nucleus. The A-vectors are rotated around the x-axis and at the same time their sense of rotation with respect to the z-axis is inverted, because the 180° pulse interchanges the spin states of the X-nucleus. Therefore, after the time 2τ refocusing is not observed (Figure 8.7a). Instead, both doublet components have a phase difference which depends on the angle ϕ and, therefore, on the time τ.

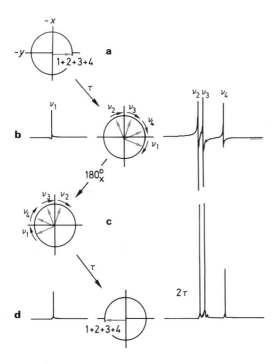

Figure 8.5 Phase correction by a spin echo experiment for four resonance lines of different Larmor frequencies: (a) situation after the excitation pulse; (b) after time τ, the individual components of the macroscopic magnetization have progressed differently in their precession around the field axis because of their different Larmor frequencies. Data detection yields a spectrum with frequency dependent phase errors; (c) a 180°_x pulse at time τ inverts the vectors and leads to a new superposition after time 2τ, the *spin echo*. The phase errors in the recorded spectrum disappear (d); (in order to observe an absorption spectrum, an additional phase correction of 180° has been applied at (d))

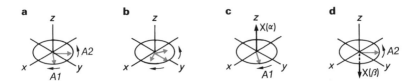

Figure 8.6 The effect of scalar spin–spin coupling J_{AX} on the transverse magnetization of the A-nucleus in the rotating coordinate system with $\omega_1 = \omega_A$, that is under "on-resonance" conditions for the A-nucleus; (a) AX-System; the effective Larmor frequencies of the vectors amount to $+J_{AX}/2$ and $-J_{AX}/2$; (b) AX_2-system; the relative frequencies of the vectors are $+J_{AX}$ and $-J_{AX}$; (c) rotation of A1 around the X(α)-spin; (d) rotation of A2 around the X(β)-spin

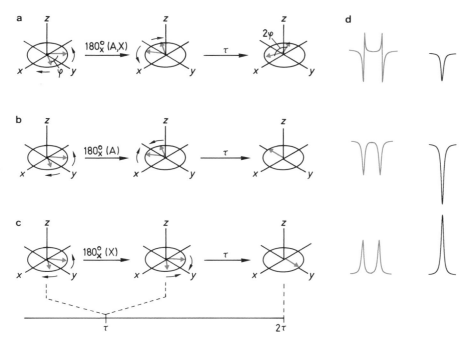

Figure 8.7 Spin echo experiments with an AX system; shown is the vector diagram for the A-nucleus;
(a) homonuclear case;
(b) heteronuclear case with a 180°(A) pulse;
(c) heteronuclear case with a 180°(X) pulse;
(d) spin echo signals without and with X-decoupling

In the heteronuclear case two variants are possible. If the A-nucleus is observed and the 180° pulse is applied in the A-region, refocusing of the A-vectors on the $-y$-axis is observed after a time 2τ (Figure 8.7b). For a 180° pulse in the X-region, however, Figure 8.7c is valid: the spin inversion for the X-nucleus interchanges the sense of rotation for the components of the A-doublet and refocusing after time 2τ is observed on the $+y$-axis. Finally, as a third alternative in the heteronuclear case an experiment can be performed where an A- and an X-pulse are applied simultaneously. The situation is then completely similar to the homonuclear case, as shown in Figure 8.7a.

If a detector on the $+y$-axis is now switched on at time 2τ, the spin echo signal of the A-nucleus is a doublet with anti-phase dispersion character in case (a). For case (b) we obtain an in-phase doublet with emission, in case (c) an in-phase doublet with absorption character. If X-decoupling is applied during A-detection, (a) yields a signal with an amplitude that depends on the angle φ that is on the time τ; (b) and (c) yield singlets in emission or absorption (Figure 8.7d).

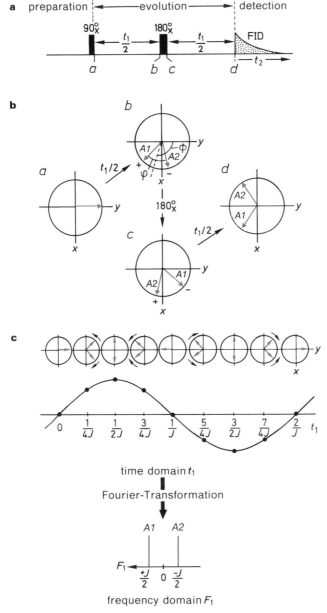

Figure 8.8 (a) Pulse sequence for two-dimensional spin echo spectroscopy (J,δ-spectroscopy); (b) vector diagram for the A nucleus of an AX system: after excitation transverse magnetization develops under the influence of Larmor frequencies and spin–spin coupling. The vectors **A1** and **A2** of the A-doublet move clockwise around the z-axis (Larmor precession, angle ϕ) and fan out (spin–spin coupling, effective Larmor frequencies $v(\mathbf{A1}) > v(\mathbf{A2})$, symbolized by $+$ and $-$, angle φ). The 180°_x pulse rotates the vectors around

Exercise 8.1   Draw the detected signal for the homonuclear spin echo experiment without and with X-decoupling, as shown in Figure 8.7a, for angles $\alpha = 90°$, $180°$ and $270°$.
Exercise 8.2   Derive in analogy to Figure 8.7 diagrams for spin echo experiments which use a $180°_y$ pulse.
Exercise 8.3   Redraw the diagram shown in Figure 8.7 for the 'off-resonance' situation with $v(A1) > v(A2) > v_0$.

3. Two-dimensional spin echo spectroscopy: the separation of the parameters δ and J

The first two-dimensional experiment which will be discussed in more detail provides us with the possibility to observe the two most important parameters of n.m.r. spectroscopy, resonance frequencies δ and scalar coupling constants J separately on two distinct frequency axes. In the resulting 2D n.m.r. spectrum the coupling appears only in the F_1 domain, while the chemical shift is confined to the F_2 domain. This result is achieved with the pulse sequence

$$90°_x - t_1/2 - 180°_x - t_1/2 - \text{FID}(t_2) \qquad (8.2)$$

which is illustrated in graphical form in Figure 8.8a. It starts with signal excitation by a $90°$ pulse, followed by the evolution time t_1, which is divided by a $180°_x$ pulse. Finally, the signal is detected in t_2.

For a simple two-spin system of the AX-type with resonance frequencies v_a and v_x and the scalar coupling J_{AX} it is possible to describe this experiment within the framework of the classical Bloch vector picture already used in Chapter 7 for the illustration of the effect of r.f. pulses on the macroscopic magnetization M (Figure 8.8b). The important new aspect, as compared to the simple FT n.m.r. experiment, is the evolution time t_1. During this time interval, transverse magnetization is affected by the mechanisms discussed in Section 2.

The pulse sequence described by equation 8.2 is nothing else than a spin echo experiment, and the rules developed above apply: defocusing of the transverse magnetization by field inhomogeneity and the effects of chemical shifts are eliminated; the effect of spin–spin coupling, however, remains. Thus, during the time interval t_1 — if we neglect true transverse relaxation — the spin system is subject only to the influence of scalar spin–spin coupling. For the vectors A1 and A2 of the

the x-axis and interchanges the spin states of the X-nucleus and as a consequence the effective Larmor frequencies for the vectorsA1 and A2. In the second half of the evolution time the effects of Larmor precession and of field inhomogeneity (not shown here) are eliminated and A1 and A2 are symmetrically oriented with respect to the y-axis. They show a phase error of $\pm \Delta\varphi$; (c) phase error $\Delta\varphi$ of one component of the A-doublet relative to the y-axis during the evolution time t_1 and second Fourier transformation

A-doublet a phase error results at the end of the evolution time which is dependent on t_1. The signals detected in a series of experiments with different t_1 values are, therefore, phase modulated. The modulation frequency is just half the frequency of the spin–spin coupling, $J_{AX}/2$, and can be determined by a second Fourier transformation of the data in the frequency domain F_1 (Figure 8.8c).

We now turn to the detection time t_2 and the frequency domain F_2 of the 2D spectrum. The receiver signal decays as usual with a time constant T_2^*. It contains, however, the resonance frequencies *and* the frequencies of the spin–spin couplings, because during t_2 both are effective. Therefore, our experiment so far did not eliminate the frequencies of the scalar coupling constants from the F_2 domain. After the second Fourier transformation the two-dimensional data matrix thus looks like the contour diagram shown in Figure 8.9a: the spin multiplets are aligned along diagonals, which are inclined with respect to F_2 by 45°. The elimination of the coupling information from F_2 is, however, achieved in a very simple manner. We just tilt the data matrix along the F_2 axis (Figure 8.9b). The projection of the 2D signals on the frequency axes F_1 and F_2 now yields spectra which contain only coupling or chemical shift information and the separation of the two parameters J and δ is complete (Figure 8.9c,d).

3.1   APPLICATIONS OF HOMONUCLEAR ^{1}H J,δ-SPECTROSCOPY

The n.m.r. experiment described above is known as *J-resolved* or *J,δ-spectroscopy*. Its main field of application is in the analysis of crowded spectra, where spin multiplets of different protons strongly overlap. These type of spectra are observed for mixtures of structurally related compounds or for alicyclic systems with numerous, strongly coupled CH_2 protons. J,δ-spectroscopy can then be used to separate the different spin multiplets, which paves the way for the determination of spin–spin coupling constants and chemical shifts.

An example for the analysis of a mixture is given in Figure 8.10 (p. 286) with a spectrum of *n*-butylbromide and *n*-butyliodide. In the 1D spectrum of the mixture only the triplets of the CH_2 groups adjacent to the halogen atoms are resolved at 400 MHz. In contrast, the contour diagram of the 2D spectrum shows all multiplets separated, even the triplets of the CH_3 groups, and the F_1 traces of the 2D data matrix yield well resolved line patterns ready for analysis.

An interesting aspect of 2D J,δ-spectra is the possibility to produce '^{1}H-

---

Figure 8.9   Data treatment during homonuclear J,δ-spectroscopy; (a) data matrix after twofold Fourier transformation; (b) data matrix after tilting along F_2; (c) comparison of a conventional 1D n.m.r. spectrum and a 2D J,δ-spectrum with separate frequency axes for chemical shifts and spin–spin coupling constants; (d) 400 MHz-^{1}H-n.m.r. spectra of *n*-butylbromide: 1D spectrum and 2D J,δ-spectrum as stacked and as contour plot; recording parameters: F_2/t_2: sweep width 1602.5 Hz, 4K data points, 32 scans, relaxation delay 11 s; F_1/T_1: $\Delta t_1/2 = 10$ ms $\hat{=}$ 25 Hz sweep width, 64 experiments $\hat{=}$ 64 data points, zero filling to 256 data points yielded a digital resolution of 0.2 Hz/Pt; total measuring time 8 h

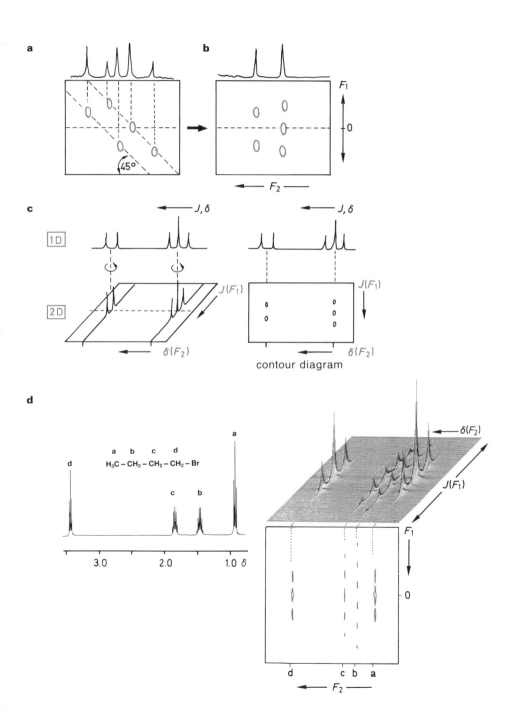

286

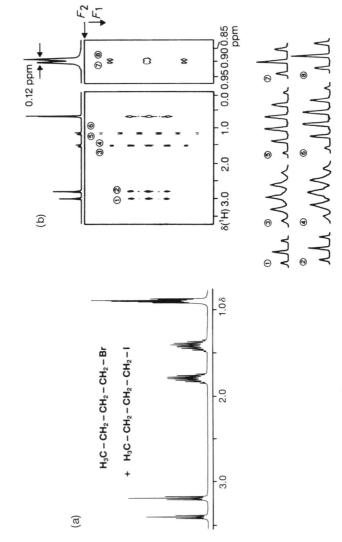

Figure 8.10 (a) 400 MHz ^{1}H 1D n.m.r. spectrum of a mixture of n-butyl bromide and *n*-butyl iodide; (b) contour diagram of the 2D-J,δ spectrum with F_2 projection and F_1 cross-section (methyl signals also on expanded scale; after Ref. 1)

decoupled' ^{1}H n.m.r. spectra by F_1 axes projections of the data (Figure 8.10b). This allows us to determine the chemical shifts of all ^{1}H resonances directly without multiplet analysis.

The most important applications of homonuclear J,δ-spectroscopy can be expected where strong multiplet overlap is observed for weakly coupled spin systems. This situation is met quite frequently with natural products or biomolecules which contain alicyclic partial structures with a large number of methylene groups. Sometimes severe spectral overlap is also met for stereoisomers, where the chemical shifts may differ only slightly. Besides E,Z-isomers, mixtures of diastereomers, as well as racemic or partially resolved optically active samples in chiral solvents (formation of diastereomeric solvation complexes) belong to this group. ^{1}H J,δ-spectroscopy can thus be a valuable analytical tool in the field of stereoselective synthesis.

Coupling constants to heteronuclei like ^{19}F or ^{31}P are treated in 2D ^{1}H J,δ-spectra like resonance frequencies, because the $180°_x$ ^{1}H pulse does not affect the spin states of these nuclei. The effect of such couplings on transverse proton magnetization is, therefore, eliminated at the end of the t_1 interval (see Figure 8.7b, p. 281). The line splitting due to ^{1}H,X couplings then appears on the F_2 axis, as shown in the spectrum of Figure 8.11c. This example also demonstrates the advantage of spin echo spectroscopy: small line widths through elimination of field inhomogeneity.

Singlet signals of solvents are not modulated by coupling and in the 2D J,δ experiment they appear on the shift axis with the coordinates $F_1 = 0$, F_2. Using special techniques, like signal saturation in the preparation time, they can be eliminated.

3.2 PRACTICAL ASPECTS OF J,δ-SPECTROSCOPY

Because ^{1}H,^{1}H spin–spin coupling constants are generally smaller than 15 Hz, the width of spin multiplets seldom exceeds 30 Hz. A spectral window of 50 or ± 25 Hz is, therefore, in most cases sufficient for the F_1 domain in J,δ-spectroscopy. If quadrature detection is used, the Nyquist frequency in F_1 then amounts to 25 Hz and the dwell time, that is the t_1 increment, is 20 ms. With 64 t_1 experiments a t_1(max) value of 1.28 s results, and the digital resolution in F_1 is 0.78 Hz/Pt. It can be improved by zero-filling before Fourier transformation. An improvement through a larger number of t_1 increments is not practicable, because the resulting increase in evolution time would strongly diminish the signal intensity in t_2 due to transverse relaxation effects. As the data shown in Figure 8.9d demonstrate, the long relaxation delay contributes strongly to the overall experimental time of J,δ-spectroscopy. Similarly, if high resolution in F_1 is needed in order to resolve small couplings, the increased number of t_1 increments leads to long measuring times.

The signals originally detected in homonuclear J,δ-spectroscopy with the

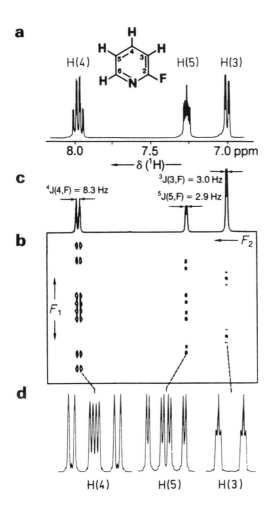

Figure 8.11   Homonuclear J-resolved 2D ^{1}H n.m.r. experiment for H(3), H(4) and H(5) in 2-fluoropyridine (1.4 M in acetone-d_6); (a) one-dimensional ^{1}H n.m.r. spectrum; (b) 2D contour plot; (c) F_2 projection; (d) traces parallel to F_1. The F_2 frequency axis contains the ^{1}H,^{19}F-coupling constants, the F_1-axis shows only the ^{1}H,^{1}H couplings. The traces along F_1 show ^{19}F-decoupled ^{1}H multiplets, while the F_2 projection shows the '^{1}H-decoupled' ^{1}H spectrum, which now displays the ^{1}H,^{19}F couplings. The following experimental data were used: Δt_1 = 26.4 ms   (this   corresponds   to SW_1 = 37.9 Hz); SW_2 = 606 Hz;   data   matrix $F_1 \times F_2 = 128 \times 1$K; sine filter functions in F_1 and F_2; 16 scans for each t_1 experiment; 10 s relaxation delay; total measuring time 7.2 h. The ^{1}H,^{1}H coupling constants are $J(3,4)$ = 8.26, $J(3.5)$ = 0.76, $J(3.6)$ = 0.76, $J(3.6)$ = 0.76, $J(4,5)$ = 7.23, $J(4,6)$ = 2.02,   and $J(4,6)$ = 4.92 Hz (after Ref. 1)

coordinates $F_2 \pm J/2$ and $F_1 \pm J/2$ are transformed by the data treatment discussed above to signals at F_2, $F_1 \pm J/2$, that is the $F_1 = 0$-axis divides the spectrum. Zero-filling (see above) and the use of filter functions (see p. 278) lead to improved lineshapes. A number of artefacts in 2D-J,δ-spectra, which are a consequence of imperfect pulse lengths (phantom or ghost signals), can be eliminated by the phase cycle EXORCYCLE, which we will not, however, discuss.

The largest disadvantage of J,δ-spectroscopy must be seen in the relatively long measuring times (see above), which are usually larger than that of other 2D n.m.r. methods. Furthermore, strongly coupled spin systems show a number of artefacts which prevent a simple analysis of the spin multiplets. Today, 2D J,δ-spectroscopy is thus used less frequently.

4. The COSY experiment—Two-dimensional ^{1}H,^{1}H shift correlations

The *COSY experiment*—also known as *Jeener experiment* after its inventor, the Belgian physicist Jeener—is certainly the most important measuring technique of two-dimensional n.m.r. spectroscopy. Here, both frequency axes contain chemical shifts and so-called *cross peaks* indicate which nuclei are spin–spin coupled. The experiment is based on scalar spin–spin coupling, as is J,δ-spectroscopy discussed in the preceding section. The determination of spin–spin coupling constants, however, is usually not the major object of COSY spectroscopy, which is primarily used in order to obtain structural information via the spin connectivities revealed by the cross peaks. In this sense, the COSY experiment is the two-dimensional equivalent to the one-dimensional selective spin–spin decoupling experiment.

The COSY pulse sequence contains only two 90° pulses, separated by the evolution time t_1 (Figure 8.12):

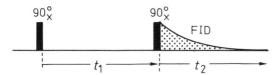

Figure 8.12 Pulse sequence for a COSY spectrum

For an AX-system, the 2D spectrum consists of *diagonal signals*, centred at coordinates v_A,v_A and v_X,v_X, as well as of off-diagonal signals, so-called *cross peaks*, centred at v_A,v_X and v_X,v_A. The cross peaks correlate the chemical shifts v_A and v_X and indicate scalar coupling between the A and the X nucleus (Figure 8.13a). This leads to a direct assignment of adjacent protons and yields important information about molecular structure (Figure 8.13b). Since in a COSY-spectrum chemical shifts are measured in F_1 and F_2, these spectra can be called δ,δ-spectra, in analogy to the classification of J-resolved spectra as J,δ-spectra.

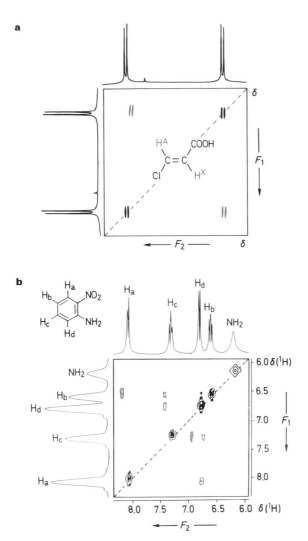

Figure 8.13 (a) COSY spectrum of an AX system; (b) COSY spectrum of *o*-nitroaniline with cross peaks for vicinal neighbours and projections on both frequency axes; δ values in ppm. The NH$_2$ resonance shows no cross peaks. Measurement time 2 h (after Ref. 2)

Besides the Larmor frequencies the frequency axes F_1 and F_2 of a COSY spectrum also contain the frequencies of the homonuclear scalar spin–spin interactions because during the evolution time both parameters, chemical shifts and scalar couplings, operate. Under conditions of high resolution a fine structure is therefore observed for the diagonal as well as for the cross peaks of a COSY spectrum from which scalar spin–spin coupling constants can be extracted (see

Section 4.1). The intensity distribution within the cross peaks, however, follows rules that differ from those that govern the intensities of the multiplets in 1D-spectra. Furthermore, in the case of crowded spectra, signal overlap can lead to the elimination of cross peak components. The assignment of line splittings to certain spin–spin coupling constants in more complex 2D spectra, therefore, is by no means trivial and it is not surprising that for the majority of practical applications the structural information which comes from the correlation of chemical shifts is by far the most important aspect.

A correct interpretation of the COSY pulse sequence on the basis of the Bloch vector model is not feasible and the systematic development of the spin physics behind this experiment must be postponed until the discussion of the basic theory in Section 5 of this chapter. With a simple, qualitative picture the origin of the diagonal and the cross peaks in a 2D COSY spectrum can be, however, rationalized. For this purpose let us first look at the singlet signal of a nucleus A, which is not spin–spin coupled. The first 90° pulse of the COSY sequence produces transverse A magnetization $M(A)$ which rotates around the z-axis. The second 90° pulse moves the y-part of $M(A)$ into the negative z-direction, while the x-part remains in the x,y-plane for detection. The intensity of the detected signal depends on the orientation of the vector $M(A)$ at the end of the evolution time which is determined by the Larmor frequency v_A. The signal amplitude is thus modulated with this frequency during t_1 and double Fourier transformation yields in both dimensions the frequency v_A, that is a diagonal signal.

Exercise 8.4   Illustrate the situation discussed above in graphical form (because scalar coupling is absent, Bloch vector pictures can be used).

In the presence of spin–spin coupling the second 90° pulse not only affects transverse magnetization, but also leads to population changes for the various transitions in the spin system. In this way magnetization is exchanged between all nuclei which are mutually coupled and their signals are amplitude-modulated in a series of t_1 experiments by the frequencies of the neighbouring nuclei. This mechanism leads to the cross peaks of the 2D-spectrum at v_i,v_j and v_j,v_i.

In comparing the COSY experiment with the one-dimensional decoupling experiments discussed in Chapter 7, we note that 2D spectroscopy yields all information about coupled nuclei in the spectrum of interest with only one experiment. The two-dimensional method is thus superior to its 1D equivalent and in the meantime indispensible, not only for complicated spectra, where it also has a big advantage with regard to the time necessary for a complete spectral analysis, but also for routine applications.

Practical applications of COSY experiments are further illustrated in Figure 8.14 with two examples. In Figure 8.14a the multiplets in the spectrum of the mixture of n-butylbromide and n-butyliodide, already discussed in Section 3.1, are assigned by COSY spectroscopy. The information obtained is not available from the J,δ-

spectrum in Figure 8.10 (p. 286). It is found that the high field as well as the low
field components of the overlapping multiplets are correlated by cross peaks. Due to
low digital resolution a decision in the case of the methyl group signals, where the
chemical shift difference amounts to only 0.12 ppm or 48 Hz, requires the enlarged
spectrum (b). Since the COSY-experiment only yields relative assignments, the
absolute assignment must be based on the known shift data for alkyl halides (p. 72).
According to this information the high field signals belong to *n*-butyliodide. Figure
8.14c finally shows convincingly the assignment of the protons of an [18]annulene.
In both spectra only cross peaks based on vicinal ^{1}H,^{1}H-coupling constants were
detected.

4.1 SOME EXPERIMENTAL ASPECTS OF 2D-COSY SPECTROSCOPY

In contrast to J,δ-spectroscopy, a considerably larger spectral window in the F_1-
dimension must be chosen in δ,δ-spectroscopy, because the sweep width in F_1 is
now determined by the chemical shift scale of a nucleus and not by the width of a

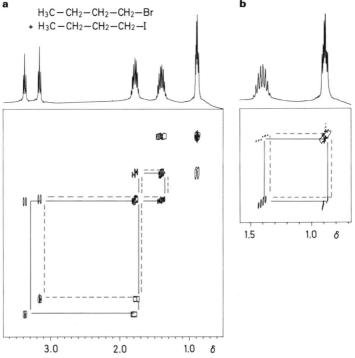

Figure 8.14 (a) 400 MHz ^{1}H,^{1}H COSY spectrum of a mixture of *n*-butylbromide and *n*-
butyliodide (ca. 1:1, see Figure 8.10); the sweep width was 3.18 ppm; 1K data points in F_2;
128 t_1-experiments, measuring time 2.5 h; data treatment with sine functions in both (b)
enlarged high field section without data treatment (c) ^{1}H,^{1}H COSY spectrum of 9,11-
bisdehydrobenzo[18]annulene; measuring time 4 h (after Ref. 3).

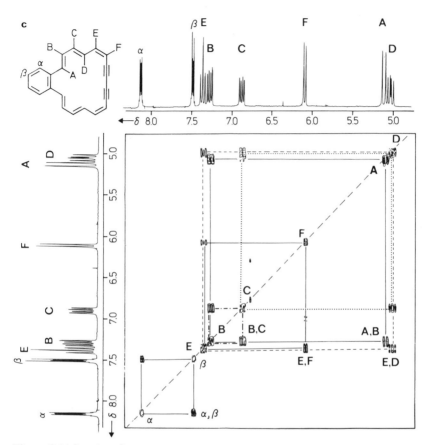

Figure 8.14 (*continued*)

spin multiplet. Since for a correlation experiment lower resolution can be tolerated and the spectra are often processed in the absolute value mode, 64 or 128 t_1 increments suffice in most cases for the detection of cross peaks. As a consequence, the digital resolution in a COSY spectrum is normally lower in F_1 than in F_2, which is seen quite clearly in Figure 8.13b.

The aspect of digital resolution is more critical if scalar coupling constants are to be determined from the fine structure of the cross peaks. It is then necessary to use *phase-sensitive* detection. For a simple AX-system Figure 8.15 (p. 295) shows the result of a COSY experiment which was processed in the phase-sensitive mode. All lines appear on both frequency axes, and a different phase behaviour for diagonal and cross speaks is found, which are detected in dispersion and in absorption or emission, respectively. In case of low digital resolution, cross peaks can thus be eliminated.

As already mentioned, problems related to the signal phase present difficulties in

2D n.m.r. spectroscopy, because double Fourier transformation in principle produces four components for the frequency function $S(F_1,F_2)$: S(real,real), S(imaginary,imaginary), S(real,imaginary), and S(imaginary,real). Their separation is obviously more complicated than the separation of real and imaginary part in 1D n.m.r. The same is true for quadrature detection which presents no problem for the time domain t_2 (see p. 252), but is principally more difficult in the t_1 domain, where it is impossible to measure simultaneously two signals. Here, a number of tricks must be used which will be discussed more closely on page 318 after we have introduced further details of the COSY experiment. For the moment we just note that three different procedures are available, where the older one uses a special phase cycle, but does not produce pure signal phases. The 2D spectrum thus has to be recorded in the absolute value mode. In order to achieve a sufficient signal separation for the resulting broad lines, filter functions for resolution enhancement are applied before data transformation. The two other procedures—developed independently by States, Haberkorn and Ruben (we use the acronym SHR) and by Marion and Wüthrich (known by the acronym TPPI, cf. p. 318)—yield pure signal phases and are clearly superior to the older method. They should always be used if an analysis of the cross peak fine structure is desirable. More recent developments, however, which are based on field gradients (cf. Section 6), are even more attractive, because they achieve quadrature detection within less experimental time.

A simple relation between cross peak intensity and the magnitude of the scalar spin–spin coupling involved does not exist. As can be shown theoretically (see p. 310 f.), cross peaks develop according to

$$S(\omega_A, \omega_X) = \sin(\pi J t_1)\sin(\pi J t_2) \tag{8.3}$$

They can be enhanced by filter functions which have a maximum after the interval $1/2J$. This procedure not only yields an intensity increase as compared to the intensity of the diagonal peaks, but also allows, within certain limits, a selection with respect to the magnitude of J. Sometimes, cross peaks can be identified only after a careful examination of the 2D-spectrum with contour diagrams at various intensity levels. For a detailed analysis of COSY spectra cross sections through the 2D data matrix parallel to both frequency axes are very useful. They often yield more reliable information than the contour plots, because in the resulting pseudo-1D spectra cross peaks can be distinguished more easily from the noise. The following factors can be responsible for failures to detect cross peaks and should be kept in mind: small coupling constants, cross peak elimination because of low digital resolution, inadequate filter functions or acquisition times which are too short with respect to the time development of cross peak magnetization.

4.2   ARTEFACTS IN COSY SPECTRA

An artefact in 2D COSY spectra are the so-called *axial peaks* which appear parallel to the F_2 dimension at $F_1 = 0$ (Figure 8.16a, p. 296). They result from longitudinal

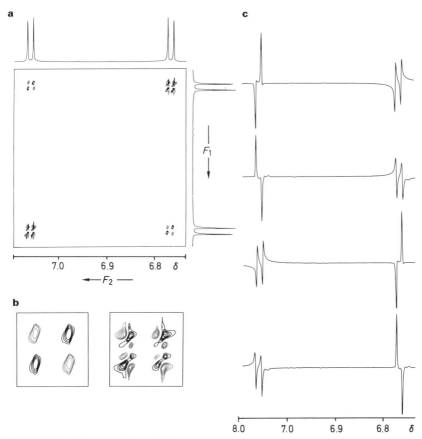

Figure 8.15   Phase-sensitive COSY spectrum of the AX-system of 2,3-dibromothiphene ($J = 5.8$ Hz) with complete resolution of all signals; (a) total spectrum; (b) enlargement of the cross and diagonal peaks; (c) traces through the two-dimensional data matrix parallel to F_2

relaxation during the evolution time. The z-magnetization which is generated in this way is transformed into transverse magnetization by the second 90° pulse of the COSY sequence. This magnetization is not amplitude modulated and after Fourier transformation appears at $F_1 = 0$. Using a simple phase cycle by alternating the phase of the second 90° pulse, they can be eliminated. Together with the CYCLOPS cycle for quadrature detection in t_2 (see p. 254), the simplest phase cycle for the COSY experiment thus amounts to: *(000) (020) (111) (131) (222) (202) (333) (313)**.

---

* For the explanation of this code see the Table on p. 254.

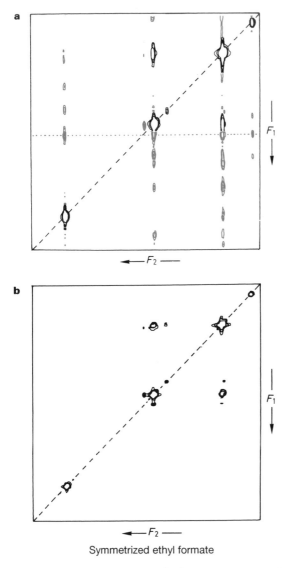

Symmetrized ethyl formate

Figure 8.16 (a) Artefacts in a ^{1}H,^{1}H COSY spectrum (red signals): axial peaks parallel to the F_2 axis at $F_1 = 0$ ($\cdots$) and t_1 noise parallel to the F_1 axis; (b) the same spectrum after symmetrization of the data matrix

---

Exercise 8.5   Draw vector pictures in order to explain the origin of axial peaks

---

2D COSY spectra very often suffer from the phenomenon of t_1 *noise*. Various instrumental instabilities during the measurement (for example instable pulse power or field/frequency lock) cause during a series of t_1-experiments a statistical, noise-like modulation of the 2D signals. This leads to noise which extends parallel to F_1 at the frequency of the F_2 signals (Figure 8.16). The intensity of t_1-noise is proportional to the intensity of the respective 1D signal and, therefore, particularly large in the case of solvent peaks. By careful adjustment of all instrumental parameters these artefacts can be minimized. Elimination is possible by *symmetrization*, a software operation with the 2D data matrix which takes advantage of the symmetry of the COSY spectrum. Data points which do not appear in symmetrical pairs with respect to the diagonal are thereby eliminated. In many cases, however, false cross peaks are produced by this technique. For instance, if two t_1 noise signals have just the coordinates i,j and j,i, they will give rise to an artificial cross peak. This technique must thus be executed with great care.

The time necessary for a routine COSY spectrum is in the case of modern high field instruments relatively short and amounts to usually less than one hour with standard samples. Even shorter measuring times have become possible by the application of gradient enhanced COSY spectroscopy, a new technique which will be described in Section 5.8.3. In any case, of course, if high digital resolution in F_1 is desirable, a lengthening of the experiment is unavoidable.

4.3   MODIFICATIONS OF THE JEENER PULSE SEQUENCE

The COSY pulse sequence has been modified with different intentions. The important variants, given in Figure 8.17, will be discussed in the following section.

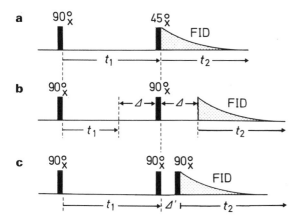

Figure 8.17   Modifications of the COSY-experiment: (a) COSY-45 for the reduction of diagonal signals; (b) long-range COSY (COSY-LR) to emphasize small couplings; the delay Δ is in the ms-region; (c) COSY with double quantum filter (COSY-DQF) to eliminate singlet signals; the delay Δ is an interval in the μs-region

4.3.1   COSY-45

If the frequency difference between coupled nuclei in the 1D spectrum is small, the COSY cross peaks appear close to the diagonal and overlap with diagonal peaks occurs. In such cases the pulse sequence COSY-45 is useful where the second 90° pulse is replaced by a 45° pulse:

$$90°_x - t_1 - 45°_x, \text{ FID } (t_2) \tag{8.4}$$

As can be shown theoretically (see p. 313), this reduces the intensities of the diagonal signals with respect to the intensity of the cross peaks. Possible overlaps can thus be diminished or even avoided. Furthermore, the smaller pulse angle introduces a nonsymmetry in the cross peaks. Normally, these are arranged in a square-like fashion. Smaller pulse angles lead to a rhombic distortion, where the orientation of the large diagonal depends on the relative sign of the coupling constants. Thus, COSY spectroscopy allows the determination of the relative signs of scalar spin–spin interactions.

4.3.2   Longrange COSY (COSY-LR)

With organic compounds proton spectra are normally dominated by geminal and vicinal coupling constants. These interactions with magnitudes between 5 and 15 Hz also determine the COSY spectra. If *correlations via long-range couplings* are of interest, special provisions have to be made. For small couplings, due to the relatively slow development of cross peak magnetization (see equation 8.3, p. 294), the relation

$$t_1(\text{max}) = t_2 = T_2 \tag{8.5}$$

should be satisfied. This can be achieved by the introduction of a fixed delay Δ in the evolution and detection time, that is before and after the second pulse of the COSY sequence:

$$90°_x - t_1-, \ \Delta, \ 90°_x, \ \Delta, \ t_2 \tag{8.6}$$

Exercise 8.6   Consider why equation (8.5) cannot be satisfied by an increase of the t_1-increment.

With T_2 values of 0.2–0.6 s for protons the value for Δ amounts to 50–500 ms. In order to illustrate this experiment, Figure 8.18a shows again a COSY spectrum for *o*-nitroaniline. Contrary to the COSY-90 experiment, where the cross peaks arise from vicinal couplings (Figure 8.13b, p. 290), cross peaks for ^4J coupling constants between meta-protons are now observed; 125 ms were chosen for the delay Δ. As one sees, a number, but not all of the vicinal correlations are also detected. With unknown compounds it is, therefore, advisable to measure first a COSY-90 or COSY-45 spectrum in order to identify the vicinal and geminal correlations, before

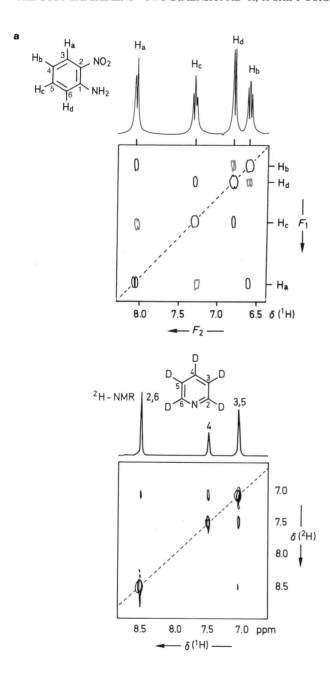

Figure 8.18 (a) 400 MHz ^{1}H,^{1}H COSY spectrum for *o*-nitroaniline with pulse sequence (8.6) and $\Delta = 125$ ms; cross peaks for ^4J-correlations in red (after Ref. 2); (b) 61.42 MHz ^{2}H,^{2}H COSY spectrum for [D$_5$]pyridine (after Ref. 4)

sequence (8.6) can be used for the verification of long-range correlations by a careful comparison of these spectra.

   In this context it is interesting to note that cross peaks can be detected even if the 1D spectrum does not show any line splittings, that is when the coupling is in the order of the halfwidth of the 1D signal. This aspect is of practical importance in several cases. Examples are spectra of organic ligands of paramagnetic complexes where even the vicinal interactions in the 1D spectrum are not resolved due to severe paramagnetic line broadening. Figure 8.18b shows another application of pulse sequence (8.6), where the extremely small couplings between deuterons, which are not resolved in the 1D ^{2}H spectrum are detected via the cross peaks in the ^{2}H,^{2}H COSY experiment. From the pyridine data (p. 522) one calculates vicinal interactions in the order of 0.1 Hz by the equation $J(^2H,^2H) = (v_D/v_H)^2 \, J(^1H,^1H)$ (cf. equation (6.1), p. 218).

Exercise 8.7   Figure 8.19 shows the COSY-90 and the COSY-LR ^{1}H n.m.r. spectrum of naphthobiphenylene-dianion ((a) and (b), respectively). Assign the signals 1–6 to the protons H_a–H_f.

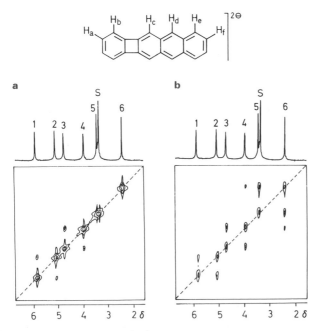

Figure 8.19   400 MHz ^{1}H,^{1}H COSY spectra of naphthobiphenylene-dianion; S = solvent signal (after Ref. 5)

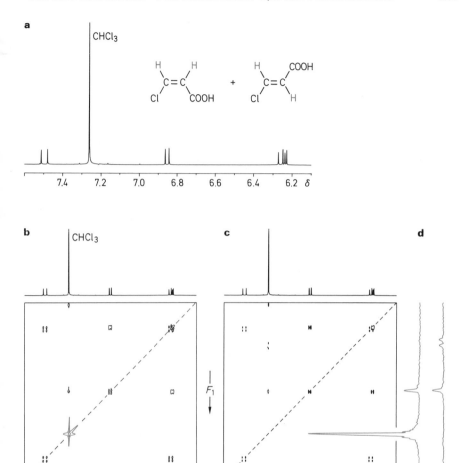

Figure 8.20   400 MHz ^{1}H n.m.r. spectra of a mixture of Z- and E-2-chloroacrylic acid in CDCl$_{3}$/CHCl$_{3}$; (a) 1D spectrum; (b) COSY-90 spectrum with 128 t_{1}-experiments of 32 accumulations each, measuring time 3 h; (c) COSY-DQF spectrum with 128 t_{1}-experiments of 64 accumulations each, measuring time 6 h; (d) F_{1}-traces of spectra (b) (left) and (c) (right) at the ^{1}H frequency of CHCl$_{3}$

4.3.3   COSY with double quantum filter (COSY-DQF)

For the elimination of perturbing singlet signals from solvent molecules, the COSY-sequence with double quantum filter (COSY-DQF, Figure 8.17c), discussed in this section, is particularly suitable:

$$90_{x}^{\circ} - t_{1} - 90_{x}^{\circ},\ 90_{x}^{\circ},\ t_{2} \qquad\qquad (8.7)$$

As we shall derive later, the introduction of a third 90° pulse allows us, with the help of a special phase cycle, to detect the magnetization of coupled spin systems, while the magnetization of singlet signals is suppressed. These experiments are based on the fact that only in a spin system that consists of two or more scalar coupled nuclei with different chemical shifts, can detectable double quantum phenomena arise. The double quantum magnetization associated with two nuclei A and X is selected via the phase cycle and detected after transformation into single quantum magnetization. Besides the elimination of singlet signals, this experiment also reduces the diagonal signals, which are partly eliminated because of their anti-phase character. In total, the COSY-DQF experiment has only half the sensitivity of the standard COSY-90 experiment, but this disadvantage is outweighed by the spectral simplifications obtained.

An application of the COSY-DQF experiment is shown in Figure 8.20 (p. 301) with the spectra of a sample of Z- and E-2-chloroacrylic acid, which contains $CHCl_3$. The diagonal signal of the solvent is eliminated in the COSY-DQF experiment as documented by the appropriate F_1 trace of the data matrix (Figure 8.20d).

5. The product operator formalism

The Bloch vector model describes the action of the external magnetic field B_0 and the radio-frequency field B_1 on the nuclear magnetization vector within the framework of classical physics and is only valid for isolated nuclei without spin–spin interactions. For an adequate description of pulse sequences, where scalar spin–spin coupling is important, quantum mechanical methods have to be used instead. The calculation of the effect of pulse sequences on spin systems the AX type or those of higher order is thereby based on the time-dependent Schrödinger equation. Contrary to the analysis of spin systems, treated in Chapter 5 on the basis of the Schrödinger equation for stationary states, the time dependence of the spin system under the effect of the appropriate Hamilton operator must now be taken into account. Before, only energy differences and transmission probabilities between the stationary states of the spin system were important. Even the effect of the r.f. field B_1 could be neglected. The Schrödinger equation for the double resonance experiment (p. 261) was already supplemented by this term. In the case of pulse sequences, the effects of r.f. pulses, chemical shifts, and spin–spin couplings during the evolution time as well as relaxation processes during the pulse sequence have to be included. The quantum theoretical tool available to deal with this type of situations is *density matrix theory*. Not many readers, however, will be familiar with this formalism. In addition, the application of density matrix calculations for the cases of interest in the present context is lengthy and not practical for larger spin systems. Therefore, we take advantage of a simplified procedure, introduced by Ernst and others and known as *Product Operator Formalism*. It serves to calculate observable magnetizations and to explain the spin physics of pulse sequences. It is based on Cartesian nuclear spin operators $\hat{I}_{xyz}$, already known from Chapter 5, as well as on products of these quantities. The procedure is limited in its applications to weakly coupled spin systems, that is first order spectra, and neglects all relaxation effects. In the following section we shall develop its basic principles and later on apply the method to the COSY sequence and several of its modifications. All our discussions are related to the rotating frame of reference.

5.1   THE PHENOMENON OF COHERENCE

Before we discuss the basic principles of the product operator formalism, let us first introduce the important phenomenon of *coherence* which plays a central role in pulse n.m.r. In principle, a coherence between two spin states corresponds to the transition in the n.m.r. energy level diagram, discussed in Chapter 5, and thus to transverse magnetization. However, the term is more general, because it describes all possible mechanisms for the exchange of spin population between different states, in particular transitions which cannot be observed directly in the experimental spectrum. For example, relaxation transitions with $\Delta m_T = +2$, as well as the double and zero quantum or combination lines belong to this group. Through such pathways spin population can be exchanged as we shall discuss in Chapter 10 in connection with the nuclear Overhauser effect, but observable n.m.r. signals do not necessarily result in a direct way. During certain pulse sequences 'invisible' coherences of this type play an important role. Only coherences which obey the quantum chemical selection rules, however, can be directly detected.

Coherences can arise between all eigenstates which belong to the same irreducible representation of the symmetry group of the nuclear spin system. Coherences between eigenstates of different symmetry are forbidden. Within the framework of density matrix theory, coherences correspond to the non-diagonal elements σ_{kl} between the eigenstates $|k\rangle$ and $\langle l|$. This means that the state function $\varphi(t)$ of the system is a coherent superposition of these and other eigenstates and, therefore, no eigenfunction of the time-dependent Hamilton operator. In the field of nuclear magnetic resonance coherences between more than two states, however, are not important. The order of a coherence, p_{kl}, corresponds to the difference Δm of the magnetic quantum numbers of the respective states: $p = 0$ for zero quantum coherences and longitudinal z-magnetization, $p = 1$ for one quantum coherences, $p = 2$ for double quantum coherences, to mention only the most important. During a pulse sequence all coherences can be excited, but only coherences of the order $p = 1$ are detectable.

An important aspect of modern pulse n.m.r. is the transformation of coherences during a pulse sequence and a change of their order. The notion of coherence order is then replaced by the term coherence level, which also has a sign. The sign results from the raising and lowering operators $\hat{I}^+$ and $\hat{I}^-$, respectively, which have been introduced in Chapter 5 (p. 141, see also Section 5.3). For example transverse x-magnetization of the coherence order, $p = 1$, represented by the nuclear spin operator $\hat{I}_x$ as described below, can be expressed, following equation (8.11a) (p. 305), by $\hat{I}^+$ and $\hat{I}^-$. A 90° pulse, which produces transverse x-magnetization, thus leads to coherences of the level $+1$ (for $\hat{I}^+$) and -1 (for $\hat{I}^-$). Double quantum

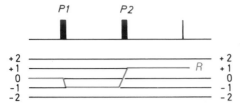

Figure 8.21   Coherence level diagram of a pulse sequence with two pulses *P1* and *P2* and the receiver *R* at the level +1

coherences are characterized by the products $\hat{I}^{+}\hat{I}^{+}$ and $\hat{I}^{-}\hat{I}^{-}$ and, therefore, the coherence levels +2 and −2 exist. A different sign of the coherence level indicates a difference in the sense of evolution for the respective operator, that is a different rotational sense in the coordinate system.

The fate of coherences during a pulse sequence is best illustrated by a *coherence level diagram*, as shown in Figure 8.21. It documents the *coherence transfer pathway* during the pulse sequence, which starts always at the zero level corresponding to longitudinal *z*-magnetization. The first pulse of a sequence produces coherences of the order $p = 1$. If a certain coherence shall be detected finally as a signal, the coherence transfer pathway must end at the level +1, which in our convention (cf. legend to Figure 8.22, p. 307) is the coherence level of the receiver.

5.2   THE OPERATOR BASIS FOR AN AX SYSTEM

The complete set of operators which can be derived for a spin system of N spin $\frac{1}{2}$ nuclei on the basis of the density matrix theory contains 4^N components. For a two-spin system of AX type, to which we will limit ourselves here, the following operators result:

$$
\begin{aligned}
&\hat{I}_x(A)\ \hat{I}_y(A)\ \hat{I}_z(A) \\
&\hat{I}_x(X)\ \hat{I}_y(X)\ \hat{I}_z(X) \\
&2\hat{I}_x(A)\hat{I}_x(X)\ \ 2\hat{I}_x(A)\hat{I}_y(X)\ \ 2\hat{I}_x(A)\hat{I}_z(X) \\
&2\hat{I}_y(A)\hat{I}_x(X)\ \ 2\hat{I}_y(A)\hat{I}_y(X)\ \ 2\hat{I}_y(A)\hat{I}_z(X) \\
&2\hat{I}_z(A)\hat{I}_x(X)\ \ 2\hat{I}_z(A)\hat{I}_y(X)\ \ 2\hat{I}_z(A)\hat{I}_z(X)
\end{aligned}
\tag{8.8}
$$

As one sees, nuclear spin operators for the individual nuclei appear as single operators and as products. The 16th operator, not shown above, is the unity operator.

The importance of the one-spin operators was already demonstrated in Chapter 5: $\hat{I}_z(A)$ and $\hat{I}_z(X)$ represent longitudinal, $\hat{I}_x(A)$, $\hat{I}_x(X)$ and $\hat{I}_y(A)$, $\hat{I}_y(X)$ transverse A- and X-magnetization, respectively. Within the framework of the product operator formalism these operators have two functions. On the one hand they correspond to the coherences of the spin system, on the other they behave as operators in the true sense and transform these coherences. This is known as *coherence transfer*. For the transfer pathways, coherence selection rules exist, for example:

(1) Coherence transfer can only occur between states of the same symmetry.
(2) In a weakly coupled spin system coherence can be exchanged between different spins only in the presence of scalar coupling.

5.3 ZERO AND MULTIPLE QUANTUM COHERENCES

The more general term *coherence* provides the basis for a discussion of magnetization components which cannot directly be observed. To characterize these contributions, we use the raising and lowering operator, $\hat{I}^+$ and $\hat{I}^-$, respectively, which have already been defined in Chapter 5 (p. 149):

$$\hat{I}^+ = \hat{I}_x + i\hat{I}_y \tag{8.9a}$$

$$\hat{I}^- = \hat{I}_x - i\hat{I}_y \tag{8.9b}$$

Application of these operators to spin functions yields the spin functions of the next higher or next lower magnetic quantum number n, for example

$$\hat{I}^+\beta = [\hat{I}_x + i\hat{I}_y]\beta = \tfrac{1}{2}\alpha - i^2\tfrac{1}{2}\alpha = \alpha \tag{8.10}$$

Therefore, the term raising and lowering operator. A double quantum coherence where two nuclei change their spin orientation at the same time is then characterized by the products $\hat{I}^+\hat{I}^+$ or $\hat{I}^-\hat{I}^-$, whilst products of the form $\hat{I}^+\hat{I}^-$ or $\hat{I}^-\hat{I}^+$ describe zero quantum coherences. In the energy level diagram of an AX system these coherences arise between the states $\alpha\alpha$ and $\beta\beta$ and $\alpha\beta$ and $\beta\alpha$, respectively (cf. Table 5.1, p. 156).
From equations (8.9a) and (8.9b) one obtains on the other hand

$$\hat{I}_x = \tfrac{1}{2}(\hat{I}^+ + \hat{I}^-) \tag{8.11a}$$

$$\hat{I}_y = \tfrac{1}{2i}(\hat{I}^+ - \hat{I}^-) \tag{8.11b}$$

For the operator products with two transverse components the following equations result:

$$2\hat{I}_x(A)\hat{I}_x(X) = \tfrac{1}{2}[\hat{I}^+(A)\hat{I}^+(X) + \hat{I}^+(A)\hat{I}^-(X) + \hat{I}^-(A)\hat{I}^+(X) + \hat{I}^-(A)\hat{I}^-(X)] \tag{8.12a}$$

$$2\hat{I}_y(A)\hat{I}_y(X) = -\tfrac{1}{2}[\hat{I}^+(A)\hat{I}^+(X) - \hat{I}^+(A)\hat{I}^-(X) - \hat{I}^-(A)\hat{I}^+(X) + \hat{I}^-(A)\hat{I}^-(X)] \tag{8.12b}$$

$$2\hat{I}_x(A)\hat{I}_y(X) = \tfrac{1}{2i}[\hat{I}^+(A)\hat{I}^+(X) - \hat{I}^+(A)\hat{I}^-(X) + \hat{I}^-(A)\hat{I}^+(X) - \hat{I}^-(A)\hat{I}^-(X)]$$

$$(8.12c)$$

$$2\hat{I}_y(A)\hat{I}_x(X) = \tfrac{1}{2i}[\hat{I}^+(A)\hat{I}^+(X) + \hat{I}^+(A)\hat{I}^-(X) - \hat{I}^-(A)\hat{I}^+(X) - \hat{I}^-(A)\hat{I}^-(X)]$$

$$(8.12d)$$

All these terms contain double and zero quantum contributions, namely the products $\hat{I}^+(A)\hat{I}^+(X)$ or $\hat{I}^-(A)\hat{I}^-(X)$ for $\Delta m_T = 2$ (spin change in the same sense) and $\hat{I}^+(A)\hat{I}^-(X)$ and $\hat{I}^-(A)\hat{I}^+(X)$ for $\Delta m_T = 0$ (spin change in the opposite sense). Through a linear combination of these terms pure double or zero quantum coherences result:

Double quantum coherence:

$$\tfrac{1}{2}[2\hat{I}_x(A)\hat{I}_x(X) - 2\hat{I}_y(A)\hat{I}_y(X)] = \tfrac{1}{2}[\hat{I}^+(A)\hat{I}^+(X) + \hat{I}^-(A)\hat{I}^-(X)] = DQ_x \quad (8.13a)$$

$$\tfrac{1}{2}[2\hat{I}_x(A)\hat{I}_y(X) + 2\hat{I}_y(A)\hat{I}_x(X)] = \tfrac{1}{2i}[\hat{I}^+(A)\hat{I}^+(X) - \hat{I}^-(A)\hat{I}^-(X)] = DQ_y \quad (8.13b)$$

Zero quantum coherence:

$$\tfrac{1}{2}[2\hat{I}_x(A)\hat{I}_x(X) + 2\hat{I}_y(A)\hat{I}_y(X)] = \tfrac{1}{2}[\hat{I}^+(A)\hat{I}^-(X) + \hat{I}^-(A)\hat{I}^+(X)] = ZQ_x \quad (8.13c)$$

$$\tfrac{1}{2}[2\hat{I}_y(A)\hat{I}_x(X) - 2\hat{I}_x(A)\hat{I}_y(X)] = \tfrac{1}{2i}[\hat{I}^+(A)\hat{I}^-(X) - \hat{I}^-(A)\hat{I}^+(X)] = ZQ_y \quad (8.13d)$$

5.4   THE EVOLUTION OF OPERATORS

During a pulse sequence operators are transformed, that is they are time dependent due to the action of several factors. Within the framework of our model, three aspects are of importance:

(1) the effect of r.f. pulses,
(2) the effect of Larmor precession, that is chemical shifts, and
(3) the effect of scalar spin–spin coupling.

How these different factors act on various magnetization components will be derived with the simple example of the Cartesian operators $\hat{I}_x$, $\hat{I}_y$, $\hat{I}_z$ (Figure 8.22). Let us start for this purpose with the z-operator $\hat{I}_z$. This operator can be treated as classical magnetization. It is, therefore, not surprising that $\hat{I}_z$ is transformed through a 90°_x pulse into $\hat{I}^*_y$ (Figure 8.22). This transformation can be written as follows:

$$\hat{I}_z \xrightarrow{\;90^\circ \hat{I}_x\;} \hat{I}_y \qquad (8.14)$$

The term of the respective Hamiltonian operator, which is responsible for the transformation described, is written above the arrow and is called the *propagator*.

---

* For the convention chosen here compare the legend to Figure 8.22.

Note that the propagator also contains an operator. The operators thus appear, as already mentioned, in active as well as in passive form.

90° pulses in other directions of the coordinate system have similar effects:

$$\hat{I}_z \xrightarrow{90°\hat{I}_y} -\hat{I}_x \tag{8.15}$$

$$\hat{I}_z \xrightarrow{90°(-\hat{I}_x)} -\hat{I}_y \tag{8.16}$$

$$\hat{I}_z \xrightarrow{90°(-\hat{I}_y)} \hat{I}_x \tag{8.17}$$

On the other hand, a $90°_x$ pulse on the operators $\hat{I}_x$ and $\hat{I}_y$ yields:

$$\hat{I}_x \xrightarrow{90°\hat{I}_x} \hat{I}_x \tag{8.18}$$

$$\hat{I}_y \xrightarrow{90°\hat{I}_x} -\hat{I}_z \tag{8.19}$$

If we choose a different pulse angle $\alpha < 90°$ it follows:

$$\hat{I}_z \xrightarrow{\alpha\hat{I}_x} \hat{I}_z \cos \alpha + \hat{I}_y \sin \alpha \tag{8.20}$$

These equations hold for A as well as X magnetization. For clarity, we have dropped the index here. Figure 8.22 shows the relations discussed again in graphical form.

Exercise 8.8   Formulate the effect of 90° pulses in the $+y$- and $-y$-direction on $\hat{I}_x$ and $\hat{I}_y$.

Since the effect of r.f. pulses is momentary (pulse duration is neglected), the propagators used so far do not contain the evolution time t_1. This changes if we now turn our attention to the effect of Larmor precession. The propagator for this case is

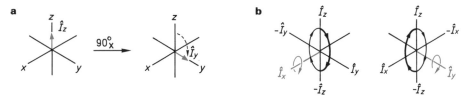

Figure 8.22   The effect of r.f. pulses on the operators $\hat{I}_z$, $\hat{I}_x$ and $\hat{I}_y$. We choose here the coordinate system in agreement with Chapter 7 and the effect of the B_1-field for nuclei with positive γ-factor is predicted by the left-hand rule (see p. 235). Transverse magnetization then rotates clockwise: $+M_y \rightarrow +M_x \rightarrow -M_y \rightarrow -M_x$. In this respect, there is no unique convention in the literature and the original publication (Ref. g) employs the right-hand rule and anti-clockwise rotation of transverse magnetization. The reader must be aware of differences in signs, therefore, if he or she compares results. The same applies to the choice of the receiver level, where we use +1 while others use −1.

simple. It must contain the Larmor frequency of the respective nucleus, the evolution time t_1 and the operator $\hat{I}_z$, since transverse magnetization rotates around the z-axis. Indeed, we can derive the propagator for the Larmor precession as well as the propagator for scalar spin–spin coupling from the Hamilton operator, equation (5.10), introduced in Chapter 5 (p. 140). For weakly coupled spin systems of first order and with the relation $\omega = 2\pi\nu$, a simple rearrangement of terms transforms equation (5.10) into equation (8.21) (remember that for an AX system only the z-operator is responsible for scalar spin–spin interaction; compare equation (2.9), p. 32):

$$\hat{H} = \sum_i \omega_i \hat{I}_z(i) + \sum_{i<j}\sum \pi J_{ij} 2\hat{I}_z(i)\hat{I}_z(j) \tag{8.21}$$

If the evolution time t_1 is now added to the various terms, the propagators for Larmor precession become $\omega_A t_1 \hat{I}_z(A)$ and $\omega_X t_1 \hat{I}_z(X)$, respectively, and for the spin–spin coupling $\pi J_{AX} t_1 2\hat{I}_z(A)\hat{I}_z(X)$; $\omega_A t_1$, $\omega_X t_1$ and $2\pi J_{AX} t_1$ are angles in radiants.

The operators are then transformed under the action of Larmor precession, remembering equation (8.20), according to:

$$\hat{I}_x(A) \xrightarrow{\omega_A t_1 \hat{I}_z(A)} \hat{I}_x(A)\cos\omega_A t_1 - \hat{I}_y(A)\sin w_A t_1 \tag{8.22a}$$

$$\hat{I}_y(A) \xrightarrow{\omega_A t_1 \hat{I}_z(A)} \hat{I}_y(A)\cos\omega_A t_1 + \hat{I}_x(A)\sin w_A t_1 \tag{8.22b}$$

These transformations, which are illustrated in graphical form in Figure 8.23a, can also be regarded as z-pulses. It is immediately clear that the $\hat{I}_z$ operator is invariant with respect to the Larmor propagator.

Finally, let us study the effect of scalar spin–spin coupling. Again, the effect is restricted to the x- and y-operators and with the propagator derived above, the following transformations, illustrated in Figure 8.23b, result[*]:

$$\hat{I}_x(A) \xrightarrow{\pi J_{AX} t_1 2\hat{I}_z(A)\hat{I}_z(X)} \hat{I}_x(A)\cos(\pi J_{AX} t_1) - 2\hat{I}_y(A)\hat{I}_z(X)\sin(\pi J_{AX} t_1) \tag{8.23a}$$

$$\hat{I}_y(A) \xrightarrow{\pi J_{AX} t_1 2\hat{I}_z(A)\hat{I}_z(X)} \hat{I}_y(A)\cos(\pi J_{AX} t_1) + 2\hat{I}_x(A)\hat{I}_z(X)\sin(\pi J_{AX} t_1) \tag{8.23b}$$

$$2\hat{I}_x(A)\hat{I}_z(X) \xrightarrow{\pi J_{AX} t_1 2\hat{I}_z(A)\hat{I}_z(X)} 2\hat{I}_x(A)\hat{I}_z(X)\cos(\pi J_{AX} t_1) - \hat{I}_y(A)\sin(\pi J_{AX} t_1) \tag{8.23c}$$

$$2\hat{I}_y(A)\hat{I}_z(X) \xrightarrow{\pi J_{AX} t_1 2\hat{I}_z(A)\hat{I}_z(X)} 2\hat{I}_y(A)\hat{I}_z(X)\cos(\pi J_{AX} t_1) + \hat{I}_x(A)\sin(\pi J_{AX} t_1) \tag{8.23d}$$

Accordingly, the true product operators appear in the propagator only after scalar spin–spin coupling is introduced. A further interesting and important aspect must be noted: the fanning out of the two transverse magnetization vectors of a doublet

---

[*]Note that for the calculation of the second term in equation (8.23c) and (8.23d) the relation $\hat{I}_z\hat{I}_z = \frac{1}{4}$ was used.

caused by the coupling produces — as a resultant of the counter rotating doublet components — two types of magnetizations: (1) magnetization of the same phase on the starting axis (first term in equation (8.23a,b), and (2) magnetization of opposite phase on the orthogonal axis (second term in equation (8.23a,b). These magnetization components are called *in-phase* and *anti-phase* magnetization. Both are continuously interconverted during evolution (Figure 8.23c). Only in-phase magnetization, however, contributes to macroscopic magnetization, since anti-phase magnetization cancels.

For the evolution of the double and zero quantum coherences introduced above, the following rules hold: Pure double quantum magnetization develops with the sum of the Larmor frequencies. This yields:

$$DQ_x \xrightarrow{[\omega_A \hat{I}_z(A) + \omega_X \hat{I}_z(X)]t_1} DQ_x \cos(\omega_A + \omega_X)t_1 + DQ_y \sin(\omega_A + \omega_X)t_1 \quad (8.24)$$

Pure zero quantum magnetization, on the other hand, develops with the difference of the Larmor frequencies:

$$ZQ_x \xrightarrow{[\omega_A \hat{I}_z(A) - \omega_X \hat{I}_z(X)]t_1} ZQ_x \cos(\omega_A - \omega_X)t_1 + ZQ_y \sin(\omega_A - \omega_X)t_1 \quad (8.25)$$

With respect to scalar spin–spin coupling between nuclei which contribute to the particular coherence, multiple quantum product operators are invariant.

5.5 THE OBSERVABLES

The magnetization detected in t_2 is finally the important aspect if the product operator formalism is used to describe certain pulse sequences. Consequently, we are interested to learn more about the meaning of the various operator products which

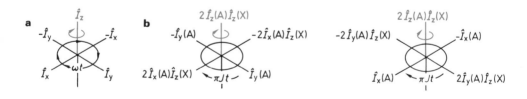

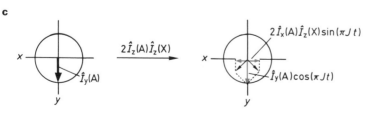

Figure 8.23 (a) Evolution under the influence of Larmor precession; (b) evolution under the influence of scalar spin–spin coupling; (c) in-phase and anti-phase magnetization

we obtain as a result of the calculations. In this respect the following rules are important:

(1) Only products that contain a single transverse component $\hat{I}_x$ or $\hat{I}_y$ yield observable signals. Examples are the one-spin operators $\hat{I}_x(A)$ or $2\hat{I}_y(A)$;

(2) Products with more than one transverse component correspond to zero or multiple quantum coherences and cannot be detected. Examples are $2\hat{I}_y(A)\hat{I}_x(X)$ or $2\hat{I}_x(A)\hat{I}_y(X)$;

(3) Products which contain one transverse component and one or several z-terms correspond to signals in anti-phase. For example, the operator product $2\hat{I}_x(A)\hat{I}_z(X)$ represents an A doublet with one absorption and one emission line, that is a phase difference of 180° between the two doublet components. The integrated intensity of such a doublet is zero.

During the various pulse experiments the operators develop under the influence of different propagators. While Larmor frequencies and scalar spin–spin couplings are important in all cases where an evolution time or fixed time delays are involved, the transformations caused by the application of r.f. pulses lead to immediate coherence transfers. A number of examples may illustrate this point:

(a) Transformation of anti-phase A-magnetization into anti-phase X-magnetization:

$$2\hat{I}_x(A)\hat{I}_z(X) \xrightarrow{90°\,[\hat{I}_y(A)+\hat{I}_y(X)]} -2\hat{I}_z(A)\hat{I}_x(X) \qquad (8.26)$$

$$2\hat{I}_y(A)\hat{I}_z(X) \xrightarrow{90°\,[\hat{I}_x(A)+\hat{I}_x(X)]} -2\hat{I}_z(A)\hat{I}_y(X) \qquad (8.27)$$

(b) Transformation of anti-phase A magnetization into zero and multiple quantum coherences:

$$2\hat{I}_x(A)\hat{I}_z(X) \xrightarrow{90°\,[\hat{I}_x(A)+\hat{I}_x(X)]} 2\hat{I}_x(A)\hat{I}_y(X) \qquad (8.28)$$

These transfers play an important role during the Jeener experiment, which we shall discuss in the following section. With the analysis of this pulse sequence we apply the relations developed so far to a specific example in order to illustrate the practical application of the product operator formalism.

5.6 THE COSY EXPERIMENT WITHIN THE PRODUCT OPERATOR FORMALISM

For the Jeener pulse sequence of the COSY experiment as applied to a homonuclear AX system

$$90°_x \text{-----} t_1 \text{-----} 90°_x,\ t_2 \qquad (8.29)$$

a product operator calculation will now be carried out. During the different steps, outlined below, the actual state of the spin system is characterized by the so-called *density operator* σ_i (i = 0, 1, 2, etc.).

(1) In the preparation period pulse excitation transforms z-magnetization (σ_0) into transverse magnetization (σ_1):

$$\sigma_0 = \hat{I}_z(A) + \hat{I}_z(X)$$

$$\downarrow 90°[\hat{I}_x(A) + \hat{I}_x$$

$$\sigma_0 = \hat{I}_y(A) + \hat{I}_y(X)$$

(2) Transverse magnetization develops during the evolution time under the influence of Larmor precession and spin–spin coupling into the state σ_2. We shall analyse these steps separately and combine in-phase and anti-phase magnetizations:

$$\sigma_1$$

$$\downarrow \omega_A t_1 \hat{I}_z(A) + \omega_X t_1 \hat{I}_z(X) \text{ (Larmor precession)}$$

$$\sigma_2 = \hat{I}_y(A) \cos \omega_A t_1 + \hat{I}_x(A) \sin \omega_A t_1 + \hat{I}_y(X) \cos \omega_X t_1 + \hat{I}_x(X) \sin \omega_X t_1$$

$$\downarrow \pi J_{AX} t_1 2\hat{I}_z(A)\hat{I}_z(X) \text{ (spin – spin coupling)}$$

$$\sigma_2 = \text{in-phase magnetization :}$$

$$[\hat{I}_y(A) \cos \omega_A t_1 + \hat{I}_x(A) \sin \omega_A t_1 + \hat{I}_y(X) \cos \omega_X t_1$$
$$+ \hat{I}_x(X) \sin \omega_X t_1] \cos(\pi J_{AX} t_1)$$

anti-phase magnetization :

$$[2\hat{I}_x(A)\hat{I}_z(X) \cos \omega_A t_1 - 2\hat{I}_y(A)\hat{I}_z(X) \sin \omega_A t_1 + 2\hat{I}_z(A)\hat{I}_x(X) \cos \omega_X t_1$$
$$- 2\hat{I}_z(A)\hat{I}_y(X) \sin \omega_X t_1] \sin(\pi J_{AX} t_1)$$

(3) The second $90°_x$ pulse, which is known as the *mixing pulse*, produces the state σ_3:

$$\sigma_2$$

$$\downarrow 90°[\hat{I}_x(A) + \hat{I}_x(X)]$$

$$\sigma_3 = \left.\begin{array}{l} -\hat{I}_z(A) \cos \omega_A t_1 \\ +\hat{I}_x(A) \sin \omega_A t_1 \\ -\hat{I}_z(X) \cos \omega_X t_1 \\ +\hat{I}_x(X) \sin \omega_X t_1 \end{array}\right\} \times \cos(\pi J_{AX} t_1)$$

$$\left.\begin{array}{l} +2\hat{I}_x(A)\hat{I}_y(X) \cos \omega_A t_1 \\ +2\hat{I}_z(A)\hat{I}_y(X) \sin \omega_A t_1 \\ +2\hat{I}_y(A)\hat{I}_x(X) \cos \omega_X t_1 \\ +2\hat{I}_y(A)\hat{I}_z(X) \sin \omega_X t_1 \end{array}\right\} \times \sin(\pi J_{AX} t_1)$$

(8.30)

As one can see, anti-phase A magnetization has been transformed into anti-phase X magnetization and vice versa. This explains the expression 'mixing pulse'.

From the terms in equation (8.30) observable magnetization, σ_3^{obs}, can now be derived if we drop, on the basis of the rules discussed in Section 5.5 (rules 1–3) the expressions which contain operator products with two transverse components:

$$\sigma_3^{\text{obs}} = [\overset{①}{\hat{I}_x(A) \sin \omega_A t_1} + \overset{②}{\hat{I}_x(X) \sin \omega_X t_1}] \cos(\pi J_{AX} t_1)$$

$$+ \overset{③}{2\hat{I}_z(A)\hat{I}_y(X) \sin \omega_A t_1} + \overset{④}{2\hat{I}_z(X)\hat{I}_y(A) \sin \omega_X t_1}] \sin(\pi J_{AX} t_1)$$

(8.31)

This magnetization then evolves during the detection time t_2, again under the influence of Larmor precession and spin–spin coupling. Remembering that only in-phase components contribute to the macroscopic magnetization, we can analyze the situation during t_2 by a straightforward application of equations (8.22) and (8.23) to the individual terms of equation (8.31). In order to facilitate this analysis we introduce the following shorthand notations: $\sin \omega_A t_1 = \sin A1$; $\sin \omega_A t_2 = \sin A2$; $\sin (\pi J_{AX} t_1) = \sin J1$; $\sin (\pi J_{AX} t_2) = \sin J2$; and similar expression for the cosine terms and ω_X, respectively.

Starting with term ①, which represents transverse A-magnetization, we have:

(1) Larmor precession:

$$\hat{I}_x(A) \sin A1 \cos J1 \xrightarrow{\omega_A t_2 \hat{I}_z(A)} \hat{I}_x(A) \sin A1 \cos J1 \cos A2 - \hat{I}_y(A) \sin A1 \cos J1 \sin A2$$

(2) spin–spin coupling:

$$\xrightarrow{\pi J_{AX} t_2 \hat{I}_z(A)\hat{I}_z(X)} \hat{I}_x(A) \sin A1 \cos J1 \cos A2 \cos J2 - \hat{I}_y(A) \sin A1 \cos J1 \sin A2 \cos J2$$

(8.32)

An analogous result is obtained for the X-magnetization, the second term in equation (8.31). This leads to the following predictions for the expected 2D spectrum: the first two terms in equation (8.31) represent A- and X-magnetization which is modulated during both the evolution time t_1 and the detection time t_2 with $\omega_A + \pi J_{AX}$ and $\omega_X + \pi J_{AX}$, respectively. After Fourier transformation this yields multiplets at $F_1 = F_2 = \omega_A$ and ω_X, respectively, which lie on the diagonal and, due to their cosine dependence on J_{AX} during t_1 and t_2, possess in-phase structure.

We now turn to terms ③ and ④ in equation (8.31). Term ③ represents anti-phase X-magnetization which develops during t_2 as follows:

(1) Larmor precession

$$2\hat{I}_z(A)\hat{I}_y(X) \sin A1 \sin J1 \xrightarrow{\omega_A t_2 \hat{I}_z(X)}$$

$$2\hat{I}_z(A)\hat{I}_y(X) \sin A1 \sin J1 \cos X2 + 2\hat{I}_z(A)\hat{I}_x(X) \sin A1 \sin J1 \sin X2$$

(2) spin–spin coupling

$$\xrightarrow{\pi J_{AX} t_2 2\hat{I}_z(A)\hat{I}_z(X)}$$

⑤   ⑥

$$2\hat{I}_z(A)\hat{I}_y(X)\sin A1 \sin J1 \cos X2 \cos J2 + \hat{I}_x(X)\sin A1 \sin J1 \cos X2 \sin J2$$

⑦   ⑧

$$+ 2\hat{I}_z(A)\hat{I}_x(X)\sin A1 \sin J1 \sin X2 \cos J2 - \hat{I}_y(X)\sin A1 \sin J1 \sin X2 \sin J2$$

$$(8.33)$$

(Note that for the calculation of the terms ⑥ and ⑧ in equation (8.33) the relationship $\hat{I}_z(A)\hat{I}_z(A) = 1/4$ was used.)

Terms ⑥ and ⑧ of equation (8.33) are the observable in-phase magnetization components which arise from the third term in equation (8.31). They have different chemical shift in t_1 and t_2 (modulation by sinA1 and cosX2 and sinA1 and sinX2, respectively). This yields after Fourier transformation a cross peak multiplet at $F_1 = \omega_A$ and $F_2 = \omega_X$ with anti-phase structure in both time dimensions (sine dependence on J_{AX}). A similar analysis for the fourth term, which represents anti-phase A-magnetization shows that it yields the symmetrical cross peak at $F_1 = \omega_X$ and $F_2 = \omega_A$.

Exercise 8.9 Develop an equation similar to equation (8.33) for the t_3 evolution of anti-phase A-magnetization (term ④ in equation (8.31)).

Because of the different modulation by pJ_{AX} (cosine and sine, respectively) there is in total a 90° phase difference between the diagonal and the cross peaks. Thus, the diagonal peaks are dispersive, while the cross peaks are detected in absorption (or emission) and *vice versa*. The experimental spectrum shown in Figure 8.15 (p. 295) confirms these relationships. If the magnitude representation of the signals is used (cf. p. 278) the phase information is completely lost. The danger of a partial cancelling of the cross peaks in the case of small couplings, however, remains.

5.7   THE COSY-EXPERIMENT WITH DOUBLE QUANTUM FILTER (COSY-DQF)

This experiment uses the following pulse sequence (cf. Figure 8.17, p. 297):

$$90^\circ_x ----t_1----90^\circ_x,\, 90^\circ_x,\, t_2 \qquad (8.34)$$

Its analysis can therefore start with the density operator σ_3 of the COSY-experiment (equation (8.30)). A further simplification results from the fact that the complete one-quantum, zero quantum and one-quantum anti-phase magnetization can be eliminated by an appropriate phase cycle (see Section 5.8). Consequently, only double quantum magnetization remains after the second 90 ° pulse. In order to select these coherences, let us first rearrange equation (8.30), where the magnetization is expressed as linear combination of pure double and zero quantum magnetization, with the help of equation (8.13a–d). This yields

$$2\hat{I}_x(A)\hat{I}_y(X)\cos\omega_A t_1 = \tfrac{1}{2}[(2\hat{I}_x(A)\hat{I}_y(X) + 2\hat{I}_y(A)\hat{I}_x(X)) - (2\hat{I}_y(A)\hat{I}_x(X)$$
$$- 2\hat{I}_x(A)\hat{I}_y(X))]\cos\omega_A t_1 \tag{8.35a}$$

$$2\hat{I}_y(A)\hat{I}_x(X)\cos\omega_X t_1 = \tfrac{1}{2}[(2\hat{I}_x(A)\hat{I}_y(X) + 2\hat{I}_y(A)\hat{I}_x(X)) + (2\hat{I}_y(A)\hat{I}_x(X)$$
$$- 2\hat{I}_x(A)\hat{I}_y(X))]\cos\omega_X t_1 \tag{8.35b}$$

We then obtain the modified density operator

$$
\begin{aligned}
\sigma_3 =\ & [-\hat{I}_z(A)\cos\omega_A t_1 + \hat{I}_x(A)\sin\omega_A t_1 - \hat{I}_z(X)\cos\omega_X t_1 \\
& + \hat{I}_x(X)\sin\omega_X t_1]\cos(\pi J_{AX} t_1) \\
& + \{\tfrac{1}{2}[(2\hat{I}_x(A)\hat{I}_y(X) + 2\hat{I}_y(A)\hat{I}_x(X)) - (2\hat{I}_y(A)\hat{I}_x(X) - 2\hat{I}_x(A)\hat{I}_y(X))]\cos\omega_A t_1 \\
& + \tfrac{1}{2}[(2\hat{I}_x(A)\hat{I}_y(X) + 2\hat{I}_y(A)\hat{I}_x(X)) + (2\hat{I}_y(A)\hat{I}_x(X) - 2\hat{I}_x(A)\hat{I}_y(X))]\cos\omega_X t_1 \\
& + 2\hat{I}_z(A)\hat{I}_y(X)\sin\omega_A t_1 + 2\hat{I}_y(A)\hat{I}_z(X)\sin\omega_X t_1\}\sin(\pi J_{AX} t_1)
\end{aligned}
\tag{8.36a}
$$

and with equations (8.13a–d)

$$
\begin{aligned}
\sigma_3 =\ & [-\hat{I}_z(A)\cos\omega_A t_1 + \hat{I}_x(A)\sin\omega_A t_1 - \hat{I}_z(X)\cos\omega_X t_1 \\
& \hat{I}_x(X)\sin\omega_X t_1]\cos(\pi J_{AX} t_1) \\
& + \{\tfrac{1}{2}[DQ_y - ZQ_y]\cos\omega_A t_1 + \tfrac{1}{2}[DQ_y - ZQ_y]\cos\omega_X t_1 \\
& + 2\hat{I}_z(A)\hat{I}_y(X)\sin\omega_A t_1 + 2\hat{I}_y(A)\hat{I}_z(X)\sin\omega_X t_1\}\sin(\pi J_{AX} t_1)
\end{aligned}
\tag{8.36b}
$$

For the double-quantum magnetization which has passed the phase cycle unhindered and which finally remains while all longitudinal, zero quantum and anti-phase single quantum terms are cancelled, we find

$$
\sigma_3^{DQ} = \{\tfrac{1}{2}(2\hat{I}_x(A)\hat{I}_y(X) + 2\hat{I}_y(A)\hat{I}_x(X))\cos\omega_A t_1 + \tfrac{1}{2}(2\hat{I}_x(A)\hat{I}_y(X)
$$
$$
+ 2\hat{I}_y(A)\hat{I}_x(X))\cos\omega_X t_1\}\sin(\pi J_{AX} t_1) \tag{8.37}
$$

The third $90°$ pulse with constant $-x$-phase then produces one-quantum magnetization:

$$
\sigma_3^{DQ} = \{\tfrac{1}{2}DQ_y\cos\omega_A t_1 + \tfrac{1}{2}DQ_y\cos\omega_X t_1)\sin(nJ_{AX} t_1) \tag{8.38a}
$$

or

$$
\sigma_4^{DQ} = \{\tfrac{1}{2}(\overset{①}{2\hat{I}_x(A)\hat{I}_z(X)} + \overset{②}{2\hat{I}_z(A)\hat{I}_x(X)})\cos\omega_A t_1 + \tfrac{1}{2}(\overset{③}{2\hat{I}_x(A)\hat{I}_z(X)}
$$
$$
\underset{④}{} + 2\hat{I}_z(A)\hat{I}_x(X))\cos\omega_X t_1\}\sin(\pi J_{AX} t_1) \tag{8.38b}
$$

σ_4^{obs} now evolves during t_2 under the action of chemical shifts and spin–spin coupling. We shall forego here a detailed treatment like the one performed above for the COSY experiment, since already by inspection of equation (8.38b) we can derive the following result: The diagonal peaks (term ① and ④, modulated by ω_A and ω_X in t_1 and t_2, respectively) are detected in both dimensions as anti-phase doublets, since the corresponding magnetization is modulated with $\sin(\pi J_{AX})$ during t_1 as well as t_2 (note the different origin of the diagonal peaks in the COSY- and the COSY-DQF experiment, cf. equation (8.30) and (8.36)). This leads, as for the cross peaks, to partial signal cancellation in the case of signal overlap if the line width is of the order of the coupling. Because the one-quantum magnetization (singlet signals) is already eliminated by the phase cycle, in total a reduction of the diagonal peaks results.

The terms ② and ③, on the other hand, which represent anti-phase X- and A-magnetization, which is modulated during t_1 with ω_A and ω_X and during t_2 with ω_X and ω_A, respectively, yield cross peaks, however, with only half of the intensity of the standard COSY experiment (cf. equation (8.31). The reduction of the diagonal peaks is, therefore, achieved only at the expense of sensitivity.

Exercise 8.10 Analyse the results of COSY experiments with variable pulse angles $\beta = 0°$, $90°$ and $180°$ for the mixing pulse with the help of the general results given in equation (8.2.3) on p. 407 of the monograph by Ernst et al. (see item 5 in the Bibliography on page 549)

5.8   PHASE CYCLES

The importance of phase cycles for the elimination of artefacts in 1D and 2D n.m.r. spectra has already been mentioned. During multi-pulse experiments phase cycles play a far more fundamental part, because signal selection, which is essential for the success of most 2D experiments, is generally achieved with the help of phase cycles. This important aspect of phase cycles is only inadequately or not at all expressed in the general diagrams which are drawn for pulse sequences. Consequently, phase cycles are often regarded as technical detail of less importance. In connection with the product operator formalism we shall, therefore, underline the vital role phase cycles play in 2D n.m.r. and demonstrate with a number of simple examples how phase cycles for multiple pulse experiments can be constructed.

For the discussion of the phase cycling technique, the coherence level diagram introduced in Figure 8.21 (p. 304) is helpful. It shows the coherence transfer pathway and allows us to determine the phases of the participating coherences for every step in a pulse sequence. The important aspect is that the various pulses not only change the coherence level but also the phase of the coherence. In this respect the following rule is important:

If the pulse phase is changed by $\Delta\theta$, those coherences, for which the pulse induces a coherence level shift Δp, change their phase ϕ by $\Delta p \times \Delta\theta$

$$\Delta\phi = \Delta p \times \Delta\theta \qquad (8.39)$$

This relation is illustrated by a number of examples in Figure 8.24a.

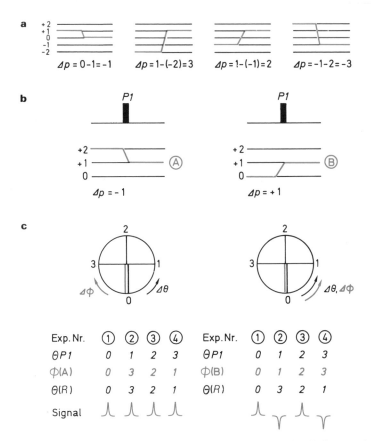

Figure 8.24 (a) Coherence level shifts; (b) coherence level diagram for two coherences which reach the receiver level +1 on different pathways; (c) analysis of the coherence and receiver phase for a phase cycle of four experiments; the phase code used has been introduced already in Chapter 7 (p. 254)

The principle of signal selection is most clearly illuminated with the simple case shown in Figure 8.24b. Assume that from two coherences A and B, which arrive at the coherence level +1 on different pathways, only one shall be detected as signal. In the case of A, double quantum magnetization is involved; in the case of B, the pulse transforms zero quantum magnetization into one quantum magnetization. Consequently, the coherence level shifts are $\Delta p(A) = -1$ and $\Delta p(B) = +1$. If in a series of four experiments the phase of the pulse *P1* is shifted by 90°, the coherence phases $\phi(A)$ and $\phi(B)$, on the basis of the rule given above (equation (8.39), shift as shown in Figure 8.24c. If we are interested in the A signals, the receiver phase must follow the phase of coherence A, $\phi(A)$. The B signals are then cancelled (Figure 8.24c). With constant receiver phase complete signal cancellation results.

In order to derive phase cycles for certain pulse sequences, the coherence pathways in the coherence level diagram must be inspected. Let us remember first that we start always at level 0 and that the first pulse *P1* produces coherences with levels $+1$ and -1. The next pulse *P2* then already excites all possible coherences for the spin system under consideration (Figure 8.25a). Phase cycling, however, puts us into the position to decide which of these coherences later reach the receiver level $+1$. The phase cycle thus has the function of a filter which can be passed only by the desired coherences.

5.8.1 The COSY Experiment

As an exercise for the application of the relations discussed so far, let us analyze the COSY experiment, equation (8.29). The appropriate coherence level diagram is shown in Figure 8.25b. After the second 90° pulse we select those coherence shifts which end at the receiver level $+1$. Other coherence shifts are not shown because they do not yield detectable magnetization.

Because of their different history, the coherences at the receiver level $+1$ have different phases. For coherence A, the coherence level shifts are $+1$ and 0, whereas coherence B has experienced shifts of -1 and $+1$. According to the rule formulated above, phase shifts for the pulses *P1* and *P2* will thus influence the coherence phases differently. Generally, pulse phase shifts of 90° are used with a code of 0, 1, 2 and 3 for relative phase differences of 0°, 90°, 180° and 270°, respectively (see also p. 254).

The effect of the pulse phase shifts for the coherence of interest can be most easily read off from a graphical diagram of the type shown in Figure 8.25c for the

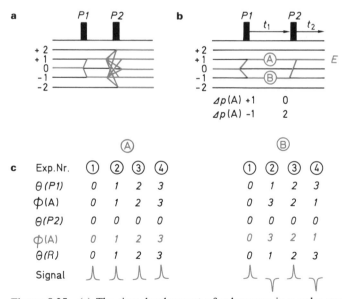

Figure 8.25 (a) The time-development of coherences in a pulse sequence; (b) coherence pathways for the COSY-90 experiment; (c) analysis of a phase cycle for the detection of coherence A in the COSY-90 experiment

coherences which arise in the COSY experiment. In the particular phase cycle shown only the phase of the first 90° pulse is changed. During four single experiments, all run with the same t_1-value, the phase then takes the values 0, 1, 2 and 3. For the coherence of level $+1$, according to Figure 8.25c, the values 0, 1, 2 and 3 result ($\Delta\phi = \Delta p \times \Delta\theta$!). Since the phase of the second 90°-pulse remains constant, the second coherence level shift has no effect ($\Delta p \times 0 = 0$). The receiver phase $\theta(R)$ is adjusted for the detection of an absorption signal with the cycle (0123). For coherence B, however, $\Delta p = -1$ and the coherence phase now follows the cycle (0321). With the given receiver phase this leads to a cancellation of the corresponding signal. On the other hand, a receiver phase cycle (0321) allows us to detect coherence B and to eliminate coherence A.

Exercise 8.11   Draw the receiver signals for both coherences if the alternative receiver phase cycle (0321) is used.

The two signals which arise in the COSY experiment are known as *antiecho* or *P-type* signal (coherence A in Figure 8.25b,c) or as *echo* (also called *coherence transfer echo*) or *N-type* signal (coherence B in Figure 8.25e,c). The abbreviations P and N result from the sense of rotation of the coherences during the evolution time which is positive ($p = +1$) or negative ($p = -1$). For the representation of COSY spectra one uses echo selection (coherence B) and the diagonal runs from the left-hand lower corner to the right-hand upper corner. Quadrature detection in F_1 requires two signals with a phase difference of 90° for each t_1 value. In the older procedure, already mentioned on page 292, these two signals are combined into one FID, which converts the amplitude modulation into a phase modulation. This is achieved with the phase cycle for *P*- or *N*-type selection discussed above. For *N*-type selection, according to Figure 8.25, the required receiver phases are (0321). The pairs of scans 1/2 and 3/4 introduce the necessary phase shift, the pairs 1/3 and 2/4 cancel axial peaks. However, this method does not lead to pure signal phases in both dimensions and the spectra have to be processed in the magnitude mode.

For true phase sensitive COSY experiments the SHR or TPPI method (p. 293) have to be used. In the former, the number of t_1 experiments is doubled and a 90° phase shift for one pulse of the COSY sequence is introduced between both data sets which are then stored as the real and imaginary part, respectively, of the t_1-FID. In the alternative approach, which is based on the Redfield quadrature detection method which we did not discuss here, the phase of one pulse is incremented in 90° steps together with t_1, hence the name *time proportional phase increment* (TPPI).

5.8.2   The COSY-DQF Experiment

During the explanation of the COSY-DQF experiment in Section 4.3.3 we have already mentioned that it is possible to eliminate zero and one-quantum magnetization, which arises after the second 90°-pulse, through an appropriate phase cycle. The final pulse is then used for the transformation of double quantum

Table 8.1 Phase cycle for the COSY-DQF
experiment

Exp.-No.	1	2	3	4	5	6	7	8
			a				b	
$\theta(P1)$	0	1	2	3	2	3	0	1
$\theta(P2)$	0	1	2	3	0	1	2	3
$\theta(P3)$	0	0	0	0	0	0	0	0
$\theta(R)$	0	2	0	2	2	0	2	0

magnetization into detectable one-quantum magnetization. For the construction of a simple phase cycle needed for the selection process, the coherence level diagram of Figure 8.26a (p. 320) applies. It shows, in contrast to the diagram in Figure 8.25b, only coherences for the levels $+2$ and -2 which were not considered before because they could not reach the receiver. Similar considerations as those used above show that a simple (0123) cycle for the $P1$ phase yields (0202) as the necessary receiver phase cycle for the detection of the original double quantum magnetization, if for $P2$ and $P3$ (0123) and (0000), respectively, are used (Table 8.1a). To illustrate this, let us follow the coherence pathway from level 0 through levels $+1$ and $+2$ to level $+1$. Coherence level shifts of $+1$, $+1$, and -1 result, the last one without consequences because $0(P3)$ is unchanged.

The coherence and receiver phases are then obtained with the relation

$$\theta(R) = (+1)\Delta\theta(P1) + (+1)\Delta\theta(P2) :$$
$$\text{Exp.} - \text{Nr. 1} : \quad (+1)(0) + (+1)(0) = 0$$
$$2 : \quad (+1)(1) + (+1)(1) = 2$$
$$3 : \quad (+1)(2) + (+1)(2) = 4 = 0$$
$$4 : \quad (+1)(3) + (+1)(3) = 6 = 2$$

Through the experiments No. 5–8 in Table 8.1b one achieves suppression of axial signals.

For the coherence pathway $0 \rightarrow (-1) \rightarrow (+2) \rightarrow (+1)$ the following applies:

$$\text{Exp.} - \text{No. 1} : \quad (-1)(0) + (+3)(0) = 0$$
$$2 : \quad (-1)(1) + (+3)(1) = 2$$
$$3 : \quad (-1)(2) + (+3)(2) = 4 = 0$$
$$4 : \quad (-1)(3) + (+3)(3) = 6 = 2$$

Therefore, this signal too does reach the receiver. The same is true for the remaining coherences with Δp-values of $+1$, -3, $+3$ and -1, -1, $+3$, with the distinction that the evolution of these coherences has a different sign.

Let us now check the cancellation of the zero- and one-quantum magnetization.

For these signal contributions the coherence level diagram of Figure 8.26b applies with four coherence pathways and Δp-values of $+1, 0, 0; +1, -1, +1; -1, +1, +1$ and $-1, 0, +2$, respectively. Because of the constant $P3$ phase again only the first two coherence level shifts are of importance. They yield the following signal phases:

Exp. – Nr. 1 : $(+1)(0) + (0)(0) = 0$ 1 : $(+1)(0) + (-1)(0) = 0$

        2 : $(+1)(1) + (0)(1) = 1$ 2 : $(+1)(1) + (-1)(1) = 0$

        3 : $(+1)(2) + (0)(2) = 2$ 3 : $(+1)(2) + (-1)(2) = 0$

        4 : $(+1)(3) + (0)(3) = 3$ 4 : $(+1)(3) + (-1)(3) = 0$

Exp. – Nr. 1 : $(-1)(0) + (+1)(0) = 0$ 1 : $(-1)(0) + (0)(0) = 0$

        2 : $(-1)(1) + (+1)(1) = 0$ 2 : $(-1)(1) + (0)(1) = 3$

        3 : $(-1)(2) + (+1)(2) = 0$ 3 : $(-1)(2) + (0)(2) = 2$

        4 : $(-1)(3) + (+1)(3) = 0$ 4 : $(-1)(3) + (0)(3) = 1$

Thus, for the chosen receiver phase cycle *0202* and *2020*, respectively (cf. Table 8.1), a cancellation of the signals transmitted through these coherence pathways is indeed achieved.

Exercise 8.12 Analyse the result for the two additional coherence pathways $0 \rightarrow (+1) \rightarrow (-1) \rightarrow (+1)$ and $0 \rightarrow (-1) \rightarrow (+1) \rightarrow (+1)$ which are not shown in Figure 8.26b.

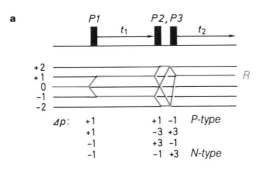

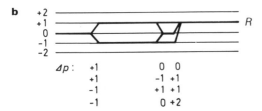

Figure 8.26 Coherence level diagrams for the COSY-DQF experiment; (a) double quantum magnetization; (b) zero- and one-quantum magnetization

6. Gradient enhanced spectroscopy

Apart from the construction of more sophisticated pulse sequences for the detection of spin correlations and magnetization transfer processes, recent experimental developments have also shown that the traditional phase cycling techniques discussed in the preceding section can be replaced in many of the important 2D experiments by the use of linear B_0 field gradients, a basic element of the n.m.r. imaging method (cf. Chapter 10).

If, after a 90_x° excitation pulse, a linear B_0 field gradient ΔG_z is applied along the z-axis to the ensemble of spins present in the n.m.r. sample tube for a time t_G, the Larmor frequencies of nuclei in different volume elements ν_i vary by $\omega + \Delta\omega_i$. Consequently, macroscopic transverse magnetization is dephased and the n.m.r. signal destroyed. This is not surprising, as we have emphasized (p. 60) that field homogeneity is a prerequisite for the detection of n.m.r. signals.

In the case of a short *gradient pulse*, where t_G is in the order of a few milliseconds, however, the signal can be recovered and a so-called *gradient echo* is detected, if within a time where diffusion is negligible the spin system is subjected to a second field gradient with the same amplitude but of opposite polarity, $-\Delta G_z$. This is true if we deal exclusively with single quantum magnetization (SQ).

Imagine now a situation where, in addition to single quantum magnetization, double quantum magnetization (DQ) is generated, as, for example, by application of two successive 90_x° pulses as in the case of a COSY experiment. A first gradient pulse ΔG_z after the 90_x° excitation pulse then changes the Larmor frequencies to $\omega + \Delta\omega$, while a second gradient pulse ΔG_z after the mixing pulse yields $\omega_{SQ} = \omega + 2\Delta\omega$ for the single quantum magnetization but $\omega_{DQ} = \omega + 3\Delta\omega$ for the double quantum magnetization (remember that DQ magnetization created by the mixing pulse rotates with twice the Larmor frequency). Selective rephasing of SQ or DQ can then be achieved by applying a reverse gradient pulse with amplitude $2\Delta G_z$ or $3\Delta G_z$, respectively.

Let us now discuss with the example of the COSY experiment how field gradients can be used for the selection of coherence pathways. On p. 317 we have shown how echo selection (coherence B) is achieved with a simple phase cycle that requires four individual experiments for one t_1 increment. Using field gradients, the same result is obtained already with only one experiment by applying pulse sequence (1) shown in Figure 8.27 (p. 332) The first gradient pulse causes dephasing of coherence B (coherence order $p = -1$) by $-\Delta G_z$ and of coherence A ($p = +1$) by $+\Delta G_z$. After the mixing pulse, coherence B is transferred to coherence level $+1$ and the second gradient pulse of the same amplitude applied before signal detection leads to a dephasing effect of $+\Delta G_z$ which exactly cancels the effect of the first gradient pulse. Thus, coherence B can be detected. On the other hand, for coherence A we have $p = +1$ during t_1 and after the mixing pulse again $p = +1$ and the dephasing amounts to $+2\Delta G_z$. The result is an elimination of coherence A and the technique allows quadrature detection in F_1 (cf. p. 318). In addition, other unwanted coherences (t_1-noise, axial peaks) are also dephased, thereby improving the quality

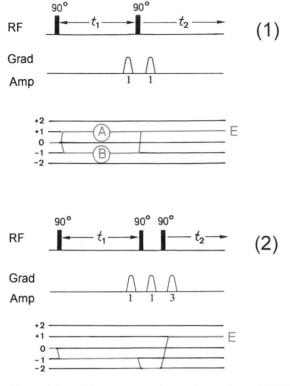

Figure 8.27 Pulse sequences for gradient enhanced COSY and COSY-DQF spectroscopy ((1) and (2), respectively)

of the spectrum. We note further that coherence selection by the field gradient technique occurs for each FID and does not rely on the addition or subtraction of signals as required by the conventional phase cycling technique. The method is thus less sensitive to hardware instabilities because dephasing of unwanted signals occurs even if the gradients are not perfectly adjusted.

In quantitative terms, for the effect Φ of a linear B_0 field gradient pulse on a particular coherence we have

$$\Phi = zp\gamma\Delta G_z t_G \tag{8.40}$$

where z is the distance (in cm) of an individual spin from the gradient origin, p is the coherence order, γ the gyromagnetic ratio (in rad T^{-1} s^{-1}), ΔG_z the gradient strength (in T cm^{-1}), and t_G (in s) the gradient pulse time. The requirement for the detection of the desired magnetization obviously is $\sum \Phi = 0$.

Coherence selection is also important for the COSY-DQF experiment and can be achieved with the gradient technique by pulse sequence (2) (Figure 8.27). In order to see how this sequence works we follow the coherence pathway $0 \rightarrow (-1) \rightarrow (-2) \rightarrow (+1)$ for the N-type signal. Application of the first gradient pulse leads to a dephasing proportional to $-\Delta G_z$ which increases after the second gradient pulse to $-3\Delta G_z$. The desired echo is finally generated by a refocusing gradient $+3\Delta G_z$ which is applied after single quantum magnetization has been re-established by the third 90°_x pulse. The P-type signal (see Fig. 8.26a, p. 320), on the other hand, dephases with $+6\Delta G_z$.

Exercise 8.13 Discuss the consequences of pulse sequence (2) for the remaining double quantum coherences as well as for the zero- and one-quantum coherences.

In the meantime, the gradient technique has been installed in the majority of pulse sequences for homo- and heteronuclear shift correlations (see also Chapter 11) and, as compared to the phase cycling technique, improves these experiments with respect to the suppression of artefacts as well as to the measuring time necessary and thus leads to a better signal-to-noise ratio.

7. Universal building blocks for pulse sequences

As we discuss more advanced pulse sequences in Chapters 10 and 11, we shall see that the spin echo experiment (p. 237) is repeatedly used as an important building block that refocuses inhomogeneity and chemical shift effects. Similarly, in a heteronuclear sequence a simple 180° X- or A-pulse can be introduced at the centre of a time interval in order to decouple the X from the A nuclei and vice versa. In the following, we discuss briefly four other important building blocks frequently used in modern pulse n.m.r. spectroscopy. A graphical representation of these experiments is given in Figure 8.28 (p. 324).

7.1 CONSTANT TIME EXPERIMENTS: ω_1-DECOUPLED COSY

In the COSY-90 sequence (Figure 8.12, p. 289), homonuclear coupling operates during t_1 and t_2. Accordingly, splittings due to spin–spin coupling appear in both frequency dimensions, F_1 and F_2. Homonuclear decoupling in F_1 requires that J-modulation of transverse magnetization during t_1 is eliminated. This is achieved by introducing a non-stationary 180° pulse in a constant time interval, Δ, between the two 90° pulses:

$$90^\circ\text{-----}t_1/2\text{-----}180^\circ\text{-----}t_1/2\text{-----}|\Delta - t_1, \ 90^\circ, \ \text{FID} \ (t_2)$$

$$|\xleftarrow{\hspace{4cm}} \Delta \xrightarrow{\hspace{4cm}}| \tag{8.41}$$

Since the 180° pulse effects both the A and the X nucleus, the evolution of coupling within the constant time interval Δ is unaffected and identical for every t_1 experiment. Hence, there is no signal modulation by J-coupling. On the other hand, chemical shift effects are refocused during t_1, but evolve during the remaining

a)
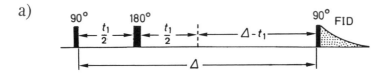

b)
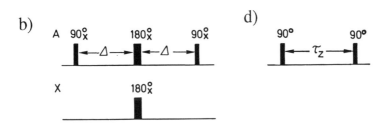

d)

c)
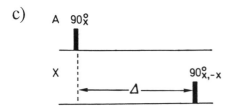

Figure 8.28 Building blocks for modern pulse sequences: (a) the 'constant time' experiment for ω_1-decoupled COSY spectroscopy; (b) the BIRD pulse; (c) low-pass filter; (d) z-filter

interval $\Delta - t_1$, which is the true evolution time. Clearly, by shifting the 180° pulse, $\Delta - t_1$ is incremented. The result is a ω_1-decoupled COSY spectrum which yields in F_1 a '^{1}H-decoupled ^{1}H spectrum', that is a singlet for every proton resonance. Resolution in F_1 is thus greatly improved and the ^{1}H chemical shifts are immediately available.

7.2   BIRD PULSES

For a heteronuclear AX system it is often important to separate transverse magnetization due to coupled spins from magnetization due to uncoupled or weakly coupled spins. For this purpose the following pulse sandwich can be used:

$$90^\circ_x\text{-----}\Delta\text{-----}180^\circ_x(A), \ 180^\circ_x(X)\text{-----}\Delta\text{-----}90^\circ_{-x} \qquad (8.42)$$

If the delay Δ is set equal to $1/2J(A,X)$, coupled A-magnetization $\hat{I}_y(A)$ evolves to form anti-phase A-magnetization $2\hat{I}_z(X)\hat{I}_x(A)$. The 180° pulses change the sense of rotation and after the second Δ delay this magnetization is refocused along the $+y$-axis and transferred into $+z$-magnetization by the 90°_{-x} pulse. Uncoupled or only weakly coupled A-magnetization, on the other hand, evolves according to the following scheme:

$$
\begin{array}{ccccc}
90^{\circ}_{x} & \Delta & 180^{\circ}_{x} & \dot{\Delta} & 90^{\circ}_{-x} \\
\hat{I}_z \longrightarrow \hat{I}_y \longrightarrow \hat{I}_y \longrightarrow & -\hat{I}_y \longrightarrow & -\hat{I}_y \longrightarrow & -\hat{I}_z &
\end{array}
\tag{8.43}
$$

and is thus selectively inverted. This sequence is known as 'bilinear rotation operator' or BIRD pulse.

7.3 LOW-PASS FILTER

Another building block for heteronuclear situations which yields a separation of coupled from non-coupled or weakly coupled magnetization employs two 90° pulses separated by a delay Δ:

$$
90^{\circ}_{x}(A)\text{-----}\Delta\text{-----}90^{\circ}_{x,\,-x}(X)
\tag{8.44}
$$

For $\Delta = 1/2J(A,X)$, anti-phase magnetization which evolved after the first 90° pulse is transformed into double quantum magnetization which changes sign with the phase change for the second 90°(X) pulse. Addition of two experiments destroys this magnetization, while that of uncoupled A spins is essentially unchanged and that of weakly coupled spins only slightly reduced because evolution to anti-phase magnetization is here much slower.

7.4 z-FILTER

An efficient way to select a desired magnetization component and purging a pulse sequence from undesired coherences is the so-called z-filter. It consists of a 90° pulse pair separated by a delay τ_z:

$$
90^{\circ}\text{-----}\tau_z\text{-----}90^{\circ}
\tag{8.45}
$$

The first pulse is used to transform a desired magnetization into z-magnetization. During the following delay τ_z all remaining transverse components oscillate and are effectively eliminated if experiments performed with different τ_z values are coadded. The desired magnetization, which was stored on the z-axis during τ_z, is transformed into transverse magnetization suitable for detection by the second 90° pulse.

8. Homonuclear shift correlation by double quantum selection of AX systems—the 2D-INADEQUATE experiment

After being familiar with the phenomenon of double quantum coherence, another correlation experiment based on double quantum magnetization will now be discussed. It was originally developed for ^{13}C n.m.r. spectroscopy in order to facilitate the recognition of neighbouring ^{13}C pairs and is known as INADEQUATE (*i*ncredible *n*atural *a*bundance *d*ouble *q*uantum *t*ransfer) experiment. It can be applied rather straightforwardly also to ^{1}H n.m.r. spectra where, because of the high natural abundance of protons, it does not suffer from low sensitivity.

The basic idea of the INADEQUATE pulse sequence is to use a two-dimensional experiment in order to separate overlapping, weakly coupled two-spin systems of the AX type. The separation then yields a unique assignment. Since in a chain of n.m.r. active nuclei each spin, with the exception of those at chain ends, has two neighbours and, therefore, participates in two different AX spin systems, the identification of the AX systems reveals the spin connectivities in a particular molecular structure.

Each AX-system is characterized by its Larmor frequencies v_A and v_X as well as by the double quantum frequency $v_{DQ} = v_A + v_X - 2v_0$ (v_0 = transmitter frequency). On a frequency axis $F_1 = v_{DQ}$ it can thus be distinguished from all other systems with different v_{DQ}-values, if in a two-dimensional spectrum the F_2-axis contains the Larmor frequencies v_A and v_X, and the F_1 frequency axis the double quantum frequency v_{DQ}.

In principle, the experiment can be performed with the pulse sequence equation (8.46):

$$90^\circ_x \text{---} 1/4J_{AX} \text{---} 180^\circ_x \text{---} 1/4J_{AX} \text{---} 90^\circ_x \text{---} t_1 \text{---} 90^\circ_x,\ t_2 \qquad (8.46)$$

We shall use this sequence in order to illustrate the spin physics behind the experiment more closely. For practical applications a number of modifications are necessary, which will be discussed later.

An analogy between the INADEQUATE and the COSY-DQF experiment may be seen if the pulse sandwich of the preparation period, 90°, 1/4J, 180°, 1/4J, 90°, which is in essence a spin echo sequence followed by a 90° pulse, is regarded as equivalent to the first two 90° pulses of the COSY-DQF sequence. During the evolution time t_1, which follows, double quantum magnetization evolves that is converted to observable single quantum magnetization by a 90° pulse as read pulse. A more detailed analysis is possible within the framework of the product operator formalism.

Let us start with longitudinal A and X magnetization and the 90°-excitation pulse:

$$\sigma_0 = \hat{I}_z(A) + \hat{I}_z(X)$$
$$\downarrow\ 90^\circ[\hat{I}_x(A) + \hat{I}_x(X)]$$
$$\sigma_1 = \hat{I}_y(A) + \hat{I}_y(X)$$

The effect of the delay 1/4J_{AX}, 180°_x, 1/4J_{AX} can easily be visualized with the help of the classical Bloch vector picture: the components of the A and the X doublet fan out and will be turned around the *x*-axis by the 180°_x pulse. Because this pulse is non-selective, it exchanges the spin states of the A as well as the X nucleus. Consequently, the doublet components continue to fan out until they are in anti-phase on the *x*-axis after the second 1/4J_{AX} interval. In this way pure anti-phase A and X magnetization is produced and phase differences which result from differences in the Larmor frequencies are eliminated by the 180° pulse. Accordingly, the density operator σ_2 results:

$$\sigma_2 = 2\hat{I}_x(A)\hat{I}_z(X) + 2\hat{I}_z(A)\hat{I}_x(X)$$

Exercise 8.14   Derive σ_2 with the help of equation (8.23b) (p. 308).

The next 90_x° pulse yields

$$\sigma_3 = 2\hat{I}_x(A)\hat{I}_y(X) + 2\hat{I}_y(A)\hat{I}_x(X)$$
$$= (1/i)[\hat{I}^+(A)\hat{I}^+(X) - \hat{I}^-(A)\hat{I}^-(X)]$$

that is pure double quantum magnetization of the order $p = 2$ and $p = -2$, which develops during the evolution time t_1 according to equation (8.24) with the sum of the Larmor frequencies $v_A + v_X$. The frequency axis F_1 thus contains the double quantum frequencies of the AX systems. Transformation of the double quantum magnetization in detectable transverse magnetization is achieved by the last pulse:

$$\sigma_3$$
$$\downarrow 90°[\hat{I}_x(A) + \hat{I}_x(X)]$$
$$\sigma_4 = -2\hat{I}_x(A)\hat{I}_z(X) - 2\hat{I}_z(A)\hat{I}_x(X)$$

which generates anti-phase A and X magnetization, detectable as an anti-phase doublet at v_A and v_X, respectively. Using magnitude representation, positive signals are obtained.

The result of a 2D-INADEQUATE experiment is shown in Figure 8.29 for the mixture of E- and Z-2-chloroacrylic acid in $CDCl_3/CHCl_3$ discussed already on p. 301. Both AX-systems are clearly separated and the centres of both spectra fall on a line with the inclination 2, since their F_2 frequency is $(v_A + v_X)/2$, their F_1 frequency, however, is $v_A + v_X$. As in the COSY-DQF experiment the solvent signal (one-quantum magnetization) is eliminated.

As compared to equation (8.40), the pulse sequence of the 2D-INADEQUATE experiment is somewhat modified for practical applications. A 135° readpulse eliminates unwanted signals which arise in cases where the A and/or the X nucleus are further coupled to a third nucleus. In addition, for any t_1-value a second sequence is applied where a phase shift of 90° for the detected signal results. In this way quadrature detection in F_1 is achieved. The phase cycle of the INADEQUATE experiment is relatively complicated, since it combines the selection of double quantum magnetization with the CYCLOPS cycle for quadrature detection and the suppression of axial signals. As in other cases, signal selection is simplified by the use of gradient pulses.

In comparison to the standard COSY experiment which yields practically the same correlation information, the 2D-INADEQUATE experiment has the advantage that

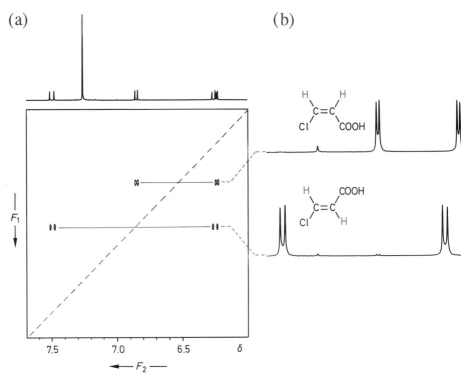

Figure 8.29 2D INADEQUATE ^{1}H,^{1}H n.m.r. spectrum of a mixture of Z- and E-2-chloroacrylic acid at 400 MHz (olefinic region); (a) contour diagram; (b) F_2-traces of the two AX-systems. For the $1/4J$ delay (23 ms) the average of the two vicinal ^{1}H,^{1}H-coupling constants (J_{cis} = 8.4, J_{trans} = 13.4 Hz) was used

diagonal signals are absent. Its dependence on the $1/4J_{AX}$ delays may be seen as a disadvantage. In addition, small coupling constants will lead to long measuring times. However, the experiment is not very critical with respect to the correct choice of J_{AX} and AX systems with similar couplings can be detected simultaneously without difficulties. In the case of larger spin systems, where more than two protons are involved (for example AMX-systems), magnetization transfer to so-called passive spins, only weakly coupled to the A or X nucleus, yields additional signals. As mentioned above, a variation of the readpulse angle eliminates these peaks. In general, satisfactory results for ^{1}H,^{1}H INADEQUATE spectra can always be expected if the structure of interest is dominated by vicinal coupling constants.

The usefulness of the 2D-INADEQUATE experiment will finally be demonstrated with two examples. In Figure 8.30 we show how the pulse sequence can be used for spectral analysis in the case of mixtures, where strong signal overlap prevents the

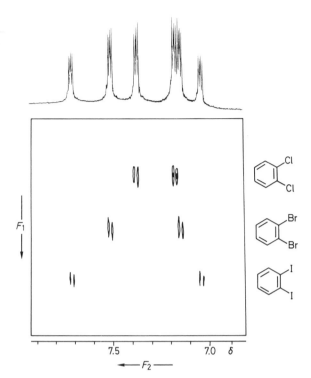

Figure 8.30   Contour diagram of the 400 MHz 2D-INADEQUATE ^{1}H n.m.r. spectrum of a mixture of o-dichloro-, o-dibromo-, and o-diiodobenzene with 1D spectrum. For the 2D experiment the following parameters were used: 64 t_1 increments, 32 scans each; sweep width in F_1 500 Hz, in F_2 1 kHz, $1/4J$-delay 40 ms (optimized for the N-parameter of ca. 6 Hz); 2 s relaxation delay, digital resolution in F_1 1.95 Hz/pt, in F_2 7.8 Hz/pt; measuring time 2.7 h (after Ref. 6)

assignment and the identification of subspectra. In the present case spectra of the AA'XX'-type, as are found in ortho- or para-disubstituted benzenes, have been separated by a 2D-INADEQUATE experiment in form of a 'spin-chromatography'. The N-parameter (cf. p. 186), which is approximately 6 Hz, was chosen to adjust the $1/4J_{AX}$ delay.

The possibility of evaluating vicinal proton connectivities for spectral assignments is finally demonstrated in Figure 8.31 with the results of a 2D-INADEQUATE experiment for the ^{1}H n.m.r. spectrum of adenosine, where the assignment of the ribose protons, which have vicinal coupling constants of the order of 3.0–9.7 Hz, is achieved. With an $1/4J$-delay of 83 ms, which corresponds to a coupling of 3 Hz, all neighbouring protons can be recognized (Figure 8.31b). Only the signals due to the geminal protons 5'-H and 5"-H are not detected. The coupling amounts here to

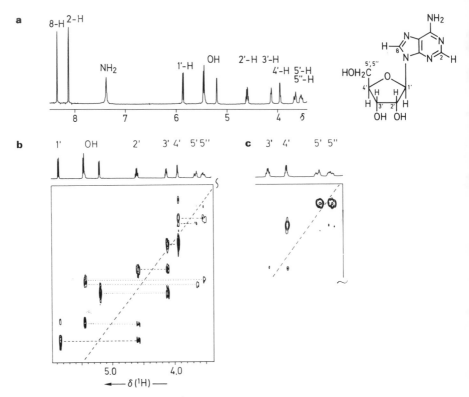

Figure 8.31 2D-INADEQUATE ^{1}H n.m.r. spectrum of adenosine at 400 MHz; solvent DMSO-d_6; (a) 1D spectrum; (b) 2D-INADEQUATE spectrum in the ribose region with $1/4J = 83$ ms ($\triangleq 3$ Hz coupling); **H−C−C−H** correlation − − − − ; **H−C−O−H** correlation $\cdots$. For one F_2 frequency one finds up to 3 correlation signals; (c) part of the 2D-INADEQUATE spectrum with $1/4J = 50$ ms ($J = 5$ Hz)

12.7 Hz and a second experiment with $1/4J = 50$ ms had to be performed (Figure 8.31c). It is interesting to note that the assignment of the OH signals is also feasible, because in the solvent dimethylsulphoxide the hydroxyl protons exchange slowly and couple with the protons at the ribose ring.

9. Three-dimensional NMR experiments

A consistent extension of the principle of two-dimensional n.m.r. spectroscopy has led in recent years to the introduction of three-dimensional and even four-dimensional experimental techniques which are important and valuable tools for the analysis of complicated spectra of biological macromolecules (peptides, proteins, nucleic acids). The basic philosophy behind such experiments will be illustrated in this section with a short description of a 3D experiment.

Through the introduction of a second evolution time in a known two-dimensional pulse sequence, during which the spin system develops under the action of a different Hamilton operator, it is possible to generate a three-dimensional spectrum after a three-fold Fourier transformation of the experimental data. The resulting signals are characterized by three different frequency parameters in the frequency space F_1, F_2, F_3:

$$90°, \quad t_1, t_2, t_3 \rightarrow S(t_1, t_2, t_3) \xrightarrow{\text{1. FT}} S(F_1, t_2, t_3) \xrightarrow{\text{2. FT}}$$

$$S(F_1, F_2, t_3) \xrightarrow{\text{3. FT}} S(F_1, F_2, F_3) \qquad (8.47)$$

This principle is illustrated with the simple example of a heteronuclear J-resolved three-dimensional ^{13}C spectrum for a CHD-group. The three frequency parameters of interest, which are to be separated, are the chemical shift of the ^{13}C nucleus, $\delta(^{13}C)$, and the spin–spin coupling constants $^1J(^{13}C,^1H)$ and $^1J(^{13}C,^2H)$. Consequently, a J,J,δ-spectrum results.

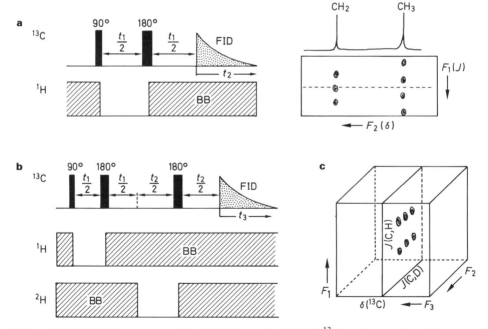

Figure 8.32 (a) Pulse sequence for the heteronuclear 2D J,δ-^{13}C-spectrum and contour diagram for a CH_2 and CH_3 group (schematically); (b) pulse sequence for heteronuclear 3D-J,J,δ ^{13}C spectrum; (c) results for a CHD-group in the frequency space F_1, F_2, F_3: doublet splitting by $J(^{13}C,^1H)$ in F_1, triplet splitting by $J(^{13}C,^2H)$ in F_2 chemical shift $\delta(^{13}C)$ in F_3

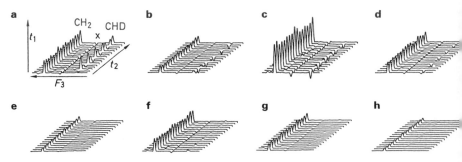

Figure 8.33 Results of 3D J,J,δ-^{13}C experiment for a mixture of diphenylmethane and diphenylmethane-d$_1$ after single Fourier transformation; aliphatic region with resonances of the CH$_2$ and CHD group (X artefact of the pulse transmitter). Shown are the pseudo-2D spectra $S(t_1,t_2,F_3)$ which are obtained after Fourier transformation with respect to t_3 for various t_1 and t_2 values. As expected, only the ^{13}C signal of the CHD group shows modulation in t_2 and t_1. The signal of the CH$_2$ group shows amplitude modulation only in the t_1-dimension (after Ref. 7)

Let us focus first on the "simple" heteronuclear 2D J,δ-spectrum. As with the homonuclear case (cf. p. 283), a ^{13}C spin echo experiment is performed. ^{13}C,^{1}H coupling must be eliminated during half of the evolution time, since otherwise no signal modulation during t_1 would result (see Figure 8.7, p. 281). The experiment is, therefore, performed with gated ^{1}H-decoupling (Fig. 8.32a). Because the 180° pulse only affects the ^{13}C nucleus, the effects of the Larmor frequencies during t_1 are completely eliminated. Therefore, only the ^{13}C,^{1}H coupling constants are detected in F_1, while F_2 contains the chemical shifts, $\delta(^{13}C)$.

For the proposed 3D J,J,δ-spectrum one can now proceed according to Figure 8.32b. ^{13}C magnetization is modulated during t_1 and t_2 with the ^{1}H- and ^{2}H-decoupler, respectively. After the first Fourier transformation with respect to t_3, which is now the detection time, one obtains the 2D spectra shown in Figure 8.33. Here the high field ^{13}C signal of the CHD group clearly shows the modulation of its signal amplitude with respect to the time axes t_2 *and* t_1. The low field signal comes from a ^{13}CH$_2$-group and, as expected, shows amplitude modulation only with respect to t_1 (compare the decoupler sequence of Figure 8.31b). After complete Fourier transformation with respect to all three time domains one expects finally six points in the F_1,F_2,F_3 frequency space which are shown schematically in Figure 8.32c.

It is evident that these types of experiments need long measuring times since *two* evolution times, t_1 and t_2, must now be incremented. With a minimum of 16 t_1 and t_2 values already $16^2 = 256$ experiments are necessary. Nevertheless, the development of powerful data processing systems paved the way for growing use of 3D experiments which combine several 2D experiments, for example homo- and heteronuclear shift correlations or NOESY and COSY pulse sequences in order to analyze complicated ^{1}H spectra with strong signal overlap. From the practical point of view especially 3D shift correlation experiments are important. 3D n.m.r.

spectroscopy has thus become an attractive and powerful technique for special applications, mainly in the field of biopolymers, not withstanding the long measuring times which require excellent hardware and field stability.

10. References

1. H. Günther and P. Schmitt, *Kontakte (Merck)*, **1985** (2), 3.
2. H. Günther and D. Moskau, *Kontakte (Merck)*, **1986** (2), 41.
3. R. Benn and H. Günther, *Angew. Chem.*, **95**, 381 (1983); *Angew. Chem. Int. Ed. Engl.*, **22**, 350–380 (1983).
4. D. Moskau and H. Günther, *Angew. Chem.*, **99**, 151 (1987); *Angew. Chem. Int. Ed. Engl.*, **26**, 1212–1220 (1987).
5. H. Hausmann, *PhD Thesis*, University of Siegen, 1991.
6. D. Schmalz, *PhD Thesis*, University of Siegen, 1989.
7. J. R. Wesener, *PhD Thesis*, University of Siegen, 1985.

RECOMMENDED READING

J. K. M. Sanders and B. K. Hunter, *Modern NMR Spectroscopy—A Guide for Chemists*, Oxford University Press, Oxford, 1987.

A. E. Derome, *Modern NMR Techniques for Chemistry Research*, Pergamon Press, Oxford, 1987.

G. E. Martin and A. S. Zektzer, *Two Dimensional NMR Methods for Establishing Molecular Connectivity—A Chemist's Guide to Experiment Selection, Performance, and Interpretation*, VCH Publishers, New York, 1988.

W. S. Brey (Ed.), *Pulse Methods in 1D and 2D Liquid-Phase NMR*, Academic Press, New York, 1988.

R. R. Ernst, G. Bodenhausen and A. Wokaun, *Principles of Nuclear Magnetic Resonance in One and Two Dimensions*, Clarendon Press, Oxford, 1987.

Review articles

(a) R. Benn and H. Günther, Modern Pulse Methods in High Resolution NMR Spectroscopy, *Angew. Chem.*, **95**, 381 (1983); *Angew. Chem. Int. Ed. Engl.*, **22**, 350 (1983).
(b) W. E. Hull, Experimental Aspects of Two-dimensional NMR, in W. R. Croasmun and R. M. K. Carlson (Eds.), *Two-Dimensional NMR Spectroscopy, Methods in Stereochemical Analysis*, Vol. 9, VCH Publishers, New York, 2nd ed. 1994, p. 67.
(c) G. A. Gray, Introduction to Two-dimensional NMR-Methods, as (b), p. 1.
(d) G. A. Morris, Modern NMR Techniques for Structure Elucidation, *Magn. Reson. Chem.*, **24**, 371 (1986).
(e) D. L. Turner, Multiple Pulse NMR in Liquids, in F5, **16**, 311 (1984).
(f) D. L. Turner, Basic Two-dimensional NMR, in F5, **17**, 281 (1985).
(g) O. W. Sørensen, G. W. Eich, M. H. Levitt, G. Bodenhausen and R. R. Ernst, Product Operator Formalism for the Description of NMR Pulse Experiments, in F5, **16**, 163 (1984).
(h) H. Kessler, M. Gehrke and C. Griesinger, Two-Dimensional NMR Spectroscopy: Background and Overview of the Experiments, *Angew. Chem.*, **100**, 507 (1988); *Angew. Chem. Int. Ed. Engl.*, **27**, 490 (1988).

(i) J. Keeler, Phase Cycling Procedures in Multiple Pulse NMR Spectroscopy of Liquids, in P. Granger and R. K. Harris, (Eds.), *Multinuclear Magnetic Resonance in Liquids and Solids*, Kluwer, Dordrecht, 1990.

(j) K. R. Williams and R. W. King, The Fourier Transform in Chemistry—NMR Part 3. Multipulse Experiments, *J. Chem. Educ.*, **67**, A93 (1990); Part 4. Two-Dimensional Methods, *J. Chem. Educ.*, **67**, A125 (1990).

(k) G. M. Clore and A. M. Gronenborn, Application of Three- and Four-Dimensional Heteronuclear NMR Spectroscopy to Protein Structure Determination, in F5, **23**, **43** (1991).

(l) C. Griesinger, 3D-NMR Spectroscopy in High Resolution NMR, in I. Bertini, H. Molinari, N. Niccolai (Eds.), *NMR and Biomolecular Structure*, VCH Publishers, New York, 1991.

9 THE INFLUENCE OF DYNAMIC EFFECTS ON ¹H NUCLEAR MAGNETIC RESONANCE SPECTRA

As we have mentioned in the introduction and in the previous chapter, nuclear magnetic resonance spectroscopy can be applied to study fast reversible reactions because the line shape of n.m.r. signals is sensitive to chemical exchange processes if these affect the n.m.r. parameters of the nucleus in question. The n.m.r. spectra of many compounds are therefore temperature dependent. In the following the physical basis of this phenomenon, known today as dynamic n.m.r., will be discussed and applications in organic chemistry will be illustrated with reference to specific examples.

1. The exchange of protons between positions with different Larmor frequencies

Let us consider at the outset the classical example of *N,N*-dimethylformamide. The C$-$N bond between the carbonyl group and the nitrogen atom in this compound has significant double bond character as is represented by the contribution of structure (b) to the resonance hybrid. In the lowest energy planar conformation the protons of the two methyl groups are in different chemical environments and therefore they have different resonance frequencies, v_A and v_B. Internal rotation around the N$-$CO bond leads to an intramolecular exchange of the methyl groups (a) $\rightleftharpoons$ (a)′; however, because of the high energy barrier to rotation (about 88 kJ mol^{-1} or 21 kcal mol^{-1}) the exchange frequency at room temperature is low. The residence time of the methyl groups in positions *cis*- or *trans* to the carbonyl group is thus relatively long and consequently two separate resonance signals are observed (see Fig. 3 of the Introduction).

If the temperature is raised these signals broaden and finally, at temperatures above 120 °C, coalesce to a single line. Apparently, at the elevated temperatures the ability to differentiate between the *cis*- and the *trans*-methyl groups is lost.

A similar temperature-dependent phenomenon is observed for the proton resonance signals of 2-methyloxepine (**137**). If we single out the absorption of the methyl group the change in the n.m.r. spectrum as a function of temperature is clearly established, as is shown in Figure 9.1. This variation with temperature can be rationalized on the basis of the valence isomerization between 2-methyloxepine and

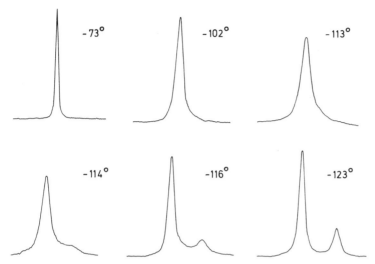

Figure 9.1 ¹H n.m.r. absorption of the methyl protons in 2α-methyloxepine (**137**) as a function of temperature (°C)

2-methylbenzene oxide (c ⇌ d), which is fast at room temperature and which causes a periodic change in the chemical environment and thus in the Larmor frequency of the methyl protons. Here, where the energy barrier is only about 30 kJ mol⁻¹ (7 kcal mol⁻¹) the methyl resonances for the two isomers can be observed only at very low temperature. Furthermore, one finds an intensity difference for these signals since both isomers are of different energy and therefore are present in different concentrations.

A further important type of temperature-dependent variation is illustrated in the spectrum of methanol (Figure 9.2). At −65 °C the splitting pattern expected for an AX_3 system is observed. The fine structure caused by spin–spin coupling becomes less sharp and finally vanishes at higher temperatures as the result of increased line broadening until, at 37 °C, sharp signals, that however show no splitting, are again observed. In addition to the loss of spin–spin splitting we note that the chemical shift difference between the OH and the CH_3 resonances decreases: the OH resonance moves to higher field since hydrogen bonds dissociate with increasing temperature.

The common cause of the effects described above is the very small energy difference that exists between magnetic sites of different Larmor frequency. In order to measure such small differences, the life time, τ, of the nuclei in each site must be sufficiently long. According to the uncertainty principle, the lower limit for τ is given by

$$\tau\delta v \approx 1/2\pi \qquad (9.1)$$

where $\delta v = \Delta E/h$ is the frequency difference $v_A - v_B$ involved. If τ becomes too small, a time-averaged spectrum is observed.

In the case of dimethylformamide, equation (9.1) is violated if the frequency of

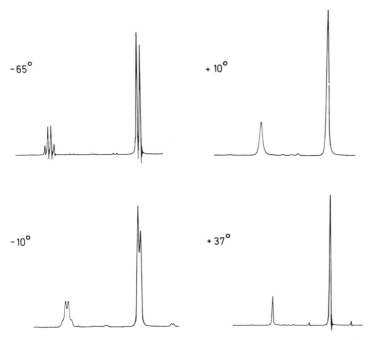

Figure 9.2 The proton resonance spectrum of methanol as a function of temperature (°C)

internal rotation increases with temperature causing the life time of the methyl protons in positions of different Larmor frequency to decrease. The same is true for the interconversion of the valence tautomers 2-methyloxepine and 2-methylbenzene oxide, where fast reaction rates at room temperature shorten the life time below the acceptable limit. Similarly, in the case of methanol, the observation of spin–spin splitting corresponds to a detection of very small energy differences. The OH proton is bound to OCH$_3$ residues that have different total spin ($+\frac{3}{2}$, $+\frac{1}{2}$, $-\frac{1}{2}$, $-\frac{3}{2}$) and different effective Larmor frequencies result. In order to resolve the corresponding energy differences the life time of the OH proton at each site must satisfy equation (9.1) and the intermolecular exchange through hydrogen bonds must be slow. Otherwise, the differentiation between individual signals disappears and only a time-averaged spectrum is recorded.

The ability of the n.m.r. spectrometer to detect such phenomena has an obvious analogy in its likeness to a camera loaded with a film of low sensitivity. In order to obtain a sharp picture, we need a long exposure time and therefore our object, the nucleus in question, must have a relatively long residence time in a given environment.

Infrared and ultraviolet spectroscopy, in comparison, operate substantially faster and time averages of spectral parameters are usually not observed. The explanation for this difference lies in the fact that the individual bands in these spectra represent distinct energies. Thus a difference in wavenumbers of 10 cm^{-1} is equivalent to a ΔE of 119.7 J mol^{-1} ($28.6 \text{ cal mol}^{-1}$) while, as already mentioned above, a difference of 100 Hz at 60 MHz corresponds to a value of only about $4.2 \times 10^{-7} \text{ J mol}^{-1}$ ($10^{-7} \text{ cal mol}^{-1}$).

1.1 THE QUANTITATIVE DESCRIPTION OF DYNAMIC NUCLEAR MAGNETIC RESONANCE

For establishing a quantitative description concerning the correlation between the line shape of the n.m.r. spectrum on the one hand and the mechanism and kinetics of dynamic processes of the type described above on the other, we must find a relationship between the life time of the protons in positions A and B and the line shape of the nuclear magnetic resonance signal. This is possible on the basis of the Bloch equations which describe the shape of the resonance signal as a function of frequency, v, and transverse relaxation time, T_2. Since, as noted on p. 233, chemical exchange processes represent an effective mechanism for transverse relaxation, their influence on the n.m.r. line shape is a function of T_2.

For the present case—a dynamic process that causes a change of Larmor frequencies $v_a \rightleftharpoons v_b$—the Bloch equations, which in their usual form allow only the calculation of n.m.r. signals at v_a or v_b, must be modified. In addition to the normal relaxation effects, time-dependent changes of the magnetization at each site are caused by the chemical equilibrium

$$A \underset{k_B}{\overset{k_A}{\rightleftharpoons}} B$$

Thus, the x,y-magnetization at site A, M_A, is increased through arriving nuclei by an amount proportional to $k_B M_B$. On the other hand, because of the departing nuclei, it suffers a loss proportional to $k_A M_A$. An analogous situation develops at site B. The explicit consideration of these terms finally leads, as we will show in more detail in the Appendix, to the following equation for the *line shape g(v)* of the resonance signal

$$g(v) = [(1 + \tau\pi\Delta)P + QR]/(4\pi^2 P^2 + R^2) \tag{9.2}$$

where

$$P = (0.25\Delta^2 - v^2 + 0.25\delta v^2)\tau + \Delta/4\pi$$
$$Q = [-v - 0.5(p_A - p_B)\delta v]\tau$$
$$R = 0.5(p_A - p_B)\delta v - v(1 + 2\pi\tau\Delta)$$

in which τ represents $\tau_A\tau_B/(\tau_A + \tau_B)$. The symbols above have the following meanings:

τ_A, τ_B	Average life times of the nuclei in positions A and B, respectively (s).
p_A, p_B	Molar fractions of components A and B, respectively.

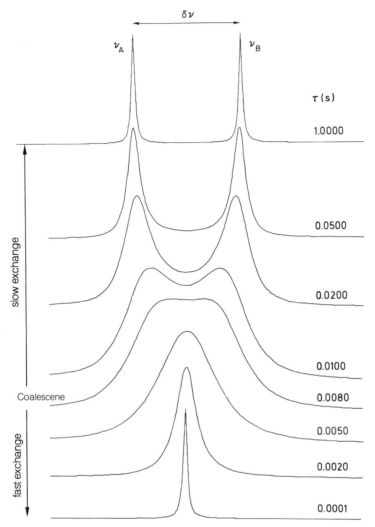

Figure 9.3 Theoretical nuclear magnetic resonance spectra for an exchange process $A \rightleftarrows B$ as a function of the parameter τ

$\delta v = v_A - v_B$ Difference between the resonance frequencies of the nuclei in positions A and B (Hz)

Δ Width at half-height (Hz) of the signal in the absence of exchange $(\tau \to \infty)$ in which case Δ_A would be set equal to Δ_B for simplicity.

v Variable frequency (Hz).

For the *average life times* τ_A and τ_B the relationships in equation (9.3) also hold:

$$\tau_A = \tau/p_B = 1/k_A \text{ and } \tau_B = \tau/p_A = 1/k_B \qquad (9.3)$$

In Figure 9.3 the result of a calculation according to equation (9.2) is represented for a typical case of an exchange process. With the data $\delta v = 30$ Hz, $p_A = p_B = 0.5$ and $\Delta = 1.0$ Hz one obtains different line shapes, $g(v)$, for different τ values. For large values separate signals at v_A and v_B are observed. This area is called the region of *slow exchange*. At the *coalescence point* the two signals merge into a broad band and in the region of *fast exchange* the spectrum becomes a singlet with a normal line width that is recorded at a frequency $(v_A + v_B)/2$. In the presence of fast exchange, each n.m.r. parameter $\tilde{P}$ is an average value according to

$$\tilde{P} = p_A P_A + p_B P_B \qquad (9.4)$$

where P_A and P_B are the parameters in question at positions A and B, respectively. In general, for n positions

$$P_{obs} = \sum_i p_i P_i (i = 1, 2, n) \qquad (9.5)$$

1.2   RELATIONSHIPS TO REACTION KINETICS

The significance of equation (9.2) must be seen in the fact that it opens up the possibility of determining rate constants if a reversible chemical reaction is accompanied by a change in the Larmor frequency of one or more nuclei. The order of magnitude of the rate constants that can be determined using this technique lies between 10^{-1} and 10^5 s^{-1}. The dynamic processes of interest are first-order reactions and are characterized by energy barriers of between 20 and 100 kJ mol^{-1} (5 and 25 kcal mol^{-1}). In general, the kinetics of these processes are too fast to be determined by using classical methods. As we shall show, especially in Section 2 of this chapter, our knowledge of the internal dynamics of molecules has been expanded tremendously by means of n.m.r. spectroscopy. Most of these processes and others that we shall discuss are different from the chemical reaction in the usual sense where the irreversible conversion of a reactant to a product through the agency of some reagent occurs. They can rather be classified in a broad sense as structural changes brought about by thermal energy, as *conformational transformations, inversions of configuration,* and *valence isomerizations*. The results obtained are therefore of interest both to chemistry and molecular physics.

In the application of n.m.r. spectroscopy for the determination of reaction rates — a field known as *dynamic NMR spectroscopy (DNMR)* — one compares the experimental spectrum measured at a specific temperature with a series of theoretical spectra calculated for different values of τ, that is, according to equation (9.3), for different rate constants. By the trial and error matching of the spectra the correct value of k can be found. This procedure is not feasible without the use of a computer and can then even be performed iteratively by changing one or more of the line shape parameters. Use is also made of the possibility of plotting the results of the computer calculation in the scale of the experimental spectrum. Thus one obtains spectra as are shown in Figure 9.3.

In the specific case of a kinetic investigation the n.m.r. spectrum of the spin system of interest is measured over as wide a range of temperature as possible. In the region of slow exchange, in which the spectrum is no longer sensitive to the kinetic effect, the values of δv and Δ, the frequency difference of interest and the natural line width of the resonance lines, are determined. One then makes the assumption, which sometimes is not justified, that these values are temperature independent and uses them for the calculations at higher temperature according to equation (9.2). The rate constants determined by the comparison of theoretical with experimental spectra for a series of different temperatures are then substituted in the *Arrhenius equation* (equation (9.6)). An *Arrhenius plot* yields the *Arrhenius activation energy*, E_a, and the *frequency factor*, A, for the reaction under investigation. Thereby it is assumed that both quantities are temperature independent. Also of general interest are the activation parameters of transition state theory, the enthalpy of activation, $\Delta H^{\ddagger}$, and the entropy of activation, $\Delta S^{\ddagger}$, which can be calculated by using equations (9.7) and (9.8), respectively:

$$\ln k = -\frac{E_a}{RT} + \ln A \tag{9.6}$$

$$\Delta H^{\ddagger} = E_a - RT \tag{9.7}$$

$$\Delta S^{\ddagger} = R\left[\ln\left(\frac{hA}{\kappa k_B T}\right) - 1\right] \tag{9.8}$$

where T is the absolute temperature (K), R is the universal gas constant $(8.31 \ \text{J K}^{-1})$, k_B is the Boltzmann constant $(1.3805 \times 10^{-23} \ \text{J K}^{-1})$, h is Planck's constant $(6.6256 \times 10^{-34} \ \text{J s})$ and κ is the so-called transmission coefficient, which is usually set equal to 1. This means that every transition state is going on to the product.

From the relation

$$\Delta G^{\ddagger} = \Delta H^{\ddagger} - T\Delta S^{\ddagger} \tag{9.9}$$

the free energy of activation at a specific temperature can also be obtained. E_a, $\Delta H^{\ddagger}$ and $\Delta G^{\ddagger}$ are measured in kJ mol^{-1} and $\Delta S^{\ddagger}$ in J K^{-1} mol^{-1}.

Another possibility for the evaluation of the free energy of activation is based on the well known *Eyring equation*:

$$k = \kappa \frac{k_B T}{h}(-\Delta G^{\ddagger}/RT) \tag{9.10}$$

or

$$\Delta G^{\ddagger} = RT[23.76 - \ln(k/T)] \tag{9.11}$$

or

$$\ln(k/T) = 23.76 - (\Delta H^{\ddagger}/RT) + (\Delta S^{\ddagger}/R) \tag{9.12}$$

Here one plots $\ln(k/T)$ against $1/T$. A straight line is obtained, with a slope of $-\Delta H^{\ddagger}/R$ and an intercept of $(23.76 + \Delta S^{\ddagger}/R)$. From these, the free activation parameters can be calculated. Unlike the data obtained directly through the use of equations (9.7) and (9.8), the activation parameters calculated using the Eyring equation are independent of temperature and these should be used in the comparisons of the energy profiles of analogous reactions.

If the components of the equilibrium are not of equal energy, the temperature dependence of the equilibrium constants, K, in the region of slow exchange must be determined by integration of the appropriate signals. One then calculates the enthalpy and the entropy differences according to *van 't Hoff*:

$$\ln K = -\Delta G^{\circ}/RT = -\Delta H^{\circ}/RT + \Delta S^{\circ}/R \tag{9.13}$$

In the calculation of the spectra the molar fractions of the components of the equilibrium must be considered. If they are very different, that is, if there is a large free energy difference between the two systems, the coalescence of the two resonance signals often is not well defined. The exchange process is then apparent only from a transient line broadening of the resonance signal of that component which is present in excess.

Finally, we should mention that spin echo experiments (see p. 237) also have been used for the investigation of dynamic processes on the basis that the amplitude of the echo is related to transverse relaxation. The shortening of T_2 by means of chemical exchange is not important in this connection if the residence time of the nuclei in positions with different Larmor frequencies is large compared with the delay time, τ, between pulses. However, for fast reactions or larger pulse intervals the amplitude of the echo decreases more rapidly than in the case of unperturbed transverse relaxation. If one determines this additional decrease as a function of τ one can calculate the rate constant at the temperature at which the experiment is run. The method has the advantage that it is applicable over a large temperature range and is not affected by spin–spin interactions. However, so far it has found only a few applications.

1.3   APPROXIMATE SOLUTIONS AND SOURCES OF ERROR

The relatively complex form of equation (9.2) and the reluctance to use a complete line shape analysis have made popular a number of approximate solutions. The best known of these is the relation

$$k_{coal.} = \pi \delta v / \sqrt{2} = 2.22 \delta v \tag{9.14}$$

which gives the rate constant at the coalescence temperature, $T_c(K)$, for the exchange between two equally populated sites ($p_A = p_B$). This enables one to use the Eyring equation to make a quick evaluation of the *energy barrier* for the process at this temperature. Making the appropriate substitution, one obtains

$$\Delta G\ddag = RT_c[22.96 + \ln(T_c/\delta v)](\text{J mol}^{-1}) \tag{9.15}$$

According to equation (9.11) separate signals at v_A and v_B are observed for $k/\delta v > 2.22$. If, on the other hand, $k/\delta v < 2.22$, an average signal is recorded at $(v_A + v_B)/2$. It must be emphasized that the value of the energy barrier thus obtained is temperature dependent because of the entropy term ($\Delta G^{\ddag} = \Delta H^{\ddag} - T\Delta S^{\ddag}$).

Since, in general, the coalescence temperature is different for different dynamic processes, a comparison of the $\Delta G^{\ddag}$ values is reasonable only if the entropy of activation for each of the processes under consideration is zero. To the extent that it is experimentally feasible a complete line shape calculation for a kinetic measurement over a larger temperature range is always preferable. Of course, this requires a greater expenditure of time and effort.

Additionally, attempts have been made to use characteristic variations of the line shape such as the separation of the maxima before the coalescence or the ratio of the intensities in the maximum or minimum for a simpler evaluation of k values. These approaches are, however, subject to systematic errors so that the results obtained are less reliable. More suitable is the use of the additional line broadening caused by the exchange, the so-called 'exchange broadening', Δ_e. This can be obtained from the observed line width, Δ_{obs}, by subtracting the natural line width, Δ_o, and the broadening due to inhomogeneity, Δ_I:

$$\Delta_e = \Delta_{obs.} - \Delta_o - \Delta_I \tag{9.16}$$

The following equations based on equation (9.2) for approximate values of the rate constant then can be written:

Slow exchange (for the signal in position A):

$$\frac{1}{\tau_A} = k_A = \pi\Delta_e \tag{9.17}$$

Fast exchange (for the average signal):

$$\frac{1}{\tau_A} = k_A = 4\pi p_A p_B^2 \delta v^2/\Delta_e \tag{9.18}$$

The applications of equations (9.17) and (9.18) are, however, limited to relatively narrow ranges that lie at the upper and lower ends of the temperature interval for which a variation in the line shape is observed.

In addition to the uncertainty that arises from an incorrect calculation of the line shape, the n.m.r. method for determining rate constants is susceptible to a series of *systematic errors* that sometimes cannot be eliminated. Especially critical are those cases in which the parameter δv amounts to only a few hertz. The variation in the line shape then falls within a very narrow temperature region and is therefore not as well defined as it is in the case of more widely separated signals. The occasional temperature dependence of the parameters δv and $\Delta_{obs.}$ mentioned above is also a source of error in these determinations. In polar solvents and in systems with heteroatoms and polar substituents one can count on association through hydrogen bonding to cause a temperature-dependent variation of the shift difference. If possible, the temperature dependence of δv and $\Delta_{obs.}$ should therefore be determined in the region of slow exchange in order to allow extrapolations to higher temperature where fast exchange occurs. Attempts can also be made to obtain δv and $\Delta_{obs.}$ together with the k value for the specific temperature directly from the line shape. To do this one matches the experimental spectrum with calculated spectra obtained by the iterative variation of all three parameters.

Also to be considered is the fact that all effects that cause an additional line broadening, such as partial saturation of the resonance line or field inhomogeneity simulate too high a value of k in the region of slow exchange and too low a value of k in the region of fast exchange. As a result the calculated activation energy is too low. Through the observation of signals of protons that are not involved in the exchange process, for example the signal of the internal standard, these errors can, to some extent, be eliminated. One must be aware, however, that the relaxation times and thus the natural line widths of signals from different substances and even of those arising from different protons within the same molecule are not necessarily equal. In many cases further complications arise through *spin–spin splittings* that make a correct interpretation difficult. Generally, then, one must consider very carefully which factors may affect the results and how errors can be avoided. By means of careful experimental technique errors in the activation energy can usually be limited to about 2 kJ mol^{-1} (0.5 kcal mol^{-1}) and can be even less than that in favourable cases. In a number of systems with relatively high energy barriers measurements using both the methods of classical kinetics and of n.m.r. spectroscopy have been possible and satisfactory agreement between the results was obtained.

Between 1956 and 1969 the values determined for the energy of activation for the internal rotation of dimethylformamide cited at the outset of this chapter rose from 29 $\pm$ 12 kJ mol^{-1} (7 $\pm$ 3 kcal mol^{-1}) to 118 $\pm$ 8 kJ mol^{-1} (28.2 $\pm$ 2 kcal - mol^{-1}). The increase in the magnitude of the value and its precision over the years illustrates the achievements of research in improving experimental techniques and our understanding of the sources of possible errors. Not only the small shift difference between the methyl proton signals (10 Hz at 60 MHz) but also the polar structure of the molecule and the long-range coupling with the formyl proton make this system a difficult 'test case' for the n.m.r. method. Today, the accepted values for the energy of activation, E_a, and the frequency factor, A, determined with the neat liquid are 85.8 $\pm$ 0.8 kJ mol^{-1} (20.5 $\pm$ 0.2 kcal/mol^{-1}) and (5 $\pm$ 3) $\times$ 10^{12}, respectively.

1.4 MORE COMPLEX EXCHANGE PHENOMENA

Equation (9.2) holds, as should be emphasized again, only for the simplest case of an exchange process, namely, for the periodic change of the Larmor frequency of a nucleus between two values v_A and v_B in the absence of scalar spin–spin coupling. In practice, however, dynamic processes that involve different modes of exchange are frequently encountered. For example, the exchange between more than two Larmor frequencies can occur if equilibria of the type

$$A \rightleftharpoons B \rightleftharpoons C \quad \text{or} \quad A \rightleftharpoons B \rightleftharpoons C$$
$$\searrow D \nearrow$$

are involved. In addition, processes are known that lead to the magnetic equivalence of specific nuclei, that is, cause the transition from an AB to an A_2 system. Furthermore, all *intermolecular* exchange processes that lead to the collapse of spin–spin multiplets can not be investigated by means of equation (9.2).

Specific examples of these different processes will be discussed in Section 2 of this chapter. The theoretical expressions that enable one to calculate the line shapes in these cases as well cannot, however, be discussed in detail. In spite of this, we do not wish to forego the demonstration of the many possible applications of variable-temperature n.m.r. spectroscopy. Also, because of these manifold applications, we must emphasize the importance of analysing the dynamic process in question carefully so that the correct theoretical method can be chosen for the interpretation of the experimental data. The power of modern data processing and software packages like 'DNMR' of Binsch and Stephenson then allow, even in the case of relatively complicated exchange processes, the calculation the complete line shape of the spectrum.

1.5 APPLICATION OF DOUBLE RESONANCE EXPERIMENTS TO THE DETERMINATION OF RATE CONSTANTS

In the presence of a chemical exchange process of the type $A \overset{\ddagger}{\rightleftharpoons} B$, as has been discussed in detail in the preceding chapter, the rate constant, k, in the region of slow exchange can be determined by double resonance experiments if separate signals for the resonances at v_A and v_B are observed. By irradiating v_A with the second field the spin distribution of the A nucleus is perturbed. This perturbation is now transmitted through the reaction $A \rightarrow B$ to the resonance signal at v_B, assuming that the longitudinal relaxation time T_{1A} is not too short relative to the rate of the reaction.

The result of such an experiment is shown in Figure 9.4 for the OH resonance of 2-hydroxyacetophenone (**138**) in the presence of salicylaldehyde (**139**) with which there is an exchange of hydroxyl protons. If the field B_2 is turned on ($\downarrow$) at v_{OH} (**139**) the intensity of the signal at v_{OH} (**138**) begins to decrease exponentially until the magnetization reaches the new equilibrium value $M_z^A(t \rightarrow \infty)$. Turning off the

138   **139**

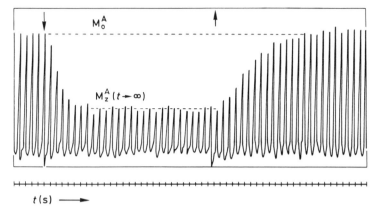

M_0^A

$M_z^A(t \rightarrow \infty)$

$t\,(s)\ \longrightarrow$

Figure 9.4 Variation of the intensity of the hydroxyl proton signal of 2-hydroxyacetophenone (**138**) as a function of the time, t, after turning on and off a B_2 field with the frequency of the OH resonance of salicylaldehyde (**139**) (after Ref. 1)

perturbing field ($\uparrow$) leads, likewise exponentially, to a restoration of the original equilibrium magnetization M_0^A.

The theoretical treatment of the phenomenon yields the following result: the lifetime τ_{1A} of a spin state in the position A depends on the longitudinal relaxation time T_{1A} in this position and the rate constant k_A. We can therefore write

$$1/\tau_1 = 1/T_{1A} + 1/\tau_A \tag{9.19}$$

For the new equilibrium magnetization it holds that

$$M_z^A(t \rightarrow \infty) = M_0^A(\tau_{1A}/T_{1A}) \tag{9.20}$$

while previously (we shall omit the derivation here) $M_z(t)$ was expressed as

$$M_z^A(t) = M_0^A[C\exp(-t/\tau_{1A}) + \tau_{1A}/T_{1A}] \tag{9.21}$$

With equation (9.20) it follows that

$$M_z^A(t) = M_0^A C\exp(-t/\tau_{1A}) + M_z^A(t \rightarrow \infty) \tag{9.22}$$

and

$$M_z^A(t) - M_z^A(t \rightarrow \infty) = M_0^A\ C\exp(-t/\tau_{1A}) \tag{9.23}$$

or

$$\ln[M_z^A(t) - M_z^A(t \rightarrow \infty)] = -t/\tau_{1A} + \text{const.} \tag{9.24}$$

Therefore, if one plots the observable $\ln[M_z^A(t) - M_z^A(t \rightarrow \infty)]$ against the time, t, one can obtain τ_{1A} as the slope of the line and with equations (9.20) and (9.19) the desired value of τ_A since the ratio $M_z^A(t \rightarrow \infty)/M_z^A$ can be determined experimentally (cf. Figure 9.4). One proceeds analogously for the determination of τ_B by irradiating the signal at site A if the two sites A and B are unequally populated.

The double resonance process just described represents a useful extension of the application of n.m.r. spectroscopy to the measurement of rate constants. It is applicable in the region of slow exchange where the line shape of the spectrum is insensitive to the dynamic process being investigated. It also represents an elegant method of identifying nuclei that are involved in chemical exchange or for analysing qualitatively the dynamic behaviour of a molecule. For example, it was successfully applied in detecting the conformational flexibility of [18]annulene at room temperature. Irradiating the resonance of the "internal" protons led to a distinct decrease in the intensity of the "external" protons (cf. p. 367), thereby establishing chemical exchange between these positions.

In Fourier transform n.m.r. spectroscopy the experiment can be performed with the DANTE technique (cf. page 250) or any other method, which allows the selective inversion of one resonance line. As in the inversion-recovery experiment for T_1 determinations, the spectrum is then measured after a certain delay time with a 90° pulse. During the delay, magnetization transfer by a chemical exchange operates. Varying the delay, the rate constants k can be calculated as above from the time dependence of the ratio $M_z^A(t)/M_z^A(t \to \infty)$.

1.6   TWO-DIMENSIONAL EXCHANGE SPECTROSCOPY (EXSY)

An interesting alternative for the one-dimensional double resonance experiment described above is two-dimensional exchange spectroscopy, EXSY (*ex*change *s*pectroscopy). With the pulse sequence

$$90°\text{-----}t_1\text{-----}90°_x\text{-----}t_M\text{-----}90°_x, \ \text{FID}(t_2) \tag{9.25}$$

cross peaks are observed for dynamic systems in the region of slow exchange between the resonances of those nuclei which are exchanging their Larmor frequencies. The mechanism of this experiment can be explained on the basis of a classical Bloch vector picture as shown in Figure 9.5. The first $90°_x$ pulse produces transverse magnetizations $M_y(A)$ and $M_y(X)$ which develop during the evolution time t_1 according to their Larmor frequencies (Figure 9.5a,b). At the end of the evolution time all resonance signals of the particular spectrum have different phases, in other words, they are labelled with their Larmor frequency. The second $90°_x$ pulse produces longitudinal z-magnetization, which has, depending on the vector orientation, positive or negative sign and different magnitude (Figure 9.5c). During the mixing time t_M which follows, the dynamic process induces magnetization transfer. The amplitude of the transverse magnetization produced by the third $90°_x$ pulse and detected in t_2 depends therefore on the evolution time t_1 as well as on the efficiency of the magnetization transfer. The transfer rate during the mixing time t_M, which is in the order of 1–2 s, is a function of the rate constants of the particular dynamic process, but also of the magnitude of the z-magnetization which is present

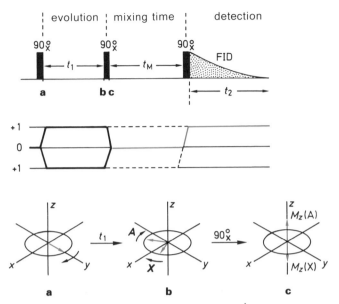

Figure 9.5 Pulse sequence for two-dimensional ^1H exchange spectroscopy with coherence transfer and vector diagram; only the z-components of $M(A)$ and $M(X)$ are shown in (c).

at $t_M = 0$ (Figure 9.5c). As a consequence, magnetization transfer is t_1-dependent and the signals of exchanging nuclei are amplitude-modulated which leads to cross peaks in the 2D spectrum.

The application of an EXSY experiment is demonstrated in Figure 8.6 with the 2D ^1H exchange spectrum for the methyl resonances of 2,3-dimethylanthraquinone-9-cyanimine. Here, the inversion process at the CN double bond leads to an equilibrium between two isomers A and B:

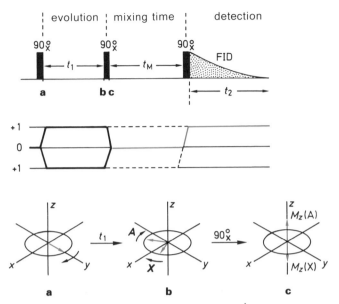

Consequently, in the region of slow exchange four different environments result for the methyl groups a–d, which can, however, exchange their Larmor frequencies only pairwise. The 2D exchange spectrum measured at $-33\,°C$ (Figure 9.6) yields the information that magnetization is transferred between signals 1 and 4 as well as 2 and 3. Accordingly, in the region of fast exchange at room temperature one observes only one signal, since the average signals $(1 + 4)/2$ and $(2 + 3)/2$ are superimposed.

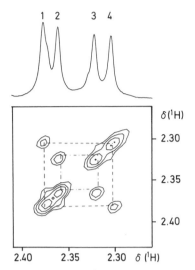

Figure 9.6 2D ^{1}H exchange spectrum (EXSY spectrum) of 2,3-dimethylanthraquinone-9-cyanimine in CD$_2$Cl$_2$ at -33 °C; $t_M = 1$ s, measuring time 6.3 h (after Ref. 2)

The pattern of the cross peaks observed in the 2D exchange spectrum is characteristic for the mechanism of the exchange process. Further applications of this experiment will be discussed in Sections 2.3 and 2.4. In addition, it is important to mention that the volume intensity of the cross peaks yields information about the rate constants of the various transfer processes. For the simple case of slow exchange between two equally populates sites ($p_A=p_B$) and in the absence of scalar spin–spin coupling the rate constants can be calculated from the intensity ratio of diagonal to cross peaks, I_D and I_C, respectively, and the mixing time t_M according to equation (9.26):

$$I_D/I_K \simeq (1 - kt_M)/kt_M \quad \text{or} \quad k \simeq 1/[t_M(I_D/I_K + 1)] \tag{9.26}$$

2D-EXSY spectroscopy has, therefore, a great potential for the investigation of dynamic processes in solution.

Within the product operator formalism (Chapter 8.5) the EXSY experiment for an AX system without scalar spin–spin coupling can be described by the following equations:

$$\hat{I}_z(A) + \hat{I}_z(X) \xrightarrow{90°[\hat{I}_x(A)+\hat{I}_x(X)]} \hat{I}_y(A) + \hat{I}_y(X)$$

$$\xrightarrow{\omega A t_1 \hat{I}_z(A)+\omega_X t_1 \hat{I}_z(X)} \hat{I}_y(A) \cos \omega_A t_1 + \hat{I}_x(A) \sin \omega_A t_1 + \hat{I}_y(X) \cos \omega_X t_1$$
$$+ \hat{I}_x(X) \sin \omega_X t_1$$

$$\xrightarrow{90°[\hat{I}_x(A)+\hat{I}_x(X)]} -\hat{I}_z(A) \cos \omega_A t_1 + \hat{I}_x(A) \sin \omega_A t_1 - \hat{I}_z(X) \cos \omega_X t_1 + \hat{I}_x(X) \sin \omega_X t_1$$
$$\tag{9.27}$$

As is expected on the basis of the pulse sequence equation (9.25), equation (9.27) also contains terms which have been derived for the COSY sequence (p. 310 f.), with the exception of those contributions which arise from scalar spin–spin coupling. During the mixing time t_M the dynamic process leads to an exchange of the new z-magnetization:

$$\hat{I}_z(A) \cos \omega_A t_1 \rightleftharpoons \hat{I}_z(X) \cos \omega_X t_1 \qquad (9.28)$$

Part of the magnetization which precesses during t_1 with ω_A will be modulated during t_2 with $\cos \omega_X t_2$ and *vice versa*. This leads to cross peaks at ω_A,ω_X and ω_X,ω_A. The transverse magnetization $\hat{I}_x$ present at the beginning of the mixing time is partly lost through transverse relaxation and partly transformed into transverse magnetization of the other nucleus by the exchange process. It will be eliminated through the phase cycle which selects only those coherences which have the order zero during the mixing time (Figure 9.5). The same is true for COSY-type cross peaks which result if the nuclei under exchange are scalar coupled. The zero quantum magnetization also present in these cases can be eliminated through various modifications of the pulse sequence, for example through a 180° pulse during the mixing time.

Finally, it is important to note that the pulse sequence of equation (9.25) is also used to record two-dimensional nuclear Overhauser spectra. This aspect will be treated in detail in Chapter 10, where we also address the question how cross peaks that arise from chemical exchange can be distinguished from those that arise through nuclear Overhauser effects.

1.7 MEASUREMENTS OF FIRST ORDER RATE CONSTANTS BY INTEGRATION

Our survey would be incomplete without mentioning that n.m.r. spectroscopy can also be employed in following slow irreversible reactions. The integration of individual signals is employed to trace the course of a transformation

$$A \rightarrow B$$

by monitoring the changes in concentrations of A or B as a function of time. In this way the concentration ratio [A]/[B] can be calculated directly or, by using an internal standard of known concentration, the increase in [B] or the decrease in [A] can be determined. As an example, the course of the conversion of the dichlorocarbene adduct of cyclooctatetraene (**A**) into the 1,2-dihydroindene derivative (**B**) is illustrated in Figure 9.7. As the spectra show, the resonance signals at $\delta 4.8$ and $\delta 3.7$ assigned to H_a, H_b and H_c in **B** appear as the reaction progresses. At the same time, the relative intensity of the singlet at $\delta 2.4$, which is due to the tertiary proton of **A**, decreases. The reaction was run in n.m.r. tubes in a thermostat at 80 °C and every hour a tube was removed and the reaction quenched at −70 °C. The [A]/[B] ratio could then be determined by integration of the n.m.r. spectrum. Its dependence on

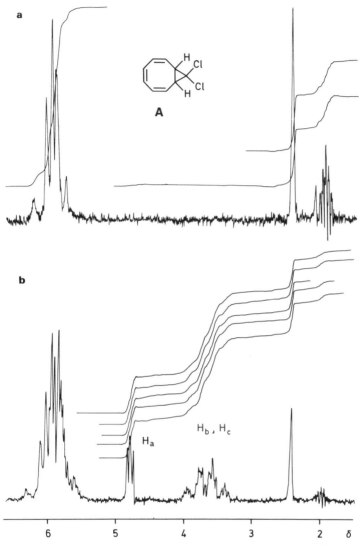

Figure 9.7 Proton resonance spectra of the rearrangement $\mathbf{A} \rightarrow \mathbf{B}$ in acetonitrile-d$_3$; (a) starting material; (b) reaction mixture after heating for 6 h at 80 °C

the reaction time led to a rate constant of $k(80\ °C) = 6 \times 10^{-5}\ s^{-1}$ for the transformation in acetonitrile as solvent and thence by use of equation (9.11) to a free energy of activation of 115.4 kJ mol^{-1} (27.6 kcal mol^{-1}).

2. The internal dynamics of organic molecules

As mentioned earlier, we want to deal with the *conformational mobility* of organic molecules and with *valence isomerizations* in this section. The examples chosen should give an indication of the multitude of phenomena that can be studied by means of n.m.r. spectroscopy. Thus we shall be more concerned with the description of the experimental observations and their qualitative significance than with quantitative aspects. Nevertheless, we shall make mention of the calculated activation parameters (E_a, $\Delta H^{\ddagger}$, $\Delta S^{\ddagger}$,, and $\Delta G^{\ddagger}$,) in order to provide the reader with a feeling for the energy requirements of the individual processes that without exception are *intramolecular*.

2.1 HINDRANCE TO INTERNAL ROTATION

2.1.1. Bonds with partial double bond character

As with dimethylformamide, other molecules having a group with the general structure

$$D-C\overset{\displaystyle A}{\underset{\displaystyle \diagdown}{\diagup\!\!\!\diagup}} \quad \text{or} \quad D-N\overset{\displaystyle O}{\diagup\!\!\!\diagup}$$

exhibit hindered rotation around the D—C or the D—N bond. In these compounds D is an electron-donating group, such as nitrogen or an aromatic π system, and A is an electron acceptor such as oxygen or sulphur. Temperature-variable n.m.r. spectra are therefore observed for thioamides, carbamates, furfuraldehyde and benzaldehyde, and with protonated ketones as well as with nitrosamines and nitrites in which the oxygen serves as the electron donor. In Table 9.1 the activation parameters of a few compounds that have been measured by means of n.m.r. spectroscopy are given.

In several systems the dynamic process causes two protons to change from an AB system into an A_2 system. One such case is encountered in N-methyl-2,4,6-trinitroaniline (140) in which the benzene nucleus acts as an electron acceptor because of the three nitro substituents. As the resonance structure (140a) suggests, one conformation in which the N—CH$_3$ bond lies in the plane of the aromatic nucleus is favoured by resonance stabilization.

The asymmetric substitution of the amino nitrogen leads to the low temperature AB system observed for the two aromatic protons since, because of the steric interaction with the methyl group, one of the two *ortho*-nitro groups probably has its plane perpendicular to the plane of the benzene nucleus. With the more rapid

Table 9.1 Examples of hindered rotation with activation energies in kJ mol^{-1} and (in parentheses) kcal mol^{-1}

Structure	Value		Structure	Value
(thioformamide)	E_a ~105		(diaryl)	$\Delta G^{\ddagger}_{298}$ 57
(dimethylcarbamate)	$\Delta G^{\ddagger}_{270}$ 67		(nitrosamine)	E_a 96
(benzaldehyde)	$\Delta G^{\ddagger}_{146}$ 33		(nitroso aniline)	$\Delta H^{\ddagger}$ 62
(furaldehyde)	$\overleftrightarrow{E}_a$ 50 / $\overleftrightarrow{E}_a$ 46		(nitrite)	E_a ~42

140   **140 a**

rotation of the methylamino group at higher temperatures the two protons become equivalent and the spectrum degenerates to an A$_2$ system (Figure 9.8). It seems worthwhile to discuss this example in more detail because, as already mentioned, neither equation (9.2) nor the approximate methods derived from it can be used for the interpretation of the n.m.r. line shape. If we describe the two nuclei as we have done before through the product functions $\alpha\alpha$, $\alpha\beta$, $\beta\alpha$, and $\beta\beta$ the exchange process leads to the result that the $\alpha(1)\beta(2)$ state is transformed into the $\beta(1)\alpha(2)$ state. The functions $\alpha\beta$ and $\beta\alpha$ are now acceptable only when there is no interaction between the two nuclei. That is, however, not the case, since the two nuclei couple with one another. The line shape must therefore be derived on the basis of quantum mechanical theory, a procedure that we cannot discuss in detail here. Let it merely be

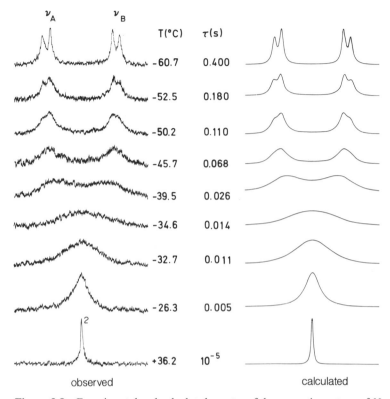

Figure 9.8 Experimental and calculated spectra of the aromatic protons of *N*-methyl-2,4,6-trinitroaniline as a function of temperature (after Ref. 3)

noted that even for this case an explicit expression for the shape of the spectrum as a function of the exchange rate can be obtained by means of which the theoretical spectra shown in Figure 9.8 were calculated. The relation

$$k_{coal.} = \pi \sqrt{(\nu_A - \nu_B)^2 + 6J_{AB}^2} / \sqrt{2}$$

can be used as an approximation at the coalescence point so that the rate constant at the coalescence temperature can be obtained rather easily.

2.1.2 Substituted Ethanes

It has been known for a long time that the rotation around the carbon–carbon bond in ethane is hindered. The first evidence for this came from the finding that the entropy of the compound was lower than that expected on the basis of theoretical calculations assuming free rotation. The molecule must therefore surmount an energy barrier in the conversion of the staggered conformation (a) into its rotational isomer (a'). The transition state for such a process can be represented by the eclipsed conformation (b).

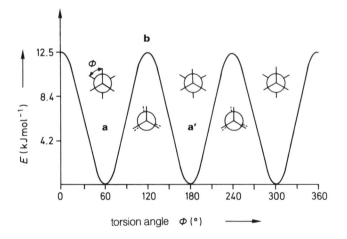

The energy profile for a full rotation as a function of the torsional angle ϕ is shown in Figure 9.9; the height of the barrier as measured by microwave spectroscopy is about 12 kJ mol^{-1} (3.0 kcal mol^{-1}). This hindrance to rotation has its origin in electronic repulsions, as shown by MO calculations. It is also known that the replacement of one or more hydrogen atoms by groups with larger steric requirements increases the height of the barrier, and on this basis one can conclude that steric interactions play a dominant role in restricting rotation in substituted ethanes.

With respect to this problem it is of considerable significance that additional experimental data have been obtained by using n.m.r. spectroscopy. Thus, data concerning the stable conformations of substituted ethanes were gathered from vicinal coupling constants and their dihedral angle dependence (p. 115). In addition, the barrier to rotation in a series of systems has been measured by analysing the temperature dependence of the spectra. For these investigations ^{19}F n.m.r. spectroscopy was used almost exclusively because here large chemical shift differences δv are observed.

Figure 9.9 Energy profile for rotation around the carbon–carbon bond in ethane

In the present context it is of interest to mention that in a number of cases the hindrance to rotation of a tertiary butyl group is so large that at low temperatures the methyl groups become nonequivalent. Thus, in the case of butylcycloheptane, for example, and also in other t-butyl-substituted cycloalkanes, a 2:1 doublet is observed for the t-butyl protons. This observation is consistent with the staggered conformation (141) in which only two methyl groups are equivalent. The barrier to rotation at the coalescence point (− 126 °C) has been determined to be 31.0 kJ mol^{-1} (7.4 kcal mol^{-1}). It was found further that this barrier is a function of the cycloalkane ring size and reaches a maximum of 32.6 kJ mol^{-1} (7.8 kcal - mol^{-1}) in t-butylcyclooctane.

Compound 142, even at room temperature, exhibits a doublet with an intensity ratio of 2:1 for the methyl resonances and fine structure arising from direct fluorine–hydrogen coupling. At 134 °C the signals coalesce and a barrier to rotation of 77.4 kJ mol^{-1} (18.5 kcal mol^{-1}) has been found.

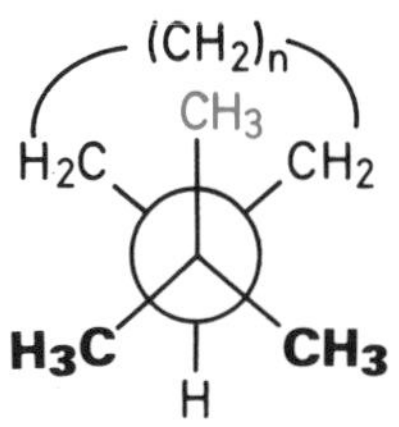

141 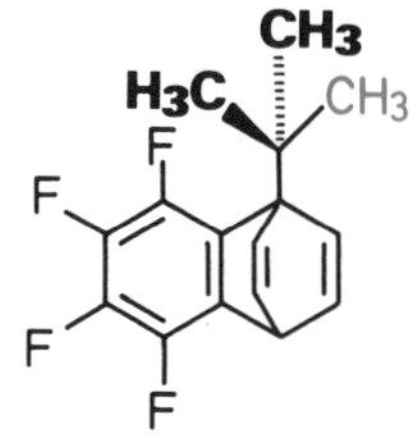   142

2.2 INVERSION OF CONFIGURATION

As was shown above, because of its lone pair of electrons, a nitrogen atom plays a very important role in molecules that exhibit hindrance to internal rotation. Another phenomenon that can be closely studied by means dynamic n.m.r. is the *inversion of configuration*, which has been well established in the case of ammonia by microwave spectroscopy. It is also known as pyramidal atomic inversion:

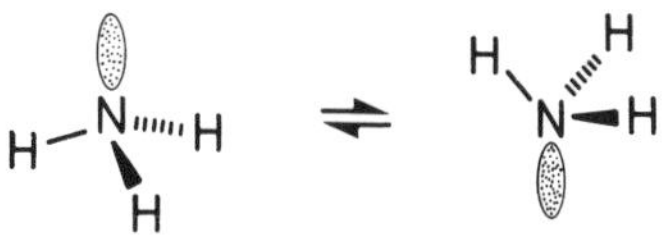

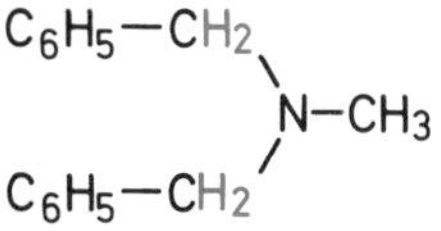

143

That inversion occurs in dibenzylmethylamine 143 has been elegantly demonstrated by using n.m.r. spectroscopy and, without resorting to low-temperature measurements, the activation energy for inversion at room temperature has been determined. According to our explanations in Chapter 4 we would expect the methylene protons in 143 to give rise to an AB system due to their diastereomeric environment. However, because of rapid inversion at room temperature, the signals degenerate into a singlet. On the other hand, if the measurement is carried out in acidic solution with a pH less than 2.0 the amine exists in the protonated form and the inversion process is slowed down. The CH_2 protons in the salt are diastereotopic and give rise to an AB system. Under these conditions the equilibrium below, in which the colour change indicates an inversion of nitrogen, can be formulated.

$$(C_6H_5-CH_2)_2\overset{\oplus}{\underset{H}{N}}-CH_3 \; \rightleftharpoons \; (C_6H_5-CH_2)_2N-CH_3 \; + \; H^{\oplus}$$

$$\updownarrow \text{Inversion}$$

$$(C_6H_5-CH_2)_2\overset{\oplus}{\underset{H}{N}}-CH_3 \; \rightleftharpoons \; (C_6H_5-CH_2)_2N-CH_3 \; + \; H^{\oplus}$$

The inversion therefore can take place only in the amine and its rate depends on the concentration of the free amine and thus on the pH of the solution. The protonation and deprotonation of the amine is also a rapid equilibrium reaction in which the configuration of the amine does not change. The fact that spin–spin coupling of the N—H proton with the protons of the N-methyl group vanishes at lower acid concentrations (pH = 2.0) before the methylene protons become enantiotopic provides experimental evidence for this additional process. A further increase in the pH then causes the methylene protons to become magnetically equivalent. The expression for the rate constant, k', that characterizes the exchange of the methylene protons and that can be derived from the line shapes of the n.m.r. spectra measured at different pHs is

$$k' = k_{\text{inv.}}[\text{Amine}]/([\text{Amine} + \text{Salt}])$$

The ratio of the concentrations at different pHs can be calculated from the known pK_a value of the amine. The rate constant for inversion is then obtained graphically by plotting k' against the concentration ratio. This gives $k_{\text{inv.}} = (2 \pm 1) \times 10^5 \text{ s}^{-1}$, which corresponds to a $\Delta G^{\ddagger}$ value of about 42 kJ mol^{-1} (10 kcal mol^{-1}). In a later study diastereotopic methylene protons in dibenzylamine were observed directly at $-155\,°C$.

In *cyclic* amines nitrogen inversion has also been detected. For example the AA′BB′ system of the ring protons in N-ethylaziridine (**144**) coalesces to a singlet at $108 \pm 5\,°C$. The energy barrier for this process is about 81 kJ mol^{-1} (19.4 kcal mol^{-1}). In the case of N-chloroaziridine (**145**) the coalescence temperature lies above the temperature at which the sample decomposes ($> 180\,°C$); the barrier to inversion must therefore be fairly high. This barrier-raising effect of chlorine substitution — which results from the higher p-character in the C–Cl bond due to the $-I$-effect of the halogen and a stabilizing effect of the pyramidal ground state structure — led in the case of 7-chloro-7-azabi-

cyclo[4.1.0]heptane (**146**) to the actual separation of the diastereomeric invertomers
146a/146b.

144   **145**

146 a   **146 b**

In cyclic diazines, such as 2,3-diazabicyclo[2.2.1]hept-5-ene (**146**), evidence for a
consecutive inversion of the two nitrogen atoms has been obtained. The equilibrium
147a $\leftrightarrows$ **147b** is characterized by a barrier of $\Delta G^{\ddagger}_{264} = 60.7$ kJ mol^{-1}
(14.5 kcal mol^{-1}). Figure 9.10 shows the time-averaged spectrum recorded at
57 °C and the superposition of the identical spectra of the two enantiomers obtained
at -39 °C. It is clear that at the lower temperature the molecule has lost the C_s
symmetry that was effective on the n.m.r. time scale at 57 °C. While the AB system
of the bridge protons ($\delta 1.7$ and $\delta 2.1$) remains unaffected, as expected, separate
absorptions are observed for the methylene protons ($\delta \approx 2.5$) and also for the
bridgehead and the olefinic protons ($\delta \approx 4.0$ and 6.5, respectively). Of further
interest is the fact that the vicinal and allylic coupling constants between olefinic and
bridgehead protons differ as a result of the different configurations of the two
nitrogen atoms.

147 a   **147 b**

In the diaziridine **148**, an analogous inversion process requires more than
96 kJ mol^{-1} (23 kcal mol^{-1}). This increase in energy barrier obviously results from
the change in ring size since in the three-membered ring the planar transition state

for inversion is expected to be of higher energy than in larger ring systems because of the smaller internal C−N−C bond angle.

On the other hand, the inversion barrier decreases if substituents on the nitrogen lead to a flattening of the pyramidal configuration ($sp^3 \rightarrow sp^2$) through electronic interactions with the nitrogen lone pair. Thus in compound **149** a value for $\Delta G^{\ddagger}_{231}$ of only 49.8 kJ mol^{-1} (11.9 kcal mol^{-1}) is observed.

148   **149**

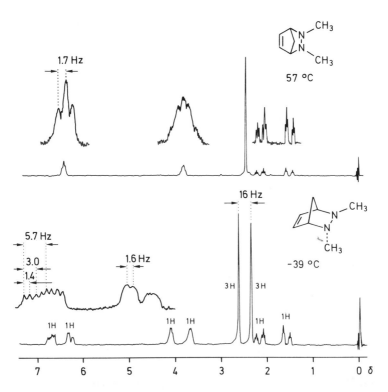

Figure 9.10 Temperature dependence of the proton resonance spectrum of 2,3-diazabicyclo[2.2.1]hept-5-ene (**147**) (after Ref. 4)

Relative to *inversion* and *hindered rotation* associated with the nitrogen atom, it is interesting that in cyclic systems the ring size influences the energy barrier for the two processes differently. The activity energy for inversion in the series **150–152** (R = Cl or CH_3) becomes smaller as a result of reduced destabilization in the transition state while the barrier to rotation in the same series (R = COOR) increases. Here the ground state is stabilized in the larger ring systems because the conjugative interaction is favoured by an increase in the C−N−C bond angle that allows a change in nitrogen hybridization from sp^3 towards sp^2.

150   **151**   **152**

Trialkyloxonium salts are isoelectronic with amines and inversions of configuration at the oxygen atom have also been detected in compounds of this type. For example, the isopropyloxonium ion of ethylene oxide (**153**) exhibits a singlet for the methylene proton of the ring at 40 °C that at −70 °C becomes an AA′BB′ multiplet. Similar inversion processes have been observed in phosphorous and arsenic compounds.

153

2.3   RING INVERSION

Nuclear magnetic resonance spectroscopy has also significantly contributed to our understanding of the conformational mobility of saturated and unsaturated ring systems.

The classical example, the inversion of the cyclohexane ring, has been investigated extensively. The observed activation parameters are $\Delta H^{\ddagger} =$ 42.5 ± 0.4 kJ mol^{-1} (10.8 ± 0.1 kcal mol^{-1}) and $\Delta S^{\ddagger} = 11.71 ± 2.1$ J K^{-1} mol^{-1} (2.8 ± 0.5 cal K^{-1} mol^{-1}). They characterize the transition of the chair conformation (a) into a high-energy conformation such as the boat or the twist boat (b, c) and finally into the equivalent chair conformation (a′). The energy profile has the form illustrated in Figure 9.11.

The transitions between the boat and the twist boat conformers are effected without encountering angle deformation simply by rotation around carbon–carbon single bonds. Such conformational transitions that are generally distinguished by low energy barriers are called *pseudo rotations*.

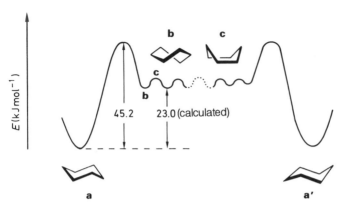

Figure 9.11   Energy profile for the ring inversion of cyclohexane

The transmission coefficient, κ, for the reaction (a) $\rightarrow$ (a') must be assumed to be $\frac{1}{2}$ since the probabilities that an individual molecule is transformed from conformation (b) or (c) to (a) or to (a') are identical. In other words, the rate constants k_{ab} and k_{ac} must be twice as large as k_{aa}.

Experimentally, one takes advantage of the different chemical shifts of the cyclohexane protons in the axial and equatorial positions (cf. p. 80). Especially simple spectra are obtained for cyclohexane-d_{11} if one eliminates the spin–spin interaction between the proton and the deuterium atoms by spin–spin decoupling. The result of such an experiment is shown in Figure 9.12 (for a description of spin–spin decoupling see Chapter 7). In substituted cyclohexanes corresponding ring inversion processes have also been detected. In Table 9.2 the results for several monosubstituted cyclohexanes obtained by means of n.m.r. experiments at about -80 °C are given. At this temperature all systems are in the region of slow exchange and the equilibrium constant can be determined by integration of the separate resonance signals for the axial and equatorial methine protons (cf. Figure 9.13a for bromocyclohexane). From the equation $-\Delta G^\circ = RT \ln K$ the difference in free energy of the two conformations, known as the A value of the substituent, is obtained. It is positive if the substituent favours the equatorial position. Also included in Table 9.2 are A values for alkyl groups that were determined by other methods. It is interesting to note that the A values for the halogens do not follow the order of increasing van der Waals radii (F, 0.135 nm; Cl, 0.180 nm; Br, 0.195 nm; and I, 0.215 nm). Increased carbon–halogen bond length and polarizability reduce non-bonded interactions in the cases of axially disposed iodine and bromine atoms. However, the $-OCD_3$ and the $-SCD_3$ and also the alkyl groups exhibit the expected gradation in A values.

From the data for chlorocyclohexane it can be estimated that at -150 °C the half-life of each isomer should be several hours. By fractional crystallization in an n.m.r. tube it was actually possible at this temperature to prepare a solution of the isomer with equatorial chlorine. As the n.m.r. spectrum of this solution (Figure 9.13c) when

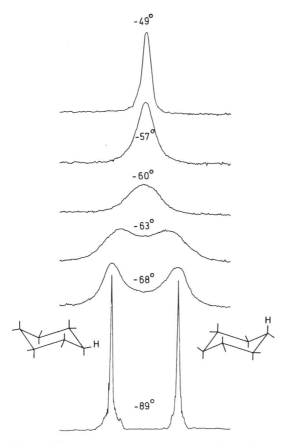

Figure 9.12 Proton resonance spectrum of cyclohexane-d_{11} as a function of temperature (after Ref. 5)

Table 9.2 Conformational energies of dierent substituents on cyclohexane

Substituent	A value		Temperature (°C)	Equilibrium constant
C≡N	1.00[a]	0.24[b]	−79	1.88
F	1.15	0.276	−86	2.10
I	1.96	0.468	−80	3.39
Br	1.99	0.476	−81	3.48
Cl	2.21	0.528	−81	3.99
OCD$_3$	2.29	0.547	−82	4.23
SCD$_3$	4.48	1.07	−79	16.1
CH$_3$	7.1	1.7		
C$_2$H$_5$	7.5	1.8		
i-C$_3$H$_7$	8.8	2.1		
C(CH$_3$)$_3$	21	5		

[a] kJ mol^{-1} [b] kcal mol^{-1}

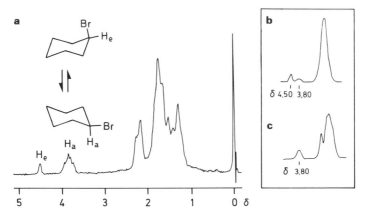

Figure 9.13 Proton resonance spectra of monosubstituted cyclohexanes: (a) equilibrium mixture of axial and equatorial bromocyclohexane at $-100\,^{\circ}$C (after Ref. 6); (b) mixture of axial and equatorial chlorocyclohexane at $-150\,^{\circ}$C (after Ref. 7); (c) pure equatorial chlorocyclohexane at $-150\,^{\circ}$C (after Ref. 7)

compared with that of an incompletely separated mixture (Figure 9.13b) indicates, the enrichment of the equatorial isomer was essentially complete. This represented the first preparative separation of conformational isomers of monosubstituted cyclohexanes.

In the case of larger cycloalkanes, dynamic processes can also be analysed by n.m.r. spectroscopy, albeit due to the greater mobility of these ring systems the barriers are often below the limit of detection. A relatively simple process of ring inversion has been found for the methylene chain in the paracyclophane **153**. Formally a ten-membered ring is involved, however, the partial structure of the aromatic ring reduces the flexibility and thereby the number of the possible dynamic processes. As a consequence, the intra-molecular motion is confined to the inversion process of the methylene chain, **154a** $\rightleftharpoons$ **154b**. The kinetics of this process can be traced most easily by the line shape changes of the AB-spectrum of the aromatic protons, which degenerates in the region of fast exchange to an A_2 system (Figure 9.14a, see also p. 355). The Eyring-diagram (Figure 9.14b) yields activation parameters of $\Delta H^{\ddagger} = 43.3$ kJ mol^{-1} (10.3 kcal mol^{-1}), $\Delta S^{\ddagger} = -51$ J K^{-1} mol^{-1} (12.2 cal K^{-1} mol^{-1}), and $\Delta G^{\ddagger}(298) = 58.5$ kJ mol^{-1} (14.0 kcal mol^{-1}).

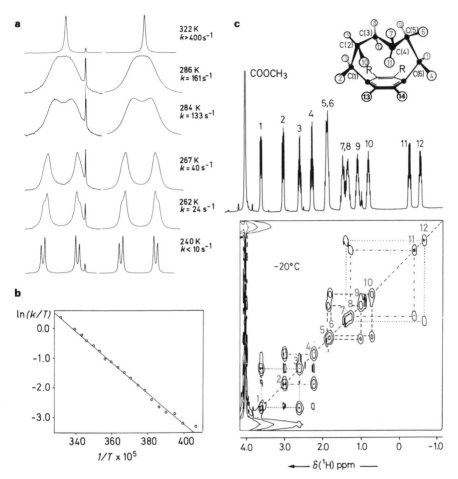

Figure 9.14 Ring inversion in [6]paracyclophane-8,9-dimethylcarboxylate **154**; (a) line shape changes of the AB spectrum of the aromatic protons H^{13} and H^{14}; experimental (left) and calculated (right); (b) Eyring-diagram; (c) 400 MHz ^{1}H 2D-exchange spectrum at -20 °C recorded with the pulse sequence equation (9.25); mixing time $t_M = 1$ s (after Ref. 8)

We have cited this example also because two-dimensional exchange spectroscopy in the slow exchange region yields clearcut results with respect to the conformational changes which occur in the CH_2 chain. As seen in Figure 9.14c, cross peaks are observed between those methylene protons which become identical if a rotation of the C(3)–C(4) group around the fixed positions C(2) and C(5) is assumed: 7/12; 8/11; 5/9; 6/10; 1/3; 2/4. The assignment of the ^{1}H n.m.r. spectrum, which forms the basis for this analysis, was again established by two-dimensional methods.

Of the many dynamic processes involving cyclic olefins that have been studied we shall mention only two. In cyclohexene, ring inversion has the effect of making the methylene protons H_a and H_b equivalent. The activation energy for this interconversion is 22.6 kJ mol^{-1} (5.4 kcal mol^{-1}). It is similar to that for cycloheptatriene [25 kJ mol^{-1} (6 kcal mol^{-1})], which according to electron diffraction measurements, exists in a tub conformation. At $-150\,°$C the non-equivalence of the methylene protons can be detected while at room temperature both protons are equivalent because of the fast equilibrium indicated below:

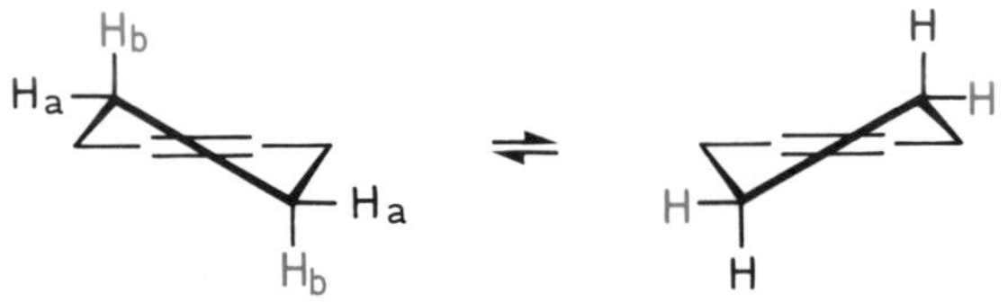

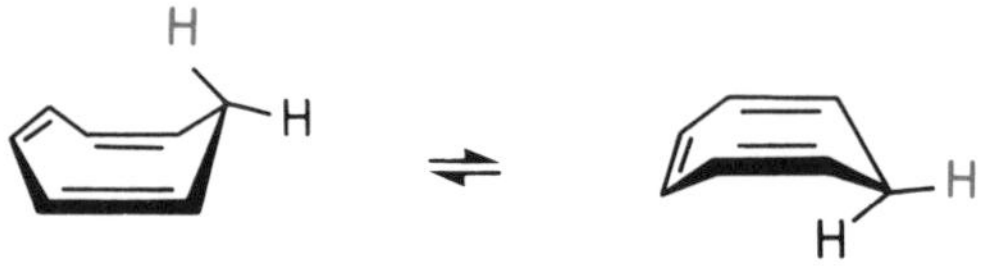

Interesting conformational equilibria have been observed and quantitatively investigated in the case of certain annulenes. Two examples may be mentioned here. Figure 9.15 shows the temperature-dependent spectra of [18]annulene. The spectrum measured at $-60\,°$C corresponds to expectations for a planar diatropic $4n + 2\pi$ system since resonances are found at $\delta - 3.0$ for the six internally directed protons and at $\delta\ 9.3$ for the twelve externally directed protons. Increasing the temperature leads to line broadening and finally a coalescence at $40\,°$C that is indicative of an exchange of internal and external protons. The signal at $\delta\ 5.4$ recorded at $100\,°$C lies, as expected, near the statistical average of $\delta\ 5.29$. The exchange process is effected in this case by rotations around carbon–carbon bonds, a phenomenon that is normally not expected in an aromatic molecule. However, in [18]annulene it is possible because the size of the ring confers unusual flexibility on the carbon skeleton. In this connection it is interesting that the first n.m.r. measurements on [18]annulenes led to puzzling results since, because of line broadening due to exchange at normal temperatures around $30\,°$C, no "satisfactory" spectrum could be recorded and the possibility of conformational transformations was not even suspected at that time.

Complicated conformational transformations and bond isomerizations have been elucidated in the case of [16]annulene. At $-30\,°$C the spectrum of this compound consists of a singlet of $\delta\ 6.74$. The detailed analysis of the temperature-dependent line shape down to $-130\,°$C leads to the following conclusions:

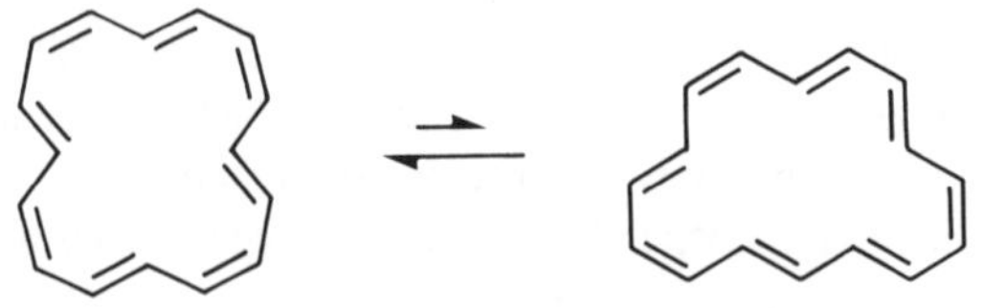

a   b

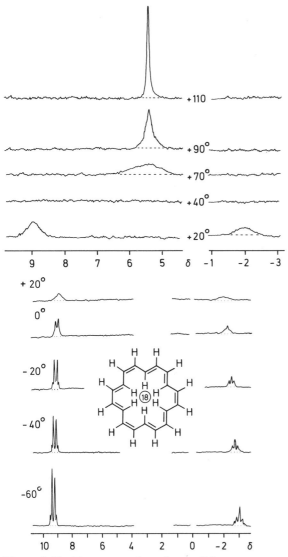

Figure 9.15 Temperature dependence of the proton resonance spectrum of [18]annulene (after Ref. 9)

(1) [16]Annulene exists in two conformations, (a) and (b), that are in fast reversible equilibrium. At −60 °C the equilibrium constant is $K = (p_A/p_B) = 2.9$.

(2) Each conformation possesses an internal mobility such that, in connection with a simultaneous double-bond shift, all protons can successively assume the four or sixteen magnetically different positions, respectively, in the two structures. Isomer (a) thereby passes through 8 and isomer (b) through 32 identical conformations that are differentiated only by the exchange of certain protons.

Noteworthy in this case is the fact that the existence of conformation (b) was demonstrated by the finding that the experimental spectra could not be matched with spectra obtained by line shape calculations without assuming a contribution of (b) to the equilibrium.

2.4   VALENCE TAUTOMERISM

A fascinating area of organic chemistry, the development of which would be difficult to imagine without n.m.r. spectroscopy, is that of fast, reversible *valence tautomerism* or *valence isomerization*. Such a process was observed for the first time by means of n.m.r. spectroscopy in the case of 3,4-homotropilidene (**155**). The reaction involves a degenerate Cope rearrangement in which the cyclopropane bond between C-3 and C-4 migrates to the position between C-1 and C-6 via a 3,3-sigmatropic shift:

155   **a**   **a'**

Since the starting material and the product are completely identical with respect to their chemical and physical properties, this kind of rearrangement has been termed an *identity* or *'invisible'* reaction. The equilibrium (a) ⇆ (a') is likewise called *isodynamic*. Through n.m.r., however, a process such as this becomes 'visible' and thus experimentally detectable, since the spectrum allows us to identify the chemical environment of an individual nucleus in the molecule and to trace possible changes in this environment. Thus, in the course of the rearrangement, C-7 proceeds from an allylic to a cyclopropyl position. Similarly the other atoms, with the exception of those in positions 2 and 5, experience an exchange of their environments and since this change occurs very rapidly at room temperature homotropilidene is called a *fluxional molecule*.

As can be seen immediately, n.m.r. spectroscopy is the method of choice for the detection of such valence tautomerism since the spectrum is sensitive to a change in the Larmor frequency of the nuclei in question. The transition (a) ⇆ (a') manifests itself in the region of fast exchange through the chemical equivalence of the protons in positions 7 and 8 as well as those in positions 1, 3, 4, and 6. The spectrum of the 'frozen' structure can be first observed after line broadening in the region of slow exchange at temperatures below −30 °C.

For the mechanism of the rearrangement 2D EXSY spectroscopy yields important information. According to the results of investigations using different techniques, among them the analysis of the ^{1}H n.m.r. spectrum, 3,4-homotropilidene exists in the ground state in a *transoid* boat conformation (**155b**). If the COPE-rearrangement would proceed through the corresponding *transoid* transition state **155c**, the

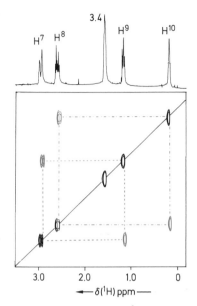

Figure 9.16 400 MHz ^{1}H 2D-EXSY spectrum of 3,4-homotropilidene (**155**) in the region slow exchange at -7 °C (aliphatic region). The cross peaks between the resonances of the cyclopropane protons H^3,H^4 and the olefinic protons H^1,H^6 are not shown. The 1D spectrum was recorded at -30 °C.

exchange of the methylene protons should follow the scheme H$^9 \rightleftharpoons$ H^8 and H$^{10} \rightleftharpoons$ H^7:

The 2D-EXSY spectrum (Figure 9.16), however, shows cross peaks between the protons H^9 and H^7 as well as between H^{10} and H^8. Consequently, the rearrangement takes place via the *cisoid* transition state **155e**. The valence tautomerism is thus preceded by a conformational change **155b** ⇌ **155d**.

This result is also in agreement with a stereochemical analysis of this concerted reaction based on molecular orbital considerations.

The subsequent development of the idea of degenerate Cope rearrangements led to the conception of bullvalene (**156**), the synthesis and n.m.r. spectrum of which

156

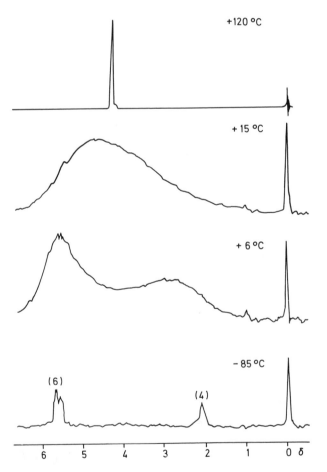

Figure 9.17 Temperature dependence of the proton resonance spectrum of bullvalene (**156**) (after Ref. 10)

created much interest. In this compound a series of valence isomerizations of the type just described leads at 80 °C to the equivalence of all protons, as is evidenced by a singlet in the n.m.r. spectrum (Figure 9.17). Lowering the temperature leads to line broadening and finally, at -80 °C, a spectrum that is consistent with the frozen structure **156** (the resonances of the cyclopropane protons and the tertiary proton are by chance superimposed). An activation energy of 49 kJ mol^{-1} (11.8 kcal mol^{-1}) has been derived for the process responsible for the averaging of all proton chemical shifts from the analysis of the temperature-dependent line shape.

A classical problem of organic chemistry is the question of whether or not cycloheptatriene (**157a**) does participate in a fast and reversible equilibrium with its valence tautomer norcaradiene (**157b**). By using n.m.r. spectroscopy it has been possible to detect norcaradiene cycloheptatriene equilibria in a series of derivates. The case involving 7-cyano-7-trifluoromethylcycloheptatriene (**157c**) was the first example.

157a   **157b**   **157c**

Exact data for one such reaction have been determined in the case of dicarbomethoxycycloheptatriene (**157**), the low-temperature spectrum of which is shown in Figure 9.18. Signals of the α-protons of both isomers can clearly be seen as the doublet at $\delta 5.72$ (**158**, $J_{\alpha\beta} = 9.5$ Hz) and the broadened singlet at $\delta 2.86$ (**159**). Three signals can be assigned to the protons of the ester methyl groups. Those of lower intensity ($\delta 3.74$ and 3.50) arise from the *exo*- and *endo*-carbomethoxy groups in norcaradiene (**159**) and the more intense singlet ($\delta 3.64$) has been assigned to the equivalent methyl groups in **158**. At room temperature a time-averaged spectrum is observed.

158   **159**

Since the components of the equilibrium above are not of equal energy, the temperature dependence of the equilibrium constant had to be determined by integration of the α-proton signals in the region of slow exchange. The enthalpy of the norcaradiene was found to be 673 $\pm$ 21 J mol^{-1} (161 $\pm$ 5 cal mol^{-1}) lower than that of the cycloheptatriene, while an entropy difference of

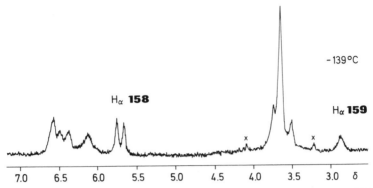

Figure 9.18   Proton resonance spectrum of the fluxional system **158** ⇆ **159** in the region of slow exchange (after Ref. 11). The x's identify spinning side bands

11.9 ± 0.1 J K^{-1} mol^{-1} (2.85 ± 0.03 cal K^{-1} mol^{-1}) favoured the triene. The entropy factor is thus responsible for the fact that at room temperature the free energy of the triene is less than that of the norcaradiene: $\Delta G^{\ddagger}_{298} = -2.88$ kJ mol^{-1} (-0.69 kcal mol^{-1}). A line shape analysis leads to values of 29.4 ± 0.6 kJ mol^{-1} (7.02 ± 0.15 kcal mol^{-1}) for the Arrhenius activation energy and $10^{11.9 \pm 0.2}$ for the frequency factor of the reaction **159**→**158**.

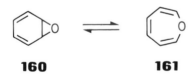

160   **161**

In case of the heterocycle oxepine (**161**) an equilibrium with the isomeric benzene oxide has been detected for the parent compound. Here the bicyclic structure is about 7.1 kJ mol^{-1} (1.7 kcal mol^{-1}) more stable while the entropy of the oxepine is about 42 J K^{-1} mol^{-1} (10 cal K^{-1} mol^{-1}) larger. The activation energies for the forward and reverse reactions (**160** → **161** and **161** → **160**) are 38.0 kJ mol^{-1} (9.1 kcal mol^{-1}) and 30.1 kJ mol^{-1} (7.2 kcal mol^{-1}), respectively.

Another interesting variation of valence tautomerism is the phenomenon of *double bond* or *π-bond shift* that has already been mentioned in the case of [16]annulene. Such a process was first detected in a very elegant manner in the case of cyclooctatetraene. The situation here is complicated by the fact that in addition to the bond shift a ring inversion occurs. It has been possible, however, to analyse both processes separately by using 1-(dimethylhydroxymethyl)-2,3,4,5,6,7-hexadeuter-ocyclooctatetraene.

H3C CH3
C
OH
H

a

HO CH3
C CH3
H

b

CH3 CH3
C
H OH

c

H
OH
C
H3C CH3

d

In this compound two pairs of structural isomers [(a)/(b) and (c)/(d)] and two pairs of conformational isomers [(a)/(c) and (b)/(d)] are in equilibrium with one another as a result of the following processes:

(1) ring inversion: (a) ⇆ (c) and (b) ⇆ (d);
(2) double bond shift: (a) ⇆ (b) and (c) ⇆ (d);
(3) double bond shift and ring inversion: (a) ⇆ (d) and (b) ⇆ (c).

At −35 °C all of the transformations are sufficiently slow and the resonance of the olefinic proton appears as two equally intense sifinglets at $\delta 5.80$ and $\delta 5.76$ that can be assigned to the isomers (a) and (c) and (b) and (d), respectively. As one would expect, because of the chirality of the non-planar olefinic system, two signals, separated by about 0.05 ppm, are observed for the methyl protons.

At higher temperature the signals of the methyl protons and also those of the olefinic proton degenerate to singlets. The coalescence temperatures for the changes are −2 and 41 °C, respectively. This difference shows that the transition from diastereotopic to enantiotopic methyl groups is achieved only through a ring inversion that confers an effective plane of symmetry on the molecule on the n.m.r. time scale. At higher temperatures the double bond isomerization accounts for the time averaging of the environment of the olefinic proton and the processes (a) ⇆ (b) and (c) ⇆ (d) now also contribute to the magnetic equivalence of the methyl protons.

These observations lead to the conclusion that a lower activation energy is necessary for the ring inversion than is required for the double isomerization, especially since the parameter δv is smaller for the exchange (a) ⇆ (b) or (c) ⇆ (d) than for (a) ⇆ (c) and (b) ⇆ (d). If the free energy of activation for both processes were the same, the coalescence of the olefinic proton signals would have been observed first because, according to equation (9.14), a smaller δv value requires a smaller k value and hence a lower temperature for coalescence.

A *quantitative* evaluation of $\Delta G^{\ddagger}$ at $-2\ ^\circ$C leads to a value of 61.5 kJ mol^{-1} (14.7 kcal mol^{-1}) for ring inversion and 71.5 kJ mol^{-1} (17.1 kcal mol^{-1}) for double bond isomerization. Assuming for both processes a planar transition state characterized in one case by **162** and in the other by **163** it can be concluded that the structure **163** with a delocalized $4n$ π-system is destabilized by about 8 kJ mol^{-1} (2 kcal mol^{-1}) with respect to the olefinic structure **162** with localized double bonds; **163** therefore is *antiaromatic*.

Interesting with respect to this observation is the finding that in 1,6:8,13-anti-bis-(methano)-[14]annulene (**164**) a double bond isomerization also takes place. The noteworthy point is that here, for the first time, we encounter such a process in a

162   **163**   **164**

$4n + 2\pi$ system that formally satisfies the Hückel rule with 14 π-electrons. Of course, this rule also assumes the coplanar arrangement of the carbon atoms of the molecular perimeter. In the case of **164** this requirement obviously is not fully met. The twisting of individual carbon–carbon bonds—according to models, especially those in the centre fragments, C-6 to C-8 and C-13 to C-1—gives rise to a considerable barrier to conjugation. The delocalization energy is therefore not sufficient to stabilize the resonance hybrid. The spectrum shown on p. 93 and recorded at $-138\ ^\circ$C is accordingly temperature dependent. After line broadening and coalescence, only two singlets for the olefinic protons and an AB system for the protons of the methylene bridge are observed at room temperature. Characteristic of the internal dynamics of this molecule is the observation that the singlet of the protons in positions 7 and 14 is temperature independent.

The rate constants for the valence tautomerization **164a** $\leftrightarrows$ **164b** can be derived from the temperature-dependent variation of the methylene proton resonances. Figure 9.19 shows several typical experimental and calculated spectra for different temperatures. The activation energy and the frequency factor were calculated to be 30 kJ mol^{-1} (7.1 kcal mol^{-1}) and $10^{12.2}$, respectively. If one assumes, as we did in the case of cyclooctatetraene, the delocalized structure **165** to be the transition state, one obtains the following energy profile:

The reaction can of course, be formulated alternatively as the result of electrocyclic processes.

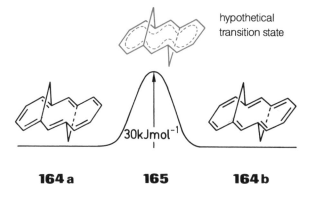

164 a   **165**   **164 b**

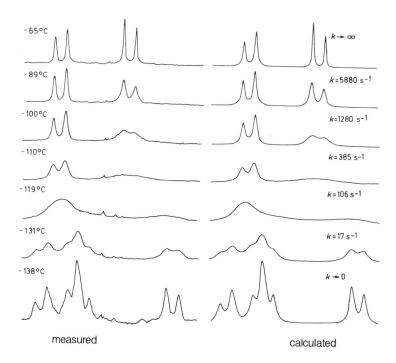

measured   calculated

Figure 9.19 Temperature dependence of the absorption of the methylene protons in 1,6;8,13-anti-bis(methano)[14]annulene (after Ref. 12). Since the molecule effectively possesses C_{2h} symmetry only one AB system is observed in the region of fast exchange. At -138 °C the methylene groups are no longer equivalent and one finds two AB systems that are partially superimposed.

2.5 DYNAMIC PROCESSES IN ORGANOMETALLIC COMPOUNDS AND CARBOCATIONS

Considerable progress in developing an understanding of the structure of Grignard reagents has been made through the application of n.m.r. techniques. For example, the ¹H n.m.r. spectrum of allylmagnesium bromide **166** consists of a quintuplet at $\delta 6.2$ and a doublet at $\delta 2.5$ in an intensity ratio of 1:4, indicating the presence of an AX_4 system. This observation is consistent with a rapid equilibrium involving the two isomers **166a** and **166b**.

$$\overset{1}{H_2C}=\overset{2}{CH}-\overset{3}{CH_2}-MgBr \quad \rightleftharpoons \quad BrMg-\overset{1}{CH_2}-\overset{2}{CH}=\overset{3}{CH_2}$$

166 a   **166 b**

In the case of 3,3-dimethylbutylmagnesium chloride (**167**) the spectrum has been observed to be temperature dependent. At $-53\ °C$ an AA'BB' system is obtained while at $32\ °C$ one observes a spectrum of the A_2X_2 type (Figure 9.20).

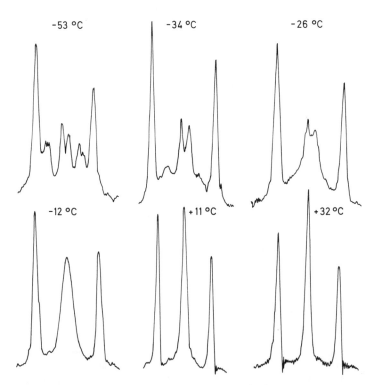

Figure 9.20 Temperature-dependent variations in the proton resonance spectrum of 3,3-dimethylbutylmagnesium chloride; AA' portion (after Ref. 13)

For a simplified explanation we assume that at low temperature the compound exists predominantly in the *trans* conformation **168**. The observation of an AA'XX' system is consistent with this conclusion. The transition to an A_2X_2 system means then that at high temperatures the parameter L becomes zero. This indicates that for the spin–spin interaction between the A and the X protons only one coupling

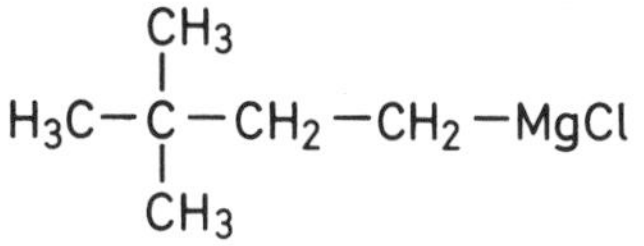

167

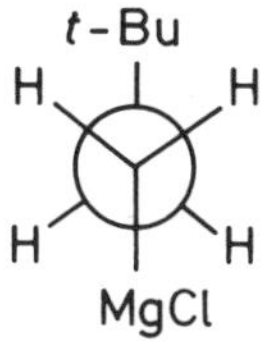

168

constant exists. The protons A and A' and X and X' must therefore be magnetically equivalent, that is they exchange magnetic sites with one another. The most plausible mechanism for such an exchange is the inversion of the carbanion (**169** ⇆ **170**), that is isoelectronic with an amine. The activation energy for this process was determined from the temperature dependent line shape of the spectrum to be 48 ± 8 kJ mol^{-1} (11 ± 2 kcal mol^{-1}).

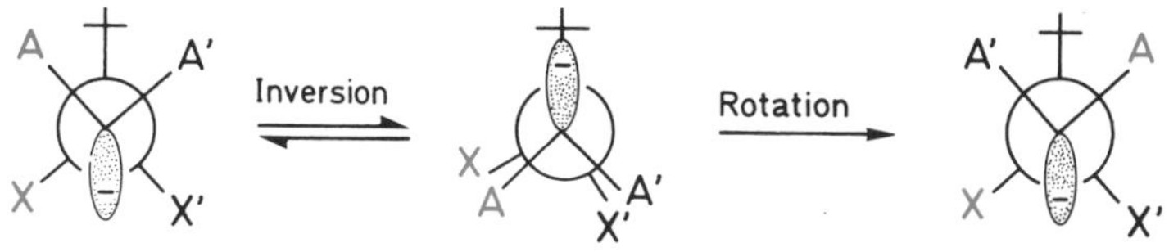

169

170

In *metal carbonyl complexes of olefins* interesting exchange phenomena involving the metal carbonyl moieties have been observed. Thus, the Fe(CO)$_4$ complex of tetramethylallene (**171**) at room temperature exhibits only a singlet (Figure 9.21). At -60 °C, however, the spectrum consists of three signals with an intensity ratio of 1:1:2, as would be required for a 'frozen' structure. An exchange of the Fe(CO)$_4$ group is therefore responsible for the simple room temperature spectrum. The intramolecular nature of this tautomerism has been confirmed by the fact that in the fast-exchange region signals for uncomplexed excess tetramethylallene can be observed separately.

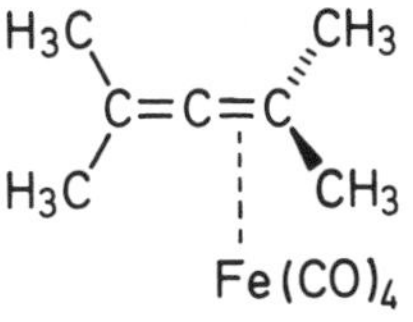

171

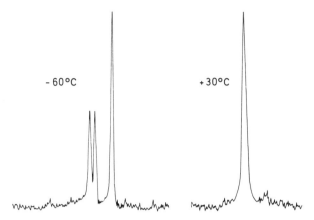

-60°C   +30°C

Figure 9.21 Proton resonance spectra of tetramethylallene iron tetracarbonyl at different temperatures (after Ref. 14)

Analogous processes have also been reported in cyclic olefins. Thus with methylcyclooctatetraeneirontricarbonyl (**172**) and the N-carbomethoxyazepineirontricarbonyl (**173**) the indicated equilibria with free energies of activation of 33 kJ mol^{-1} (8 kcal mol^{-1}) and 65 kJ mol^{-1} (15.5 kcal mol$^-$), respectively, have been observed. In **172** the exchange of the irontricarbonyl group is accompanied by a double bond shift that can not take place in **173**.

172

173

Deeper insight into the mechanisms of proton shifts in carbocations has also developed as the result of applications of n.m.r. spectroscopy. Thus, *hexamethylbenzene* in hydrofluoric acid forms the *σ-complexed cation* that is the accepted intermediate in electrophilic aromatic substitution. The n.m.r. spectrum of this cation (**174**) at room temperature indicates that there is a rapid migration of the added proton around the ring (Figure 9.22). One observes a doublet ($J = 2.1$ Hz) for the

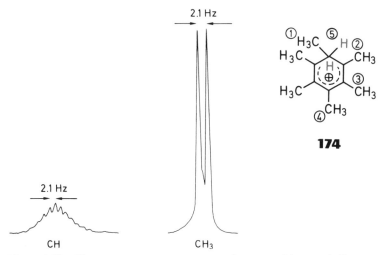

Figure 9.22 The proton resonance spectrum of protonated hexamethylbenzene (**174**) in the region of rapid exchange (after Ref. 15)

methyl resonances and a multiplet for the tertiary proton that, as a result of its rapid exchange, is effectively coupled to all of the eighteen methyl protons. Therefore, the observed splitting, according to equation (9.5), is an average of the coupling constants observed in the low-temperature spectrum of the 'frozen' structure **174**; $J_{15} = 6.8$ Hz, $J_{45} = 3.5$ Hz, $J_{25} = 1$ Hz and $J_{35} = 0$ Hz:

$$J_{\text{obs.}} = (J_{15} + J_{45} + 2J_{25})/6$$

The comparison of experimental and calculated line shapes in the temperature range between -105 and $55\,°C$ leads to an activation energy of 55.7 ± 3.3 kJ mol^{-1} (13.3 ± 0.8 kcal mol^{-1}) for the proton migration.

In an analogous case, namely that of the heptamethylbenzonium ion, even the mechanism of the exchange process can be clarified in this way. The theoretical spectra expected for a series of 1,2-shifts (**175**) and a statistical exchange via an intermediate π-complex (**176**) were both calculated. Only in the first case did the calculated and the experimental spectra agree.

177   **178**

One problem of carbocation chemistry that has attracted much interest for many years, the structure of the norbornyl cation (**177**), was also investigated by dynamic n.m.r.

The ion was prepared from the fluoride (**178**) in a solution of SbF$_5$ and SO$_2$F$_2$ and exhibited the spectrum displayed in Figure 9.23a. The presence of three signals in the intensity ratio of 4:1:6 that coalesce at higher temperatures to a singlet, allows the conclusion that even at -113 °C no 'frozen' structure exists. Wagner–Meerwein rearrangements and 2,6-hydride shifts (see Scheme) cause the protons in positions 1,

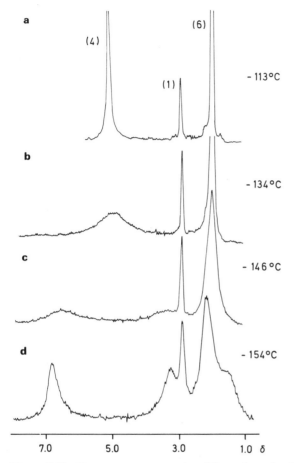

Figure 9.23 Proton resonance spectra of the norbornyl cation as a function of temperature (after Ref. 16)

2, and 6 and also those in positions 3, 5, and 7 to become equivalent under these conditions. The proton in position 4 does not participate in this exchange process and the cation possesses C_{3v} symmetry on the n.m.r. time scale. At -113 °C the 2,3-hydride shift, however, is slow so that the enthalpy of activation of 45.2 ± 2.5 kJ mol^{-1} (10.8 ± 0.6 kcal mol^{-1}) derived from the line shape changes of the spectrum at higher temperatures (-100 to 3 °C) can be assigned to this process.

If one uses a mixture of SbF$_5$, SO$_2$ClF, and SO$_2$F$_2$ it is even possible to record spectra down to -158 °C. By further cooling, the signal at lowest field splits into two singlets of equal intensity. In addition, the signal at highest field develops a shoulder while the signal of the proton in the 4-position remains unchanged throughout. The low-temperature spectrum obtained in this way (Figure 9.23) is best rationalized on the basis of a protonated nortricyclene structure (**179**), that is, a non-classical carbocation with a five-coordinate carbon atom. The bond to C-2 and C-6 can be described as a three-centre bond and the three-fold axis of symmetry has been replaced by a symmetry plane. Neither the classical structure **177** nor the assumption that the Wagner–Meerwein rearrangements are still fast even at this low temperature can explain this spectrum. For the second dynamic process an activation energy of 24.7 ± 0.8 kJ mol^{-1} (5.9 ± 0.2 kcal mol^{-1}) has been derived on the basis of spectral changes from a line shape calculation according to equation (9.2).

179

Finally, we want to discuss the rearrangement of acyloxonium salts proceeding as neighbouring group participation. In the case of the cation **180** obtained by treating glycerine triacetate with antimony pentachloride the n.m.r. spectrum indicates a fast reversible rearrangement into the isomer **181**. At room temperature, with acetonitrile-d$_3$ as solvent, the methyl protons appear as two singlets at $\delta 3.35$ and $\delta 2.16$, the line widths of which increase upon heating. Finally, the two signals coalesce at 105 °C and a value of 78.2 kJ mol^{-1} (18.7 kcal mol^{-1}) was determined for $\Delta G^{\ddagger}$.

In cyclic compounds a neighbouring group reaction of this kind leads to analogous valence isomerizations and in the case of the cation **182** a participation of all the acetoxy groups in a series of identity reactions can take place.

180   **181**

182

3. Intermolecular exchange processes

The exchange processes discussed so far were without exception *intramolecular*. In the remaining section of this chapter we want to mention briefly a few exchange phenomena that are characterized by their *intermolecular* nature. These include almost all proton transfer reactions such as those already mentioned in our discussion of methanol. The collapse of spin–spin multiplets is in many cases of diagnostic value. Table 9.3 gives the results of several investigations on alcohols, amines, and related substances. The processes involved are exclusively second-order reactions, in contrast to the first order reactions treated previously.

Of special interest to the chemist are keto–enol equilibria that involve both intermolecular and intramolecular proton transfer and that can also be detected and studied by means of n.m.r. spectroscopy. In the spectrum of acetylacetone (Figure 9.24) one clearly sees the signals of both the keto form [$\delta(CH_3)2.2$. $\delta(CH_2)3.7$] and the enol form [$\delta(CH_3)2.0$, $\delta(=CH)5.7$], indicating that the rate of the equilibrium reaction is slow on the n.m.r. time scale ($k < 10^{-1}$). With the addition of base or by raising the temperature, the reaction rate increases and line broadening and finally coalescence to time-averaged spectrum is observed.

Table 9.3 Intermolecular proton transfer reactions and their second-order rate constants ($1 \text{ mol}^{-1} \text{ s}^{-1}$) as determined using n.m.r. spectroscopy

Reaction				k
H_2O	$+ H_3O^+$	$\leftrightarrows H_3O^+$	$+ H_2O$	10^{10}
H_2O_2	$+ H_3O^+$	$\leftrightarrows H_3O_2^+$	$+ H_2O$	2×10^7
CH_3OH	$+ H*OH$	$\leftrightarrows OCH_3OH*$	$+ HOH$	3
CH_3OH	$+ H_3O^+$	$\leftrightarrows CH_3OH_2^+$	$+ H_2O$	10^8
RSH_2^+	$+ RSH$	$\leftrightarrows RSH$	$+ RSH_2^+$	10^2
RS^-	$+ RSH$	$\leftrightarrows RSH$	$+ RS^-$	6×10^5
R_3NH^+	$+ H_2O$	$\leftrightarrows R_3N$	$+ H_3O^+$	10^{-2}
R_3NH^+	$+ R_3N$	$\leftrightarrows R_3N$	$+ R_3NH^+$	10^8
R_2NH	$+ OH^-$	$\leftrightarrows R_2N^-$	$+ HOH$	5×10^6
$(CH_3)_3PH^+$	$+ (CH_3)_3P$	$\leftrightarrows (CH_3)_3P$	$+ (CH_3)_3PH^+$	10^2
$H_5C_6-C\equiv C-$ H	$+ OH^-$	$\leftrightarrows H_5C_6-$ $C\equiv C^-$	$+ H_2O$	5×10^2

From the relative intensities of the signals it is clear that at room temperature the enol form predominates in the equilibrium mixture. Integration shows that the mixture contains 86% enol and 14% ketone. The strong deshielding of the enol hydroxyl proton is a consequence of intramolecular hydrogen bond formation. The magnetic equivalence of the methyl groups of the enol form, provided it is not accidental, indicates a rapid intramolecular exchange of the hydrogen atoms between the two oxygen functions.

183

Intramolecular exchange processes are also encountered in *organometallic* chemistry. Figure 9.25 shows the temperature dependence of the n.m.r. spectrum of trimethylaluminium in toluene. At -55 °C two separate signals corresponding to the dimeric structure **183** are observed. These signals can be assigned to the methyl bridges and to the terminal methyl groups. The equivalence of all of the methyl groups at higher temperature can be rationalized on the basis of the following mechanism:

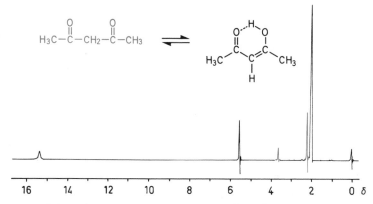

Figure 9.24 The proton resonance spectrum of acetylacetone at room temperature in carbon tetrachloride

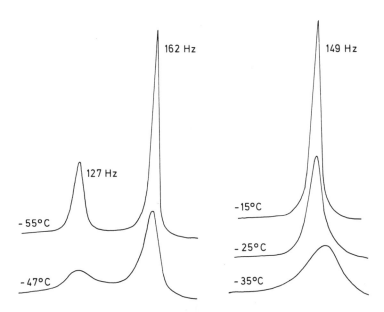

Figure 9.25 The temperature dependence of the proton resonance spectrum of trimethylaluminium (in toluene) (after Ref. 17)

The activation energy for the exchange was determined to be 65.3 kJ mol^{-1} (15.6 kcal mol^{-1}). The intermolecular course of the process was confirmed by the fact that in the presence of trimethylgallium not only the individual signals of the methyl groups of trimethylaluminium and trimethylgallium coalesced but at higher temperatures even the two average signals of these two compounds merged with one another.

In other cases the coupling constants between the proton and the metal of the organometallic compound has been used as an indicator for a dynamic process. For example, naturally occurring cadmium consists of a mixture of isotopes in which there are ^{111}Cd (13%) and ^{113}Cd (12%) nuclei each with a spin of $I = \frac{1}{2}$ in addition of ^{112}Cd nuclei with a spin of $I = 0$. In the proton resonance spectrum of dimethylcadmium one therefore observes satellites that are caused by spin–spin interaction with the cadmium (Figure 9.26). As in the case of the splitting of the hydroxyl proton signal in methanol, this fine structure disappears when individual methyl groups are exchanged between different cadmium atoms. It has been observed experimentally that with increasing temperature the Cd satellites at first broaden and finally disappear under the principal signal of the methyl resonance. The fact that one can still easily observe the ^{13}C satellites after the disappearance of the Cd satellites clearly indicates that the exchange involves entire methyl groups and not individual protons.

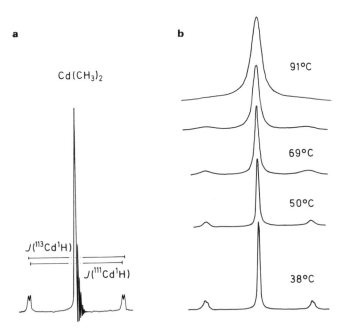

Figure 9.26 (a) The proton resonance spectrum of dimethylcadmium in toluene (after Ref. 17); (b) Temperature dependence of the proton resonance spectrum of pure dimethylcadmium (after Ref. 18)

4. The influence of ^{14}N nuclei on proton resonances

Finally at this point, we wish to discuss the effects that can be caused in proton resonance spectra by the presence of ^{14}N nuclei in the molecule under study.

As mentioned earlier (p. 10), the ^{14}N nucleus has a spin quantum number of $I = 1$ and, as a result of its non-spherical charge distribution, a quadrupole moment that is of significance for its relaxation behaviour (cf. p. 230). Interactions of the quadrupole moment with the electric field gradients of the electron cloud of the molecule in question represent a relaxation mechanism by which the ^{14}N nucleus can exchange its energy with the lattice. If relaxation is fast, an N–H proton 'sees' only a time average of the possible spin states $m_I(^{14}N) = 1$, 0, and -1 and the expected splitting into a 1:1:1 triplet vanishes. This amounts to a 'decoupling' of both nuclei.

As a rule the relaxation times of the ^{14}N nuclei in organic compounds are of such a magnitude that they do not lead to a complete disappearance of line splittings. The result is usually a more or less extensive line broadening for the ^{1}H signals due to ^{14}N, ^{1}H interactions over one or two bonds (^{14}N–^{1}H, ^{14}N–C–^{1}H). Therefore, if one wishes to measure ^{14}N–^{1}H couplings the quadrupolar relaxation must be suppressed. This can be achieved by *increasing the temperature* or by making the electronic environment of the ^{14}N nucleus symmetrical. Accordingly, the proton resonance spectrum of the ammonium ion ($^{14}NH_4^+$) consists of a 1:1:1 triplet with sharp lines (cf. Exercise 2.9, p. 39/40). ^{14}N–^{1}H coupling is also observed in isonitriles, from which it can be concluded that the electric field gradient of the electron cloud at the nitrogen in these compounds is small.

If, on the other hand, one wishes to observe H–H coupling in a molecule such as pyridine without the perturbing effect of ^{14}N, quadrupolar relaxation must be increased. This can often be accomplished by *lowering the temperature*. In addition, heteronuclear double resonance is used to decouple the ^{14}N nucleus (see figure). The perturbing effect of ^{14}N can also be eliminated by substituting ^{15}N with $I = \frac{1}{2}$ for ^{14}N, but this of course requires expensive syntheses.

The relationships described can further be complicated by the fact that N–H protons participate in exchange processes via hydrogen bond formation. In many cases this also leads, as mentioned on p. 23, to line broadening. Furthermore, the exchange provides an additional mechanism that, as in the case of methanol (p. 338), can lead to a disappearance of ^{14}N–^{1}H or ^{1}H–^{1}H coupling constants.

Since the rate of proton exchange is increased by raising the temperature or by the addition of catalysts such as H_2O, OH^-, or H_3O^+ the temperature must be decreased for the elimination of these effects. Likewise, by using dimethyl sulphoxide as solvent the proton exchange can be suppressed to the extent that coupling constants of N–H protons can be measured.

388

REFERENCES

5. References

1. S. Forsén and R. A. Homan, *J. Chem. Phys.,* **39**, 2892 (1963).
2. H. Günther and H. E. Mons, unpublished results.
3. J. Heidberg, J. A. Weil, G. A. Janusonis and J. K. Anderson, *J. Chem. Phys.,* **41**, 1033 (1964).
4. J. E. Anderson and J. M. Lehn, *J. Am. Chem. Soc.,* **89**, 81 (1967).
5. F. A. Bovey, *NMR Spectroscopy,* Academic Press, New York, 1969.
6. *Japan Electron Optics Lab. News,* No. 4 (1970).
7. F. R. Jensen and C. H. Bushweller, *J. Am. Chem. Soc.,* **91**, 3223 (1969).
8. H. Günther, P. Schmitt, H. Fischer, W. Tochtermann, J. Liebe and Ch. Wol, *Helv. Chim. Acta,* **68**, 801 (1985).
9. F. Sondheimer *et al.,* in *Aromaticity,* The Chemical Society, London, 1967.
10. G. Schrîder, J. F. M. Oth and R. Merenyi, *Angew. Chem.,* **77**, 774 (1965).
11. M. Gîrlitz and H. Günther, *Tetrahedron,* **24**, 4467 (1969).
12. E. Vogel, U. Haberland and H. GÅnther, *Angew. Chem.,* **82**, 510 (1970); *Angew. Chem. Int. Ed. Engl.,* **9**, 513 (1970).
13. G. M. Whitesides, M. Witanowski and J. D. Roberts, *J. Am. Chem. Soc.,* **87**, 2854 (1965).
14. R. Ben-Soshan and R. Pettit, *J. Am. Chem. Soc.,* **89**, 2231 (1967).
15. E. L. Mackor and C. McLean, *Pure Appl. Chem.,* **8**, 393 (1964).
16. G. A. Olah and A. M. White, *J. Am. Chem. Soc.,* **91**, 3957 (1969).
17. N. S. Ham and T. Mole, in F5, **4**, 91 (1969).
18. W. Bremser, M. Winokur and J. D. Roberts, *J. Am. Chem. Soc.,* **92**, 1080 (1970).

RECOMMENDED READING

J. Sandström, *Dynamic NMR Spectroscopy,* Academic Press, New York, 1982.

L. M. Jackman and F. A. Cotton (Eds.), *Dynamic Nuclear Magnetic Resonance Spectroscopy,* Academic Press, New York, 1975.

Review articles

(a) C. S. Johnson, Jr., Chemical Rate Processes and Magnetic Resonance, in F1, **1**, 33 (1965).
(b) G. Binsch, The Study of Intramolecular Rate Processes by Dynamic Nuclear Magnetic Resonance, in N. L. Allinger and E. L. Eliel, (Eds.) *Topics in Stereochemistry,* Vol. 3, Interscience, New York, 1968.
(c) H. Kessler, Detection of Hindered Rotation and Inversion by NMR Spectroscopy, *Angew. Chem. Int. Ed. Engl.,* **9**, 219 (1970).
(d) J. M. Lehn, Nitrogen Inversion, *Fortschr. Chem. Forsch.,* **5**, 311 (1970).
(e) J. B. Lambert, Pyramidal Atomic Inversion, in N. L. Allinger and E. L. Eliel, (Eds.) *Topics in Stereochemistry,* Vol. 6, Interscience, New York (1971).
(f) J. B. Lambert and Y. Takeuchi, *Cyclic Organonitrogen Stereodynamics,* VCH Publishers, New York,
(g) L. S. Johnson Jr. and C. G. Moreland, The Calculation of NMR Spectra for Many-Site Exchange Problems, *J. Chem. Educ.,* **50**, 477 (1973).
(h) C. S. Handloser, M. R. Chakrabarty and M. W. Mosher, Experimental Determination of pK_a Values by Use of NMR Chemical Shifts, *J. Chem. Educ.,* **50**, 510 (1973).
(i) D. E. Krahnbuehl, P. M. Metzger, D. W. Thompson and R. C. Fay, NMR Studies of the Stereochemistry and Non-Rigidity of Titanium-β-Diketonate Complexes, *J. Chem. Educ.,* **54**, 119 (1977).

(j) G. Binsch and H. Kessler, The Kinetic and Mechanistic Evaluation of NMR Spectra, *Angew. Chem. Int. Ed. Engl.,* **19**, 411 (1980).

(k) G. Willem, 2D NMR Applied to Dynamic Stereochemical Problems, in F5, **20**, 1 (1988).

(l) C. L. Perrin and T. J. Dwyer, Application of Two-Dimensional NMR to Kinetics of Chemical Exchange, *Chem. Rev.,* **90**, 935 (1990).

10 SELECTED EXPERIMENTAL TECHNIQUES OF NUCLEAR MAGNETIC RESONANCE SPECTROSCOPY

In many laboratories, nuclear magnetic resonance spectra are obtained as a daily routine for a large number of samples and the n.m.r. experiment performed with the available instrumentation, mostly for ^{1}H and ^{13}C spectra, is practically 'programmed'. If sample concentrations are similar, the experimental settings need hardly be adjusted and spectra can be obtained in an on-line fashion, especially if an automatic sample changer is installed. With this approach, obviously, only a fraction of the enormous potential behind the method is used, but this limitation is readily accepted since it guarantees a rapid return of information, mostly of an analytical nature.

On the other hand, throughout the Periodic Table there are magnetic nuclei suitable for the n.m.r. experiment and different probes for the properties of molecules and solids are thus available. A discussion of a few of these cases is included in the next chapter. In addition, a number of experimental variations are possible in n.m.r. and the physics of spin systems is as fascinating today as it was in earlier times. Nuclear magnetic resonance therefore certainly ranks among the most versatile spectroscopic techniques, and in this chapter we shall introduce the reader to those of the advanced n.m.r. techniques which have become particularly important to chemistry.

1. Superconducting magnets

Our discussion of the analysis of n.m.r. spectra has shown that the desired information concerning the chemical shifts and the coupling constants can be obtained very easily if the spectrum can be considered to be first order. Since the chemical shift is field dependent (while the coupling constants are not), complicated spectra can often be simplified by performing the experiment at higher field strength where the ratio $J/v_0\delta$ becomes sufficiently small. In addition, even without spin–spin coupling, the resolution greatly improves as still smaller chemical shift differences for groups of similar structure can be recognized. More information is thus available from the spectrum. Finally, a sensitivity enhancement is achieved by increasing B_0 since the Boltzmann distribution between the spin levels becomes more favourable (cf. equation (1.11)). It is therefore not surprising that the development of n.m.r. spectroscopy was accompanied by considerable improvements in magnet technology.

Historically, the first commercial spectrometers that were sold at the beginning of the 1950s were equipped with electromagnets of field strength of 1.0 T and a radiofrequency source of 40 MHz for ^{1}H n.m.r. Later, 60-MHz instruments*

---

* The proton resonance frequency rather than the field strength is used for the classification of n.m.r. spectrometers.

(B_0 =1.4 T) became available and by the mid-1960s the standard research n.m.r. spectrometer operated at 100 MHz and a field strength of 2.3 T. At the same time the limit for the use of conventional electromagnets was reached since with the usual ferromagnetic materials no higher degree of magnetization can be produced.

A completely new approach had to be employed for the production of more powerful magnetic fields, and this led to the development of superconducting magnets. Use was made of certain metals and alloys, for example, of niobium and zirconium, which at 4 K, that is at the temperature of liquid helium, no longer exhibit electrical resistance. Thus a significant increase in the current intensity is possible in these systems.

The principle of an n.m.r. spectrometer with a superconducting or *cryomagnet* is illustrated in Figure 10.1. The 'magnet' here consists of a coil, the solenoid S, made from a special alloy. The sample tube is located within the coil that, in turn, is immersed in liquid helium. One of the principle problems encountered in the construction of such a spectrometer was the thermal insulation of the solenoid. It

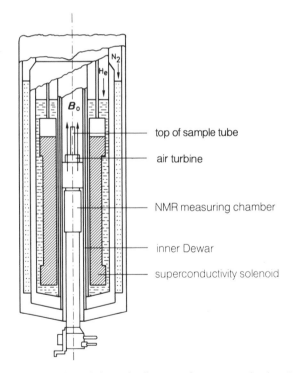

top of sample tube

air turbine

NMR measuring chamber

inner Dewar

superconductivity solenoid

Figure 10.1 Schematic diagram of a superconducting "magnet". The coil is located in a Dewar flask containing liquid helium; the Dewar flask, in turn, is cooled in liquid nitrogen. The sample tube is placed into the instrument from above. In contrast to the conventional magnets, the lines of the external magnetic field here are parallel to the long axis of the sample tube (after Ref. 1)

could be solved satisfactorily, and further progress in the technology, especially improvements in the construction of the Dewar vessel which minimize helium consumption and increase the periods between refills, has paved the way for research instruments that operate at the B_0 field strength between 4.7 and 11.8 T and ^{1}H n.m.r. frequencies from 200 to 500 MHz. For special applications, in particular work on synthetic and biopolymers, cryomagnets with field strengths even of up to 14.1 or 20.6 T with ^{1}H frequencies of 600 and 750 MHz respectively are in use. For instruments up to 500 MHz helium loss is low, typically 250 cm^3 per day, and refill times are several months. Liquid nitrogen consumption, however, is much higher and weekly refill intervals are the rule.

As an illustration for the effect that can be expected by increasing the B_0 field, the spectrum of acrylonitrile at 5.1 T/220 MHz may be cited (Figure 10.2) and compared with the 1.4 T/60 MHz spectrum of acrylonitrile shown in Figure 6.4, p. 206. In contrast to the complicated ABC system observed there, we now see an AMX system that can be analysed directly as a first-order spectrum.

In particular for biopolymers such as proteins or nucleic acids, n.m.r. investigations with high magnetic fields are indispensable. In these systems, very often a large number of subspectra from similar partial structures, such as spin systems of amino acids or ribose and desoxyribose units, strongly overlap. Only at very high magnetic fields is the separation of individual spectral regions sufficient for an analysis. Already with the relatively small coenzyme nicotinamide adenine dinucleotide (NAD$^+$) (184), which plays an important part in biological redox processes, the ^{1}H n.m.r. spectra of only two ribose units cannot be analysed at low field strength. Figure 10.3 shows the advantage of cryomagnets with high magnetic fields, where individual resonances can be distinguished at $B_0 = 14.1$ T.

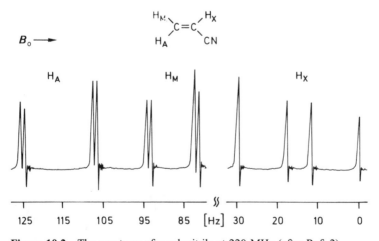

Figure 10.2 The spectrum of acrylonitrile at 220 MHz (after Ref. 2)

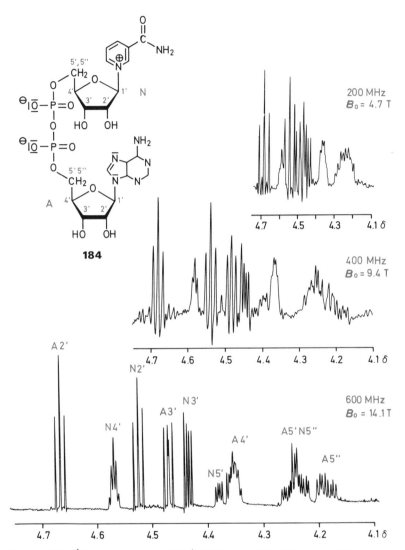

Figure 10.3 ^{1}H n.m.r. spectrum NAD$^+$ (**184**) in the high field region of the ribose protons at three different field strengths and ^{1}H frequencies. Because of the more favourable Hz/ppm ratio, the resolution increases with increasing field strength and facilitates spectral analysis (after Ref. 3)

In the investigation of dynamic effects, the use of higher field strengths likewise is of advantage. As equation (9.14) demonstrates, at the coalescence point the rate constant k is proportional to the relative chemical shift $\delta\nu$ between the Larmor frequencies in question. Fast reactions, i.e. those with large rate constants and low activation energies, can thus be investigated in a more easily accessible temperature

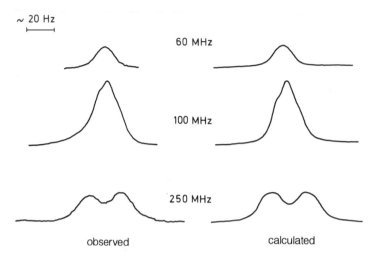

Figure 10.4 The absorption of the methyl protons of thiepine dioxide **185** at −150 °C and different frequencies (after Ref. 4)

range if it is possible to increase δv, thereby increasing $k_{\text{coal.}}$, and therefore the coalescence temperature.

The *ring inversion* of the thiepine dioxide derivative **185** offers an example for this effect. Here, the presence of diastereotopic methyl groups indicating the non-planarity of the compound was detected at −150 °C only with a 250-MHz spectrometer (Figure 10.4). Since the rate constant was the same in all three experiments because they were run at the same temperature, it is the increase in δv that is responsible for the observed line separation.

185

Finally, we should mention that for spectrometers with superconducting magnets the susceptibility correction equation (3.3) (p. 57) applies for δ values in the case of measurements made with external reference.

2. The Nuclear Overhauser Effect (NOE)

Three different phenomena are usually considered in connection with the notion of the *Overhauser effect*. Each involves a variation of the intensity of resonance signals observed in double resonance experiments. However, several different mechanisms are responsible for the effect.

In the case of the true Overhauser effect for a system consisting of nuclear spin I and an electron spin S, an increase in the intensity of the nuclear resonance signal is observed if one simultaneously saturates the electron resonance with an r.f. field of frequency v_S. This experiment may be performed on a paramagnetic solution of sodium in liquid ammonia in which the proton resonance is observed under conditions that saturate the electron resonance.

The variation of the intensities of the nuclear resonance lines that result from this experiment can be rationalized by reference to the *Solomon diagram* shown in Figure 10.5. There the eigenstates of a two-spin system IS in a magnetic field are represented. Altogether there exist four states of different energy and for their arrangement the different signs of the nuclear and electron spins were considered. Transitions for the nucleus or the electron can be stimulated by an r.f. field of the frequency v_I or v_S, respectively. Let us now consider the probability, W, for the particular relaxation transitions that are responsible for the maintenance of the Boltzmann distribution. Thus the quantities W_1 and $W_{1'}$ correspond to the probability for longitudinal relaxation of nuclear and electron spins, respectively. In addition there are also the transition probabilities W_2 and W_0 for the cases in which the nuclear and the electron spins flip simultaneously. W_2 and W_0 are of significance only when there is a spin–spin interaction between the spins I and S. Now, if the electron resonance, i.e. the transitions $(3) \to (1)$ and $(4) \to (2)$, is saturated by an oscillating field B with the frequency v_S, then the Boltzmann distribution between the states (3) and (1) as well as between (4) and (2) will be disturbed, that is, the populations of (1) and (2) will become too high while those of

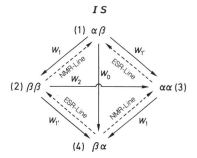

Figure 10.5   Solomon diagram for a two-spin system *IS* composed of a nuclear and an electron spin

(3) and (4) will become too low. This perturbation can be counteracted by an increased number of relaxation transitions, that is, through an increase in W_0, since in this fashion the state (1) is depopulated and the population of state (4) is increased. However, for the nuclear resonance, that is, for the transitions (4) → (3) and (2) → (1), this leads to an intensification of the absorption signal because the net effect of the process is the overpopulation of state (2) and the depopulation of state (3) since the spins are carried along the route (3) → (1) → (4) → (2). The result is a *polarization* of the nuclear spin distribution and the experiment is known as *dynamic nuclear polarization*.

It is important for the preceding case that the relation $W_0 \gg W_2$ is satisfied. An electron spin thus can flip only when a nucleus simultaneously changes its spin orientation in the opposite direction. In this case the relaxation is produced predominantly through a time-dependent *scalar* spin–spin coupling. In the previously mentioned solution of sodium in liquid ammonia the unpaired electrons are solvated by ammonia molecules. A fast exchange of these molecules between the solvation shell of different paramagnetic centres has the result that the proton–electron coupling vanishes. It maintains its effectiveness, however, as a relaxation mechanism.

Let us consider once more the experiment we have just described but with reference to the magnitude of the energies that are exchanged with the lattice in the relaxation process. For every quantum mechanical system that is characterized by two energy levels E_p and E_q, an equilibrium is established so that the number of transitions $E_p \to E_q$ is equal to the number of transitions $E_q \to E_p$. Thus, it follows that for the eigenstates (1) and (4) of the spin system IS

$$N_\alpha n_\beta W_{\alpha\beta \to \beta\alpha} = N_\beta n_\alpha W_{\beta\alpha \to \alpha\beta} \qquad (10.1)$$

in which N_α and N_β and n_α and n_β signify the populations for the nuclei and the electrons, respectively, and $W_{\alpha\beta \to \beta\alpha}$ and $W_{\beta\alpha \to \alpha\beta}$ denote the transition probabilities. According to the Boltzmann law we have

$$\frac{N_\alpha n_\beta}{N_\beta n_\alpha} = \frac{W_{\beta\alpha \to \alpha\beta}}{W_{\alpha\beta \to \beta\alpha}} = \exp\left(\frac{-\Delta E}{kT}\right) = \exp -h(\nu_S + \nu_I)/kT \qquad (10.2)$$

Now, if the electron resonance is saturated then $n_\alpha = n_\beta$ and

$$\frac{N_\alpha}{N_\beta} = \exp[-h(\nu_S + \nu_I)/kT] \qquad (10.3)$$

Because $h\nu_S \gg h\nu_I$ the nuclear spin distribution, which normally obeys the expression

$$\frac{N_\alpha}{N_\beta} = \exp(-h\nu_I/kT) \qquad (10.4)$$

is now determined by the very much larger energy difference $h\nu_S$.

If we carry this train of thought to a spin system that consists of two nuclear spins, we arrive at the so-called *nuclear Overhauser effect (NOE)*. For this case the Solomon diagram must be modified, since now both spins have the same sign and the sequence of the states is changed (Figure 10.6). Besides the longitudinal relaxation rates W_1 for the A and the X nucleus, respectively, cross relaxation rates W_2 for the double quantum transition and W_0 for the zero quantum transition are now of importance. Both are a consequence of *dipolar* spin–spin interactions between A and X.

If the resonance of one nucleus, for example A in Figure 10.6, is now irradiated, an increase in the intensity of the X resonance occurs if $W_0 \ll W_2$. Spin population is then transported from state (3) to state (2) via states (1) and (4). Regarding the population of states (1) and (4) in a first approximation as constant, state (3) is depopulated and state (2) is overpopulated. The Boltzmann distribution for the X lines thus requires signal enhancement. Since the frequencies of W_0 transitions are in the order of Hz or kHz, while the W_2 frequencies are in the MHz range, the condition $W_0 \ll W_2$ is always satisfied for mobile liquids or solutions of low molecular weight compounds which have a low viscosity (see page 229).

The quantitative treatment of the phenomenon leads to the so-called *Solomon equation*, an expression for the increase in the z-magnetization of nucleus X, M_z^X, relative to the equilibrium magnetization M_0^X:

$$\frac{M_z^X}{M^0} = 1 + \frac{W_2 - W_0}{2W_1^X + w_2 + W_0} \frac{\gamma_A}{\gamma_X} \qquad (10.5)$$

If a pure dipole–dipole interaction exists between the two nuclei, W_2, W_1, and W_0 are in the ratio of $1:\frac{1}{4}:\frac{1}{6}$ and equation (10.5) reduces to

$$\frac{M_z^X}{M_0^X} = 1 + \frac{\gamma_A}{2\gamma_X} \qquad (10.6)$$

For the homonuclear case ($\gamma_A = \gamma_X$) a signal enhancement of 50% results. This is the maximum NOE for protons. For a ^{1}H,^{13}C spin system the effect is four times as large for ^{13}C (200%) if the ^{1}H resonance is irradiated, because here $\gamma_H/\gamma_C = 4$. The ratio $\gamma_A/2\gamma_X$ from equation (10.6) is known as nuclear *Overhauser enhancement* η and the abbreviation NOE also stands for nuclear Overhauser enhancement.

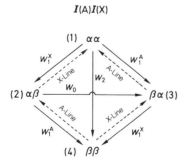

Figure 10.6 Solomon diagram for a two spin system composed of two nuclear spins

The nuclear Overhauser effect is thus an important tool for improving the sensitivity of n.m.r. measurements with less sensitive nuclei. i.e. nuclei with small γ-factors. This is standard practice in ^{13}C n.m.r. spectroscopy, for example, where ^1H broadband decoupling is applied (cf. p. 266). Problems arise, however, if the X nucleus has a negative gyromagnetic ratio, as for instance ^{15}N or ^{29}Si. The nuclear Overhauser enhancement $\gamma_A/2\gamma_X$ is then negative and the observed X signal will be inverted. Much less favourable conditions are met if competing relaxation mechanisms lead to a reduction of the nuclear Overhauser enhancement. In the limit $\gamma_A/2\gamma_X < \sim -1$ the X signal can even be eliminated. A decrease of the NOE is always to be expected if other than dipolar relaxation mechanisms are present.

2.1   ONE-DIMENSIONAL NOE EXPERIMENTS

It is important to remember that the nuclear Overhauser effect is governed by relaxation processes and does not involve scalar spin–spin coupling between the nuclei A and X. The cross relaxation rate is proportional to the factor $1/r^6$ and depends therefore on the distance between the nuclei of interest. For fast molecular motion in liquids the following equation holds:

$$W_2 - W_0 = \left(\frac{\mu_0}{4\pi}\right)^2 \frac{1}{2} \hbar^2 \gamma_A^2 \gamma_X^2 \tau_c r^{-6} \qquad (10.7)$$

where τ_c is the correlation time for molecular reorientation (cf. p. 230). The practical importance of homonuclear NOE experiments for distance measurements between protons is based on this relation. Besides the use of NOE effects for signal enhancement, this is the largest area where this technique is applied. NOE measurements are consequently valuable aids in structural research and conformational analysis when it is necessary to decide which of two nuclei, A or B, is closer in space to a third nucleus C within the same molecule. Problems of this type are encountered in connection with cis–trans isomerism about double bonds and in the conformational analysis of alicyclic compounds. Experimentally, one proceeds by irradiating each of the two resonances, in our case at ν_A or ν_B, in turn with a secondary field while simultaneously monitoring the intensity of the resonance of the nucleus C. The larger intensity increase is observed for that pair of nuclei, which has the smaller distance r, that is one uses the simplified relation

$$\eta_{AC}/\eta_{BC} = (r_{BC}/r_{AC})^6 \qquad (10.8)$$

This is valid, if for both pairs of nuclei, AC and BC, the same correlation time applies and no NOE exists between A and B. With rigid molecules, the first condition is always met.

Thus it was possible by means of nuclear Overhauser experiments to assign unequivocally the resonance of the methyl groups on dimethyl formamide (cf. p. 336 and Fig. 3 in the Introduction). Only when the methyl signal at lower field is irradiated is an intensity increase observed for the signal of the formyl proton. This absorption is thus identified as that of the methyl group that is *trans* to the carbonyl group. In other examples the configuration of the ethylidene side-chain of the alkaloid dehydrovoachalotine (**186**) was established and it could be decided whether the 2-methoxy-4,4,6-trimethyl-1,3-dioxane exists in the *cis*- or the *trans*-form (**187** or **188**, respectively). In the case of **186** an intensity increase of 26% was observed for the signal of H-15 when the resonance of the methyl group at C^{-18} was irradiated. In the case of the dioxanes the irradiation of one of the methyl groups at the 4-position in **188** led to a 12% increase in the intensity of the signal for H^2 while no such effect was observed in the case of **187**.

It should be understood that the intramolecular nuclear Overhauser effect is quenched by all influences that allow relaxation via mechanisms other than *intramolecular* dipole–dipole interactions. In particular, *intermolecular* dipole–dipole interactions must be minimized. The sample solutions should thus be free of oxygen and possibly degassed, and the solvents of choice are those that have only a few magnetic nuclei, such as CS_2 or CCl_4. In addition, the concentration of the compound under investigation should not be too high.

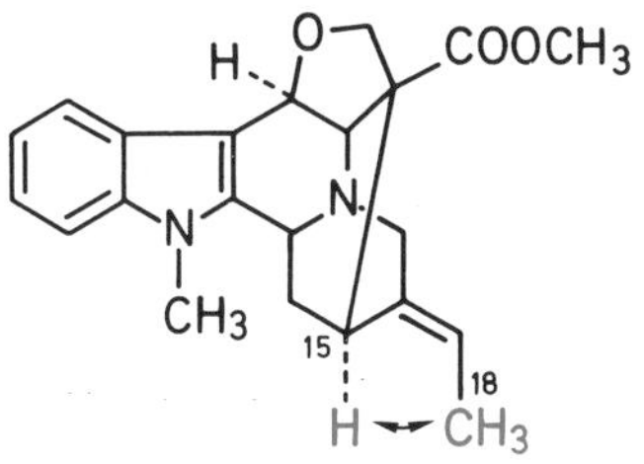

186

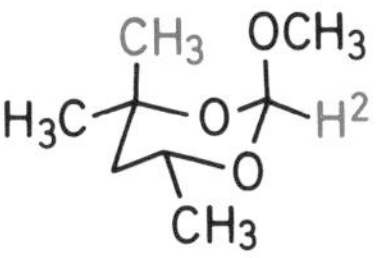

187

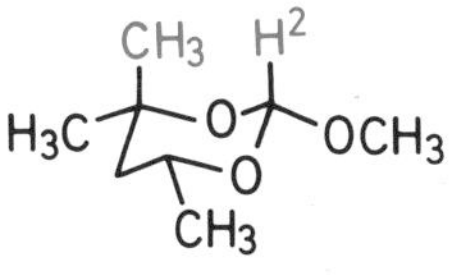

188

2.2 COMPLICATIONS DURING NOE MEASUREMENTS

A number of complications can arise during NOE measurements which shall be described shortly in the following sections. Let us first point out the difference between NOE measurements in the CW and in the FT mode. The CW NOE experiment is a steady state experiment. This means that during irradiation of the A resonance a stationary state is reached with respect to the competition between perturbation and re-establishment of the Boltzmann distribution in the spin system.

The distance dependence of the cross relaxation rate $W_2 - W_0$ determines the magnitude of the effect and the stronger the perturbation, the stronger is also the driving force to restore the original spin populations and thus the increase in the intensity of the X transitions.

In the FT method, on the other hand, the spectrum is excited by the pulse after the irradiation time t_{NOE}, during which the Overhauser effect was built up. During the detection time t_2, a transient NOE is then measured and the relation $\eta_{exp} < \eta_{max}$ is valid.

A number of complications during practical applications of NOE measurements result from the fact that in the molecules studied not only isolated two-spin systems are present, but in general a single proton has a large number of different neighbours. For the simple case of three nuclei A, B and C, for example, a linear and an angular geometry is possible:

In the first case, after A-irradiation one finds an intensity increase for nucleus B which is, however, smaller than expected. For the nucleus C, on the other hand, the intensity may have decreased. Such a finding is the result of an indirect NOE effect which is a consequence of dipolar cross relaxation between B and C. A perturbation of the Boltzmann population at B is, therefore, restored not only through an increasing number of B transitions, but also through a contribution from nucleus C. In the energy level diagram for the partial spin system BC the population differences of the B nucleus are increased as a consequence of A irradiation. This means, on the other hand, that the population differences for the C nucleus and consequently the intensity of the C resonance decreases. If a fourth nucleus D is present, the effect changes sign again and becomes positive. Indirect NOE effects of this type are generally known as *spin diffusion*.

A change of the geometry in the direction of an angular arrangement, on the other hand, can give rise to a direct NOE effect between A and C. It has a positive sign and will thus be diminished or even nullified through the negative indirect effect A–C. Consequently, even in the case of relatively short nuclear distances, an NOE effect may not be observed.

A further factor which may complicate NOE measurements is scalar spin–spin coupling. Nuclear Overhauser effect theory for strongly coupled spin systems is, as expected, complex and beyond the level of our introduction. In the case of weakly coupled spin systems, which form the majority of those cases which are studied today with high magnetic fields, NOE measurements may fail if selective population transfer, a phenomenon based on scalar spin–spin coupling to be discussed in the next section, is present. This is particularly true if irradiation is applied to spin multiplets and special precautions have to be taken in such cases to evaluate the results.

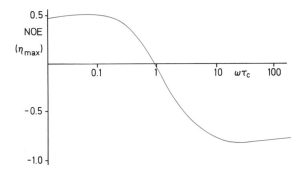

Figure 10.7 Dependence of the homonuclear NOE between protons on the product of resonance frequency, ω, and molecular correlation time, τ_c

Finally, with large molecules the dependence of the nuclear Overhauser effect on the correlation time is important. NOE effects may then completely vanish or become negative (Figure 10.7). Such situations are found for macromolecules or small molecules in viscous solvents. Equations (10.5) and (10.6) are, therefore, valid only under the condition $\omega\tau_c \ll 1$, which is always true for small and isotropic molecules (MG < 500) in solutions of low viscosity (extreme narrowing condition). In the negative region of the NOE effect we have for the cross relaxation $W_0 > W_2$, because during slow molecular motions frequencies in the kHz region dominate the relaxation process. According to Figure 10.6, upon A irradiation spin population is now transported from state (4) to state (1) via states (2) and (3), which results in a depopulation of state (4) and an overpopulation of state (1). Consequently, the X lines have an inverse Boltzmann distribution and yield negative (emission) signals. With large molecules it is, furthermore, impossible to assume a unique molecular correlation time τ_c, since intramolecular dynamic processes must be added and the motion of individual parts of the molecule are not necessarily correlated. Furthermore, large molecules seldom behave isotropically.

2.3 NOE DIFFERENCE SPECTROSCOPY

Experimentally, a nuclear Overhauser measurement for the estimation of relative proton distances is performed most easily with the help of FT n.m.r. difference spectroscopy. With this technique, the time signal $S(t_2)$ is accumulated for experiments which are alternatively run with and without irradiation at the ^1H resonance of interest. The data are stored in two different blocks of the computer memory and the difference is obtained before or after Fourier transformation. The Overhauser enhancement remains as the *NOE difference spectrum* (Figure 10.8). This form of NOE spectroscopy is very sensitive and allows the detection of very small intensity differences (as low as ca. 3%), because the signals from experiments with and without irradiation are recorded practically under identical conditions. In

404

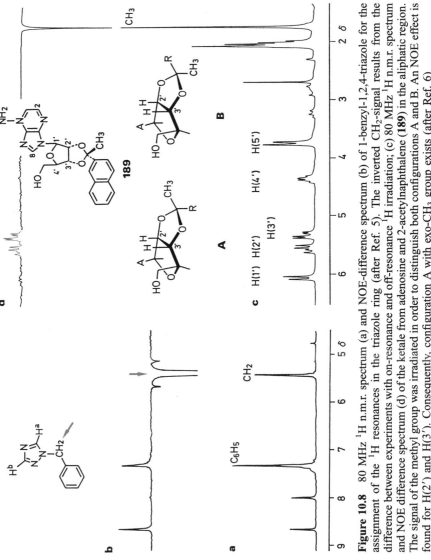

Figure 10.8 80 MHz ^{1}H n.m.r. spectrum (a) and NOE-difference spectrum (b) of 1-benzyl-1,2,4-triazole for the assignment of the ^{1}H resonances in the triazole ring (after Ref. 5). The inverted CH$_2$-signal results from the difference between experiments with on-resonance and off-resonance ^{1}H irradiation; (c) 80 MHz ^{1}H n.m.r. spectrum and NOE difference spectrum (d) of the ketale from adenosine and 2-acetylnaphthalene (**189**) in the aliphatic region. The signal of the methyl group was irradiated in order to distinguish both configurations A and B. An NOE effect is found for H(2') and H(3'). Consequently, configuration A with exo-CH$_3$ group exists (after Ref. 6)

order to achieve this, the experiment without irradiation at the proton signal of interest is performed as an off-resonance experiment with the decoupler positioned outside the spectral window.

2.4 TWO-DIMENSIONAL NUCLEAR OVERHAUSER SPECTROSCOPY (NOESY)

The pulse sequence 90_x°, t_1, 90_x°, t_M, 90_x°, FID(t_2) [equation (9.22) and Fig. 9.5], which was introduced on p. 348 for 2D exchange spectroscopy (EXSY), is also used for recording two-dimensional homonuclear Overhauser (NOESY) spectra. The exchange of magnetization during the mixing time is then based on the nuclear Overhauser effect. NOESY spectra have the same structure as EXSY or COSY spectra: the 1D spectrum appears on the diagonal, the cross peaks as non-diagonal elements yield the desired information about spin correlations (Figure 10.9a). In

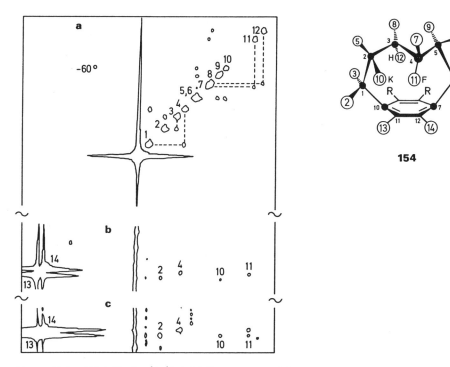

Figure 10.9   400 MHz 2D ^{1}H,^{1}H NOESY spectrum of [6]paracyclophane-8,9-dimethyldicarboxylate (**154**) at $-60\ ^\circ$C with cross peaks of different intensity. The contour diagram (a) shows strong geminal NOE effects. The less intensive cross peaks between the various CH$_2$ protons on one hand and the aromatic protons H(13) and H(14) on the other, which are characteristic for the conformation of the methylene chain, are found in diagrams (b) and (c) recorded at lower intensity levels (after Ref. 7)

those cases where exchange as well as NOE effects operate, phase sensitive spectra must be recorded to distinguish both mechanisms. In these spectra we observe for the "normal" positive NOE effect diagonal and cross peaks with alternative sign, while for exchange spectra the same sign is found. However, negative NOE effects for macromolecules also yield the same sign for diagonal and cross peaks. In such cases the ROESY experiment, discussed in Section 4.2.2, can be used to differentiate between both effects. For ROESY spectra, the sign of both types of signals is always opposite.

COSY signals, which are produced in most NOESY experiments because of the presence of geminal, vicinal or long-range $^1H,^1H$-couplings, must be eliminated as in the case of 2D exchange spectra through the phase cycle. Zero quantum magnetization, which because of its coherence order 0 passes this filter, is eliminated, as already mentioned on p. 351, through the introduction of a 180° pulse during the mixing time t_M. Alternatively, the statistical variation of t_M by about 20% serves the same purpose, since NOESY signals increase steadily, while COSY signals have a sine oscillation during t_M and, therefore, cancel. In a number of cases it is also possible to eliminate exchange signals by temperature variations. Figure 10.9 shows as an example the 2D-NOESY spectrum of [6]paracyclophane-8,9-dimethyldicarboxylate (**154**), a compound already mentioned in Chapter 9, in the methylene proton region at -60 °C. At that temperature the ring inversion is slow enough for only NOE effects to be observed. The interactions via the short distances between the geminal protons of the CH_2-groups dominate, and their closeness in space is recognized by cross peaks. It is interesting that NOE effects between the protons of the methylene chain and the protons of the aromatic ring can also be detected. Because of the larger $^1H,^1H$ distances involved, the corresponding cross peaks are considerably less intensive and appear in the contour plot only at a relatively low intensity level which already shows a lot of noise.

In case of larger molecules, NOESY spectra are a valuable tool for the assignment of partial spectra. Figure 10.10 shows such an application with the ^1H n.m.r. spectrum of the dication of benzo[*b*]biphenylene (**190**). In the ^1H n.m.r. spectrum of this compound one observes a singlet at $\delta9.85$ and two AA'BB'-systems. The NOESY spectrum yields cross peaks between the singlet and the high field BB'-part. Because the singlet can be assigned unambiguously to the protons H(9,10), the BB'-part belongs to H(5,8) and both four-spin systems are thus identified. The assignment within the low-field AA'BB' system remains to be established by other methods.

NOESY spectroscopy is today indispensable for the conformational analysis of biomolecules. It has paved the way for the three-dimensional structure determination in solution, because in addition to correlations based on spin–spin coupling via the network of the chemical bonds, direct information about distances in space can be obtained. For this purpose one determines the build-up rates of the cross peaks through a variation of the mixing time in several 2D experiments. With a known distance as reference, proton distances of the order of 0.2–1.0 nm can then be determined with an error of ca. 10%. The analysis of such NOE data for large

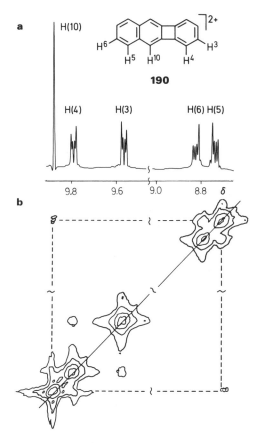

Figure 10.10 ^{1}H-2D NOESY spectrum of benzo[*b*]biphenyl dication (**190**) in SbF$_6$ at −30 °C (after Ref. 8)

molecules (MG up to 20 000), which consist of several hundred cross peaks, can only be attempted with data processing and powerful programs have been developed for this purpose.

3. Polarization transfer experiments

While the *homonuclear* Overhauser effect between protons with the possibility to derive ^{1}H,^{1}H distances is an important tool in the field of stereochemistry, the main aspect of *heteronuclear* NOE effects is the sensitivity gain for insensitive nuclei with small γ-factors (see equation (10.5), p. 399). The general importance of the latter aspect for n.m.r. spectroscopy, but also the complications that arise during NOE experiments with nuclei which have negative γ-factors, have promoted the search for alternative ways of signal enhancement. It was, therefore, of interest that experiments undertaken in order to record INDOR spectra (see p. 270) with the FT method, a new experiment for signal enhancement was discovered. The principle used here is known as *polarization transfer* or *polarization inversion*. Contrary to the nuclear Overhauser effect, these techniques rely on scalar spin–spin coupling and do not suffer from negative γ-factors. They can replace NOE measurements in all cases where a scalar spin–spin interaction between the nuclei of interest exist.

3.1 THE SPI EXPERIMENT

For a two-spin AX system of a sensitive and an insensitive nucleus, for example ^{1}H, ^{13}C or ^{1}H, ^{15}N, the equilibrium population of the energy levels and consequently the relative intensities of the A and X lines are governed by the Boltzmann law. The population difference between two states E_q and E_p is then determined by the gyromagnetic ratio of the particular nucleus which changes its spin state during the transition $E_p{\rightarrow}E_q$. For states which are connected by the transitions of the sensitive nucleus (A, large γ), a larger population difference results than for those which belong to the transitions of the insensitive nucleus (X, small γ) (Figure 10.11). With

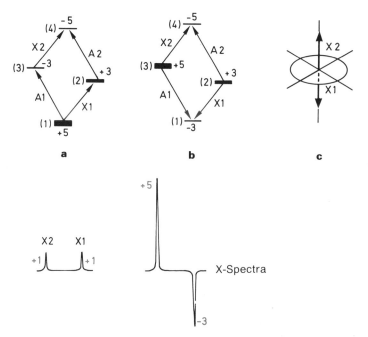

Figure 10.11 Perturbation of the Boltzmann distribution in an AX spinsystem through a selective 180°(A) pulse and resulting X spectra; (a) equilibrium state; (b) perturbed Boltzmann distribution of the spin population after a selective 180° pulse on the line A1 of the sensitive A nucleus; (c) vector diagram for the X magnetization after population inversion

$E = \pm \gamma h B_0/4\pi$ (see p. 2) and $h B_0/4\pi = p$, we obtain for the energy of the states $\beta\beta$, $\beta\alpha$, $\alpha\beta$ and $\alpha\alpha$ of a two-spin system

$$E_{\beta\beta} = -(\gamma_A + \gamma_X)p$$
$$E_{\beta\alpha} = -(\gamma_A - \gamma_X)p$$
$$E_{\alpha\beta} = (\gamma_A - \gamma_X)p$$
$$E_{\alpha\alpha} = (\gamma_A + \gamma_X)p$$

and in the case of a ^{1}H,^{13}C pair because of $\gamma(^1$H$)/\gamma(^{13}$C$) = 4$ the relative population numbers 5, 3, -3, and -5, as shown in Figure 10.11a. The relative intensities of the n.m.r. lines are directly proportional to the corresponding differences, that is $I(^{13}$C$):I(^1$H$) = 4:16$ or $1:4$ (note that the different natural abundance of the A- and the X-nucleus is not considered here).

A selective population inversion for an A line, for example A1, which interchanges the populations of the connected spinstates, then leads to the energy level diagram of Figure 10.11b. It shows for the X lines increased absorption (X2) or emission (X1) and the relative intensities $+5$ and -3, respectively. The total intensity found before for the sensitive nucleus is now observed for the absolute intensity of the lines of the insensitive nucleus. This phenomenon is called *polarization transfer*. The intensity increase achieved for the X nucleus corresponds to the ratio γ_A/γ_X. It is fully developed for the absorption line X2 ($I = 1 + \gamma_A/\gamma_X$). For the emission line X1, on the other hand, one obtains $I = 1 - \gamma_A/\gamma_X$, which still leads to an improved signal-to-noise ratio because generally $\gamma_A/\gamma_X > 1$. It is important to note, that, contrary to the nuclear Overhauser effect (equation (10.6)), the results are independent of the sign of γ_X. For negative γ_X-values the emission and absorption lines are just interchanged. On the other hand one must remember that the experiment does not yield a net effect, because the integrated intensity, $I(X1) + I(X2)$, is unchanged and with ^{1}H decoupling during ^{13}C detection an X line of the relative intensity 2 is observed.

Experimentally, population inversion can be achieved by a selective 180°_x pulse on one of the A lines. In practice this is done most simply for a ^{1}H,^{13}C pair with the proton decoupler which is adjusted in the CW mode at the frequency of one of the ^{1}H lines, that is one of the ^{13}C satellite lines in the ^{1}H-spectrum. If complete population inversion is not achieved, one speaks of selective *population transfer*. The experiment is, therefore, known under the acronyms SPI (*s*elective *p*opulation *i*nversion) or SPT (*s*elective *p*opulation *t*ransfer). Figure 10.12 shows the experimental result for the ^{1}H,^{13}C spinsystem of chloroform, which was the first published example.

Exercise 10.1 Describe the results for a 180° pulse on line A2.

The SPI experiment has importance for signal assignments as well as for the measurement of insensitive nuclei like ^{13}C, ^{15}N or ^{29}Si, where ^{1}H, ^{19}F or ^{31}P can

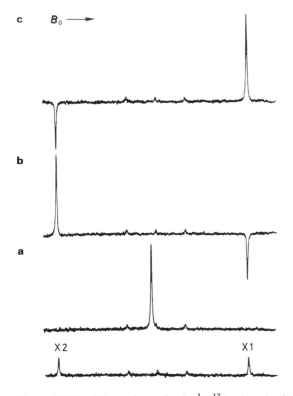

Figure 10.12   SPI experiment for the ^{1}H,^{13}C spin pair of chloroform; (a) ^{1}H decoupled and
^{1}H coupled ^{13}C n.m.r. signal; (b) spectrum after inversion of the ^{1}H line A1 (high-field ^{13}C
satellite; population inversion according to Figure 10.11b); (c) spectrum after inversion of the
^{1}H line A2 (low-field ^{13}C satellite; after Ref. 9)

		a					n			**b**				
			1				0			1				
		1		1			1		-3		5			
		1	2	1			2		-7	2	9			
	1	3	3	1			3	-11	-9	15	13			
	1	4	6	4	1		4	-15	-28	6	36	17		
1	5	10	10	5	1		5	-19	-55	-30	-50	65	21	
1	6	15	20	15	6	1	6	-23	-90	-105	20	135	102	25

Figure 10.13   Line number and relative intensities for the X multiplet of an A_nX group
(A = ^{1}H) at normal Boltzmann distribution (a) and after selective population inversion for one
A line (b)

serve as sensitive A nuclei. The intensity increase which is obtained for a first order AX spin system of spin-$\frac{1}{2}$ nuclei can be judged from a comparison with Pascal's triangle which describes the normal intensity behaviour of first order spin multiplets (Figure 10.13). An important limitation of the SPI-method must be seen, however, in the fact that only one line can be inverted at a time and sensitivity enhancement is limited to a particular A,X spin pair. For structures with several insensitive nuclei of interest the experiment has to be repeated for each X nucleus.

Finally, SPI experiments can be used as the INDOR experiment in CW n.m.r. spectroscopy for the determination of the relative signs of scalar coupling constants and for spectral analysis.

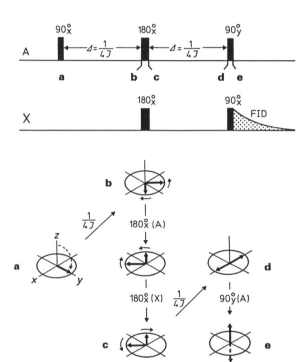

Figure 10.14 Pulse sequence of the INEPT method for an AX system (e.g. A = ^{1}H, ^{19}F, or ^{31}P; X = ^{13}C, ^{15}N, or ^{29}Si). The vector diagram shows only the A magnetization in the rotating frame ($\nu_0 = \nu_A$). After 90°_x excitation (a) the transverse magnetization of the nucleus A is modulated by spin–spin coupling to the less sensitive nucleus X. After time $\tau = 1/4J$ a phase difference of 90° exists between both doublet vectors (b). A 180°_x pulse in the A as well as in the X frequency region leads to the vector diagram (c) so that after time 2τ state (d) is reached. A 90°_y pulse now inverts the magnetization of one proton line (e). This corresponds to selective population inversion. The polarization of the spin system is detected by a 90°_x pulse in the frequency region of the less sensitive nucleus, whose lines show emission or enhanced absorption. Refocusing is achieved by an additional spin echo sequence, for a doublet $1/4J - 180^\circ_x(A,X)-1/4J$. The X resonance can then be detected as a positively polarized doublet or — with simultaneous ^{1}H decoupling during the detection period — as a singlet

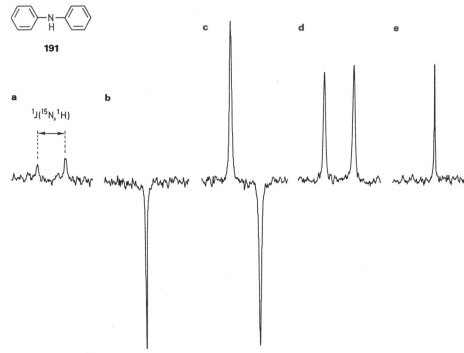

Figure 10.15 ^{15}N n.m.r. signal (40.53 MHz) in diphenylamine (**191**): (a) without NOE effect; (b) with NOE effect; (c) INEPT signal; (d) INEPT signal after refocussing with a spin echo sequence $\Delta_2 = 1/4J$, 180_x° (A,X), $1/4J$; (e) as (d), however, with ^{1}H decoupling; the coupling constant $^1J(^{15}N,^1H)$ amounts to 89 Hz; note that the intensity of the signal is now reduced due to the negative NOE effect.

3.2   THE INEPT-PULSE SEQUENCE

With respect to practical applications of polarization transfer experiments for the measurement of insensitive X nuclei important progress was made after it was recognized that polarization transfer can be achieved also non-selectively through a suitable pulse sequence. For this purpose the INEPT sequence (INEPT = insensitive nuclei enhanced by polarization transfer), shown in Figure 10.14, was introduced. It is illustrated by a series of vector diagrams in Figure 10.14a–e (p. 411).

Important elements of the INEPT sequence are the modulation of transverse magnetization of the sensitive nucleus (A) by scalar coupling to the insensitive nucleus (X) and the simultaneous application of two 180_x° pulses in the A and X region. The vector arrangement reached for the doublet components of the A resonance after the evolution time 2Δ can be transformed by a 90_y° pulse into an arrangement typical for a spin system with selectively inverted Boltzmann distribution. As shown above, in the energy level diagram this corresponds to a population inversion over one A line and leads to X line polarization (cf. Figure 10.11c).

Table 10.1 Enhancement factors η_{NOE} and η_{INEPT} for nuclear Overhauser and INEPT experiments, respectively, with $\{^1H\}X$ pairs

X	^{11}B	^{13}C	^{15}N	^{29}Si	^{57}Fe	^{103}Rh	^{109}Ag	^{119}Sn	^{183}W
$\eta_{NOE}{}^a$	2.56	2.99	−3.94	−1.52	16.48	−16.89	−9.75	−0.41	13.02
$\eta_{INEPT}{}^b$	3.12	3.98	9.87	5.03	30.95	31.77	21.50	2.81	24.04

^{a} Note that the observed intensity is equal to $1 + \eta_{NOE}$
^{b} For ^{19}F or ^{31}P as polarization source (A nucleus) the data for η_{NOE} and η_{INEPT} are reduced by the factor 0.941 $[(\gamma(^{19}F)/\gamma(^1H)]$ and 0.407 $[\gamma(^{31}P)/\gamma(^1H)]$

The most important aspect of the INEPT method is the fact that it allows a much larger intensity increase for insensitive nuclei than the nuclear Overhauser effect. Furthermore, negative γ-factors are no disadvantage because polarization transfer is governed by the ratio γ_A/γ_X, while nuclear Overhauser enhancement is determined by the sum $1 + \gamma_A/2\gamma_X$ (equation (10.6), p. 399). In addition, cumulative effects can arise during polarization transfer experiments in case of degenerate lines. The theoretical enhancement factors for both methods, which, quite naturally, are only approximated in practice, are collected in Table 10.1. A comparison of the experimental results is given in Figure 10.15 with the $^1H,^{15}N$ nuclear spin pair of diphenylamine.

The practical application of INEPT spectroscopy is documented in Figure 10.16 with the ^{15}N n.m.r. spectrum of 9-methylpurine. The evolution time of the INEPT pulse sequence is based on one A,X coupling and, therefore, has to be estimated if $J(A,X)$ is not known. In the ^{15}N n.m.r. spectroscopy of unsaturated heterocycles the geminal $^{15}N,^1H$-couplings with magnitudes of ca. 10 Hz ($\Delta = 1/40$ s = 25 ms), for example, are well suited for detection. If in addition NH- or NH_2-groups are present,

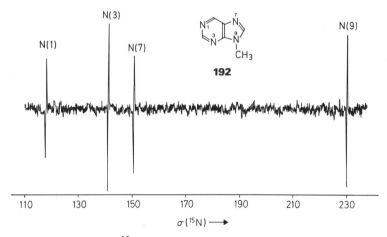

Figure 10.16 INEPT ^{15}N n.m.r. spectrum of 9-methylpurine (**192**); measuring frequency 40.53 Mz, 6293 transients, spectral width 6024 Hz, measuring time 4.75 h. The σ-scale refers to external $CD_3{}^{15}NO_3$ (after Ref. 10)

a second experiment with $\Delta = 1/4\,^1J(^{15}N,^1H)$ is necessary. The Δ delay then amounts to $1/250$ s = 40 ms.

Typical for simple INEPT spectra without ^{1}H-decoupling is the zero intensity of the central line of an uneven multiplet as well as the inversion of half of the multiplet lines (NH: $-1, +1$; NH$_2$: $-1, 0, +1$; NH$_3$: $-1, -1, +1, +1$). This results from the phase cycle used, which eliminates the original X magnetization. Only magnetization generated by polarization transfer is detected. The integrated total intensity of an INEPT multiplet is, therefore, zero (compare the net effect in the SPI experiment). ^{1}H decoupling can be used if the X signals are refocussed by an additional spin echo sequence. Without ^{1}H decoupling, "normal" multiplets with positive phases for all lines are observed, while ^{1}H decoupling leads to a singlet (see Figure 10.15). For AX, A$_2$X and A$_3$X spin systems (for example CH-, CH$_2$-, CH$_3$-groups) different echo times Δ_2 have to be used. This paves the way for signal selection and assignment. This aspect is treated in more detail in Chapter 11 (p. 474). In general, a prolongation of the pulse sequence leads to a reduction in signal intensity due to relaxation effects. Therefore, the simple pulse sequence without A decoupling, as shown in Figure 10.14, is often the best choice in order to measure insensitive nuclei.

An important factor for the success of a polarization transfer experiment is the relaxation time of the sensitive nucleus. Since nuclei like ^{1}H or ^{19}F have considerably shorter relaxation times than, for example, ^{13}C, the repetition rate for data accumulation can be much higher than for direct measurements of the insensitive nucleus. This shortens the time necessary to obtain spectra of insensitive nuclei, in the case of ^{15}N n.m.r. measurements by a factor of 2–3. Interesting INEPT experiments have also been performed for different metal nuclei. In these cases protons (for ^{103}Rh and ^{109}Ag) but also ^{31}P nuclei (for ^{57}Fe, ^{103}Rh and ^{183}W) have been used as polarization sources. Of great importance are also $\{^1H\}^{29}$Si and $\{^1H\}^{119}$Sn experiments. Complications arise if strongly coupled ^{1}H spin systems are present. The complex relaxation behaviour then leads very often to a failure of the INEPT experiment.

A detailed discussion of the DEPT sequence, a polarization transfer experiment used today for spectral editing of ^{13}C signals, will be given in Chapter 11.

4. Rotating frame experiments

4.1 SPINLOCK AND HARTMANN–HAHN CONDITION

During all one- and two-dimensional experiments discussed so far the static magnetic field B_0 pointed along the z-direction of the laboratory as well as of the rotating frame of reference. In the following we shall describe experimental techniques where effectively the direction of the static field in the rotating frame, at least for a short period of time, is changed. These techniques are based on a famous idea introduced by Hartmann and Hahn in the area of solid state n.m.r. which has found in recent years new applications also in high-resolution n.m.r. of liquids.

The object of the Hartmann–Hahn experiment was to improve the detection of an insensitive X nucleus of low natural abundance, for example ^{13}C, by magnetization transfer from a sensitive A nucleus of high natural abundance, for example ^1H. The strategy developed for this purpose can be understood best if we introduce the concept of spin temperature. Following equation (1.11) (p. 6), the temperature T_s of a spin system depends on the population ratio N_α/N_β. An ensemble of sensitive nuclei with high population excess in the ground state has, therefore, a low spin temperature, while an ensemble of sensitive nuclei with only a small population excess has a high spin temperature. This differentiation is possible since the spins are only weakly coupled to their surroundings, the lattice, while strong coupling exists within the spin system due to dipolar interactions (long T_1 and short T_2 times). For ^1H and ^{13}C spins the relation $\gamma(^1H)/\gamma(^{13}C) = 4$ leads to a ratio $T_s(^{13}C):T_s(^1H) = 4:1$.

If we now succeed in establishing a thermal contact between the 'hot' and the 'cold' spin reservoir, the ensemble of the insensitive nuclei should be cooled at the expense of the ensemble of the sensitive nuclei. The consequence would be a high population difference for the X nuclei and, therefore, a sensitive X resonance. How can thermal contact be achieved? Energy exchange between spins proceeds for homonuclear spin systems in solids by a flip-flop mechanism. As discussed in Chapter 7 (p. 231), spin–spin interaction leads to an exchange of energy quanta γB_0 and a change of spin orientation for one nucleus is accompanied by the opposite change for its neighbour. For heteronuclear AX spin systems we have, however,

$$\gamma(A)B_0 \neq \gamma(X)B_0 \tag{10.9}$$

and since all parameters in relation (10.9) are fixed, there is no possibility of overcoming this inequality and achieving magnetization transfer.

In the rotating coordinate system, however, a different situation arises. Here we are able to meet the condition

$$\gamma(A)B_1(A) = \gamma(X)B_1(X) \tag{10.10}$$

by adjusting the two B_1 amplitudes. In the case discussed above the equation

$$\gamma(^1H)B_1(^1H) = \gamma(^{13}C)B_1(^{13}C) \tag{10.11}$$

results, which is also known as the *Hartmann–Hahn condition*. Now energy can be exchanged by cross polarization between both spin reservoirs.

In practice, the experiment starts with a 90°_x ^1H pulse which is immediately followed by a shift of the B_1-field from the x- to the y-axis (Figure 10.17). Thus, B_1 is parallel to the magnetization vector $\mathbf{M}$ and the protons behave in this new field as before in the B_0 field: they precess with their Larmor frequency $\omega(^1H) = \gamma(^1H)B_1$ around the y-axis, in other words, they are locked in the y-direction of the rotating frame. This state of the spin system is called *spinlock*.

If we now adjust the B_1 field in the ^{13}C frequency region according to equation (10.11), the z-components of both magnetizations oscillate with the same frequency and an energy transfer from the hot to the cold spin reservoir is possible. In order to achieve this, the spinlock time must be of the order of the proton spin lattice relaxation time.

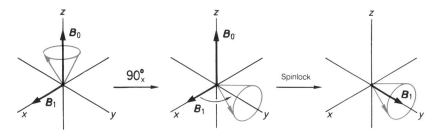

Figure 10.17   Illustration of the Hartmann–Hahn spinlock experiment

4.2   SPINLOCK EXPERIMENTS IN SOLUTION

4.2.1   Homonuclear Hartmann–Hahn and TOCSY experiments

The *Hartmann–Hahn experiment*, also known as the *cross polarization (CP) experiment*, plays an important part in solid state n.m.r. spectroscopy (see Section 8, p. 440). In liquids, because of the fast molecular motion, we have $T_1 \approx T_2$ and dipolar couplings vanish. Thus, the conditions for the original spinlock experiments are not met. Nevertheless, it is possible to perform spinlock experiments in liquids with homonuclear as well as with heteronuclear spin systems. Magnetization transfer is then based on *scalar* spin–spin coupling. Such techniques can be used for homo- and heteronuclear shift correlation experiments. They are known by the acronym HOHAHA (homonuclear Hartmann–Hahn) and HEHAHA (heteronuclear Hartmann–Hahn). The physical basis for the TOCSY experiment (total correlation spectroscopy) is quite similar.

All the methods have in common that during a spinlock or mixing time the Zeeman contributions to the Hamilton operator, in other words the interactions with the static external magnetic field B_0 and thus the chemical shifts, are practically eliminated and scalar coupling between nuclei dominates. In the case of TOCSY spectroscopy, the mixing operator for complete isotropic coupling has the form

$$\hat{H}_{\mathrm{M}} \equiv \hat{H}_{\mathrm{J}} = \sum_{i<j}\sum 2\pi J_{ij}\boldsymbol{I}(i)\boldsymbol{I}(j) \tag{10.12}$$

and describes a state of *isotropic mixing* which is experimentally accessible through the application of multipulse decoupling sequences of the type of the MLEV sequence, discussed earlier. During the *spinlock* or *mixing time* (t_{SL} or t_{M}) magnetization transfer takes place (Hartmann–Hahn mixing or isotropic mixing). It reaches a maximum after a time interval equal to $1/2J$. It is important to note that the transfer proceeds beyond the next nuclei, which are directly coupled, to remote nuclei and finally progresses through the complete scalar coupled network of nuclei in the particular molecule. Two-dimensional homonuclear correlation spectra can be produced if the evolution time t_1 precedes the spinlock or mixing time as is shown in Figure 10.18 for the pulse sequences of the HOHAHA and TOCSY experiments.

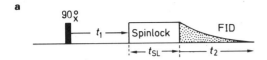

Figure 10.18 Pulse sequence for two-dimensional homonuclear HOHAHA and TOCSY experiments (a, b respectively)

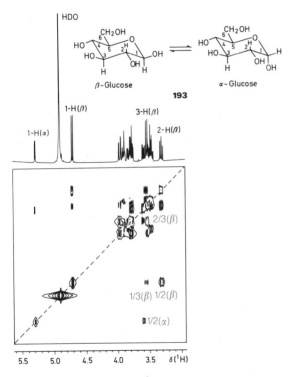

Figure 10.19 2D TOCSY ^{1}H n.m.r. spectrum of a mixture of α- and β-D-glucose (**193**) in D$_2$O with application of the MLEV sequence (mixing time 30 ms, measuring time 1.5 h); at 4.9 ppm the signal of HDO. The vicinal correlations of the anomeric protons are clearly recognized, as well as the relay signal for 1-H/3-H of β-D-glucose. The high-field portion, with its numerous cross and relay peaks, can be analysed only by use of various filter functions. Relay peaks may be suppressed by using shorter mixing times

As a comparison with the COSY sequence (Figure 8.12, p. 289) shows, the second 90_x° pulse is now replaced by the spinlock or mixing time. Here magnetization transfer takes place which is governed by the length of these delays. Short times (20–50 ms) yield primarily cross peaks of strongly coupled protons, while longer mixing times (100–300 ms) allow magnetization transfer to remote protons of the spin system. For an AMX system magnetization is then transferred from the A to the M nucleus and from there to the X nucleus. So-called *relayed spectra* result which show cross peaks from protons which are separated by more than three bonds. An advantage of the HOHAHA experiments as compared to the COSY experiment must be seen in the fact that the magnetization transfer leads to a net effect and all signals appear nearly in pure absorption. This differs also from the results of the INEPT experiments discussed above. It is, therefore, possible to record phase sensitive 2D spectra and the danger of signal elimination is removed. An analogous situation is found for TOCSY spectra. Figure 10.19 shows a two-dimensional ^{1}H,^{1}H shift correlation on the basis of a TOCSY sequence which is most frequently used in practice.

4.2.2　One-dimensional selective TOCSY spectroscopy

As already mentioned earlier, two-dimensional experiments require a considerable investment of measuring time. Alternative methods, which are less demanding in this respect, are, therefore, still attractive and a number of one-dimensional variants of several 2D experiments have been proposed as alternative n.m.r. techniques. Among them are methods with selective excitation which are most useful in those cases where a few 1D experiments are sufficient to complete the information already obtained from other sources.

From the group of selective experiments we introduce here the 1D TOCSY experiment which has a large potential for applications in structural research. It can be used to advantage in all cases where in a complicated spectrum the signal of a single proton H_i is observed separately and can be used as starting point of the magnetization transfer process. Such a situation exists, for example, for the anomeric protons in carbohydrates. The 1D TOCSY experiment consequently plays an important role in spectral analysis of oligosaccharides and oligonucleotides.

For the 1D TOCSY experiment difference spectroscopy can be used. In the first sequence a selective 180° pulse, generated by a DANTE sequence or by a GAUSS pulse inverts the magnetization of the separated proton H_i. Transfer to other protons occurs during the mixing time which follows directly after the 90° excitation pulse because the 1D experiment has no evolution time. In the second sequence the 180° pulse is used off-resonance. The difference *S1–S2* thus contains only signals which have received a magnetization transfer.

$$
\begin{aligned}
S1 &: \ 180_x^\circ \ (\text{sel, on-resonance}), \ 90_x^\circ, \ t_\text{M}, \text{FID} \\
S2 &: \ 180_x^\circ \ (\text{sel, off-resonance}), \ 90_x^\circ, \ t_\text{M}, \text{FID}
\end{aligned}
\qquad (10.13)
$$

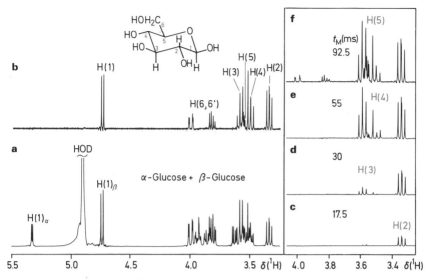

Figure 10.20 1D TOCSY ^{1}H n.m.r. spectra of a mixture of α- and β-D-glucose already used for Figure 10.19. (a) 400 MHz 1D spectrum; (b) 1D TOCSY spectroscopy with t_M = 130 ms and excitation of 1-H$_\beta$ yields the partial spectrum of the β-D-glucose; (c)–(f) 1D TOCSY spectra of β-D-glucose with increasing mixing time allow the signal assignment. The measuring times for each spectrum were 16 min. The partial spectrum of α-D-glucose with 1-H$_\alpha$ as starting point of the magnetization transfer process was not observed under the conditions used since, because of mutarotation, magnetization was transferred by chemical exchange onto the signals of β-D-glucose. The transfer in the opposite direction was not observed because of $k_{\beta \to \alpha} < k_{\alpha \to \beta}$

Alternatively to sequence (10.13), the sequence 90°(sel), t_M, FID can be used. Through the length of the mixing time the magnetization transfer is adjustable with respect to the connectivities in the spin system (close or remote), similar to the 2D experiment. If the mixing time is sufficiently long, one can produce partial spectra of individual spin systems which are superimposed in the normal 1D spectrum. An example is shown in Figure 10.20.

4.2.3 The ROESY experiment

As discussed in Section 2.2, the change of sign observed for the nuclear Overhauser effect in connection with a decrease of the molecular correlation time τ_c is a disadvantage for practical applications. It is, therefore, of general interest that NOE measurements can also be performed in the rotating frame and that under these conditions nuclear Overhauser factors are always positive. The limit for nuclear Overhauser enhancement in the rotating frame is 67.5% (Figure 10.21). Instead of the originally introduced acronym CAMELSPIN such experiments are better known by the name ROESY (rotating frame NOESY). They have gained considerable importance for the study of large molecules, in particular biological macromolecules.

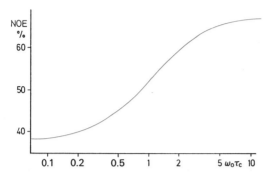

Figure 10.21 Dependence of the ROESY effect for homonuclear two-spin systems on the product of resonance frequency, ω_0, and molecular correlation time, τ_c

The pulse sequence for the 2D ROESY experiment is identical to the HOHAHA sequence in Figure 10.18 with the difference that during the spinlock time magnetization transfer now proceeds via dipolar interactions. It is immediately clear that in such an experiment complications can arise through HOHAHA signals (see Section 4.2.1) which are either detected in addition to the ROESY signals or, because of their different phase, can lead to an elimination of ROESY signals. In the first case a differentiation is possible by using phase sensitive spectra, because only HOHAHA signals have the same (positive) phase as the diagonal signals. Compared to NOESY spectra, ROESY experiments have the advantage that spin diffusion is less pronounced. However, double transfer processes of the type HOHAHA-ROESY or ROESY-HOHAHA can lead to further signals which have a different phase behaviour than that described. The analysis of 2D ROESY spectra may thus sometimes be difficult. In the one-dimensional version of the ROESY experiments two pulse sequences are used alternatively: (1) 90°, spinlock, FID, and (2) $180°_{sel}$, 90°, spinlock, FID. The selective 180° pulse in the second sequence starts the cross relaxation process, the difference (2)–(1) yields the ROE effects. With respect to the measuring time, the same advantages apply as discussed above for the 1D TOCSY experiment. An application of 1D ROESY difference spectroscopy is shown in Figure 10.22.

5. Chemically induced dynamic nuclear polarization (CIDNP)

Intimately connected with the relaxation effects responsible for the nuclear Overhauser effect is a process that has become known as 'chemically induced dynamic nuclear polarization' (CIDNP). Since its discovery it has developed into a powerful technique for the investigation of radical reactions.

The starting point was the observation that in the thermal decomposition of dibenzoyl peroxide using cyclohexanone as a solvent the resonance signal of the benzene produced temporarily appeared as an emission line (Figure 10.23).

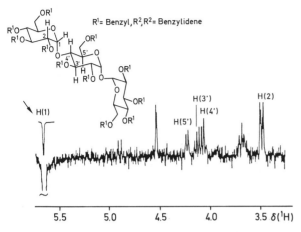

Figure 10.22 1D ROESY difference spectrum of a trisaccharide. The ROESY signals for the protons of neighbouring hexose units yield informations about the ring connections (note that rotation around the glycoside bond is possible; after Ref. 11)

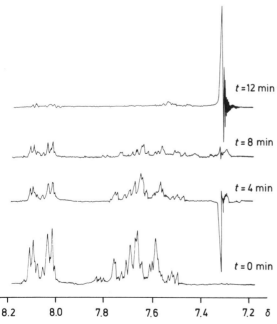

Figure 10.23 ^{1}H n.m.r. spectra during the decomposition of dibenzoyl peroxide in cyclohexanone; the decomposition starts at $t = 0$ (after Ref. 12)

It could be ascertained that the resonance line was that of a diamagnetic species that accordingly must have had an inverse spin population. The theory of this phenomenon has been developed to the extent that the effect can be attributed to an intermediate radical pair, $(R^\cdot, {}^\cdot R)$, that exists within a solvent cage. Within the electron–nuclear spin system of this radical pair relaxation and singlet–triplet transitions lead to a perturbation of the Boltzmann distributions and thereby to a polarization of the nuclear spin system that finally manifests itself in the n.m.r. spectrum of the diamagnetic product through emission or enhanced absorption. The most important aspects of the physical chemistry behind the CIDNP effect will be discussed in the following sections.

5.1   ENERGY POLARIZATION (NET EFFECT)

On heating, benzoylperoxide decomposes into a benzoyloxy- and a phenyl-radical with the formation of CO_2

After the decomposition, both fragments stay together for a certain time in the solvent cage and the spin orientation of the electrons which were paired in the peroxide bond before bond breaking corresponds to the singlet state. This radical pair is known as a correlated radical pair. Because the decomposition takes place in the magnetic field B_0, the singlet state of the radical pair in the moment of its formation can be described by a vector model as shown in Figure 10.24a. One electron spin (S_1) is oriented parallel, the other (S_2) anti-parallel with respect to the B_0-field. The phase angle between the spin vectors amounts to 180 ° and the total spin, therefore, is zero. Because of the interaction between B_0 and S_1 and S_2, respectively, a precession of the electron spins around the field axis starts, completely analogous to the Larmor precession of the nuclear spins discussed in Chapter 7. Since the radicals R_1 and R_2 have different structures, the Larmor precession frequencies of the electrons, ω_S, is also different. For ESR, ω_S is related to the Bohr magneton μ_B and the external magnetic field B_0 by equation (10.14), where g is a typical constant for a particular radical, the so-called Landé- or g-factor.

$$\omega = g\frac{\mu_B B_0}{h} \qquad (10.14)$$

During a short period of time, because of the different Larmor frequencies ω_1 and ω_2, the phase angle between S_1 and S_2 changes (Figure 10.24b). Eventually a situation is reached where both spins precess with the same phase (Figure 10.24c). This corresponds to the triplet state T_0 where the addition of the x,y-components of S_1 and S_2 results in a total spin angular momentum of 1. The interaction of the electron spins with B_0 has thus induced a singlet-triplet transition for the radical pair.

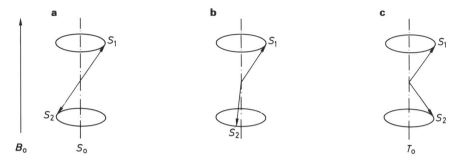

Figure 10.24 Spin states of a radical pair in the magnetic field B_0; (a) singlet state S_0 with total spin 0; phase angle 180°; (b) different precession frequencies of both electron spins lead to a change of the phase angle; (c) triplet state T_0 with total spin 1; the phase angle is now 0 °

The situation is a little more complex if we consider scalar coupling between the electron spin and the protons of the phenyl ring in the phenyl radical. As a consequence and in analogy to scalar coupling between nuclear spins, discussed earlier, the effective magnetic field at the electron and, therefore, its Larmor frequency changes. In a phenyl radical the coupling constants between the electron and the protons are positive. Because the nuclear spins can be oriented anti-parallel or parallel relative to the direction of B_0, we must anticipate two varieties of phenyl radicals: those, where B_0 is enhanced through the coupling (type I), and those where B_0 is attenuated through the coupling (type II). Because of the resonance condition, equation (1.10) the relation $\omega_I > \omega_{II}$ holds. Radicals of type I thus reach the triplet state earlier than those of type II.

The life time of a radical pair is limited by diffusion and chemical reactions with the solvent and other compounds that are present. Furthermore, recombination is possible (Figure 10.25), however, only for the singlet state. In the present case, therefore, radicals of type II, which have a longer life time in the singlet state, will recombine more effectively than those of type I. The latter have a good chance to leave the radical pair by escape out of the solvent cage. Through diffusion processes

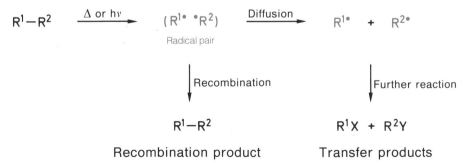

Figure 10.25 Possible reactions of a radical pair ($R^{1\cdot}$, R^2) produced by thermolysis or photolysis of R^1–R^2

these phenyl radicals get in contact with solvent molecules and the *escape* or *transfer* product benzene is formed by hydrogen abstraction. The important point is that the protons of the so-formed benzene have a non-equilibrium Boltzmann distribution because for radicals of type I the nuclear spin orientation was anti-parallel to the external field B_0. The nuclear spin system in the transfer product is thus polarized in such a way that the excited state has the higher population. Consequently, an n.m.r. emission line is observed at ω_H (benzene).

The surprising aspect of the CIDNP experiment obviously is the result that for the chemical reaction between phenyl radicals and cyclohexanone in the magnetic field B_0 a selection has taken place. Only phenyl radicals with a certain spin polarization have reacted.

5.2 ENTROPY POLARIZATION (MULTIPLET EFFECT)

The effect described above is known as *energy polarization* or *net effect*. It is recognized by increased absorption or emission. In addition, a second phenomenon exists, known as *entropy polarization*.

During the thermal or photochemical decomposition of diphenyldiazomethane (**194**) in methylphenylacetate (**195**) one observes, for example, the lines of the AB-system of the tertiary protons in the product methyl-2,3,3-triphenylpropionate (**196**) simultaneously in absorption and emission (Figure 10.26).

$$(C_6H_5)_2CN_2 \xrightarrow{\;h\nu \text{ or } \Delta\;} (C_6H_5)_2C\colon$$

194

$$(C_6H_5)_2C\colon \;+\; C_6H_5-CH_2-COOCH_3 \longrightarrow \left[(C_6H_5)_2\overset{H}{C}\cdot \quad \cdot\overset{H}{C}-COOCH_3 \atop \qquad\qquad C_6H_5 \right]$$

195

$$(C_6H_5)_2CH-CH-COOCH \atop \qquad\qquad C_6H_5$$

196

This *multiplet effect* results apparently from a situation where the nuclear spin levels of the AB-system with the total spin $m_T = 0$ are stronger populated than those with total spin $m_T = +1$ or -1. In order to understand this result we note that the Larmor frequencies of the electron spins of many organic radicals because of similar g-factors are practically identical. This is also true in the present case. Singlet–triplet interchange can then be achieved only if coupling between nuclear and electron spins induces a $\Delta\omega_S$ for the radicals in question.

As is known from the results of other investigations, diphenylcarbene formed by thermal decomposition of diphenyldiazomethane exists in the triplet state. The radical pair ($R_1^{\cdot}$ $^{\cdot}R_2$) formed by proton abstraction from **195**, therefore, also exist in the triplet state. For the AX nuclear spin system we consider the four eigenstates $\alpha\alpha$, $\alpha\beta$, $\beta\alpha$ and $\beta\beta$. If the ESR hyperfine coupling constants **a** for both radicals are also identical, the effect of spin coupling between nuclei and electrons on the Larmor frequencies of the electrons in the radical pair is eliminated for the nuclear states $\alpha\alpha$ and $\beta\beta$, respectively, since both radicals are effected in the same way. These radical pairs thus stay in the triplet state and have a good chance to recombine.

For radical pairs with the nuclear spin combinations $\alpha\beta$ and $\beta\alpha$ on the other hand, a Larmor frequency difference $\Delta\omega_S$ is induced because for the state $\alpha\beta$ we have $\omega_1 = +\frac{1}{2}\mathbf{a}$ and $\omega_2 = -\frac{1}{2}\mathbf{a}$ while for the state $\beta\alpha \frac{1}{2}\mathbf{a}$ is used with different sign. Radicals of this pair can thus recombine after the triplet–singlet transition and a product results where the nuclear spins populate the levels $\alpha\beta$ and $\beta\alpha$, respectively. The resulting spectrum, shown in Figure 10.26, is drawn again schematically in Figure 10.27a together with the corresponding Boltzmann distribution in the energy level system for the AX system.

Figure 10.27b shows, on the other hand, the result of an analogous reasoning for the case that a radical pair is formed in the singlet state. The reader may easily convince himself or herself that in the recombination product the nuclear spin levels $\alpha\alpha$ and $\beta\beta$ with total spin $m_T = \pm 1$ should then be stronger populated.

CIDNP spectroscopy is thus based on the finite lifetime of the radical pair during which spin–spin interactions between electrons and nuclei induce changes of the electronic spin states. Since recombinations, that means bond formations, can arise only out of singlet states, there is a selection for the consecutive reactions which

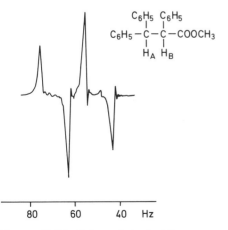

Figure 10.26 Spin polarized AB-spectrum of the ethane protons in methyl-2,3,3-triphenylpropionate after the thermal decomposition of diphenyldiazomethane in methylphenylacetate at 140 °C (after Ref. 13)

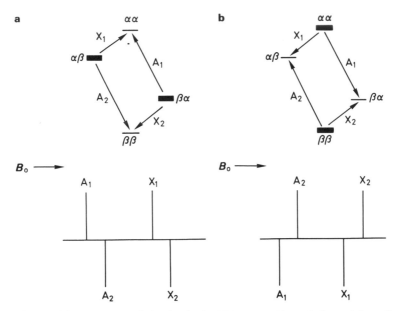

Figure 10.27 Entropy polarization in the AX system: (a) result for a triplet radical pair; (b) result for a singlet radical pair. The relative population of the nuclear spin levels is indicated by the light or heavy bar

results in non-equilibrium Boltzmann distributions for the nuclear spins in the products.

5.3   THE KAPTEIN RULES

From our discussion it has become clear that the CIDNP effect is an important tool for investigations of reaction mechanisms in radical and photochemistry. In order to interpret the experimental results correctly, a good knowledge of the theoretical background is necessary. A number of important questions can, however, already be answered by a simple interpretation of the observed spectra on the basis of the *Kaptein rules*. According to these rules the sign of the expected polarization is calculated as a product of the signs of a number of factors which characterize the properties of the radicals and products. The effect of energy polarization (net effect, $+$ for absorption, $-$ for emission) is then described by the following equation:

$$\Gamma_N = \mu\varepsilon\Delta g\mathbf{a}_i \tag{10.15}$$

and for entropy polarization (multiplet effect, $+$ for the relation emission/absorption, $-$ for absorption/emission) we have:

$$\Gamma_M = \mu\varepsilon\mathbf{a}_i\mathbf{a}_j J_{ij}\sigma_{ij} \tag{10.16}$$

The physical constants which appear in these equations and which are only used with their sign, have the following meaning:

μ characterizes the spin state of the correlated radical pair:
  + for a triplet educt
  − for a singlet educt
ε characterizes the product
  + for a recombination product
  − for a transfer product
Δg stands for the sign of the difference $g_i - g$ of the Landé factors of the corresponding radicals where g_i belongs to the radical of the observed nucleus i
a_i sign of the ESR-hyperfine coupling constant of radical fragment i
J_{ij} sign of the n.m.r.-coupling constant
σ_{ij} characterizes situation where the nuclei i and j are in the same or in different radical fragments:
  +, if i and j are in the same radical fragment
  −, if i and j are in different radical fragments

The signs of **a** and J_{ij} must be determined independently by ENDOR- and n.m.r. investigations not discussed here.

As can be easily understood, only one unknown factor can be determined through the application of equations (10.15) and (10.16). For example, if for a process where the multiplet effect is observed the signs of **a**, J_{ij}, and σ_{ij} are known, the spin state of the correlated radical pair can be derived. Very often this is an important piece of information for chemistry. For the formation of **196** one obtains the following result. The ESR hyperfine coupling constants of the radicals involved as well as the vicinal $^1H,^1H$ coupling constants of the product have positive signs, the nuclei i and j stem from different radical fragments ($\sigma_{ij} = -1$) and the experiment shows the relation A/ E ($\Gamma_M = -1$). It then follows $(-) = (\mu)(+)(-)$; μ is, therefore, positive, and the educt existed in the triplet state.

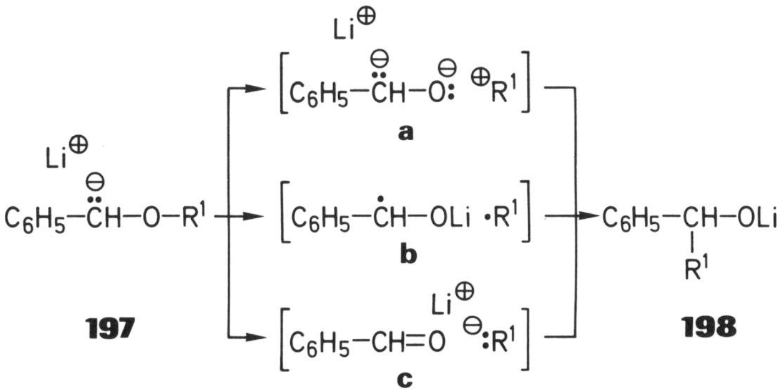

The application of the CIDNP effect is not restricted to photolysis or thermolysis reactions. CIDNP effects have also been observed during reactions of organolithium compounds and were used to study reaction mechanisms in organometallic chemistry. The method thus provides opportunities for the study of reaction mechanisms. For example, the *Wittig ether rearrangement*, in which an α-methylated alkylbenzyl ether (**197**) is transformed into the corresponding alkyl phenyl carbinol (**198**), can be rationalized using either an ionic or a radical pair mechanism. While the path (a) could be excluded relatively easily on the basis of a variety of experimental data, it was considerably more difficult to decide between the alternatives (b) and (c) using conventional methods. The observation that in the rearrangement of lithium t-butyl benzyl ether the signal of the tertiary proton in the corresponding lithium 1-phenyl-2,2-dimethylpropanolate appeared as an emission signal was a direct indication that the reaction can proceed via path (b). Whether alternative mechanisms, in our case (c), can be excluded completely becomes available from investigations of this kind only if the experiment is quantitatively evaluated. The observation of a CIDNP effect alone is not a sufficient criterion for the intermediacy of radicals, since, because signal enhancements by a factor of 10–10^3 are possible, it may arise from a side-reaction.

6. Nuclear magnetic resonance spectroscopy of paramagnetic materials

6.1 CONTACT SHIFTS

In Chapter 7 we alluded to the fact that paramagnetic impurities markedly accelerate the longitudinal relaxation of protons and therefore hinder the measurement of n.m.r. spectra. It follows that materials that themselves are paramagnetic cannot, in general, be investigated by using n.m.r. spectroscopy. However, there are cases in which a paramagnetic compound such as a radical anion of the an aromatic hydrocarbon ($R \doteq$) exists in a low concentration in the presence of the corresponding diamagnetic species (R). Here an intermolecular exchange of the lone electron will result and the paramagnetic spin density is distributed over a large number of molecules with the effect that it is *'diamagnetically diluted'*. If, in addition, electron spin relaxation is fast, it becomes possible to observe the n.m.r. signals of the system.

$$R^{\cdot\ominus} + R_* \rightleftharpoons R + R_*^{\cdot\ominus}$$

They are broadened in contrast to those of pure diamagnetic compounds, but in spite of this they may be easily recognized.

It is significant that the interaction between electron and nuclear spin that manifests itself in electron spin resonance (e.s.r.) spectroscopy as the hyperfine splitting of the e.s.r. signal, here merely leads to a shift of the n.m.r. signal to higher or lower field relative to the signal of the same proton in the diamagnetic compound. This effect is known as *contact shift*.

In e.s.r. spectroscopy the hyperfine coupling constant **a** for the scalar interaction between electron and nuclear spin that leads to the line splitting in the e.s.r. spectrum is given by the relation

$$\mathbf{a} = \mathbf{Q}\rho \tag{10.17}$$

where Q is a proportionality factor of about -25 G and ρ is the unpaired spin density at the carbon atom under consideration; ρ is a dimensionless quantity and **a** is thus expressed in gauss. The spin density, ρ_μ, at the centre μ of a π system can be calculated through the relation $\rho_\mu = c_{\kappa\mu}^2$ from the coefficient $c_{\kappa\mu}$ of the wave function ψ_κ at the centre μ; ψ_κ is that molecular orbital in which the unpaired electron resides. The coefficients in turn can be obtained using Hückel MO calculations where the molecular orbitals are expressed as linear combinations of the $2p_z$ atomic orbitals at the carbon atoms:

$$\Psi_\kappa = \sum_{\mu=1}^{n} c_{\kappa\mu}\phi_\mu \qquad (10.18)$$

One would now expect that the multiplet structure of the electron resonance in the case of a two-spin system of the AX type, as represented here by the nuclear and electron spin, would also lead to a splitting in the n.m.r. spectrum. There are two reasons why this is not the case. The first is the fast spin relaxation of the electrons and the second is the rapid exchange of electrons occurring between radical anions (R$\dot{-}$) and diamagnetic molecules (R) in the solution. As in the case of methanol (p. 338), time averaging results and the line splitting vanishes since the electron interacts with a large number of nuclei in different spin states. The average n.m.r. line should then assume the same position as the corresponding resonance signal of the diamagnetic compound. However, as the phenomenon of the contact shift shows, this is not observed. The reason for this lies in the different populations of the two electronic eigenstates. Since the energy difference $h\nu_S$ is substantially larger than the corresponding value $h\nu_I$ in n.m.r., the low energy level ($m_S = +\frac{1}{2}$) is much more highly populated and contributes with a greater weight ($N_{+1/2} > N_{-1/2}$) to the time average of ν according to

$$\tilde{\nu} = N_{+1/2}\nu_{+1/2} + N_{-1/2}\nu_{-1/2} \qquad (10.19)$$

The magnitude of the contact shift is field dependent and satisfies the expression

$$\frac{\Delta B}{B_0} = \frac{\mathbf{a}\gamma_e^2\hbar}{4\gamma_p kT} \qquad (10.20)$$

where γ_e and γ_p signify the magnetogyric ratio of the electron and the proton, respectively, k is the Boltzmann constant and T is the absolute temperature. Moreover, there is a dependence on the concentration of the paramagnetic molecules. As one can see, the equation above allows us to determine the sign of the hyperfine coupling constant, **a**, information that is not available from the e.s.r. spectrum alone.

As an illustration, Figure 10.28 shows the n.m.r. spectra of 1-propylnaphthalene (**199**) for different concentrations of the corresponding radical anion. It can be seen that the hyperfine coupling constants for the methyl protons and the protons of the β-methylene group have opposite signs since the CH$_3$ lines are shifted to lower field while the CH$_2$ lines are shifted to higher field. Moreover, the spectra confirm that the

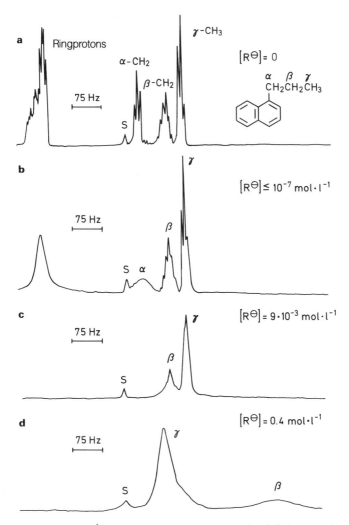

Figure 10.28 ¹H n.m.r. spectrum of 1-propylnaphthalene in the presence of different concentrations of the corresponding radical anion [R·]; in tetrahydrofuran—the absorption at s is that of incompletely deuterated solvent (after Ref. 14)

magnitude of the contact shift, as expected according to the equation above, is proportional to the magnitude of the constants **a** (cf. formula **199**). A more detailed consideration of the relaxation processes shows further that the line width is also proportional to the magnitude of **a**. In addition, it is a function of the factor $1/r^6$, where r is the distance of the nucleus under consideration from the radical centre. In our case this is the $2p_z$ orbital of the corresponding carbon atom of the naphthalene ring. The lines of the ring protons and the methylene protons are therefore especially strongly shifted and broadened so that they can no longer be observed in the

$$\overset{\alpha \quad \beta \quad \gamma}{CH_2CH_2CH_3}$$

199

$a_\alpha = + 267.0 \mu T$
$a_\beta = - 21.2 \mu T$
$a_\gamma = + 6.4 \mu T$

spectrum (c). The values of the hyperfine coupling constants determined in this experiment are indicated next to formula **191**.

An additional advantage of the method is its *sensitivity*, which enables us to determine very small hyperfine coupling constants that cannot be measured by using e.s.r. spectroscopy directly. In favourable cases it has even been possible to observe the nuclear magnetic resonance spectra of *radicals* and to determine the e.s.r. hyperfine coupling constants from the magnitude and signs of the contact shifts.

6.2 PSEUDOCONTACT SHIFTS—SHIFT REAGENTS

In addition to the mechanism of the interaction between electron and nuclear spins already mentioned, a second possibility of mutual influence can be detected using n.m.r. spectroscopy. This mechanism, known as *pseudocontact interaction*, leads to a shift of the n.m.r. line and operates when a strongly anisotropic paramagnetic centre is present in the molecule. The unpaired electrons in the valence orbitals of the rare earth metals, for example, possess such anisotropic properties. The influence on the proton resonance is the result of a dipolar interaction between the magnetic moments through space. Its magnitude is proportional to the expression $(3 \cos^2 \theta - 1)/r^3$, where r is the distance between the nucleus under consideration and the site of the paramagnetism and θ is the angle between the effective symmetry axis of the paramagnetic moment and the distance vector, r, to the nucleus. For example, the dipole field of an axially symmetrically paramagnetic centre then has the following form:

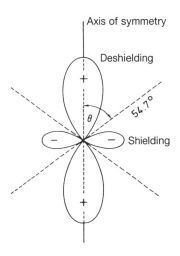

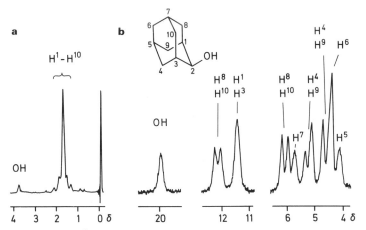

Figure 10.29 ^{1}H n.m.r. spectrum of 2-adamantanol: (a) normal spectrum; (b) spectrum in the presence of tris(dipivaloylmethanato)europium(III) (after Ref. 5)

The pseudocontact interaction may be observed in both *crystals* and *solutions*. Thus, it has been found that the coordination of hydroxyl or amino groups with europium(III) causes a significant paramagnetic shift of the proton resonances of the corresponding alcohols or amines if one adds the complex formed from Eu(III) and the β-diketone (**200**) (the tris(dipivaloylmethanato)-Eu(III) complex [Eu(DPM)$_3$]) to the solution of the substance under investigation. On the basis of the angle factor (3 cos^2 θ − 1), shifts to both lower and higher field can be observed. Complexes with praseodymium or ytterbium show similar effects. These substances, known as *shift reagents*, have found wide application, since they allow complex spectra in which numerous proton resonance signals overlap to be significantly simplified. The effects is in certain ways comparable to that which one attains by the application of higher B_0 fields, i.e. with superconducting magnets. Especially with saturated compounds, a considerable increase in information is achieved through the use of shift reagents, as is shown in Figure 10.29 for the case of 2-adamantanol. Here the resonances of all of the protons and the geminal coupling constants, which have the largest values of all spin–spin interactions in this compound, can be unravelled. It is fortunate that the line broadening caused by the paramagnetic moment is in the case of lanthanides relatively small.

$$H_3C \diagdown \diagup CH_3 \quad H_3C \diagdown \diagup CH_3$$

200

The effect described originates from complex formation between the shift reagent, where the lanthanide cation still has free coordination sites, and the substrates. The observed spectra are the time average for free and complexed substrate. Individual

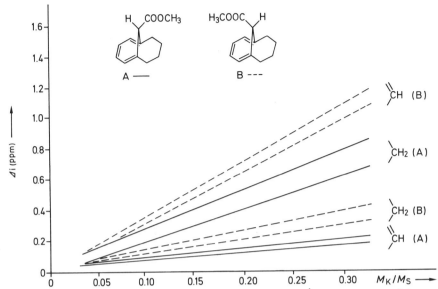

Figure 10.30 Eu(FOD)$_3$-induced pseudo contact shifts in the ^{1}H n.m.r. spectra of the exo/endo-isomers of 11-carbomethoxy-bicyclo[4.4.1]undeca-1,3,5-triene; M_K/M_S = molar ratio Eu(FOD)$_3$/substrate (after Ref. 17); complexation is achieved with the ester group

lanthanides differ with respect to the sign of their dipole field. For example, for Eu(III) we find a situation as shown on p. 431, while for Pr(III) the signs are reversed. The magnitude of the resulting shift is strongly dependent on the basicity of the complexing group in the substrate which decreases in the following sequence:

$$-NH_2 > OH > C{=}O > COOR > CN$$

Acidic groups very often lead to a decomposition of the lanthanide complex and can, thus, in general not be investigated (see, however, Table 10.2). In addition, due to the lack of complex formation, the spectra of saturated and unsaturated hydrocarbons cannot be simplified by the application of shift reagents. However, it proved possible to use silver salts as auxiliary reagents in order to observe lanthanide induced shifts for olefinic and aromatic compounds. The proper signals of the groups in the shift reagents are generally not observed because they are shifted in the opposite direction. Furthermore, because of the small concentration of the shift reagent, these signals are of low intensity (see also Table 10.2). For the induced shifts of the n.m.r. signals (^{1}H or ^{13}C) in the substrate, Δ_i, which are known as LIS-values (LIS = *l*anthanide *i*nduced *s*hift), the following relation holds for axial symmetry:

$$\Delta_i = K\frac{3\cos^2\theta - 1}{r_i^3} \tag{10.21}$$

Here K is an empirical constant typical for the complex under consideration. On the basis of the geometry factor structural data (angles, distances) can be derived from

Table 10.2   Most common shift reagents and their properties (after Ref. 16)

Anion	Cation

DPM⁻ (= dipivalomethanato)
FOD⁻ (= 1,1,1,2,2,3,3-heptafluoro-7,7-dimethyl-4,6-octandionato)

Chiral Ligand

DFHD⁻   (= 1,1,1,5,5,6,6,7,7,7-decafluoro-2,4-heptandione)
FACAM⁻ (= 3-trifluoroacetyl-D-campherato)

Properties			
DPM			FOD
good solubility in organic solvents	(+)	(+) (+)	excellent solubility in organic solvents, high concentration possible, good resolution can be achieved
good complex formation with strong bases	(+)	(+)	strong complex formation with strong bases
weak complex formation with weak basis	(−)	(+)	good complex formation even with weak bases
t-butyl signal close to TMS, in most cases at higher field	(+) (+)	(−)	t-butyl signal can appear in the n.m.r. region of aliphatic protons; substrate conformation not much affected
substrate conformation weakly influenced	(+)	(−)	substrate conformation can be strongly changed
unstable against weak acids, phenols, carboxylic acids	(−)	(+)	stable against weak acids

the experimentally observed Δ_i-values for various nuclei in the ligand (^{1}H, ^{13}C), provided that contact and non-axial contributions to the LIS-data are absent.

The aspect of quantitative analysis of lanthanide induced shifts is of considerable interest for conformational analysis. In practice, however, very often a qualitative empirical analysis is already sufficient to solve a stereochemical problem by application of shift reagents. For example, Figure 10.30 shows the resonance shifts measured for the olefinic and the methylene protons of the syn- and anti-isomer of 11-carbomethoxy-bicyclo[4,4,1]undeca-1,3,5-triene. They are sufficiently different to allow an assignment of the stereochemistry at the bridge carbon. Due to the different distances of the ester protons on one side and the olefinic and methylene protons on the other we find for isomer A $\Delta(CH_2) > \Delta(=CH)$ and for B $\Delta(CH_2) < \Delta(=CH)$.

Lanthanide complexes with chiral ligands are used quite frequently to determine the optical purity of enantiomeric mixtures. As in the case of diastereomers which are formed by solvation with optically active solvents (see p. 207), one observes different n.m.r. signals for diastereomeric complexes of D- and L-species, if a shift reagent with a unique stereochemistry of the ligand is used. Complexes of various lanthanides with different ligands are commercially available. Those ligands and cations used most frequently are collected in Table 10.2 with their most relevant properties.

7. Nuclear magnetic resonance of partially oriented molecules

As was mentioned earlier, Brownian motion of molecules in the liquid phase causes the direct dipole–dipole coupling between individual nuclear moments in a sample to be reduced to zero and thus dipolar interactions only contribute to relaxation. However, it has been discovered that certain substances when used as solvents within specific temperature ranges are able to restrict the Brownian motion of solute molecules and enforce upon them a specific orientation within the solvent. Such substances are called *liquid crystals*, and here we shall be concerned in particular with the *nematic phases*. These materials exhibit, even in the liquid state, in a

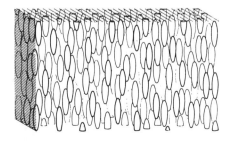

Figure 10.31 The structure of a nematic phase

temperature range of about 20–40 °C above the melting point, an ordered structure resembling in a remote way that of a solid. Finally, at the *clearing point*, their behaviour becomes similar to that of other liquids. The nematic phases, in particular, show predominantly a parallel arrangement of the individual molecules (Figure 10.31), the structures of which are rod-like in shape. substances of this class are compounds such as 4,4′-dimethoxyazoxybenzene (**201**) or *N*-(ethoxybenzylidene)-4-butylaniline (**202**).

For n.m.r. it is of significance that a solute molecule dissolved in the nematic phase will assume a preferred orientation within the liquid crystal. Obviously, the position of the solute molecule is not fixed and it can undergo translational and rotational motion. These motions are, however, not completely free as in a normal isotropic liquid, but restricted by the surrounding structure of the liquid crystal. The degree of ordering that results for the solute is relatively low, but it is sufficient to induce dipole–dipole coupling between protons when the n.m.r. spectrum of a compound is measured in such an anisotropic environment. *Intermolecular* interactions, on the other hand, are precluded as before because of the diffusion of the solute molecules through the solvent. They are encountered only in true solids.

The dipolar interaction between the nuclear spins leads now to an additional line splitting in the spectrum. This is shown in Figure 10.32 for the simple case of the A_2 system of the protons of methylene chloride. In isotropic liquids we would obtain a singlet, but here, using **202** as a solvent, a splitting of 4000 Hz is observed.

In order to understand this finding, let us recall the energy level diagram for the A_2 case (p. 148). By the application of symmetry functions, we obtained an antisymmetric state and three symmetric eigenvalues linked through the degenerate transitions $E_2 \rightarrow E_1$ and $E_4 \rightarrow E_2$ (Figure 10.33a). The interaction between two nuclear dipoles μ_1 and μ_2 separated by a distance r_{12} now causes the eigenstates of the spin system to be either stabilized or destabilized. The energy of the interaction is given by the expression*

$$E = \frac{\mu_1\mu_2}{r_{12}^3} - \frac{3(\mu_1 r_{12})(\mu_2 r_{12})}{r_{12}^5} \tag{10.22}$$

---

* For brevity, the factor $\mu_0/4\pi$ is omitted for the present discussion

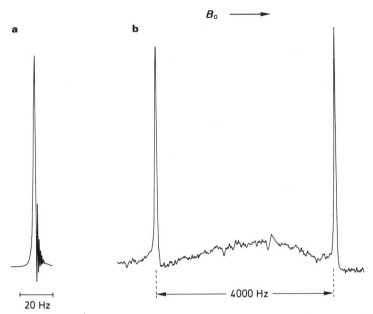

Figure 10.32 ^{1}H n.m.r. spectrum of methylene chloride in (a) isotropic and (b) anisotropic environments. The nematic phase was provided by compound **202**

where r_{12} is the distance vector between the two nuclei as defined in Figure 10.33c. Using the relation (1.2), equation (10.22) is transformed into the quantum mechanical Hamiltonian operator (10.23):

$$\mathscr{H}_{12} = \gamma_1\gamma_2\hbar^2 \frac{\hat{\mathbf{I}}_1\hat{\mathbf{I}}_2}{r_{12}^3} - \frac{3(\hat{\mathbf{I}}_1 r_{12})(\hat{\mathbf{I}}_2 r_{12})}{r_{12}^5}$$

If we apply this operator to the symmetric wave functions $\alpha\alpha$, $(\alpha\beta + \beta\alpha)/\sqrt{2}$, and $\beta\beta$, the detailed calculation, assuming $\mu_1 = \mu_2$ (that is, for equivalent nuclei), leads to a destabilization of states (1) and (4) by the amount

$$\Delta E_{1,4} = \tfrac{1}{4}\gamma^2\hbar^2(3\cos^2\theta - 1)r_{12}^{-3} \qquad (10.24)$$

and a stabilization of state (2) by the amount

$$\Delta E_2 = \tfrac{1}{2}\gamma^2\hbar^2(3\cos^2\theta - 1)r_{12}^{-3} \qquad (10.25)$$

The angle θ, as shown in Figure 10.33, is the angle between the distance vector, r_{12}, and the direction of the external magnetic field, B_0. Resonance then occurs relative to ν_0 at

$$\nu = \nu_0 \pm \frac{3}{4}\gamma^2 \frac{h}{4\pi^2}(3\cos^2\theta - 1)r_{12}^{-3} \qquad (10.26)$$

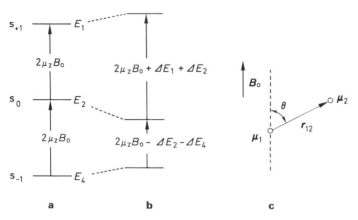

Figure 10.33   The energy level diagram of an oriented A_2 system

or

$$v = v_0 \mp \tfrac{3}{2} D_{12} \tag{10.27}$$

if one introduces the general expression

$$D_{ij} = \frac{-h\gamma^2}{4\pi^2} \tfrac{1}{2} (3\cos^2 \theta - 1) r_{ij}^{-3} \tag{10.28}$$

for the dipolar coupling, D_{ij}, between two protons. If one further defines the degree of orientation of the internuclear distance by the relation

$$S_{ij} = \tfrac{1}{2} (3\cos^2 \theta - 1) \tag{10.29}$$

then

$$D_{ij} = -\frac{h\gamma^2}{4\pi^2} S_{ij} r_{ij}^{-3} \tag{10.30}$$

The same result is applicable for an isolated A_2 system in solids.

As equation (10.30) shows, the line splitting depends on the distance, r_{12}, between the two nuclei. If there was a reliable independent method for calculating S_{ij}, one could obtain the internuclear distance from the experimentally determined dipolar coupling. This, however, is not the case, since in A_2 systems only one observable relates the two unknowns, S_{12} and r_{12}. This is a result of the fact that the dipolar coupling, in contrast to the scalar coupling, is not a molecular constant and the value of 4000 Hz measured above for the line splitting in methylene chloride (which corresponds to 1333 Hz for D_{12} according to equation (10.27)) is dependent through the degree of orientation upon temperature, concentration, and the nematic phase used. Therefore, it varies in practice with each new measurement.

In the case of more complicated molecules in which a larger number of independent dipolar couplings exists, one usually has enough information at hand to determine distance *and* orientation for a particular pair of nuclei from the n.m.r. spectrum. Even so, only the ratios S_{ij}/r_j are obtained which yield in turn distance ratios r_{ij}/r_{ik} but not absolute atomic distances. Since, however, bond angles can also be determined, n.m.r. spectroscopy of partially oriented molecules has developed into an important method of structure determination supplementing the well known techniques of microwave spectroscopy and electron and X-ray diffraction, especially if line splittings due to the presence of ^{13}C nuclei can be included in the analysis. These data also yield information on the carbon geometry, which is obviously of greater interest than the proton geometry alone. The most important aspect of the n.m.r. method probably is that structural information can be obtained in the liquid state.

Nuclear magnetic resonance spectra of partially oriented molecules are distinguished from the spectra measured in isotropic media by their larger spectral width and the slightly larger linewidths (see below). Since dipolar couplings exceed scalar couplings by a factor of $10-10^3$, these spectra often extend over several kHz. In addition, because the orientation parameters are strongly temperature and concentration dependent, the spectral appearance is also very sensitive to temperature and concentration. Furthermore, a unique feature of the n.m.r. spectra of partially oriented molecules is the fact, already derived above, that spin–spin coupling between magnetically equivalent nuclei becomes measurable. A particularly impressive example of this is seen in the spectrum of partially oriented benzene that shows more than 50 lines (Figure 10.34). The high symmetry of the molecule here significantly facilitates the spectral analysis which is based on the Hamiltonian operator introduced in Chapter 5. It contains, in addition to the scalar spin–spin interactions, also the dipolar couplings. One then obtains

$$
\begin{aligned}
\mathcal{H} = &\sum v_i \hat{I}_z(i) + \sum_{i<j} \sum (J_{ij} + 2D_{ij})\hat{I}_z(i)\hat{I}_z(j) \\
&+ \sum_{i<j} \sum (J_{ij} - D_{ij})(\hat{I}_x(i)\hat{I}_x(j) + \hat{I}_y(i)\hat{I}_y(j))
\end{aligned}
\tag{10.31}
$$

The resonance frequencies v_i are differentiated from those in isotropic phases by an anisotropic contribution to the shielding constant. In addition, in the Hamiltonian matrix J_{ij} is replaced by $J_{ij} + D_{ij}$ in the diagonal elements and by $J_{ij} - D_{ij}$ in the off-diagonal elements. In principle, the scalar interactions can also be determined by an analysis based on equation (10.31). However, one can simplify the problem if one obtains them from the analysis of the spectrum measured in an isotropic phase. It is of significance that by means of the n.m.r. spectra of partially oriented molecules it is possible to determine the absolute signs of the scalar coupling constants if one can predict a preferential orientation on the basis of the molecular structure. Finally, it must be emphasized that the relatively simple form of the Hamiltonian operator arises only because the *intermolecular* dipolar interactions can be disregarded as a result of rapid diffusion processes in the liquid crystal. This is not the case in solids.

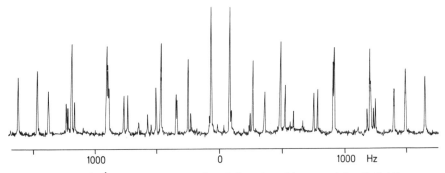

Figure 10.34 The ^{1}H n.m.r. spectrum of partially oriented benzene (after Ref. 18)

The molecules of the nematic phase are also oriented in the magnetic field. With their axis of highest magnetic susceptibility, generally also the longest molecular axis, they align parallel to the direction of the external field B_0. It is of advantage that because of the relatively high degree of order and the large number of protons, the spectra of the liquid crystal are strongly broadened and practically disappear in the noise. The spectra of the partially oriented guest molecules are thus unperturbed by the liquid crystal solvent.

As a consequence of the alignment of the liquid crystal with respect to the external field it is not possible to rotate the n.m.r. sample tube during the experiment in conventional iron magnets because rotation around the y-axis would destroy the orientation of the phase. Under these conditions, the linewidths are of the order of 5–10 Hz. In superconducting magnets used today, however, the z-axis is parallel to the rotational axis of the n.m.r. tube and rotation does not affect the ordering of the crystalline liquid. Sample rotation then leads to much narrower n.m.r. lines.

8. High-resolution solid state nuclear magnetic resonance spectroscopy

The last section has shown already that *intra*molecular dipolar couplings can complicate the n.m.r. spectra of small molecules. This effect is considerably more pronounced in solids, where in addition *inter*molecular dipolar couplings are important. Only structureless broadline spectra with halfwidths of several kHz are then observed and the information about the local environment of the individual nuclei, that is chemical shifts and scalar spin–spin couplings, is completely lost (Figure 10.35). Through a series of innovative developments, however, it was possible during the last two decades to control the mechanisms that lead to this enormous line broadening and as a result of these efforts solid state n.m.r. spectra with relatively narrow lines (10–50 Hz linewidth), which again yield chemical shift information, can be observed today. Aside from solid state n.m.r. spectroscopy of single crystals, which forms a special area not discussed here, *high resolution solid*

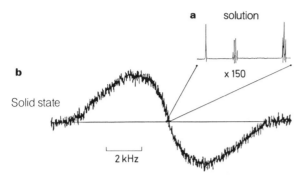

Figure 10.35 ^{1}H n.m.r. spectrum of ethanol: (a) spectrum of the liquid; (b) spectrum of a solid sample (after Ref. 19)

state n.m.r. of powdered solids has developed in recent years into a powerful tool for structural elucidations in inorganic, organic and organometallic chemistry.

Through chemical shifts and scalar spin–spin couplings which can now also be measured for solids, high-resolution solid state n.m.r. provides us with information about the bonding situation, the symmetry properties and, in particular, about the dynamic behaviour of solid state structures. In comparison to structural analysis on the basis of X rays, which requires relatively large single crystals, it is important that n.m.r. spectroscopy can be applied to microcrystalline material, crystal powders and even to amorphous materials.

8.1 THE EXPERIMENTAL TECHNIQUES OF HIGH-RESOLUTION SOLID STATE NMR SPECTROSCOPY

While in isotropic liquids the n.m.r. spectra are dominated by the Zeeman interaction with the external field B_0 ($\hat{H}_z$), the shielding constant σ ($\hat{H}_{CS}$) and by scalar spin–spin interaction J ($\hat{H}_S$), in solids additional mechanisms operate. Dipole–dipole interactions ($\hat{H}_{DD}$), quadrupolar interactions ($\hat{H}_Q$) as well as the anisotropy of nuclear shielding and scalar coupling are complicating factors which lead to severe line broadening. The Hamilton operator for both domains, the liquid and the solid, is thus defined by the following expressions:

$$\text{Solution}: \quad \hat{H} = \hat{H}_z + \hat{H}_{CS}^{iso} + \hat{H}_S^{iso} \tag{10.32}$$

$$\text{Solids}: \quad \hat{H} = \hat{H}_z + \hat{H}_{CS} + \hat{H}_S + \hat{H}_{DD} + \hat{H}_Q \tag{10.33}$$

The individual terms of these equations are of different importance and the experimental concepts for the observation of highly resolved solid state n.m.r. spectra vary. In this respect the n.m.r. active nuclei behave differently and can be divided into four groups according to the following scheme:

Nuclei with $I = \frac{1}{2}$: $\begin{cases} \text{high natrual abundance (e.g. } ^1\text{H}, ^{19}\text{F}); \text{ Group 1} \\ \text{low natural abundance (e.g. } ^{13}\text{C}, ^{15}\text{N}, ^{29}\text{Si}, ^{31}\text{P}; \text{ Group 2} \end{cases}$

$$\text{Nuclei with } I > \tfrac{1}{2} : \begin{cases} \text{integer spin (e.g. } {}^2\text{H, } {}^6\text{Li}); & \text{Group 3} \\ \text{half integer spin (e.g. } {}^{11}\text{B, } {}^{17}\text{O, } {}^{27}\text{Al}); & \text{group 4} \end{cases}$$

Most difficult are measurements for nuclei belonging to Group 1, since their high natural abundance leads to strong homonuclear dipolar interactions and homonuclear decoupling is obviously impossible. Those nuclei belonging to Groups 3 and 4 are also not without problems due to the presence of quadrupolar interactions. Therefore, mainly experiments with nuclei of Group 2 have pioneered the development of high-resolution solid state n.m.r spectroscopy with applications in chemistry. The advantage must be seen in the fact that these nuclei are diluted, either because of their low natural abundance (^{13}C, ^{29}Si) or, as in the case of ^{31}P, because the chemical structure of interest only contains one atom of this kind. In the molecular crystals of organic compounds these nuclei are then effectively isolated. Homonuclear coupling is absent and heteronuclear interactions can be removed through spin decoupling. For insensitive nuclei there is the additional option of magnetization transfer from abundant protons by cross polarization in a Hartmann–Hahn experiment. This technique paved the way for high-resolution ^{13}C n.m.r. spectroscopy in solids, which played a leading role for the development of this new branch of n.m.r. spectroscopy.

In the standard experiment for recording high resolution solid state ^{13}C n.m.r. spectra (Figure 10.36), the ^{13}C nuclei are excited by ^{1}H,^{13}C cross polarization (CP) (cf. p. 416). During signal detection high power ^{1}H-decoupling is used. In this way dipolar and scalar heteronuclear spin–spin interactions are eliminated and the signals are broadened solely by the anisotropy of the chemical shift which results from the directional dependence of the shielding constant σ in the B_0-field; σ is a tensor which can be brought into diagonal form through a suitable coordinate transformation:

$$\sigma = \begin{vmatrix} \sigma_{11} & 0 & 0 \\ 0 & \sigma_{22} & 0 \\ 0 & 0 & \sigma_{33} \end{vmatrix} \tag{10.34}$$

In solids, $\sigma_{11} \neq \sigma_{22} \neq \sigma_{33}$ or for axially symmetric molecules $\sigma_{11} = \sigma_{22} \neq \sigma_{33}$. While in liquids the chemical shift is measured as the trace of the symmetric matrix (10.34),

$$\text{tr}\sigma = \tfrac{1}{3}(\sigma_{11} + \sigma_{22} + \sigma_{33}) \tag{10.35}$$

in the case of powder spectra different σ-values are observed depending on the orientation of the individual microcrystals relative to the field direction. The experimental spectra reflect this distribution by the lineshapes shown in Figure 10.37. From these spectra, the anisotropy of the chemical shift can be determined. It gives us information about the electron distribution around the nucleus which is directly related to chemical bonding.

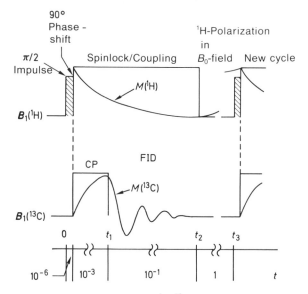

Figure 10.36 Time scale of a ^{1}H,^{13}C cross polarization (CP) experiment for solid state n.m.r. spectroscopy (after Ref. 20)

Even if the contributions of dipolar ^{1}H,^{13}C coupling to the observed spectrum are thus practically eliminated, linewidths of several kHz are the rule and especially spectra of compounds with several non-equivalent nuclei are difficult to analyse because of strong line overlap. It is only by the application of a further technique that line narrowing is sufficient in order to observe high resolution spectra: rotation of the

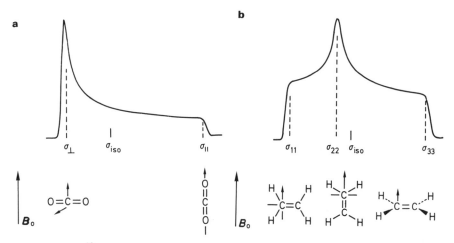

Figure 10.37 ^{13}C resonances for powder spectra of molecules with axial symmetry (a) and without axial symmetry (b)

sample cell around the magic angle (MAS = *magic angle spinning*). In this experiment, the so-called *rotor* — a small capsule of ca. 10 mm length and 4 or 7 mm diameter made from zircon oxide or boronitride and filled with the powdered solid — is rotated with a high spinning rate around an axis which makes the magic angle $\theta = 54.7\,°$ (cf. p. 9) with the axis of the external field B_0. For axial symmetric groups the time average of the chemical shift anisotropy then vanishes because of the angular term:

$$\langle \sigma A(t) \rangle \sim \tfrac{1}{3}(\sigma_{\parallel} - \sigma_{\perp})(3\cos^2\theta - 1) \tag{10.36}$$

At rotational frequencies 2–9 kHz it is possible to narrow the resonance lines to such an extent that it is indeed correct to speak of high resolution n.m.r. Combined with cross polarization we obtain CP/MAS solid state spectra (Figure 10.38).

The effect of sample rotation is shown in Figure 10.39 for the case of a ^{31}P resonance. In practice the rotational frequency accessible is not always sufficient to remove the anisotropy effect completely. Sample rotation then leads to the formation of rotational side bands which vanish only at very high spinning speeds (Figure

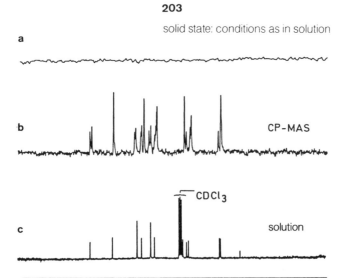

Figure 10.38 ^{13}C n.m.r. spectrum of solid 4,4′-bis[2,3-(dihydroxypropyl)oxy]benzoyl-diacetonide (**203**); (a) conditions as in solution; (b) high resolution CP/MAS solid state spectrum; (c) spectrum of a sample in solution (after Ref. 20)

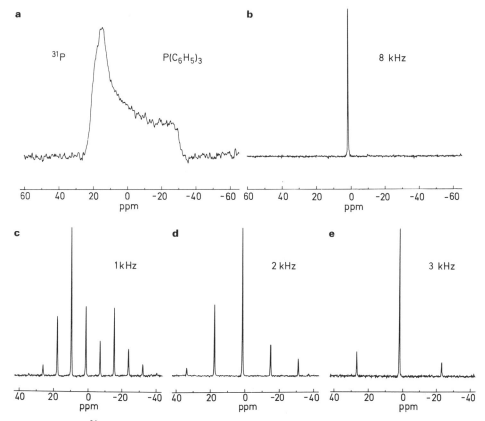

Figure 10.39 ^{31}P resonance of triphenylphosphine in the solid without (a) and with (b) MAS-effect; measuring frequency 121.5 MHz; (c)–(e) rotational side bands at different rotational frequencies

10.39c–e). While this effect complicates the spectra and the line assignment, it also has a positive aspect: one can calculate the anisotropy of the shielding tensor from the position and the intensity distribution of the side bands (Herzfeld–Berger method). Normally, these parameters, which are of interest with respect to the bonding situation of the particular atom, are obtained from single crystal studies, where the sample is measured in various orientations with respect to the B_0 field axis.

The CP/MAS technique has been used successfully for the detection of a large number of spin-$\frac{1}{2}$ nuclei, for example [29]Si, [33]S, [77]Se, [89]Y, [119]Sn, [195]Pt, [199]Hg, [205]Tl, [207]Pb. One can dispose of cross polarization and record only MAS spectra if the sample contains no protons. This applies to [29]Si spectra of silicates.

A completely different approach which yields high resolution solid state spectra has been used independently and relatively early on the basis of multi-pulse sequences. Here the time averaging of dipolar couplings is achieved through a series

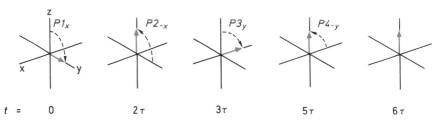

Figure 10.40   WAHUHA-pulse sequence for high resolution solid state n.m.r. spectroscopy

of r.f. pulses which is constructed in such a way that the nuclear spins practically rotate around the magic angle (spin-flip narrowing). A typical 4-pulse sequence of 90° pulses, known as a WAHUHA sequence (after its inventors *Wa*ugh, *Hu*ber, and *Ha*eberlen), has, in the rotating frame, the following form

$$PI_x - 2\tau - P2_{-x} - \tau - P3_y - 2\tau - P4_{-y} - \tau- \qquad (10.37)$$

Since homonuclear dipolar couplings can also be eliminated, the spin-flip method found recently interesting applications in connection with the MAS technique for the measurement of high resolution ^{1}H solid state spectra (CRAMPS = *c*ombined *r*otation *a*nd *m*ulti-*p*ulse n.m.r. spectroscopy of *s*olids).

For multiline spectra, as those observed for ^{13}C n.m.r of organic solids, the problem of assignment exists. Several one- and two-dimensional experiments are available for this purpose and two of them are now discussed. One is used to distinguish the resonances of quaternary carbons and CH$_3$ groups on the one hand from those of CH and CH$_2$ groups on the other. The second sequence allows the determination of signal multiplicity and thereby the number of protons bound to the carbon of interest.

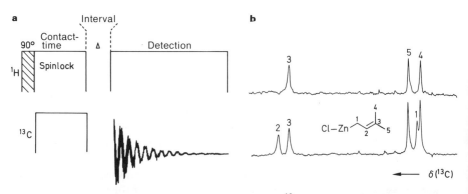

Figure 10.41   (a) Pulse sequence for the assignment of ^{13}C resonances in CP/MAS solid state n.m.r. spectra known as NQS-sequence (non-quaternary carbon suppression); (b) CP/MAS ^{13}C n.m.r. spectrum of allyl zinc chloride (below) with application of the NQS experiment (above) (after Ref. 21)

As shown in Figure 10.41a, in the first sequence one uses delayed signal detection. During the delay Δ, which amounts to 30–100 µs, proton coupling is allowed. Dipolar interactions between neighbouring $^1H, ^{13}C$ spin pairs can then lead to a decay of transverse ^{13}C magnetization and these signals are lost. Quaternary carbons are less affected since the dipolar interactions are strongly reduced with the distance between the nuclei (factor $1/r^3$). For methyl carbons, dipolar couplings to protons are eliminated even in the solid because of the high rotational frequency of the methyl groups (energy barriers < 15 kJ mol^{-1}). Signals of CH_3 carbons are thus also retained. Figure 10.41b documents the application of this technique with an example of an allyl zinc compound.

The second method is based on J,δ-spectroscopy, introduced in Chapter 8, which is now adjusted for heteronuclear spin systems (cf. p. 331). Scalar spin–spin coupling is allowed only during half of the echo time. In solid state n.m.r. this is achieved through the application of two different ^1H decoupling sequences, multipulse decoupling (elimination of dipolar couplings, see above) and high power ^1H decoupling (elimination of dipolar *and* scalar coupling). The result of such an experiment is shown in Figure 10.42 for the ^{13}C spectrum of camphor.

With quadrupolar nuclei a complicating factor in solid state n.m.r. is the splitting of the resonance by quadrupolar coupling which, in general, cannot be eliminated by magic angle spinning. The changes in the energy level diagram associated with this interaction for nuclei with integer and half integer spin are shown in Figure 10.43. As one can see, the degeneracy of individual transitions is removed and multiplet structures (doublets, triplets) arise.

A relatively simple situation is met with spin-1 nuclei, of which deuterium, because of its relatively small quadrupolar moment, has been investigated most frequently. Usually one can assume axial symmetry of the electric field gradient since deuterium appears nearly exclusively in CD bonds. The ^2H n.m.r. signal then splits according to

$$v_Q = \frac{3e^2qQ}{4h}(3\cos^2\theta - 1) \tag{10.38}$$

in a doublet. The tensor e^2qQ/h is the so-called *quadrupolar coupling constant*, the product between electronic charge e, quadrupole moment eQ and field gradient, q, at the nucleus (h = Planck's constant); θ is the angle between the direction of the external magnetic field B_0 and the main direction of the electric field gradient. For deuterons the magnitude of e^2qQ/h ranges between 170 and 200 kHz, for ^{14}N around 5 MHz and for ^{35}Cl values of up to 80 MHz can be assumed. This leads to digitizing problems (remember the Nyquist theorem!) and only for deuterons is the spectral region accessible with the standard digitizers available.

For crystal powders all orientations CD bonds relative to the field axis are possible and the resulting superposition of the resonances of individual deuterons leads to a line shape as shown in Figure 10.44a. For compounds with several anisochronic deuterons such lines are superimposed and the spectra are more complicated (Figure 10.44b).

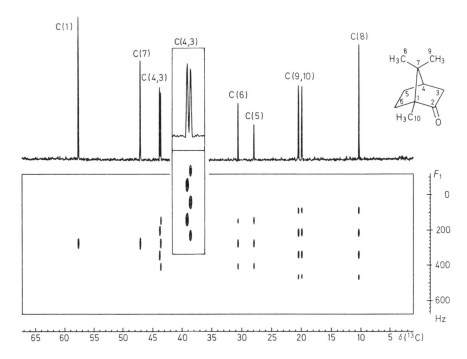

Figure 10.42 High resolution two-dimensional ^{13}C J,δ-spectrum of solid camphor; contour diagram, measuring frequency 75 MHz; the signal of the carbonyl carbon is not shown. The ^{13}C multiplets due to scalar $^{13}C,^{1}H$ coupling are clearly shown on the F_1-axis and the differentiation between resonances of quaternary carbons, CH-, CH$_2$- and CH$_3$-groups becomes possible

The line shape of the ^2H solid state resonance is, because of the angular dependence of the quadrupolar coupling (equation (10.38)) also influenced by dynamic processes in the solid. Through the analysis of such spectra important information can be gained about the dynamic properties of polymer chains. Experiments aimed at a reduction or elimination of the quadrupolar broadening of the ^2H-resonances by application of the MAS-technique are not successful because of the large spectral width. In connection with special techniques (data collection synchronized with the rotor speed, excitation of double quantum transitions) resonances of deuterons bound to aromatic and aliphatic carbon have been separated. The small chemical shift range for deuterium and the line broadening due to quadrupolar effects of higher order, however, limits application of these methods to special cases. For quadrupole nuclei with non-integral spin $I > \frac{1}{2}$ there is no quadrupolar first order line broadening for the transition between the states with $m_I = -\frac{1}{2}$ and $m_I = +\frac{1}{2}$ (see Figure 10.43b). Solid state n.m.r. spectra of nuclei like

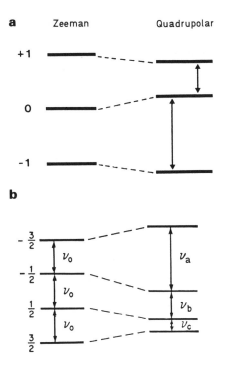

Figure 10.43 Energy level diagram for quadrupolar nuclei in solids; (a) spin $I = 1$; (b) spin $I = 3/2$

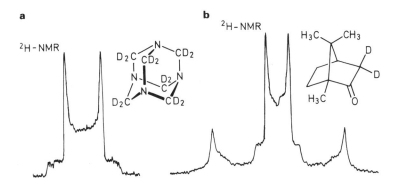

Figure 10.44 Static ^2H solid state n.m.r. spectrum of hexamethylenetetramine-d_{12} with isochronic deuterons (a) and camphor-3-d_2 with anisochronic deuterons (b) (after Ref. 19)

[11]B, [23]Na, [27]Al and [17]O have thus relatively frequently been measured as broadline spectra without MAS application. Sample rotation leads to line narrowing; however, the broadening effects of higher order cannot be eliminated. Therefore, these spectra, which in addition are complicated because of the relatively small chemical shift range for these nuclei, are difficult to analyse. Only recently, a new technique called *double rotation* has been developed to deal with these problems. One can show that if the sample is rotated around two separate axes which are perpendicular, narrow lines are indeed observed. This sophisticated technique thus constitutes a real breakthrough in solid state n.m.r. spectroscopy of quadrupolar nuclei and interesting results can be expected for the near future.

8.2   APPLICATIONS OF HIGH-RESOLUTION SOLID STATE NMR SPECTROSCOPY

From the numerous applications of the CP/MAS or MAS technique in structural research only a few typical examples can be discussed in this section. One aspect of general interest is certainly the fact that chemical shift data which are important indicators of the bonding situation in the molecules of interest are now accessible for solids. More direct information about the electronic structure of solids is thus

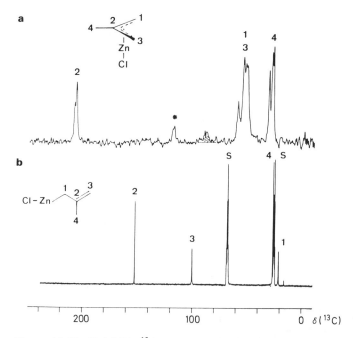

Figure 10.45   50.3 MHz ^{13}C n.m.r. spectra of 2-methallyl zinc chloride in the solid (CP/MAS) (a) and in THF solution (b) and the η^3- and η^1-structures derived from these spectra; S = solvent signals (after Ref. 22)

available. In particular charge density differences or changes can be easily recognized in the n.m.r. spectra. Furthermore, the number of the resonance lines measured is directly connected to molecular symmetry and to the number of equivalent molecules in the unit cell of the crystals.

For ^{13}C n.m.r. spectroscopy one can usually assume that the structure of a particular compound in solution and in the solid is identical if a comparison between solid state and solution n.m.r. data yields differences in the δ-values which are smaller than 5 ppm. Significant structural alterations are indicated by larger differences in the δ-values. Sometimes solvation effects, which are absent in the solid, may stabilize a certain structure in solution and spectral differences between liquid and solid state n.m.r. spectra are found. This aspect is shown in Figure 10.45 with the spectra of 2-methallyl zinc chloride where not only the position, but also the number of ^{13}C signals, clearly indicate that the solution structure is different from the structure in the solid.

Alkalides and electrides (**204, 205**, respectively) are new materials with interesting properties which are prepared from alkaline metals and crown ethers. In these systems, for example, the caesium cations are bound to the crown ether as guest ions, while the caesium anions or even free electrons are highly mobile as counter ions. For the preparation of such structures, the experimental conditions must be met very accurately and the success of the synthesis has to be controlled. For this purpose high resolution ^{133}Cs solid state n.m.r. spectroscopy is well suited because it yields for the two species $Cs^+(18\text{-crown-}6)_2/e^-$ and $Cs^+(18\text{-crown-}6)_2/Cs^-$ three well separated signals (Figure 10.46). Newly synthesized products can thus be easily tested with respect to their composition. ^{133}Cs has a natural abundance of 100% and a relatively small quadrupolar moment of -3×10^3 b and is, despite its high spin ($I = 7/2$), a suitable nucleus for these MAS n.m.r. measurements which can be performed without cross polarization.

Alkalide   Electride

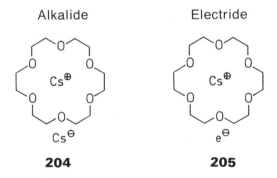

204   **205**

A related example is provided by the question of the structure of phosphorus(V)chloride, which has been solved on the basis of the charge density dependence of the ^{31}P resonance with the help of the ^{31}P MAS n.m.r. relatively easily in favour of the ionic structure $PCl_4^+[PCl_6^-]$, because two ^{31}P resonances, separated by 377 ppm, were found.

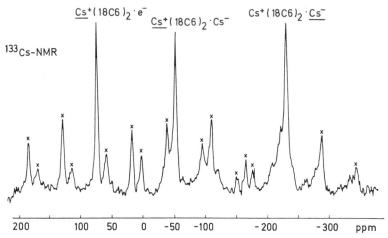

Figure 10.46 ^{133}Cs MAS n.m.r. spectrum of crystalline material from a solution of cesium and 18-crown-6 (2:3) in a polar amine/ether mixture. Rotational side bands are marked (x), the ppm scale refers to $Cs^+(H_2O)$; the line widths are 100–200 Hz (after Ref. 23)

In the field of inorganic solids and ceramic materials ^{29}Si MAS n.m.r. has become an important tool. It was found that investigations of zeolites and alumosilicates yield interesting structural information. The chemical shift is sufficiently large to distinguish the environment of different ^{29}Si nuclei with one to four aluminium atoms (Figure 10.47). Such studies are of particular interest because the X-ray method, where similar diffraction patterns are observed for Si and Al, is difficult to apply to these materials. Furthermore, it is difficult to grow crystals which are sufficiently large for X-ray investigations. With n.m.r. it has even been possible to reveal ^{29}Si–^{29}Si connectivities in zeolites using the 2D INADEQUATE experiment.

Another aspect of high-resolution solid state n.m.r. spectroscopy which attracts attention is the study of dynamic processes in solids. The basis for such investigations is the same as in liquids (see Chapter 9): the dependence of the n.m.r. line shape on the lifetime of nuclei in positions with different Larmor frequencies. Temperature dependent studies reveal the mechanisms and the activation parameters for conformational changes and other intramolecular processes by line shape analysis. Surprisingly, such investigations showed that, for example, ring inversion in cyclohexane (cf. p. 361ff.) has practically the same barrier in the solid as in the solution. The packing in the molecular crystals of organic compounds is thus not very different from solvation in solution. This high mobility even leads to so-called plastic crystals for a number of compounds, which yield high resolution n.m.r. spectra even without particular provisions. Examples are fluorcyclohexane, adamantane or P_4S_3. Finally, as in liquids, two-dimensional exchange spectroscopy can be used in solids successfully to study dynamic processes.

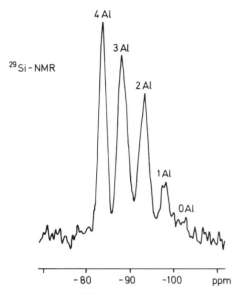

Figure 10.47 ²⁹Si-MAS-spectrum of the zeolite analcite with individual lines for different silicone environments (after Ref. 24)

9. Nuclear magnetic resonance in biology and medicine

Aside from other topics, the most spectacular applications of n.m.r. spectroscopy in the last two decades have been reported from the areas of biochemistry, biology and even medicine. In this respect new developments are of particular interest: spectroscopy with living objects (*in-vivo* n.m.r.) and n.m.r. imaging (nuclear spin tomography). Both aspects will be described in the following sections.

9.1 *In-vivo* NMR SPECTROSCOPY

The structure of *biomolecules*, an expression used for compounds which play an important role in biochemistry, was already a target of n.m.r. spectroscopic investigations in the early days of CW n.m.r. The generally complex structure and the high molecular weight of most of these compounds, however, severely limited the success of such measurements at the low field strengths available because of low sensitivity and insufficient spectral dispersion. Structural analysis of biopolymers like peptides, proteins or nucleic acids and polysaccharides were beyond the borders of the experimental tools which were at that time at the disposal of the n.m.r. spectroscopist. The introduction of cryomagnets with high magnetic fields and the modern measuring techniques of pulse Fourier transform spectroscopy today have changed this situation profoundly and n.m.r. spectroscopy is now used with great success for structural investigation of complicated biological macromolecules.

While for such measurements principally the same experimental techniques are used as in organic chemistry for the study of smaller molecules, the development of special experimental methods has opened up completely new fields for the application of n.m.r. spectroscopy in the life sciences. We are now able to study biochemical processes *in-vivo*, that is with the living object, for example a cell culture or an intact organism. While cells can be studied as cell suspensions in the n.m.r. sample tube, tissue investigations or the study of cell metabolism in perfused organs or living organisms require special transmitter and receiver coils which must be designed according to the needs of the experiment. In many cases, especially for larger objects, surface coils are used where the radiofrequency penetrates into the object and which can detect a weak signal. The particular arrangement of r.f. coils which yields optimal signal excitation and detection are, together with the object under study, placed into the magnetic field of special widebore cryomagnets which have openings up to 50 cm in diameter. In this way n.m.r. spectra can be obtained of certain areas of the object of interest at high B_0-fields. The objects studied extend from perfused organs (heart, liver) of animals to the extremities of human probants (arm, leg).

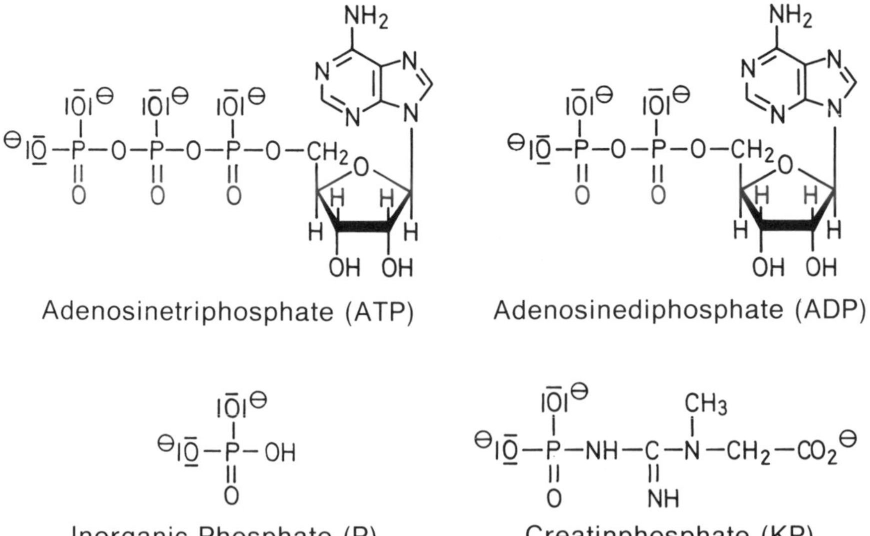

Adenosinetriphosphate (ATP)

Adenosinediphosphate (ADP)

Inorganic Phosphate (P)

Creatinphosphate (KP)

Figure 10.48 Molecules important for the energy conversion in the cell: ATP (adenosinetriphosphate, energy rich phosphate) is converted with energy production (-50 kJ mol^{-1}) by hydrolysis into ADP (adenosinediphosphate) and inorganic phosphate (P). Creatine phosphate (CP) is a storage material in equilibrium with ATP: CP + ADP $\rightleftharpoons$ ATP + creatine. It is used for ATP production if the ATP synthesis by respiration is blocked, for example in the case of insufficient blood supply

From the nuclei which are available for such investigations as sensitive n.m.r. probes, the phosphorous nucleus ^{31}P is especially well suited. It has a natural abundance of 100% and a high sensitivity $[v(^{31}P)/v(^{1}H) = 1:2]$. Protons, because of the abundance of water in living matter, are in general much too concentrated to yield simple spectra and carbon-13 is not sensitive enough due to its low natural abundance. Phosphorous atoms are part of several important products of cell metabolism. For example, the molecules shown in Figure 10.48 are important components in the reaction scheme which serves for the energy production in the cell. Since these compounds contain only few chemically different phosphorous atoms, relatively simple high resolution ^{31}P n.m.r. spectra are obtained and the assignment of the observed ^{31}P resonances to ATP and other key products, shown in Figure 10.49 (p. 456), is rather straightforward.

The excellent quality of the spectra which can be obtained (Figure 10.49) allows us to detect metabolic changes by application of ^{31}P n.m.r. spectroscopy. It has been shown, for example, that the position of the ^{31}P resonance of inorganic phosphate in the cell is a sensitive indicator for the intracellular pH value (Figure 10.49b). As shown in Figure 10.50 (p. 457) with the ^{31}P n.m.r. spectrum from a human forearm, muscle contraction under iscamic conditions, that is during insufficient blood supply, leads, aside from other characteristic changes, to a decrease in pH. After blood supply is stopped by tieing up of the arteries, the storage substance creatinephosphate is consumed. At the same time, lactate formation during glycolysis leads to a decrease in pH. After restoring the blood and thus the oxygen supply the normal state is slowly reached again.

In addition to ^{31}P, ^{13}C n.m.r. spectroscopy has been increasingly used in recent years, in combination with ^{13}C labelling, for *in vivo* n.m.r. investigations. In this way the number of biochemical reactions and pathways, which can be studied by n.m.r., has considerably grown. Of particular interest in this respect are studies of drug metabolism. Similarly, the clinical investigation of body fluids with high-field instruments is a promising research area. Because of its higher sensitivity, work in this field is based primarily on ^{1}H n.m.r. spectroscopy.

9.2   NMR IMAGING

Up to now we have dealt with n.m.r. spectra that furnish information on individual molecules, either in the liquid or—as referred to briefly in the last section—in the solid state. In this sense n.m.r. spectra may be regarded as pictures or images of molecules since the spectroscopist performs an intellectual transformation of the recorded data (chemical shifts, spin–spin coupling constants, relaxation times, etc.) into an image of the molecule, mostly the structural formula.

A different kind of *n.m.r. imaging* has been developed in recent years, where structures well above the molecular level are studied and a real two-dimensional representation of the object is derived from the spectral data. This field—also known

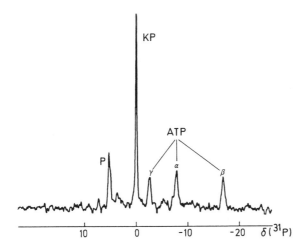

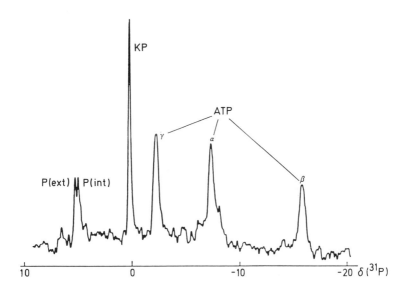

Figure 10.49 (a) ^{31}P n.m.r. spectrum of a frog muscle (73.8 MHz); P: inorganic phosphate; KP: creatine phosphate; ATP: phosphorous resonances of adenosine triphosphate (after Ref. 25); (b) ^{31}P n.m.r. spectrum (73.8 MHz) of a perfused rat heart. Two signals are observed for inorganic phosphate since the pH in the cells and in the perfusion liquid is slightly different. The α- and γ-signals of ATP also contain contributions from the α- and β-phosphorous signals of ADP, which in this experiment are not separated due to the low field strength (after Ref. 25, 26)

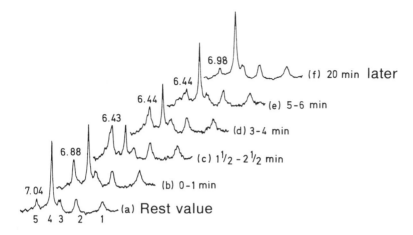

Figure 10.50 32.5 MHz ^{31}P n.m.r. spectrum of a human forearm. The first spectrum was recorded at rest, prior to exercise. Subsequent spectra (b) and (c) were recorded under ischaemic conditions with exercise in the period 0–1.5 min. After 3 min arterial flow was restored and spectra (d)–(f) show the relaxation process to normal conditions (after Ref. 25, 27)

as n.m.r. zeugmatography* or *spin mapping*—has promising prospects for applications in biology and even in medicine, and it seems appropriate to discuss these developments in order to provide information about the underlying experimental principles.

In 1973 Lauterbur first produced an n.m.r. image by applying a field gradient ΔB over the sample cell. In such an experiment, nuclei in different regions of space will experience different external fields $B_0 + \Delta B_i$ correspondingly, their resonance frequencies will differ. In other words, the field gradient introduces a chemical shift between nuclei that would be isochronous in the conventional n.m.r. experiment.

If the line width of the individual resonances is small compared to the field gradient, signals from different parts of the object—more precisely those from nuclei residing in various planes perpendicular to the gradient—can be discriminated. An intensity plot will then yield information about the spatial spin distribution.

One of the first n.m.r. images produced by this technique was that of two capillaries of 1 mm diameter each filled with H_2O and fitted into a 5-mm sample tube containing D_2O/H_2O. The construction of a two-dimensional picture of this arrangement from the ^{1}H-n.m.r. signals received is schematically illustrated in Figure 10.51a. The actual result obtained is shown in Figure 10.51b. Here the n.m.r. data were processed by a computer using image reconstruction programs that allowed the printing of a spin density map for the protons in the sample. The two capillaries are clearly recognized.

---

* From the Greak $\zeta\varepsilon\upsilon\mu\alpha$ 'that which joins together'

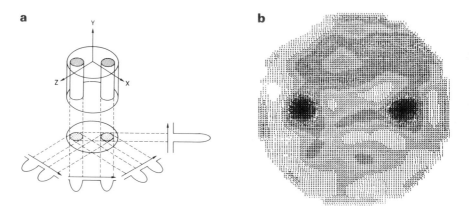

Figure 10.51 (a) The n.m.r. information obtained from two capillaries containing H_2O using four different experiments with field gradients in directions indicated by the arrows (after Ref. 28); (b) n.m.r. image produced from an arrangement shown schematically in (a); two 1 mm capillaries filled with H_2O were fitted in a cylindrical sample tube of 4.2 mm inner diameter that contained a mixture of D_2O and H_2O (from Ref. 29)

Further refinements of the imaging technique have been achieved in recent years. Using alternating field gradients along orthogonal directions, an n.m.r.-sensitive volume in space can be produced at the intersection of the three null planes of these gradients. The n.m.r. signal from this selected region is then detected, while those from other parts of the sample are rejected. Scanning the object by this sensitive point yields the data necessary for image construction. Similarly, using only two time-dependent gradients a sensitive line is produced for the detection of n.m.r. signals by FT-methods, thereby greatly reducing the time necessary for the experiment. Finally, these data are subjected to computer image processing. As an example for the progress made by these more recent developments the n.m.r. image of a cross section through an intact lemon is shown in Figure 10.52. Our earlier comparison of n.m.r. to a camera here certainly finds its justification!

It is important to emphasize that the NMR images of the internal structure of these and other objects are obtained in a non-destructive manner. The technique is therefore comparable to röntgenography. Unlike X-rays, however, the magnetic and r.f. fields used are, to our present knowledge, without harm to living cells and this makes n.m.r. imaging an important tool for biology and medicine. The construction of whole-body magnets with magnet bores of up to 50 cm has paved the way for investigations on humans and n.m.r imaging today is an indispensable aid for medical diagnosis. The enormous improvement in image quality as well as a considerable shortening of the time necessary to produce these images has placed n.m.r. imaging units today at the side of computer tomographs in the radiology departments of the big hospitals. In particular brain investigations have been

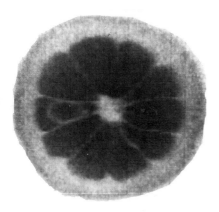 

Figure 10.52   Left: Proton n.m.r. image of the cross section of an intact lemon, diameter approximately 7 cm. The section defined in the image was about 2 mm thick. Recording time 5 minutes, resolution about 0.5 mm. The image shows strong signals from the fluid in the segments and weaker signals from the surrounding skin. The harder septum planes which partition the segments are well resolved. Right: Photograph of the same lemon cut along the selected plane after the n.m.r. image was produced. Note the pip that is also shown in the n.m.r. image (after Ref. 30)

advanced by the new n.m.r. technique which yields, as demonstrated in Figure 10.53, detailed information about brain structure and facilitates the localization of pathological regions as tumours or damaged blood vessels. A noteworthy advantage of the n.m.r. method over other techniques is the possibility to produce cross sections in every direction which is of interest. In addition, especially adapted surface coils allow more detailed investigations of certain regions of the human body. Furthermore, aside from the chemical shift other parameters, as for example relaxation times, can be used to produce an image and other sensitive nuclei like phosphorous-31 or sodium-23 can be employed. The large versatility of n.m.r. in general classifies the n.m.r. imaging method as one of the most promising techniques in radiology.

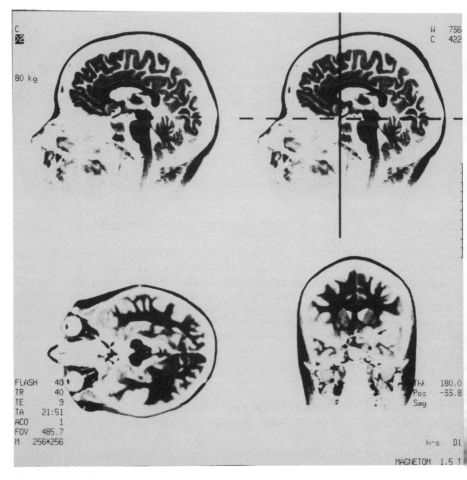

Figure 10.53   N.m.r. images of the human skull at 1.5 T (after Ref. 31)

10. References

1. Varian Associates, Palo Alto, Calif.
2. L. F. Johnson, *Anal. Chem.,* **43**, 28A (1971).
3. P. Bast, *PhD thesis*, University of Siegen 1988.
4. F. A. L. Anet, C. H. Bradley, M. A. Bradley, M. A. Brown, W. L. Mock, and J. H. McCausland, *J. Am. Chem. Soc.,* **91**, 7782 (1969).
5. W. Holzer, *Tetrahedron,* **47**, 9783 (1991).
6. H. Uzar, private communication.
7. H. Günther, P. Schmitt, H. Fischer, W. Tochtermann, J. Liebe, and C. Wol, *Helv. Chim. Acta,* **68**, 801 (1985).
8. W. Andres, *PhD thesis*, University of Siegen 1991.
9. K. G. R. Pachler, and P. L. Wessels, *J. Magn. Reson.,* **12**, 337 (1973).
10. M. Schumacher, and H. Günther, *Chem. Ber.,* **116**, 2001 (1983).

11. H. P. Wessel, G. Englert, and P. Stangier, *Helv. Chim. Acta,* **74**, 682 (1991).

12. J. Bargon, H. Fischer, and U. Johnson, *Z. Naturforsch.,* **22a**, 1551 (1967).

13. G. L. Closs, and L. E. Closs, *J. Am. Chem. Soc.,* **91**, 4549 (1969).

14. E. de Boer, and C. McLean, *Mol. Phys.,* **9**, 191 (1965).

15. *Technical Bulletin,* Nr. 4, Varian Associates, Palo Alto, Calif.

16. Review article (m), see below.

17. H. Fischer, *diploma thesis,* University of Cologne 1978.

18. A. Saupe, *Angew. Chem.,* **80**, 99 (1968).

19. C. A. Fyfe, *Solid State NMR for Chemists,* CFC Press, Guelph, 1983.

20. C. S. Yannoni, *Acc. Chem. Res.,* **15**, 201 (1982).

21. H. Grondey, *PhD thesis,* University of Siegen 1988; H. Grondey and R. Benn, unpublished.

22. R. Benn, H. Grondey, H. Lehmkuhl, H. Nehl, K. Angermund, and C. Krüger, *Angew. Chem.,* **99**, 1303 (1987); *Angew. Chem. Int. Ed. Engl.,* **26**, 1300 (1987).

23. S. B. Dawes, A. S. Ellaboudy, and J. L. Dye, *J. Am. Chem. Soc.,* **109**, 3508 (1987).

24. R. E. Wasylishen, and C. A. Fyfe, in F2, **12**, 1 (1982).

25. D. G. Gadian, *Nuclear Magnetic Resonance and its Applications to Living Systems,* Clarendon Press Oxford 1982.

26. T. H. Grove, J. J. H. Ackermann, G. K. Radda, and P. J. Bore, *Proc. Natl. Acad. Sci. USA,* **77**, 299 (1980).

27. B. D. Ross, G. K. Radda, D. G. Gadian, G. Rocker, M. Esiri, and J. Falconer-Smith, *New Eng. J. Med.,* **304**, 1338 (1981).

28. P.C. Lauterbur, *Nature,* **242**, 190 (1973).

29. P. C. Lauterbur, *Pure Appl. Chem.,* **40**, 149 (1974).

30. E. R. Andrew, private communication.

31. Siemens AG

REVIEW ARTICLES

(a) L. F. Johnson, Superconducting Magnets, see Ref. 2 above.

(b) NMR at Very High Field, J. B. Robert, Ed., F3, **25**, 1 (1991).

(c) P. D. Kennewell, Applications of the Nuclear Overhauser Effect in Organic Chemistry, *J. Chem. Educ.,* **47**, 278 (1970).

(d) G. E. Bachers, and T. Schaefer, Applications of the Intramolecular Nuclear Overhauser Effect in Structural Organic Chemistry, *Chem. Rev.,* **71**, 617 (1971).

(e) C. L. Van Antwerp, Experimental Determination of the Nuclear Overhauser Effect, *J. Chem. Educ.,* **50**, 638 (1973).

(f) J. K. M. Sanders, and J. D. Mersh, Nuclear Magnetic Double Resonance; The Use of Difference Spectroscopy, in F5, **15**, 353 (1983).

(g) A. M. Gronenborn, and G. M. Clore, Investigation of the Solution Structure of Short Nucleic Acids Fragments by Means of Nuclear Overhauser Enhancement Measurements, in F5, **17**, 1 (1985).

(h) K. E. Kövér, and G. Batta, Theoretical and Practical Aspects of One- and Two-dimensional Heteronuclear Overhauser Experiments and Selective ^{13}C T_1-Determinations of Heteronuclear Distances, in F5, **19**, 223 (1987).

(i) S. H. Pine, Chemically Induced Dynamic Nuclear Polarization, *J. Chem. Educ.,* **49**, 664 (1972).

(j) R. G. Lawler, Chemically Induced Dynamic Nuclear Polarizations, in F5, **9**, 145 (1973).

(k) R. v. Ammon, and R. D. Fischer, Shift Reagents in NMR Spectroscopy, *Angew. Chem.,* **84**, 737 (1972); *Angew. Chem. Int. Ed.,* **11**, 675 (1972).

(l) M. R. Petersen, Jr., and G. H. Wahl, Jr., Lanthanide NMR Shift Reagents, *J. Chem. Educ.,* **49**, 790 (1972).

(m) O. Hofer, The Lanthanide Induced Shift Technique. Application in Conformational Analysis, *Topics in Stereochemistry,* **9**, 111 (1976).

(n) G. R. Sullivan, Chiral Lanthanide Shift Reagents, *Topics in Stereochemistry,* **10**, 287 (1978).

(o) J. Reuben, and G. A. Elgavish, *Handbook on the Physics and Chemistry of Rare Earths*, K. A. Gschneider, Jr., and L. Eyring, Eds., North-Holland (1979).

(p) J. Reuben, Paramagnetic Lanthanide Shift Reagents in NMR-Spectroscopy: Principles, Methodology and Applications, in F5, **9**, 1 (1973).

(q) F. Inagaki, and T. Miyazawa, NMR Analyses of Molecular Conformations and Conformational Equilibria with the Lanthanide Probe Method, in F5, **14**, 67 (1982).

(r) A. Saupe, Recent Results from the Field of Liquid Crystals, *Angew Chem.,* **80**, 99 (1968); *Angew. Chem. Int. Ed. Engl.* **7**, 97 (1968).

(s) P. Diehl, and C. L. Khetrapal, NMR Studies of Molecules Oriented in the Nematic Phase of Liquid Crystals, in F3, **1**, 1 (1969).

(t) C. L. Khetrapal, A. C. Kunwar, A. S. Tracey, and P. Diehl, Nuclear Magnetic Resonance Studies in Lyotropic Liquid Crystals, in F3, **9**, 1 (1975).

(u) F. C. Nachod, and I. J. Zuckerman (1971) Eds., *Determination of Organic Structures by Physical Methods*, Vol. 4, Academic Press, New York.

(v) R. E. Wasylishen, and C. A. Fyfe, High-resolution NMR of Solids, in F2, **12**, 1 (1982).

(w) C. S. Yannoni, High-Resolution NMR in Solids: The CPMAS Experiment, *Acc. Chem. Res.,* **15**, 201 (1982).

(x) J. R. Lyerla, C. S. Yannoni, and C. A. Fyfe, Chemical Applications of Variable-Temperature CPMAS-NMR Spectroscopy in Solids, *Acc. Chem. Res.,* **15**, 208 (1982).

(y) W. S. Veeman, Carbon-13 Chemical Shift Anisotropy, in F5, **16**, 193 (1984).

(z) R. Voelkel, High-Resolution Solid State ^{13}C NMR Spectroscopy of Polymers, *Angew. Chem. Int. Ed. Engl.,* **27**, 1468 (1988).

(aa) B. Blümich, and H. W. Spiess, Two-Dimensional Solid State NMR Spectroscopy: New Possibilities for the Investigation of the Structure and Dynamics of Solid Polymers, *Angew. Chem. Int. Ed. Engl.,* **27**, 1655 (1988).

(bb) H. Saito, and I. Ando, High-Resolution Solid-State NMR Studies of Synthetic and Biological Macromolecules, in F2, **21**, 210 (1989).

(cc) P. G. Morris, NMR Spectroscopy in Living Systems, in F2, **20**, 1 (1988).

11 CARBON-13 NUCLEAR MAGNETIC RESONANCE SPECTROSCOPY

The element carbon consists of the stable isotopes ^{12}C and ^{13}C with 98.9% and 1.1% natural abundance, respectively. Only the ^{13}C nucleus has a magnetic moment and as a u,g nuclide the spin $I = \frac{1}{2}$, while the ^{12}C nucleus of the major isotope is non-magnetic. Nuclear magnetic resonance spectroscopy of carbon, which quite naturally is of great interest for organic chemistry, is, therefore, limited to the investigation of carbon-13.

The magnetic moment of carbon-13 is smaller than that of the proton by a factor of 4. Consequently, carbon-13 is less sensitive for the n.m.r. experiment than the proton. Furthermore, the low natural abundance renders its detection more difficult and ^{13}C n.m.r. spectroscopy is by far less sensitive than ^1H n.m.r. If one defines, on the basis of equations (1.5) and (1.13) (pp. 2, 12), the receptivity of a nucleus X for the n.m.r. experiment at constant B_0 field relative to the proton $D^p(X)$, according to

$$D^p(X) = \frac{\gamma_X^3}{\gamma_p^3} \frac{N_X I_X(I_X + 1)}{N_p I_p(I_p + 1)} \tag{11.1}$$

where γ is the magnetogyric ratio, N the natural abundance and I the spin quantum number, one obtains with the appropriate data $D^p(^{13}C) = 1:5700$. Quite a number of experimental developments were thus necessary in order to counterbalance this disadvantage and to pave the way for a broad application of ^{13}C n.m.r. spectroscopy. On the other hand, it must be remembered that the low natural abundance of ^{13}C also has an advantage: ^{13}C n.m.r. spectra are easily analysed because homonuclear $^{13}C,^{13}C$ spin–spin coupling constants are absent and the ^1H n.m.r. spectra of organic molecules are not disturbed by heteronuclear ^{13}C, ^1H coupling.

1. Historical development and most important areas of application

The development of ^{13}C n.m.r. spectroscopy can be divided into three stages. During the pioneer time, that is in the nineteen sixties, when only low-field spectrometers without field frequency stabilization were available, spectra had to be recorded with high sweep rates in order to avoid saturation effects (rapid passage method). Carbon-13 labelled material and large volume sample cells (diameter up to 15 mm), were applied in order to improve the low signal-to-noise ratio that results from the low natural abundance and the low sensitivity of ^{13}C.

The second period of ^{13}C n.m.r. spectroscopy started as data accumulation and proton broadband decoupling were introduced. Digital memory units could be used after the spectrometer was equipped with a field frequency lock and ^1H-decoupling led to a two-fold intensity gain: on one hand through the collapse of multiplet structures and on the other through the nuclear Overhauser effect (see Chapter 10). This is illustrated in Figure 11.1 with spectra obtained for pyridine under different experimental conditions.

A further and, without doubt, by far the most important step for the improvement of the signal-to-noise ratio was the introduction of Fourier transform spectroscopy. The possibility to accumulate a large number of spectra in a relatively short time provided the basis for routine applications of ^{13}C n.m.r. spectroscopy. Spectra could now be recorded for smaller samples and within reasonable experiment time. Today, even large molecules can be measured under conditions not very different from those of routine ^1H n.m.r., as Figure 11.2 demonstrates with the ^{13}C n.m.r. spectrum of vitamin B_{12}.

Not surprisingly, ^{13}C n.m.r. spectroscopy has since then developed rapidly into a routine method for chemical structure determinations. If one remembers that the organic chemist is interested primarily in the molecular carbon skeleton, evidently a ^{13}C spectrum then yields structural information much more directly than a proton spectrum. In particular quaternary carbons, as those of many functional groups ($C\equiv N$, $C=O$, $C=NR$ etc.), are now detectable. The amount of structural n.m.r. data has thus considerably increased. Furthermore, the large ^{13}C chemical shift range (approximately 250 ppm) and the small line width of the ^{13}C signals (0.5 Hz or less), effectively increases spectral resolution, that is signal dispersion. Carbon-13 n.m.r. is thus the method of choice for structural investigations of complex organic

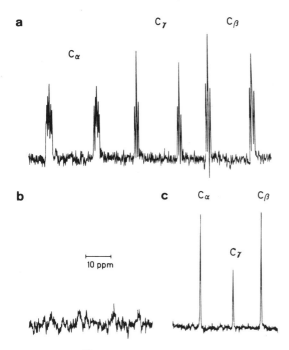

Figure 11.1 ^{13}C spectra of pyridine: (a) CW spectrum at 22.63 MHz after 64 accumulations of 320 seconds acquisition time each (after Ref. 1); (b) result of a frequency sweep experiment at 15 MHz; (c) as (b), however, with proton decoupling (after Ref. 2)

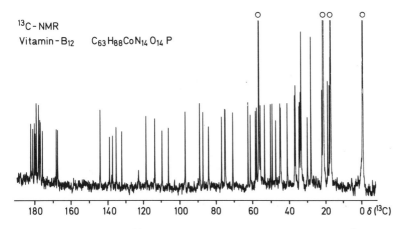

Figure 11.2 100 MHz ^{13}C FT n.m.r. spectrum of vitamin B$_{12}$ with ^{1}H-decoupling; at o signals of the reference compound trimethylsilylpropanesulphonate (TS)

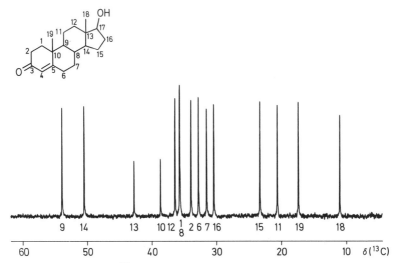

Figure 11.3   50 MHz ^{13}C n.m.r. spectrum of the steroid testosterone in the aliphatic carbon region with ^{1}H-decoupling; the resonances of C-3, C-4, and C-5 at higher frequencies are not shown

molecules such as natural products and synthetic as well as biological oligomers and macromolecules. In the field of steroids, to cite only one example, intensive investigations led to a situation where individual molecules can be recognized by their characteristic ^{13}C n.m.r. finger prints (Figure 11.3).

In this context it is important to emphasize that ^{13}C n.m.r. spectroscopy has by no means replaced proton n.m.r. spectroscopy, which is still the most widely used spectroscopy technique. Both methods are rather complementary in a very effective way. In practice, the combination of ^{1}H and ^{13}C n.m.r. data forms the basis for any structural investigation in organic and organometallic chemistry and yields in most cases sufficient information for solving the problem under study, especially if mass spectroscopic data are also available.

There are a number of general aspects of ^{13}C n.m.r. spectroscopy which require special consideration and which will now be discussed. The large chemical shift range renders ^{13}C attractive as a probe for the study of *dynamic processes*. Under the conditions of proton decoupling in most cases simple exchange systems between two sites (A $\rightleftharpoons$ B) are present which are readily analysed. Furthermore, the dynamic process very often leads to several two-sites exchange systems that allow separate lineshape analyses which increases the precision of the results. It must be remembered, however, that the lineshape changes depend on the frequency difference between the nuclei in question (in Hz). Because of its small magnetogyric ratio, ^{13}C is thus not as effective as one could expect on the basis of the chemical shift scale in ppm. Nevertheless, because of the generally greater sensitivity of ^{13}C shielding constants to structural variations, the δv values in ^{13}C n.m.r. are usually larger by a factor of 10 than the corresponding chemical shift differences in the proton spectrum.

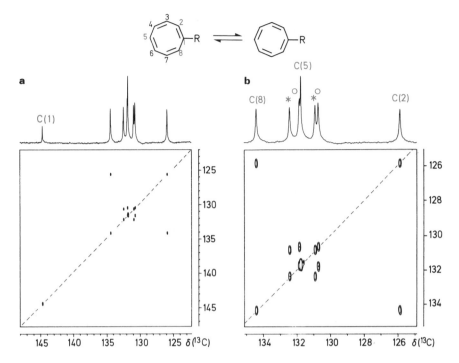

Figure 11.4   Detection of the double bond shift in *n*-butyl-cyclooctatetraene by two-dimensional ^{13}C exchange spectroscopy at room temperature; (a) total spectral range; as expected, the quaternary carbon C(1) shows no cross peaks because it does not participate in the exchange process; (b) the same is true for C(5), which gives rise to a sharp singlet in the high-field region; the signals of C(2) and C(8) are connected by cross peaks; the remaining signals have not been assigned (after Ref. 3)

In the area of dynamic effects *two-dimensional ^{13}C exchange spectroscopy* has become important. For ^{13}C nuclei, because of their low natural abundance, the application of the pulse sequence equation (9.25) (p. 348) does not lead to cross peaks which arise from homonuclear Overhauser effects. Therefore, the results are usually less ambiguous than for 2D ^{1}H exchange spectroscopy. An example of a 2D ^{13}C exchange spectrum is shown in Figure 11.4, where the double bond shift in *n*-butyl-cyclooctatetraene is detected.

Another application of significance is *^{13}C labelling*. The elucidation of reaction mechanisms in organic chemistry or biochemistry, that previously was based exclusively on the use of the radioactive isotope ^{14}C, can now be accomplished with the help of ^{13}C n.m.r. spectroscopy. As before, the synthesis of the labelled compound is unavoidable, but one is spared the often difficult and not always unequivocal degradation of the isolated reaction products since, when using ^{13}C resonance spectroscopy, the position of the carbon atom in question can be determined easily. In most applications ^{13}C enrichment is employed, but ^{13}C depletion, that is ^{12}C labelling, has also been recognized as a useful technique. In some cases, even ^{13}C double labelling may be of advantage, because then $^{13}C,^{13}C$ spin–spin coupling constants can be used for the analysis. Such experiments are of interest for mechanistic studies where one wants to prove if a certain C–C bond has been cleaved during the reaction of interest. Only for experiments which need the highest sensitivity, is labelling with radiocarbon (^{14}C) still superior to ^{13}C n.m.r.

Finally, beside chemical shifts and coupling constants ($^{13}C,^{13}C$, $^{13}C,^1H$, $^{13}C,X$ where X = ^{19}F, ^{31}P, etc.) a third ^{13}C n.m.r. parameter is of importance: the ^{13}C spin–lattice relaxation time. Through FT n.m.r., T_1 measurements are facilitated and can be carried out almost routinely. Such an experiment using the inversion-recovery technique (cf. p. 235) is shown in Figure 11.5.

Since nuclear relaxation rates ($1/T_1$) depend on a variety of molecular properties—in the case of dipolar relaxation mainly on molecular dynamics—these parameters are important sources of information on molecular motion. In general, however, their interpretation is less straightforward than is usually the case with chemical shifts or coupling constants and must be based on physical models for molecular dynamics. In general, T_1 data are less important for structural research than chemical shifts and coupling constants.

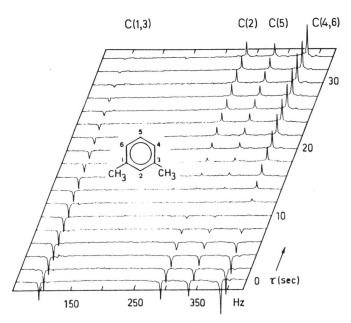

Figure 11.5 The determination of the longitudinal relaxation time, T_1, of the ^{13}C nuclei in the benzene ring of m-xylene. For each spectrum impulse sequences of 180 °-τ-90 ° were used and τ (in seconds) was increased in 2-s intervals; the following T_1 values (in s) have been determined: C(1,3) 52 ± 5; C(4,6) 16.5 ± 2; C(2) 19 ± 2; C(5) 20 ± 2 (after Ref. 4)

2. Experimental aspects of carbon-13 nuclear magnetic resonance spectroscopy

The experimental aspects of FT n.m.r. have been dealt with in some detail in Chapter 9, and the general features discussed there equally well apply to ^{13}C FT n.m.r. Spectra are obtained with the ^{13}C n.m.r. signal of tetramethylsilane as internal reference (see Section 2.2) and a heteronuclear lock system, usually employing the ^{2}H resonance of the solvent CDCl$_3$. ^{1}H broadband decoupling is applied and chemical shift determinations are thus a routine matter, as the line frequencies are printed out directly by the computer. Even if the maximum NOE (200%, see p. 399) is not always observed in practice because other relaxation mechanisms may successfully compete with dipolar relaxation, the signal-to-noise ratio is high enough to run ^{13}C n.m.r. spectra of milligram samples if high-field instruments (7.2 T or more) are available. There are a few problems, however, originating from ^{1}H decoupling which need special comment. Firstly, the elimination of line splittings prevents the measurements of ^{13}C,^{1}H coupling constants, and valuable experimental information is thus lost. Secondly, the nuclear Overhauser effect leads to intensity distortions and the integration of these spectra becomes questionable. Finally, the assignment of the n.m.r. signals to specific carbon atoms of the particular structure is by no means obvious.

2.1 GATED DECOUPLING

In order to deal with these shortcomings, special experimental techniques are available. The application of a method known as *gated decoupling* successfully handles spin–spin coupling and intensity distortions. It is based on the fact that different time scales are characteristic for the decoupling experiment and the nuclear Overhauser effect. Whereas the latter is governed by spin–lattice relaxation and thus evolves and decays within seconds, decoupling sets in or vanishes almost immediately after the B_2 field has been switched on or off. This difference can be used to advantage in two experimental sequences illustrated in Figure 11.6. In the first (Figure 11.6a), the decoupler is gated so that it is on during the delay time between different pulses and switched off during data acquisition. As a result, the spectra are not decoupled, but at the same time most of the Overhauser enhancement is retained; ^{13}C,^{1}H coupling constants can thus be measured without the necessity of sacrificing the NOE completely.

If, on the other hand, correct integrals are of interest, the second sequence of experiments (Figure 11.6b) is applied. Here, the decoupler is switched on during data acquisition, but switched off during the remaining time before the next pulse. ^{13}C,^{1}H spin–spin coupling is thus eliminated without producing nuclear Overhauser enhancement, primarily because the integrals depend on the initial value of the time domain function. In addition, or alternatively, so-called shiftless relaxation reagents can be used to eliminate an unwanted NOE. Because of their paramagnetic moment, these compounds, for example chromium acetylacetonate, provide an effective

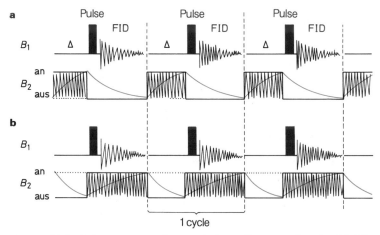

Figure 11.6 Pulse sequence for gated decoupling experiments. (a) The decoupler is switched off during data acquisition; part of the NOE enhancement is retained, ^{13}C spectra are ^{1}H-coupled. (b) The decoupler is switched on during data acquisition; the NOE is suppressed, ^{13}C spectra are decoupled. Red trace signifies NOE

mechanism for ^{13}C spin–lattice relaxation that suppresses dipolar relaxation which is the basis for the NOE (cf. p. 399). It is important, however, to ascertain that the addition of the paramagnetic substance does not affect the ^{13}C chemical shift.

2.2 ASSIGNMENT TECHNIQUES

Independent methods for the correct assignment of ^{13}C signals to the molecular structure under study were from the start of prime interest to n.m.r. spectroscopists. Many of the older procedures—like off-resonance ^{1}H-decoupling (see p. 296)—have been replaced in the meantime by modern one- and two-dimensional techniques. The most important experiments which are used today will be discussed in the following sections. In this respect, two aspects have to be distinguished: On the one hand we have methods for *spectral editing* or *multiplicity selection*, which allow the resonances of quaternary, methine, methylene, and methyl carbons to be distinguished. This was achieved before by off-resonance ^{1}H-decoupling. On the other hand, in a second stage ^{13}C signals must be assigned to certain structural elements of the molecule which calls for additional experimental methods.

2.2.1 *Multiplicity selection with the heteronuclear spin echo experiment (SEFT, APT)*

Soon after the concept of evolution time was introduced, one recognized that it can also be used to develop new one-dimensional measuring techniques. On the basis of the heteronuclear spin echo experiment, for example, an assignment technique for ^{13}C signals was established which allows us to differentiate between the resonances of quaternary carbons, (C_q), and carbons of CH, CH_2, and CH_3 groups. It is known as the APT (*attached proton test*) or SEFT (*spin echo Fourier transform spectroscopy*) experiment and was an attractive alternative to the traditional method of off-resonance ^1H decoupling. Today, polarization transfer methods (see Section 2.2.2) have replaced the SEFT or APT technique for multiplicity selection, but it seems worthwhile to describe the older techniques in some detail because the underlying principle is of general importance.

As described on p. 281, during the heteronuclear spin echo experiment for AX spin systems, the A and the X magnetization, respectively, fan out in the x,y-plane due to scalar coupling. For AX, A_2X and A_3X systems (A = ^1H, X = ^{13}C) of CH, CH_2 and CH_3 groups, characteristic vector orientations result after a certain evolution time. This fact can be used for signal selection if $^{13}C,^1H$ coupling is removed during the detection time by ^1H-decoupling. Switching on the decoupler at the end of the evolution time conserves the particular vector orientation. Because of their large values (125–250 Hz, see below), this experiment is dominated by the one-bond coupling constants and the effect of the geminal and vicinal couplings, which are considerably smaller, can be neglected.

The time development of ^{13}C magnetization and signal intensity, $I(^{13}C)$, during the delay τ under the action of $J(^{13}C,^1H)$ is shown in Figure 11.7a. For ^1H-coupled ^{13}C nuclei it is governed by equations (11.2) to (11.4):

$$CH \;:\; I + I_0 \cos(\pi J \tau) \tag{11.2}$$

$$CH_2 \;:\; I = I_0 \cos^2(\pi J \tau) \tag{11.3}$$

$$CH_3 \;:\; I = I_0 \cos^3(\pi J \tau) \tag{11.4}$$

Two τ-values are of practical interest: $\tau = 1/2J$ leads to an elimination of all ^1H-coupled resonances, and $\tau = 1/J$ leads to phase selection with positive signals for C_q and CH_2 resonances and negative signals for CH and CH_3 resonances. The corresponding vector pictures are shown in Figure 11.7b and the pulse sequences of the SEFT experiment is given in Figure 11.7c. The $180^\circ_x(^{13}C)$ pulse of the spin echo experiment eliminates the effects of chemical shifts and transverse relaxation and the ^1H-decoupler determines the evolution of ^{13}C magnetization.

A practical application of this experiment is given in Figure 11.8 with a spectrum of 4-*tert*-butylcyclohexanone. The proton decoupled ^{13}C n.m.r. spectrum of this compound (Figure 11.8b) shows five signals of which two are nearly degenerate. The traditional off-resonance ^1H-decoupling experiment (Figure 11.8a) yields unambiguous results for the methine resonance, C(4), for one CH_2 resonance and for the signal of the quaternary carbon. However, for the closely spaced signals at highest field strong overlap prevents an assignment. The SEFT experiment, first run with $\tau = 1/2J$, confirms the resonance of the quaternary carbon (Figure 11.8c). A second experiment with an evolution time of $\tau = 1/J$ then shows additional signals for C(4) (negative phase), C(2) (positive phase) and at high field clearly two signals: One with negative and one with positive phase. Accordingly, these two resonances must belong to a CH_2 and a CH_3 group.

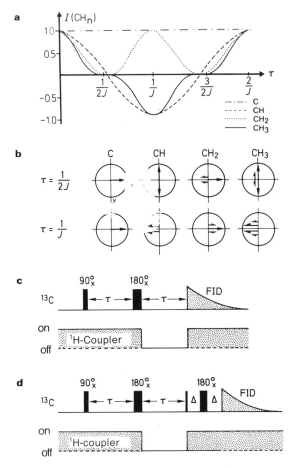

Figure 11.7 (a) Time dependence of the ^{13}C magnetization for CH, CH$_2$ and CH$_3$ groups modulated by scalar $^{13}C,^{1}H$ coupling; relaxation effects are neglected and the magnetization of quaternary carbons is not changed; (b) vector orientation for the evolution delays $\tau = 1/2J$ and $1/J$; (c) pulse sequence of the SEFT experiment; (d) the APT sequence, $\Delta \sim 1$ ms

Because 90° pulses require relatively long relaxation delays, the performance of the SEFT experiment can be improved if smaller excitation pulse angles are used. It is then necessary, however, to align the remaining ^{13}C z-magnetization again along the $+z$-axis by a short spin echo sequence after the evolution time and before signal detection (Figure 11.7d). For this experiment the acronym APT (attached proton test) has been coined.

Exercise 11.1 Verify the effect of the additional spin echo sequence in the APT experiment with the help of vector diagrams.

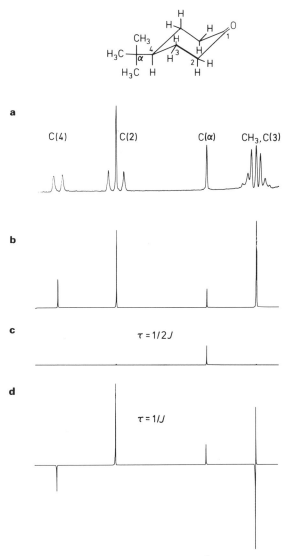

Figure 11.8 Signal assignment for the ^{13}C n.m.r. spectrum of 4-*tert*-butylcyclohexanone; (a) *off-resonance* ^{1}H-decoupling; (b) ^{1}H broadband decoupling; (c) SEFT experiment with $\tau = 1/2J$; (d) SEFT experiment with $\tau = 1/J$

Additional SEFT or APT experiments with evolution delays around $1/2J$ are necessary to distinguish between resonances of CH and CH$_2$ groups. This is one of the reasons why polarization transfer experiments have replaced the spin echo technique for ^{13}C assignments. The time-consuming two-dimensional heteronuclear J,δ experiment, which was quite popular some years ago for multiplicity selection, has been replaced by these techniques, on the other hand, for sensitivity reasons.

2.2.2 Polarization transfer experiments

With the development of polarization transfer pulse sequences INEPT and DEPT (see p. 411) new methods for ^{13}C assignment became available which have the additional advantage of signal enhancement. As shown in Figure 11.9, the evolution of the magnetization of CH, CH$_2$, and CH$_3$ carbons during the INEPT sequence differs and the maxima are observed at different Δ_2 values, a fact which can be used for signal selection. For example, a value of $1/2J(^{13}C,^1H)$ allows us to detect the magnetization of CH groups, while that of CH$_2$ and CH$_3$ groups is eliminated. On the other hand, a Δ_2 value of $3/4J(^{13}C,^1H)$ yields positive signals for resonances of CH and CH$_3$ groups and negative signals for CH$_2$ groups.

Today, in practical applications of polarization transfer experiments for resonance assignments the DEPT sequence, shown in Figure 11.10, is usually preferred. It yields, after three $1/2J$ delays, multiplets with uniform phase and with the application of ^{1}H-decoupling singlet signals for all types of ^{13}C resonances. The pulse angle θ of the last A pulse can be optimized for individual groups in order to allow signal selection.

Within the product operator formalism (p. 302 ff.) the DEPT sequence, applied to a CH fragment and neglecting the 180° pulses, yields the following expressions:

$$\hat{I}_z(^1H) \xrightarrow{\; 90°\hat{I}_x(^1H)\;} \hat{I}_y(^1H) \xrightarrow{\; 1/2J\;} 2\hat{I}_x(^1H)\hat{I}_z(^{13}C)$$

$$\xrightarrow{\; 90°\hat{I}_x(^{13}C)\;} 2\hat{I}_x(^1H)\hat{I}_y(^{13}C) \xrightarrow{\; 1/2J\;} 2\hat{I}_x(^1H)\hat{I}_y(^{13}C) \qquad (11.5)$$

$$\xrightarrow{\; 90°\hat{I}_y(^1H)\;} 2\hat{I}_z(^1H)\hat{I}_y(^{13}C) \xrightarrow{\; 1/2J\;} \hat{I}_x(^{13}C)$$

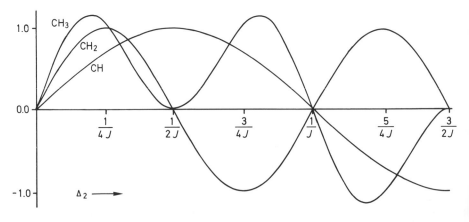

Figure 11.9 Time dependence of ^{13}C magnetization modulated by ^{13}C,^{1}H coupling in the refocussing period Δ_2 of the INEPT sequence

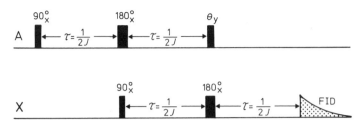

Figure 11.10 The DEPT pulse sequence (DEPT = distortionless enhancement by polarization transfer); the angle of the last pulse θ_y can be varied

The most important aspects are the development of pure anti-phase ^1H magnetization in the first $1/2J$ delay, the transformation of this magnetization into heteronuclear double quantum magnetization by the $90°(^{13}C)$ pulse, and the transformation of this magnetization into anti-phase ^{13}C magnetization, which is refocussed in the final $1/2J$ delay. The central $1/2J$ delay results from the necessity to perform in the ^1H as well as in the ^{13}C domain a spin echo sequence in order to remove relaxation and chemical shift effects.

On the basis of Figure 11.9 the relations 11.6–11.8 can be derived for the refocussing of ^{13}C magnetization and the editing of subspectra for CH, CH$_2$ and CH$_3$ resonances. If we define $\theta = \pi J \Delta_2$, we obtain for the last pulse of the DEPT sequence the angle $\theta(1) = \pi/4$ or $45°$ for $\Delta_2 = 1/4J$ (subspectrum S(1)), $\theta(2) = \pi/2$ or $90°$ for $\Delta_2 = 1/2J$ (subspectrum $S(2)$), and finally $\theta(3) = 3\pi/4$ or $135°$ for $\Delta_2 = 3/4J$ (subspectrum $S(3)$). The subspectrum $S(1)$ then contains all CH, CH$_2$ and CH$_3$ signals, the subspectrum $S(2)$ only the CH signals and the subspectrum $S(3)$ positive signals for CH and CH$_3$ groups and negative signals for CH$_2$ groups. Linear combinations are formulated for the selection, where the intensity differences must be taken into account:

$$S(CH) = 2S(2) \tag{11.6}$$

$$S(CH_2) = S(1) - S(3) \tag{11.7}$$

$$S(CH_3) = S(1) + S(3) - 1.414\ S(2) \tag{11.8}$$

Signals which are not part of these spectra can be assigned to quaternary carbons.

DEPT-editing of a ^{13}C spectrum is shown in Figure 11.11a. If in a simplified version $\theta_y = 135°$ is used, only the subspectrum $S(3)$ (Figure 11.11b) is obtained. In summary we can state that the assignment techniques described in Section 2.2.1 and 2.2.2 have the advantage that ^1H decoupled ^{13}C signals are detected and the information about the multiplicity is transformed into phase information. The assignment can thus be performed under the highest spectral dispersion. In addition, the intensity enhancement by NOE effects or polarization transfer are conserved. Because of the J-dependence of the evolution time, however, in many cases two different experiments are necessary since the $^1J(^{13}C,^1H)$ values for sp, sp^2, and sp^3 carbons differ considerably (see below).

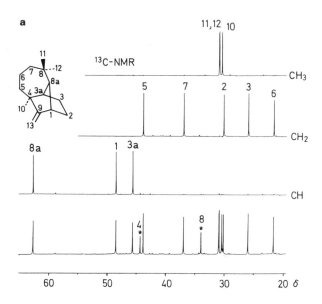

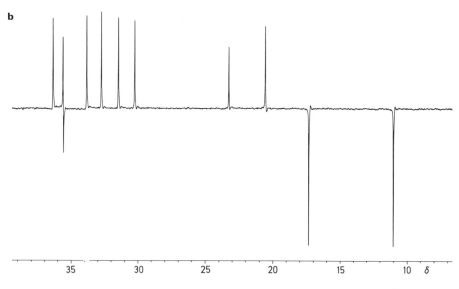

Figure 11.11 (a) Multiplicity selection with the DEPT sequence for the 100 MHz ^{13}C n.m.r. spectrum of the terpene longifolene (aliphatic region); $1/2J = 3.8$ ms; the signals of quaternary carbons are labelled with a star (after Ref. 5); (b) simplified DEPT spectrum with $\theta_y = 135°$ for the aliphatic region of the ^{13}C n.m.r. spectrum of testosterone (cf. Figure 11.3, p. 466)

Exercise 11.2   Verify the signal assignment given in Figure 11.3 (p. 466) with the help of the DEPT spectrum shown in Figure 11.11b.

2.2.3   Heteronuclear two-dimensional ^{1}H,^{13}C chemical shift correlation

While the methods described so far allow an assignment of ^{13}C signals with respect to their multiplicity and thus yield information about the number of attached protons, further experiments are necessary in order to derive an assignment to individual positions in a chemical structure. In this respect, two-dimensional heteronuclear ^{1}H,^{13}C chemical shift correlations are of major importance. As in the case of homonuclear shift correlations or COSY spectroscopy, discussed in detail in Chapter 8, resonance frequencies of scalar coupled nuclei, in this case ^{1}H and ^{13}C, are correlated and cross peaks with the coordinates $\delta(^1\text{H}),\delta(^{13}\text{C})$ are obtained. Most of these correlations are based on heteronuclear ^{13}C,^{1}H coupling constants over one bond. Those nuclei which show cross peaks are, therefore, direct neighbours in the particular molecule and the resonance assignment of the proton spectrum can be transferred directly to the ^{13}C spectrum and *vice versa*. In addition, correlations based on ^{13}C,^{1}H long range couplings can be used which often yield information which is crucial for the final solution of a structural problem.

Two different pulse sequences are available today for two-dimensional hetero-nuclear shift correlations between a sensitive A nucleus (^{1}H, ^{19}F, ^{31}P) and an insensitive X nucleus (^{13}C, ^{15}N): the long known polarization transfer experiment and a more recent development, which involves heteronuclear multiple quantum phenomena (Figure 11.12). In the first experiment (Sequence I, Figure 11.12a), magnitude is transferred from the sensitive nucleus, in general from the proton, to the insensitive nucleus, for example ^{13}C or ^{15}N. This magnetization transfer generates the cross peaks in the two-dimensional spectrum. The signals of the insensitive X nucleus are detected; for an example, see Figure 11.12c. The experiment is known as HETCOR (*het*eronuclear shift *cor*relation) experiment.

In the second version (Sequence II, Figure 11.12b) heteronuclear multiple quantum magnetization is produced, which develops during the evolution time and is finally transferred into detectable A magnetization. The sensitive A nucleus is used for signal detection and this method was introduced as *reverse* or *inverse* shift correlation. Today it is better known by the acronym HMQC (*h*eteronuclear *m*ultiple *q*uantum *c*oherence).

In both sequences I and II it is possible to apply a refocussing period after which decoupling can be used. This is rather simple for Sequence I, since here the protons must be decoupled. Pulse sequence II uses ^{13}C or generally X nucleus composite plus decoupling. An additional problem exists because in fact the satellites of the X nucleus in the spectrum of the sensitive A nucleus are detected. To avoid signal overlap, the main signals of the A nucleus, which originate from molecules which do not contain a magnetically active X nuclide, have to be eliminated by the phase cycle. If elimination is not complete, the spectra contain strong t_1 noise. In the case

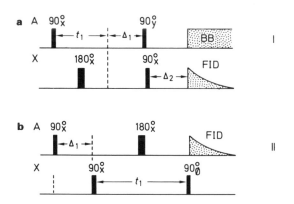

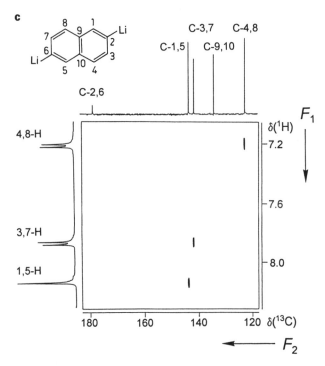

Figure 11.12 Pulse sequences for heteronuclear chemical shift correlations (HETCOR); (a) A → X polarization transfer method with X detection; the delay Δ_1 amounts to $1/2J(A,X)$; Δ_2 is the delay responsible for refocussing of the *anti*-phase X magnetization. For the $^1H,^{13}C$ experiment, the F_1 domain contains the ^1H frequencies, while the F_2 domain contains the ^{13}C frequencies; (b) 'inverse' correlation with A detection without X decoupling; the F_1 domain contains ^{13}C frequencies, while the ^1H frequencies appear on the F_2 frequency axis; (c) ^{13}C-detected $^{13}C,^1H$ shift correlation for 2,6-dilithionaphthaline in THF-d_8 based on $^1J(^{13}C,^1H)$ (pulse sequence I); spectral windows 6600 (F_2) and 500 Hz (F_1); 32 t_1 increments, 40 scans each; relaxation delay 2 s; exp. time 40 min

of X nucleus decoupling the X satellites may then coincide with the noise signals and their detection becomes difficult. For this reason X nucleus decoupling is rarely used in practice, which has the additional advantage that the $^{13}C,^{1}H$ spin–spin coupling constants can also be obtained from the 2D spectrum. One must remember, however, that the A nucleus (^{1}H) doublets are recorded in anti-phase and signal elimination may result in the case of small couplings, an important point for the long-range correlation experiment (see below).

Exercise 11.3 Develop the product operator treatment for the pulse sequences shown in Figure 11.12 for a CH fragment up to signal detection.

Because of A nucleus detection the HMQC experiment is more sensitive than experiments based on polarization transfer. Further sensitivity improvements are achieved if 'inverse' probe heads are employed, where the inner coil, which is more sensitive because of the larger filling factor, is used for A detection. The outer coil is then tuned to the X frequency. The theoretical factor of $(\gamma_A/\gamma_X)^{5/2}$ for sensitivity enhancement is not observed in practice, however, because of competing mechanisms which lead to signal losses. In any case, the success of such experiments depends on the careful adjustment of pulse angles and delays during the pulse sequence and on stable spectrometer hardware. Further improvements have been achieved by the introduction of a so-called BIRD sequence (see p. 324) which greatly reduces the signals of the isotopomers with non-magnetic X nuclides.

On the basis of the Δ_1 delay, where anti-phase A magnetization develops, and the different magnitude of one-bond and long–range $^{13}C,^{1}H$ coupling constants, the detection of one-bond or long–range correlations can be controlled in both sequences. For $^{1}J(^{13}C,^{1}H)$ values between 125 and 170 Hz, a delay of 3.3 ms is a good compromise in practice. For long–range couplings, which are in the order of 3–12 Hz, the Δ_1 values are around 25 ms. For the Δ_2 delay a value of 2 ms secures the detection of all correlations via one-bond couplings (CH, CH_2, CH_3 groups). In the case of long–range couplings these values are again considerably larger (20–50 ms). In both pulse sequences I and II the repetition rate is governed by the relative short longitudinal ^{1}H relaxation time. The relaxation delay can thus be much shorter than for one-dimensional ^{13}C experiments. Because of the interchange of the F_1 and F_2 frequency axes, one difference, however, exists: with Sequence I higher resolution can be obtained for the ^{13}C frequency axis than with Sequence II.

The general importance of heteronuclear shift correlations has initiated intensive research for improvements with respect to sensitivity and selectivity. From the various modifications which have been proposed we briefly refer to three pulse sequences which have found widespread applications. They are summarized in Figure 11.13 (p. 481).

A variant for A,X shift correlations based on long-range couplings, known as COLOC sequence (*correlation via long-range couplings*), is shown in diagram III. It is a polarization transfer experiment like Sequence I (Figure 11.12), but now the evolution time t_1 for the development of anti-phase A magnetization is incorporated into the Δ_1 delay which is flanked by the two 90° pulses, Because this delay is unchanged during t_0 incrementation, the experiment is called a 'constant time experiment' (cf. p. 323). Incrementation of the evolution time t_0 is achieved by moving the 180 °(A,X) pulse pair through the Δ_1 interval. The time τ thus varies in a series of experiments. Because of the 180°(A) pulse, which refocusses chemical shift effects at the time τ, the evolution of A magnetization which yields the frequency labelling necessary for a shift correlation is restricted to the time $\Delta_1 - \tau$. This is the true t_1 period in this experiment. From the couplings, heteronucleus A,X coupling develops during the whole Δ_1 interval, while homonuclear A coupling is eliminated from the F_1 domain (cf. p. 323). The result is a shorter pulse sequence with improved detection for fast relaxing cross peak magnetizations and with ω_1 decoupling.

Sequence IV and V are variants of the HMQC experiment (Sequence II, Figure 11.12). The BIRD pulse sandwich introduced in Sequence IV greatly improves the suppression of single quantum A magnetization from molecules with magnetically inactive X nuclei, for example ^{12}C in $^1H,^{13}C$ experiments. The 180 ° pulse refocusses chemical shift evolution and the τ-delay is tuned for the heteronuclear coupling so that at the end of this period coupled A magnetization is in anti-phase ($\hat{I}_x(A)\hat{I}_z(X)$) while uncoupled A magnetization — homonuclear A couplings neglected — points still along the y-axis ($\hat{I}_y(A)$). The $90°_{-x}$ pulse transforms this magnetization to $\hat{I}_z(A)$, while $\hat{I}_x(A)\hat{I}_z(X)$ is unaffected and proceeds to detection as in Sequence II. The neglect of homonuclear A couplings is possible because the heteronuclear A,X couplings are generally larger by a factor of 10 and dominate the evolution of transverse magnetization. The improvement achieved with Sequence IV is shown in Figure 11.14 (p. 482) with an example from a $^2H,^{13}C$ correlation experiment with ^2H detection.

Finally, Sequence V was introduced as an 'inverse' COLOC experiment. Here, the delays Δ_1 and Δ_2 are tuned to one-bond and long-range heteronuclear A,X couplings, respectively. The first 90° pulse pair, separated by the delay $\Delta_1 = 1/2^1J(A,X)$ serves as a low-pass filter (cf. p. 324) which eliminates one-bond correlations. The second 90°(X) pulse then creates after the appreciably longer delay Δ_2 (~60 ms) the desired multiple quantum coherences based on long-range couplings. Before detection, a refocusing delay Δ_2 is introduced in order to avoid signal elimination due to the initial anti-phase character of the A-magnetization in case of small couplings. It is then possible to decouple the X nucleus (^{13}C) by, for example, a composite pulse sequence. While for Sequence I the refocusing delay Δ_2 is a compromise with respect to the optimal values necessary for different $^{13}C,^1H_n$ groups (see p. 474), a unique value $1/2J(^{13}C,^1H)$ serves in Sequence C because X coupling to the A nuclei involves only one ^{13}C spin. The sequence goes by the acronym HMBC (*h*eteronuclear shift correlations via *m*ultiple *b*ond *c*onnectivities) and is today most effectively performed using gradient enhanced spectroscopy (cf. p. 321), which significantly improves the elimination of t_1-noise from the residual signals of molecules with non-magnetic X nuclei. This is demonstrated in Figure 11.15 (p. 483) with a $^{13}C,^1H$ correlation experiment for 3-fluorophenanthrene.

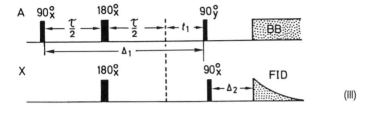

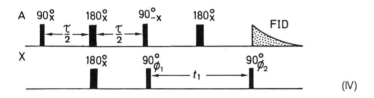

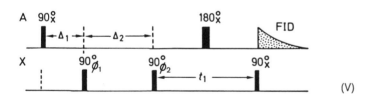

Figure 11.13   Modified pulse sequences for heteronuclear shift correlations; (III) HETCOR COLOC sequence for correlations via long-range couplings: (IV) HMQC experiment with improved elimination of single quantum A signals: (V) HMBC sequence

Since a comprehensive treatment of all aspects of heteronuclear n.m.r. experiments is clearly beyond the scope of the present introduction, the interested reader is referred for a more complete discussion to the review articles listed at the end of this chapter. We conclude our introduction with a description of four examples of ^{13}C,^{1}H chemical shift correlations in order to illustrate their application.

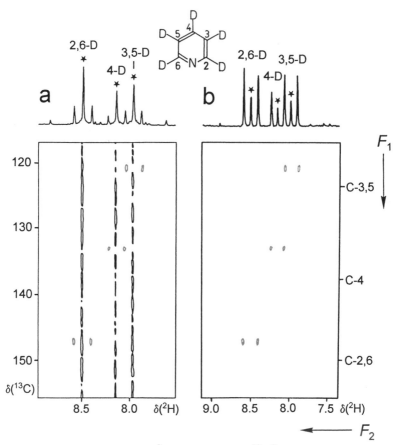

Figure 11.14 Two-dimensional ^{2}H-detected HMQC ^{13}C,^{2}H shift correlations for pyridine-d$_5$ without (a) and with (b) additional BIRD pulse (sequence IV in Figure 11.13). In (b), the $\tau/2$ delay [$=1/4\,^1J(^{13}$C,^{2}H)] was set to 10 ms corresponding to $^1J(^{13}$C,^{2}H) = 25 Hz. The starred signals are residual ^{2}H resonances of ^{12}C–^{2}H units. The cross peaks (shown in red) are split by ^{13}C,^{2}H spin–spin coupling

Figure 11.16 (p. 484) demonstrates how the assignment of the methyl resonances in the ^{13}C n.m.r. spectra of angelica and tiglic acid — acyl residues which appear quite frequently in natural products — can be achieved. Electron density considerations and stereochemical aspects are not sufficient to derive an unambiguous decision. The situation is different in the ^{1}H n.m.r. spectrum where only the protons at C(4) have a large vicinal ^{1}H,^{1}H coupling to the olefinic proton H(3). The doublet of quartets (^4J coupling between H(4) and H(5)) can, therefore, be safely assigned to H(4). The cross peaks in the 2D ^{1}H,^{13}C shift correlation experiment then yield the ^{13}C assignment which shows that the order of δ(4) and δ(5) in both compounds is different. While δ(4) is fairly constant, C(5) is considerably shielded in tiglic acid

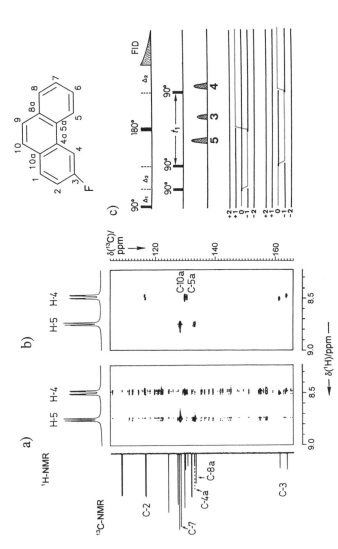

Figure 11.15 100/400 MHz $^{13}C,^1H$ HMBC correlation for 3-fluorophenanthrene; (a) with sequence IV (Figure 11.13); (b) with sequence C modified for coherence selection by linear B_0 field gradients as shown in (c). According to the coherence pathways shown and the resulting coherence orders, the desired ^1H magnetization evolves during the first $t_1/2$ interval as heteronuclear double quantum coherence with the sum $-(\omega(^1H) + \omega(^{13}C))$. After the 180 $°(^1H)$ pulse, in the second $t_1/2$ interval, we have $(\omega(^1H) - \omega(^{13}C))$ (zero quantum magnetization) and finally during detection $\omega(^1H)$. If we now consider the ratio $\gamma(^1H)/\gamma(^{13}C) = 4:1$, the gradient pulses ΔG_1, ΔG_2, and ΔG_3 shown in (c) must have the amplitudes 5, 3, and 4 in order to refocus the desired ^1H magnetization $(5 \times (-)(4 + 1) + 3 \times (4 - 1) + 4 \times 4 = 0)$. For the undesired t_1-noise from the ^{12}C molecules we have, during the same intervals, $(- 4 \times 5) + (4 \times 3) + (4 \times 4) = 8$, which leads to elimination. In both experiments, the delays Δ_1 and Δ_2 were set to 3 and 47 ms, respectively

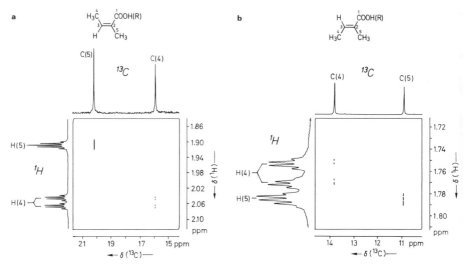

Figure 11.16   400/100 MHz 2D ^{1}H,^{13}C chemical shift correlation for angelic and tiglic acid on the basis of pulse sequence I (Figure 11.12) (after Ref. 6)

($\Delta\delta = -9.3$ ppm), apparently a consequence of steric interactions with the larger groups in geminal and vicinal position.

Because of the larger spectral dispersion of ^{13}C, a strongly overlapping ^{1}H spectrum can often be assigned by a ^{1}H,^{13}C shift correlation. However, the opposite case can also be found. Both aspects are illustrated in Figure 11.17. Thus, the signals of the methylene protons from individual CH$_2$ groups in the spectrum of the paracyclophane **154** (p. 364) are identified via their cross peaks with the same carbon resonance (Figure 11.17a). On the other hand, the nicely resolved ^{1}H n.m.r. spectrum of bisdehydrobenzo[18]annulene (p. 293), which is easily analysed on the basis of coupling constants and ring current effects, can be used to assign the closely spaced ^{13}C signals in the frequency region between 126.0 and 127.5 ppm (Figure 11.17b).

A ^{1}H,^{13}C chemical shift correlation recorded with pulse sequence II is finally shown in Figure 11.18. The cross peaks appear here as doublets, since the X nucleus was not decoupled. As with sequence I, quaternary carbons do not lead to cross peaks if Δ_1 is optimized for one-bond ^{13}C,^{1}H coupling constants.

Figure 11.17   Two-dimensional ^{1}H,^{13}C shift correlation for the paracyclophane **154** (a) and bisdehydrobenzo[18]annulene (b) (after Ref. 7, 8)

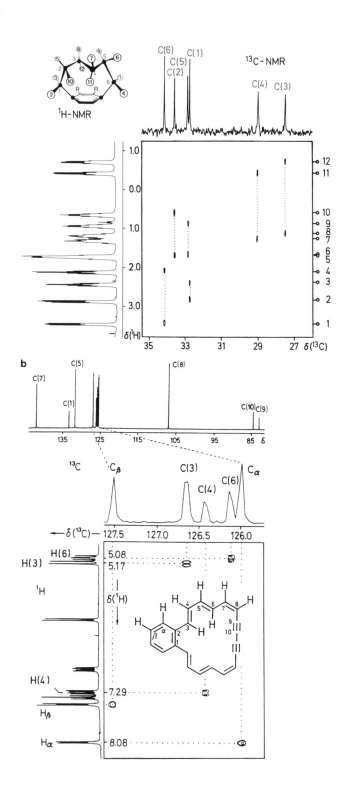

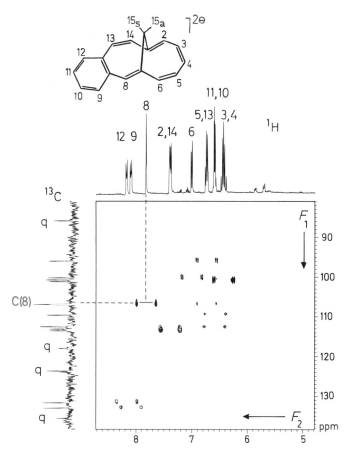

Figure 11.18   400/100 MHz ^1H,^{13}C chemical shift correlation via heteronuclear multiple quantum coherences (HMQC experiment, pulse sequence II, Figure 11.12) for the dianion of benzo{c}-1,7-methano{12}annulene with $\Delta_1 = 3.57$ ms for $^1J(^{13}C,^1H) = 140$ Hz; the cross peaks correspond to the ^{13}C satellites in the ^1H spectrum (after Ref. 9).

2.2.4   The $^{13}C,^{13}C$ INADEQUATE experiment

In Chapter 8 (p. 325ff.) we discussed the INADEQUATE experiment, a homonuclear chemical shift correlation on the basis of double quantum magnetization. This experiment was originally developed for ^{13}C n.m.r. spectroscopy and was first executed as a one-dimensional experiment. The idea was to measure $^{13}C,^{13}C$ coupling constants more easily. The experimental determination of these parameters suffers from the low natural abundance of carbon-13, which leads to only 0.1‰ probability to find two neighbouring ^{13}C nuclei in one and the same molecule. If enough sensitivity is available, satellites can be observed in the ^1H-decoupled ^{13}C n.m.r. spectrum of these isotopomers (see Chapter 6, p. 214), which accompany the

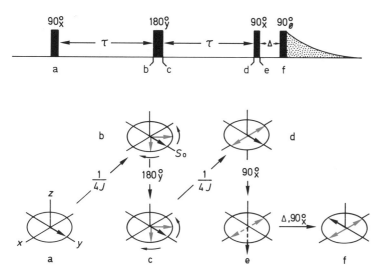

Figure 11.19 Pulse sequence of the one-dimensional INADEQUATE experiment for the detection of $^{13}C,^{13}C$ spin–spin coupling constants. The experiment is performed with ^1H broad band decoupling. In both $\Delta = 1/4J$ delays anti-phase ^{13}C magnetization develops which is transformed into homonuclear double quantum coherence by the second 90° pulse (red vectors). After a short delay (ca. 10 µs) the last pulse produces detectable single quantum magnetization. The phase program selects magnetization components which proceeded via the coherence pathway of coherence order $p = 2$. The magnetization S_0 of molecules with only one ^{13}C nucleus follows the coherence pathway with coherence order $p = 1$ and shows different phase behaviour. The 1/4J delay has to be optimized for the coupling of interest, however, deviations of up to 10% can be tolerated. With a certain τ value satellites can be detected which are separated by nJ (n = 3, 5)

200-times more intensive main signal S_0 due to molecules with only one ^{13}C. They belong to the lines of AX spin systems of $^{13}C,^{13}C$ spin pairs and can be detected best for directly bonded nuclei where the couplings are of the order of 35 to 90 Hz, depending on carbon hybridization (cf. Section 000). In the case of smaller coupling constants (geminal or vicinal spin pairs) the satellites are mostly superimposed with the main signal.

In this context, the one-dimensional INADEQUATE sequence, shown in Figure 11.19, is of great interest because it largely eliminates the strong main signal S_0. Its success depends on the phase cycle, which selects the double quantum magnetization of the $^{13}C,^{13}C$ AX systems while the single quantum magnetization of molecules with only one ^{13}C nucleus is suppressed. Spectra as that shown in Figure 11.20 are then obtained. Today the new technique of gradient enhanced coherence selection greatly facilitates this signal selection process.

Exercise 11.4 Draw a coherence level diagram for the one-dimensional INADEQUATE sequence (Figure 11.19) and propose (a) a phase cycle, and (b) the introduction of gradient pulses which allow the selective detection of $^{13}C,^{13}C$ spin systems.

An interesting and powerful extension of the 1D INADEQUATE experiment was introduced by adding the evolution time t_1, which leads to a second frequency dimension and thus a 2D experiment. This pulse sequence, already introduced with equation (8.46) (p. 326), and extensively discussed there, is used here for ^{13}C observation with ^1H broadband decoupling. During the evolution time homonuclear ^{13}C double quantum magnetization develops and the individual $^{13}C,^{13}C$ AX systems neighbouring carbon atoms in the molecule are separated along the F_1 frequency axis on the basis of their different double quantum frequencies $v_{DQ} = v_A + v_X - 2v_0$. The goal of this experiment is not the determination of $^{13}C,^{13}C$ coupling constants but rather the analysis of the carbon skeleton of a molecular structure. Neighbouring carbons are detected via their $^1J(^{13}C,^{13}C)$ coupling and this leads to a resonance assignment which is directly connected to the carbon chain in the molecule. The 2D INADEQUATE experiment would certainly be the method of choice for the analysis of complex molecular structures if its low sensitivity (see above) could be overcome. Nevertheless, many liquids and highly concentrated solutions of solid compounds with good solubility have been successfully studied.

In several cases even fairly large molecules have been investigated by the 2D INADEQUATE technique. The experiment is of special interest for systems with a greater number of identical partial structures. Strong overlap of the signals of these structural elements then results for the ^{13}C n.m.r. spectrum and the assignment

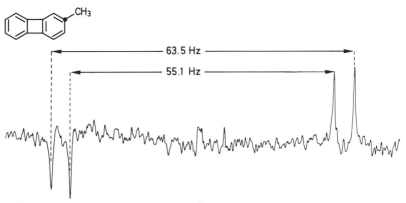

Figure 11.20 One-dimensional INADEQUATE experiment for the measurement of the $^{13}C,^{13}C$ coupling constants of the methylated β-carbon of 2-methylbiphenylene. With a delay $\Delta = 1/4J = 4.5$ ms ($\stackrel{\triangle}{=} J = 55.6$ Hz) the couplings to C-1 and C-3 have been detected. A separate measurement for the CH_3 signal yields $^1J(\text{C-2},CH_3) = 44.2$ Hz; note the good suppression of the S_0 signal in the centre of the spectrum (after Ref. 10)

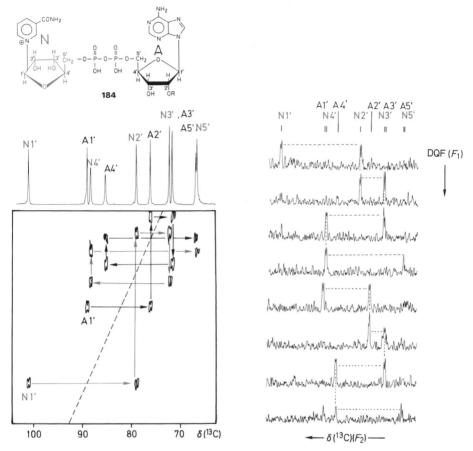

Figure 11.21 Result of a 2D INADEQUATE experiment for the assignment of the ribose
^{13}C n.m.r. signals of nicotinamide-adenine-dinucleotide (NAD, **184**); (a) contour plot with the
assignment of the ribose spectra; (b) F_1 traces of the two-dimensional data matrix. For each
double quantum frequency the corresponding AX system is observed (after Ref. 11)

becomes difficult. Such conditions are met, for example, with nucleic acids and
related compounds with their repeating ribose or desoxyribose units. This is
demonstrated with the example of NAD (**184**), where sufficiently concentrated
solution can be prepared in D_2O (Figure 11.21).

2.2.5 Assignment techniques with selective excitation

As other two-dimensional techniques, those described in the preceding section for
the assignment of ^{13}C resonances have the disadvantage of long measuring times, in
the case of the INADEQUATE method even more so because of the low natural
abundance of carbon-13. As in ^{1}H n.m.r. resonance, alternative one-dimensional

experiments are therefore of interest if selective excitation is feasible. Such experiments rely on the DANTE sequence or on GAUSS pulses (Chapter 7, p. 250). They are especially useful if a number of informations on the molecular structure is already known and only a few assignments are in question. A complete 2D experiment would then duplicate much of the information already available and the expense in instrument time would not be justified. Two examples will illustrate this idea.

With a selective INEPT sequence, for instance, where a GAUSS pulse at a certain ^1H resonance v_i is used for the 90_y° (^1H) pulse, ^{13}C nuclei adjacent to H_i can be detected and assigned. As in the case of the selective ^1H-decoupling experiment one has to irradiate both ^{13}C satellites in the ^1H spectrum. The selectivity achieved depends on the relative magnitude of the $^{13}C,^1H$ coupling constants involved. Because of the rule $^1J \gg {}^nJ$ ($n > 1$) it is generally sufficient in order to distinguish next neighbours from remote neighbours (Figure 11.22).

In an analogous way, instead of a 2D INADEQUATE experiment, a selective 1D INADEQUATE experiment can be constructed if in the pulse sequence shown in Figure 11.1 the third 90° pulse is replaced by a selective GAUSS pulse. Detectable anti-phase magnetization is then produced only for the selectively excited ^{13}C nucleus and its direct neighbours and one observes only AX systems where these nuclei participate. The acronym SELINQUATE was coined for this experiment.

2.2.6 Alternative assignment techniques

Aside from the methods described so far, which were exclusively of spectroscopic origin, other techniques exist which rely on arguments based on chemical structure. In the first place we mention here $^{13}C,^1H$ spin–spin coupling constants, discussed in more detail later in this chapter. The structural dependence of these parameters often yields unequivocal assignments. For example, methine and methylene groups in three-membered rings are immediately recognized by their large $^{13}C,^1H$ couplings over one bond, which are much larger than the corresponding couplings in open-chain compounds. In three-membered rings we find values around 160 Hz, while for strainless ring systems like cyclohexane or aliphatic chains values around 125 are observed.

In many cases ^1H-coupled ^{13}C spectra yield first-order multiplets and such a straightforward application of coupling information for assignment purposes is illustrated with the spectrum of the 2-thio-oxotetrahydropteridine dianion shown in Figure 11.23. Here three pairs of carbon atoms can be distinguished by the multiplicity of their ^{13}C n.m.r. signals: the quaternary carbons C(2) and C(4) show no splitting, the carbons C(9) and C(10) show vicinal coupling to H(7) and H(6), respectively, whereas C(6) and C(7) show line splitting due to $^1J(^{13}C,^1H)$ and $^2J(^{13}C,^1H)$ (doublets of doublets). Of course, the assignment within each pair of carbons must be based on independent arguments from different sources.

Among chemical methods used for the assignment of ^{13}C n.m.r. signals there are shifts induced by the addition of shift reagents, by changing the pH of the solution

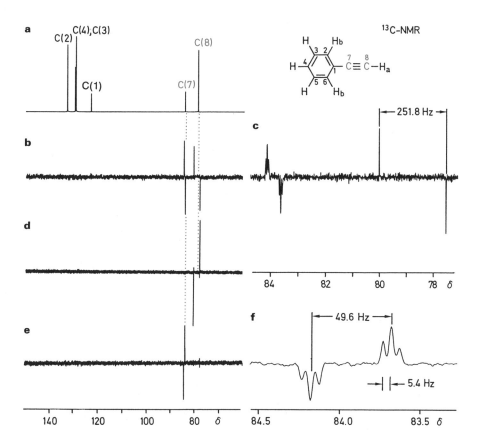

Figure 11.22 Selective INEPT experiments in the spectrum of phenyl acetylene; (a) 100 MHz $^{13}C\{^{1}H\}$ spectrum of phenyl acetylene (10% in acetoned-d_6); (b) INEPT experiment (see Figure 10.14, p. 411) with selective $90^\circ_x(^{1}H)$ pulse at the H_a resonance; the delay $1/4J(^{13}C,^{1}H)$ of the INEPT sequence was optimized with 1 ms for the ^{1}J coupling C(8),H_a which amounts to 250 Hz. Since the ^{2}J coupling between C(7) and H_a in phenylacetylene is unusually large (49.6 Hz), a polarization transfer is also observed for C(7). Both carbons at the triple bond can, therefore, be distinguished if the coupling pattern of their signals is compared (see spectrum (c) and (f)). For the more distant carbons of the benzene ring the relation $^{n}J \ll {^{1}J}$ is valid and these signals are not detected. (c) Signals of C(7) and C(8) from spectrum (b) enlarged; one also observes the vicinal coupling between C(7) and H_b; (d), (e) INEPT experiments with selective $90^\circ_x(^{13}C)$ read pulse (see Figure 10.14) at the C(8) and C(7) resonance and optimized INEPT delay (1 and 5 ms), respectively; (f) spectrum (e) enlarged

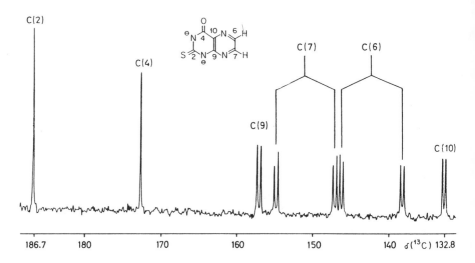

Figure 11.23 First-order splittings due to ^{13}C,^{1}H spin–spin coupling in the ^{13}C n.m.r. spectrum of 2-thio-4-oxotetrahydropteridine dianion (after Ref. 12)

causing protonation or deprotonation, or by solvent effects. Finally, if a simple and unequivocal route to specific deuteration is available, ^{2}H labelling yields the desired information. Owing to ^{13}C,^{2}H coupling, 1:1:1 triplet structures are observed for the carbon resonances, most pronounced for the directly substituted carbon [remember, however, that $J(^{13}C,^{2}H) = J(^{13}C,^{1}H)/6.5144$, as discussed on p. 218]. In addition, characteristic isotope effects on chemical shifts are observed and these are also of diagnostic value. The method is best explained by using an example. Figure 11.24 shows the olefinic ^{13}C resonances of 1,6-indane oxide deuterated specifically in position 4. From the triplet observed in the high-field absorption it is immediately clear that this resonance belongs to C(3) and C(4). We note that $\delta(3)$ and $\delta(4)$ are different owing to the isotope effect that shifts $\delta(4)$ to higher field by 0.3 ppm as compared with the shift in the undeuterated compound. For $\delta(3)$, two bonds away, this effect is much smaller (0.1 ppm). The isotope effect also discriminates between the resonances of C(2) and C(5), the former at lower field broadened owing to an unresolved $^{3}J(^{13}C,^{2}H)$ coupling.

Finally, in the case of complicated structures such as those of natural products, T_1 measurements are sometimes used to assign different ^{13}C resonances. As will be briefly discussed later, spin–lattice relaxation in organic molecules depends on C,H distances, and relaxation times T_1 for the quaternary carbon atoms are therefore considerably longer than those for carbon atoms substituted by hydrogen. An example is given with the data of diphenylacetylene (T_1 in seconds):

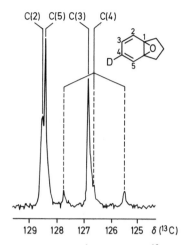

Figure 11.24 ¹H-decoupled ¹³C n.m.r. spectrum of 1,6-indane oxide specifically deuterated at C(4) for the assignment of the ¹³C resonances (after Ref. 13)

3. Chemical shifts

The chemical shifts of ¹³C resonances in organic molecules span a region about 250 ppm, including extreme values to high and low field (tetraiodomethane and carbonium ions, respectively) of even 650 ppm. A general survey is presented in Figure 11.25.

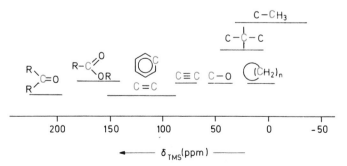

Figure 11.25 δ scale (ppm) of the ¹³C resonances in organic compounds

As in ^{1}H n.m.r. the δ-scale for carbon-13 can be divided into subregions for the resonances of aliphatic, olefinic, and acetylenic carbon atoms. Carbonyl carbon atoms are the most strongly deshielded and their resonances form a separate region at lowest field. In earlier work carbon disulphide served simultaneously as reference compound and solvent, but later the ^{13}C resonance of tetramethylsilane (TMS) was introduced and is today accepted as internal reference. This has the advantage that the majority of ^{13}C δ-values are positive, as are the $\delta(^1H)$ values. A more detailed chemical shift diagram, a collection of ^{13}C chemical shifts from organic compounds, and a Table of ^{13}C resonances of important solvents and references are included in the Appendix.

3.1 THEORETICAL MODELS

For the discussion of the correlation between $\delta(^{13}C)$ and structure we remember equation (2.4) (p. 17) where the shielding constant σ is given as the sum of three terms: the local diamagnetic and paramagnetic contributions and the effect of neighbouring groups:

$$\sigma = \sigma_d + \sigma_p + \sigma' \qquad (11.9)$$

As for other heavy nuclei, ^{13}C chemical shifts are determined mainly by variation of σ_p and, to a lesser extent, by σ_d. Neighbouring group effects that are so well known in ^{1}H n.m.r. are of only minor importance.

Substituent-induced changes of σ_d for a particular carbon atom C_i can be assessed in a semi-empirical manner through an equation similar to the Lamb formula for the free atom:

$$\sigma_d^i = \frac{\mu_0}{4\pi} \frac{e^2}{3m_c} \sum_{j \neq i} Z_j R_{ij}^{-1} \qquad (11.10)$$

where Z_j is the atomic number of nucleus j, R_{ij} is the internuclear distance, and the other terms are well known constants. This contribution is of particular importance in the case of heavy atoms, where increased shielding is observed with increasing atomic number. For the halogens, this effect is most pronounced for iodo substitution and has become known as the *heavy atom effect* (Figure 11.26).

For the paramagnetic contribution, early theoretical considerations led to the expression

$$\sigma_p^i \approx -\frac{1}{\Delta E} \left(\frac{1}{r_i^3} \right)_{2p} \sum_{j \neq i} Q_{ij} \qquad (11.11)$$

where ΔE is a mean electronic excitation energy, r_i is the average radius of the carbon $2p_z$ orbitals, and Q_{ij} is a bond order term that originates from the presence of π-bonds.

As it turns out, structural changes usually effect all of the individual contributions to σ_p and only in a few cases does a predominant influence of one component justify a separate treatment. Nevertheless, equation (11.11) can still be used as a guide for

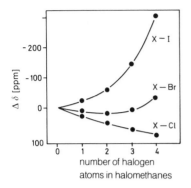

Figure 11.26   The effect of halogens on ^{13}C shielding (after Ref. 14)

the interpretation of a number of prominent features observed for carbon shielding, since it allows a discussion of experimental data in terms of chemical significance. It has to be pointed out, however, that the calculation of chemical shifts—and equally that of spin–spin coupling constants—has made considerable progress in recent years and a number of powerful quantum chemical procedures are available today for this purpose. One of the most successful approaches is provided by the IGLO method (IGLO = *i*ndividual *g*auge for *l*ocalized *o*rbitals) and such data have already been used for structure determinations. Again, the limited scope of our introduction prevents us from a detailed discussion of these developments and the interested reader is referred to the review articles listed at the end of this chapter.

Going back to the discussion of equation (11.11), we note that with respect to the ΔE dependence of σ_p, except for $\pi \rightarrow \pi^*$ excitations, which are excluded owing to symmetry considerations, all other transitions ($\sigma \rightarrow \pi^*$, $\pi \rightarrow \sigma^*$, $\sigma \rightarrow \sigma^*$, $n \rightarrow \pi^*$, and $n \rightarrow \sigma^*$) are important, the largest contribution usually arising from the transition of lowest energy.

The increasing shift of ^{13}C resonances in the series alkanes–alkenes–carbonyl compounds is thus not unexpected. For the last group of compounds, a linear correlation between the wavelength of the $n \rightarrow \pi^*$ transition and the chemical shift of the carbonyl resonance has even been found. The bond order term superimposes additional changes that are most pronounced for the central carbon of allene and the ^{13}C resonances of alkynes:

H_3C-CH_3	$H_2C=CH_2$	$H_2C=C=CH_2$	$HC\equiv CH$
δ 5.9	123.3	74.8 213.5	71.9

For the latter the theory yields $\Sigma_{j \neq i} Q_{ij} = 0$ as in alkanes, whereas for allene $\Sigma_{j \neq i} Q_{ij} = 0.8$.

Perhaps the most important contribution to σ_p, at least the one that is most frequently supported by experimental observation, is the $(r_i^{-3})_{2pz}$ term, which can be related to the charge density. Partial negative charge thus leads to an increase in r_i (orbital expansion) and consequently to a diminution in σ_p, and increased shielding results. For partial positive charge, the opposite reasoning holds, orbital contraction producing a deshielding effect. This charge dependence of [13]C n.m.r. chemical shifts was recognized in the early stages of the technique and led to the development of an empirical relation based on the ^{13}C resonance of benzene and aromatic ions and the corresponding π charge density changes, $\Delta\rho$:

$$\Delta\sigma = 160 \, \Delta\rho \qquad (11.12)$$

An extended diagram of this correlation is given in Figure 11.27.

Equation (11.12) is analogous to the similar correlation derived for protons (equation 4.2). Note, however, that the origin of the two correlations is different, since in the case of protons, changes in the local diamagnetic term are responsible.

A great variety of data have been subjected to regression analysis based on equation (11.12) and consequently various proportionality constants, ranging from 60 to 360, have been derived. This is not so surprising if we remember the mutual dependence of the different terms in equation (11.11). Also, different methods of calculating the charge density changes — pure π-electron calculations as well as those including σ-electrons — have been applied. Numerical calculations for estimating $\Delta\sigma$ or $\Delta\rho$ using such a simple equation as (11.12) with a particular proportionality constant are therefore restricted to certain classes of compounds.

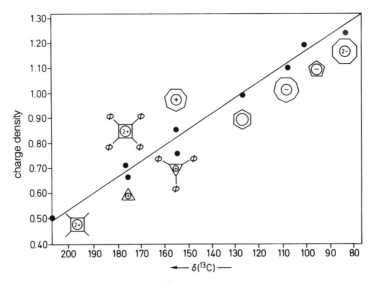

Figure 11.27 Correlation of ^{13}C chemical shifts and π-electron densities in aromatic systems (after Ref. 15)

The charge density dependence of ^{13}C resonances also forms the basis of the fact that resonance structures are often helpful in rationalizing ^{13}C chemical shifts. The following examples serve to illustrate this point:

δ 2.5 194.0
CH₂=C=O
Ketene

$\overset{\ominus}{CH_2}-C\equiv\overset{\oplus}{O}$

δ 107.8

142.6

4,5-Dihydrooxepine

δ 149.8 128.4

2-Cyclohexenone

Furthermore, the mesomeric effect of substituents in aromatic systems documents itself through the changes induced for the ^{13}C resonance, as the data for the *para*-carbon of several substituted benzenes show:

N(CH₃)₂	OCH₃	CH₃	H	$O=\overset{CH_3}{\underset{}{C}}$	NO₂
Δδ – 11.8	–8.1	– 2.8	0	+ 4.2	+6.0

In the *ortho*-position, additional steric effects may operate.

Since charge density/chemical shift correlations also exist for protons and ^{19}F nuclei, it is not surprising that the chemical shifts of the resonance frequencies of all three nuclei are in many cases linearly related to one another. This is clearly illustrated for the ^1H and ^{13}C resonances at the *para* position of monosubstituted benzenes and the ^{19}F resonances in *para*-substituted fluorobenzenes in Figure 11.28.

Another observation related to the charge density effect is the finding that alternating and non-alternating π-electron systems are well distinguished by their ^{13}C n.m.r. spectra. Since the latter have a non-uniform charge density distribution, their ^{13}C n.m.r. spectra show a significant spread, covering a much larger shift range than that of alternating systems. An example is presented in Figure 11.29.

Finally, large shifts the ^{13}C resonance are observed in the case of protonation or deprotonation, and the study of carbanions and carbocations especially has profited a great deal from the development of ^{13}C n.m.r. To illustrate the protonation dependence of ^{13}C chemical shifts, Figure 11.30 shows the pH dependence of the ^{13}C resonance of pyridine, where the transition from the free amine to the pyridinium ion is accompanied by large chemical shifts.

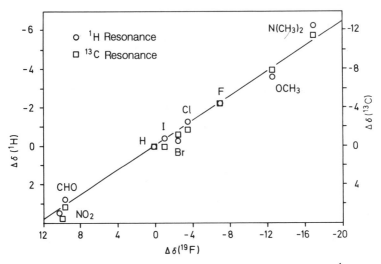

Figure 11.28 Correlation between the resonance frequencies of ^{1}H and ^{13}C *para* to substituents in monosubstituted benzenes and the ^{19}F resonance frequencies of the correspondingly *para*-substituted fluorobenzenes; $\Delta\delta$ values in ppm relative to benzene (after Ref. 16)

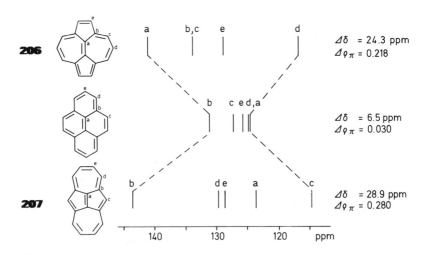

Figure 11.29 ^{13}C chemical shifts and charge density differences in pyrene and the two isomeric non-alternating hydrocarbons dicyclopentaheptalene (**206**) and dicycloheptapentalene (**207**); ΔQ_{π} was obtained from CNDO/2 calculations (after Ref. 17)

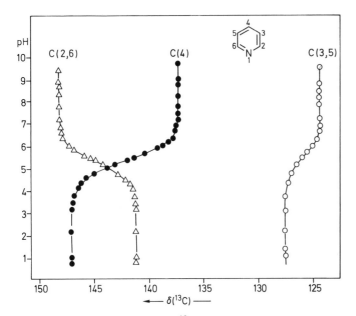

Figure 11.30 pH dependence of ^{13}C resonances of pyridine (after Ref. 18)

The interesting point to note here is that only the shifts of C-4 and C-3,5 are in accord with expectation (low field shift due to positive charge), whereas the shielding observed for the α-carbon atoms C-2,6 does not conform. It can be rationalized, however, if we remember that the protonation of the nitrogen will change the electron transition at the nitrogen from an $n \rightarrow \pi^*$ to a $\sigma \rightarrow \pi^*$ type and thus increase ΔE. If this effect dominates, σ_p will decrease, and shielding should result, as is indeed observed.

Particularly drastic shifts towards low field are encountered as expected for the ^{13}C resonance frequencies of *carbocations*. Using ^{13}C resonance spectroscopy, the distribution of the positive charge over neighbouring carbon atoms in these systems can also be studied.

As the data in Table 11.1 show, the ^{13}C resonance frequencies of the central carbon atoms in the systems with cation-stabilizing substituents such as cyclopropyl and phenyl groups are paramagnetically shifted far less than those of the positively charged carbon atoms in the simple alkyl cations, since in the former cases the charge is distributed over the substituents. The apparently anomalous paramagnetic shift in going from the dimethylphenyl to the triphenylmethyl carbenium is caused by the steric hindrance of the phenyl groups in the triphenylmethyl system. The phenyl groups in this system cannot assume a coplanar arrangement and as a consequence the conjugative delocalization of the positive charge to the substituents is reduced.

Table 11.1 ^{13}C resonances of carbenium ions (δ_{TMS})

	$^{13}\overset{\oplus}{C}$	$^{13}CH_3$
$(CH_3)_3\overset{\oplus}{C}$	328	47
$(CH_3)_2\overset{\oplus}{C}H$	318	60
$(CH_3)_2\overset{\oplus}{C}\,C_2H_5$	332	43
$(CH_3)_2\overset{\oplus}{C}{-}\triangleleft$	280	27
$C_6H_5\overset{\oplus}{C}(CH_3)_2$	254	-
$(C_6H_5)_2\overset{\oplus}{C}\,CH_3$	198	-
$(C_6H_5)_3\overset{\oplus}{C}$	211	-

Using ^{13}C resonance spectroscopy, it can also be demonstrated that fast rearrangements take place in a series of carbocations. Thus, it has been shown that the isopropyl cation (**208**) labelled with 50% ^{13}C at position 2 undergoes a rearrangement with a half-life of 1 h at $-60\,°C$ that distributes the label equally among all three carbons. This rearrangement probably takes place via a protonated cyclopropane (**209**):

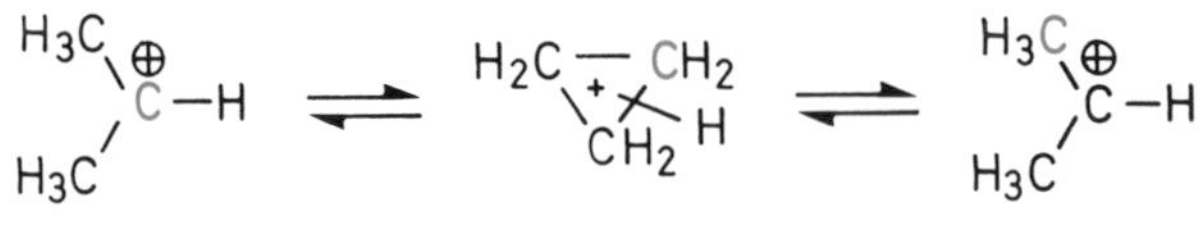

208   **209**   **208'**

Concluding this section, a short discussion of the σ' term in equation (11.9) seems appropriate. As far as diamagnetic anisotropy effects are concerned, changes in $\Delta\sigma$ depend only on the magnitude of $\Delta\chi$ and the relevant geometry. Induced shifts are therefore of the same order of magnitude as in ^1H n.m.r., usually less than 1 ppm, and in ^{13}C n.m.r. they are masked completely by the much larger changes due to σ_p and σ_d. In particular the ring current effect is practically absent for ^{13}C resonances, as is suggested by the fact that a common region for olefinic and aromatic carbons exists on the $\delta(^{13}C)$ scale. Only in carefully selected cases has it been found that variations in $\delta(^{13}C)$ can be attributed to the shielding effects of cyclic π electron systems. The detection of ring current effects is therefore only within the domain of ^{1}H n.m.r.

In contrast, in a number of cases, for example carboxylate ions and protonated amines, it has been shown that the electrical field effect contributes significantly to carbon shielding. Of course, changes in $\Delta\sigma$ due to the polarization of bonds are ultimately a consequence of changes in σ_d and σ_p, and a qualitative estimate of the electrical field effect through classical equations such as equation (4.16) merely constitutes a different approach to the same phenomenon. Similarly, the van der Waals effect, which is probably related to the γ-effect discussed below, may be treated by bond polarization models.

3.2 EMPIRICAL CORRELATIONS

Since the early days of ^{13}C n.m.r. a number of empirical chemical shift/molecular structure relations were developed which are most useful for the analysis of ^{13}C spectra. Some of them may be rationalized using the concepts discussed in the last section, but they may also be taken as empirical correlations that are more or less self-consistent.

Best known are the substituent effects observed in alkanes, where replacement of a hydrogen atom by a methyl group leads to deshielding of 9–10 ppm for the α- and β-carbon and to shielding of 2.5 ppm for the γ-carbon. Comparing the data for pentane (**211**) and the branched hydrocarbons **212** and **213** with those of butane (**210**) as a reference, these effects are clearly documented. They are fairly constant for the whole series, and an empirical additivity rule may be used to predict the chemical shift for carbons in alkane chains:

$$\delta(C_i) = B + \sum_j n_j A_j \qquad (11.13)$$

$$CH_3{-}CH_2{-}CH_2{-}CH_3$$
$$13.3 \quad 23.3$$

210

$$CH_3{-}CH_2{-}CH_2{-}CH_2{-}CH_3$$
$$13.7 \quad 22.7 \quad 34.6$$

211

$$\begin{array}{l} H_3C \\ \diagdown \\ CH-CH_2-CH_3 \\ \diagup \\ H_3C \end{array}$$
$$22.3 \quad 30.2 \quad 32.1 \quad 11.8$$

212

$$\begin{array}{l} H_3C \\ \diagdown \\ H_3C-C-CH_2-CH_3 \\ \diagup \\ H_3C \; 30.7 \qquad 9.0 \\ 29.2 \qquad 37.0 \end{array}$$

213

Here B is a constant almost equal to the chemical shift of methane ($\delta -2.3$ ppm), A_j is the chemical shift increment for α-, β-, or γ-substituents, and n_j is the number of substituents present in the particular position. By regression analysis it was

Table 11.2 Substituent increments for the empirical calculation of $\delta(^{13}C)$ values in alkanes (after Ref. 19)

$$\delta(C_k) = B + A_1 n_{kl} + S_{kl} \qquad\qquad (11.14)$$

A_1 increments S_{kl} increments

				adjacent C atom l		
		C_k	1 °	2 °	3 °	4 °
A_α	9.1	1 °	0	0	−1.1	−3.4
A_β	9.4	2 °	0	0	−2.5	−7.5
A_γ	−2.5	3 °	0	−3.7	−9.5	—
A_δ	0.3	4 °	−1.5	−8.4	—	—
A_ε	0.2					

determined that $A_\alpha = +9.1$, $A_\beta = +9.4$, $A_\gamma = -2.5$, and $B = -2.6$ ppm. More elaborate equations with correction terms for branching have been proposed and substituent increments developed for these situations are collected in Table 11.2.

Exercise 11.5 Predict the $\delta(^{13}C)$ values of 3-methylheptane with the help of equation (11.14) (Table 11.2).

Similarly, methyl substitution in cycloalkanes leads to typical shift increments that differ from those in open-chain compounds. For cyclohexane ($\delta 27.6$ ppm) the following parameters for axial and equatorial substitution have been found:

	Axial CH$_3$	Equatorial CH$_3$
α-effect	+1.4	+6.0
β-effect	+5.4	−9.0
γ-effect	−6.4	0

The differentiation between axial and equatorial substituents is of importance for conformational analysis, where the stereochemistry of different conformers can be assessed through their ^{13}C data. The two cyclohexanes **214a** and **214b**, where the different α-effects clearly characterize the axial and the equatorial isomers are illustrative.

214 a   **214 b**

In the case of alkenes, the chemical shift of the olefinic carbon C_A may be predicted using the increments below with the ethylene value (123.2 ppm) as reference. Of interest is the different sign of the β-effect (increments for C-2 and C-1′, respectively).

$$\Delta\delta: \quad -0.4 \quad +6.8 \quad +7.7 \qquad\qquad -6.5 \quad -1.6 \quad +1.1$$
$$C_\gamma - C_\beta - C_\alpha - C_A = C - C_{\alpha'} - C_{\beta'} - C_{\gamma'}$$

Of course, such schemes can be extended to other substituents and a variety of additivity rules for different classes of compounds may be found in the literature. For obvious reasons we cannot treat them in detail here and will conclude our discussion with a short summary in Table 11.3. Additional ^{13}C chemical shifts are collected in the Appendix.

One of the most frequently discussed empirical observations is the γ-effect, i.e. the shielding observed for a carbon atom if substituents are introduced in the γ-position (see above for the A_γ values for alkanes). It is not restricted to alkyl groups in alkanes, and it has been observed for other substituents and in structures such as cyclohexanes, bicyclic systems, and olefins. As the examples chosen in Table 11.4 demonstrate, upfield shifts are observed in all cases where the stereochemistry leads to van der Waals interactions of the type indicated by diagram **215**.

215

Table 11.3 Substituent-induced chemical shifts for ^{13}C resonances in alkanes, alkenes, and benzenes; $\Delta\delta$ values in ppm (after Ref. 20)

Substituent	Alkanes			Alkenes		Benzenes			
	α	β	γ	α	β	ipso	ortho	meta	para
F	70.1	7.8	−6.8	24.9	−34.3	35.1	−14.3	0.9	−4.4
Cl	31.0	10.0	−5.1	2.6	−6.1	6.4	0.2	1.0	−2.0
Br	18.9	11.0	−3.8	−7.9	−1.4	−5.4	3.3	2.2	−1.0
I	−7.2	10.9	−1.5	−38.1	7.0	−32.3	9.9	2.6	−0.4
OR	49.0	10.1	−6.2	29.4	−38.9	30.2	−14.7	0.9	−8.1
OCOCH$_3$	52.0	6.5	−6.0	18.2	−27.1	23	−6	1	−2.3
NR$_2$	28.3	−11.3	−5.1			22.4	−15.7	0.8	−11.8
NO$_2$	61.6	3.1	−4.6	22.3	−0.9	19.6	−5.3	0.8	6.0
CN	3.1	2.4	−3.3	−15	+15	−16.0	3.5	0.7	4.3
COOH	20.1	2.0	−2.8	4.2	8.9	2.4	1.6	−0.1	4.8
CHO	29.9	−0.6	−2.7	13.6	13.2	9.0	1.2	1.2	6.0
CH=CH$_2$	21.5	6.9	−2.1	14.8	−5.8	7.6	−1.8	−1.8	−3.5
C≡CH	4.4	5.6	−3.4			−6.1	3.8	0.4	−0.2
C$_6$H$_5$	22.1	9.3	−2.6	12.5	−11.0	13.0	−1.1	0.5	−1.0
CH$_3$	9.1	9.4	−2.5	12.9	−7.4	9.3	0.6	0	−3.5

Table 11.4   The γ-effect in ^{13}C n.m.r. (δ_{TMS} in ppm)

This suggests C–H bond polarization as the physical cause of the upfield shift, but it seems that further theoretical and experimental work is necessary in order to substantiate this argument. Nevertheless, the γ-effect is an important tool for stereochemical studies.

4. Carbon-13 spin–spin coupling constants

There are three important groups of spin–spin interactions in ^{13}C n.m.r.: ^{13}C,^{13}C, ^{13}C,^{1}H, and ^{13}C,X coupling constants, where X is another n.m.r.-active element. It is of interest to note that the experimental approaches used for the determination of these constants differ considerably.

If we turn first to the homonuclear ^{13}C,^{13}C coupling constants we can state that these parameters, because of the low natural abundance of ^{13}C, can be measured routinely, that is without special experimental provisions, only for ^{13}C labelled compounds. As was already pointed out in Section 2.2.4, the natural abundance of adjacent ^{13}C,^{13}C spin pairs only amounts to 0.1‰. In order to analyse such spin systems in natural abundance, long data acquisition times are necessary. The $J(^{13}$C,^{13}C) couplings between non-equivalent ^{13}C nuclei are then measured from the ^{13}C satellites in the ^{1}H-decoupled ^{13}C n.m.r. spectra. This situation prevails even after the introduction of new measuring techniques like the INADEQUATE sequence, which can be used in the majority of situations where AX- or AB-type spectra are found. Isotopomers with three ^{13}C nuclides can safely be neglected.

Secondly, there is an extensive body of experimental data concerning ^{13}C,^{1}H coupling constants, mostly those over one bond. These have been measured from ^{1}H n.m.r. spectra, as was mentioned in Chapter 3, which has the advantage of measuring the nucleus with the larger γ-factor, and more recently from ^{13}C n.m.r. spectra. Quite often, however, the spin systems observed are of higher order and the exact determination of coupling constants involves a complete analysis of the ^{1}H-coupled ^{13}C spectrum. The gated decoupling technique described in Section 2 is used to advantage here.

Finally, $^{13}C,X$ coupling constants with $X = {}^{19}F, {}^{31}P$, etc., are most easily measured from the ^{1}H-decoupled ^{13}C spectra of the appropriate compounds.

4.1 CARBON-13 COUPLING CONSTANTS AND CHEMICAL STRUCTURE

4.1.1 $^{13}C,^{13}C$ Coupling constants

From the data collected in Table 11.5, it becomes clear that $^{13}C,^{13}C$ coupling constants over one bond are sensitive parameters of the nature of the carbon–carbon bond involved. In the case of hydrocarbons, a dependence on the s-character product for the carbon orbitals ϕ_i and ϕ_j forming the C_i-C_j sigma bond has been observed:

$$^{1}J(^{13}C, {}^{13}C) = 550\, s(i)\, s(j) \qquad (11.15)$$

The basis of such a correlation is the assumption that only one of the various mechanisms that contribute to spin–spin coupling, the so-called *Fermi contact term*, dominates. This term depends on the electron density at the nucleus—hence the name 'contact term'—and consequently only on the s-orbitals involved. The following experimental observations serve to illustrate this point:

	$CH_2{=}C{=}CH_2$	$\overset{C}{\triangleright}C{=}CH_2$	$\triangleright C{=}CH_2$	
$J(^{13}C,^{13}C)$	98.7	23.2	95.2	(Hz)
$s(i)s(j)$	$\frac{1}{2}\cdot\frac{1}{3}=\frac{1}{6}$	$\frac{1}{6}\cdot\frac{1}{4}=\frac{1}{24}$	$\frac{1}{2}\cdot\frac{1}{3}=\frac{1}{6}$	

Here, the Walsh model for cyclopropane is used to derive the s-character of the particular hybrid orbitals in the three-membered ring.

4.1.2 $^{13}C,^{1}H$ Coupling constants

As for $^{13}C,^{1}H$ coupling constants, the structure dependence of the $^{1}J(^{13}C,^{1}H)$ data shown with a number of examples in Table 11.6 has been of considerable interest to chemists. In this case too an s-character dependence was found and for hydrocarbons the empirical relation

$$^{1}J(^{13}C,^{1}H) = 500\, s(i) \qquad (11.16)$$

has been derived. By analogy with equation (11.15) (note that in the present case $s(j) = 1$ for the proton 1s orbital), it relates the coupling to the fractional s-character $s(i)$ of the C,H bond involved. Since the s-character in turn is related through the equation $s(i) = 1/(1 + \alpha)$ to the hybridization parameter α of the carbon orbital sp^{α}, one can obtain information concerning the hybridization of a particular carbon atom through $^1J(^{13}C,^1H)$ measurements. Thus, for the highly strained hydrocarbon benzocyclopropene, the combination of data derived from $^1J(^{13}C,^{13}C)$ and $^1J(^{13}C,^1H)$ couplings yields the following hybridization diagram:

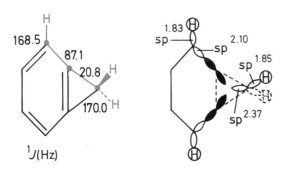

In particular, the remarkable *ring size dependence* of $^1J(^{13}C,^1H)$ in cycloalkanes is of interest and lends the coupling constant diagnostic value. Three-membered rings may therefore be identified through their large $^1J(^{13}C,^1H)$ values. For cyclopropane, the value of 161 Hz, typical for an sp^2 CH bond, is in accord with the Walsh model for the bonding situation in the three-membered ring. From equation (11.16) it follows that 32% s-character is contained in the C,H bond orbitals, leading to $sp^{2.1}$ hybrids. For the C–C bonds, on the other hand, high p-character (82%) results.

Attempts have also been made to draw conclusions from the $^{13}C,^1H$ coupling constants concerning the bond angle in the CH_2 group under consideration. However, this parameter is a better probe for the *interorbital angle*, and it must be remembered that this angle often deviates significantly from the structural angle between the internuclear axes. Also of importance is the fact that, in addition to hybridization changes, several other factors are important in determining the magnitude of the $^{13}C,^1H$ coupling constant. In particular, electronegative substituents can give rise to significant variations, as is seen in the values for the chloromethanes given in Table 11.6. Presumably changes in the effective nuclear charge of carbon are responsible for these findings. In these cases the simple correlation with the hybridization of the carbon bond orbitals breaks down.

In contrast to the strong variation of the $^1J(^{13}C,^{13}C)$ data with the bond order or bond length of the C–C bond involved, substituent effects are less pronounced. From the $^{13}C,^{13}C$ coupling constants over more than one bond the vicinal interaction is of interest. Here, a dihedral angle dependence similar to the Karplus curve for $^1H,^1H$ coupling constants (p. 115) has been found.

Table 11.5 $^{13}C,^{13}C$ coupling constants (Hz)

H_3C-CH_3	34.6	$CH_3-\overset{\displaystyle CH_3}{\underset{\displaystyle CH_3}{C}}X$	$X = CH_3$	36.9
$H_2C=CH_2$	67.6		$= NH_2$	37.1
$HC\equiv CH$	171.5		$= OH$	39.5
$C_6H_5-H_2C-CH_3$	34		$= Cl$	40.0
$C_6H_5-HC=CH_2$	70		$= Br$	40.2
$C_6H_5-\underset{\displaystyle \overset{\|}{O}}{C}-CH_3$	43	$\overset{\displaystyle H_2C}{\underset{\displaystyle H_2C}{\diagdown}}\overset{X}{\underset{Y}{\diagup}}$	$\begin{array}{cc} X & Y \\ Br & H \\ I & H \\ Cl & Cl \end{array}$	$\begin{array}{c} \\ 13.3 \\ 12.9 \\ 15.5 \end{array}$
$H_3C-C\equiv N$	57.3			

Exercise 11.6   Use the ^{13}C coupling constants measured for methylene cyclopropane (p. 505 and $^1J(C,H)_{ring} = 161.5$, $^1J(C,H)_{db} = 160.8$ Hz) and derive with the help of equations (11.15) and (11.16), on the basis of the Walsh model for three-membered rings, a hybridization diagram for this hydrocarbon.

Geminal and *vicinal* $^{13}C,^1H$ coupling constants are much smaller than the couplings over one bond. Generally, values between 1 and 12 Hz are observed. Their determination from the ^{13}C satellites in ^1H n.m.r. spectra is thus more difficult because the signals of interest are closer to the large main signal of ^{12}C molecules. In addition, it usually requires the exact analysis of complicated spin systems.

With respect to structure determinations the vicinal $J(^{13}C,^1H)$ data are of greatest interest. Like the $^3J(^1H,^1H)$ data, they depend on bond lengths and dihedral angles. For benzenoid aromatics, a correlation between $^3J(^1H,^1H)_{trans}$ and the bond length R_{CC} (in nm) or the Hückel MO π-bond order $P_{\mu\nu}$ of the central $C-C$ bond in the particular $^{13}C-C-C-^1H$ fragment, respectively, was found (Figure 11.31, p. 509). The linear relations have the form

$$^3J(^{13}C,^1H) = -404.7\,R_{CC} + 63.69 \qquad (11.17)$$

$$^3J(^{13}C,^1H) = 14.77\,P_{\mu\nu} - 2.32 \qquad (11.18)$$

The dihedral angle dependence of the $^3J(^{13}C,^1H)$ data is of interest for conformational analysis, as for example in the case of nucleotides or carbohydrates. Equations which are analogous to the well-known Karplus equation for vicinal $^1H,^1H$ coupling constants (p. 115) have been derived experimentally as well as from theoretical calculations. Those based on experimental data vary to some extent depending on the class of compounds which was used to derive these relations. For aliphatic hydrocarbons we have

$$^3J(^{13}C,^1H) = 4.50 - 0.87\,cos\phi + 4.03\cos 2\phi \qquad (11.19)$$

an equation shown in graphical form in Figure 11.32 (p. 509). As in the case of the vicinal $^1H,^1H$ coupling one finds $^3J(0°) < {}^3J(180°)$ and in the case of olefins $^3J_{trans} < {}^3J_{cis}$.

Table 11.6 ^{13}C,^{1}H coupling constants (Hz) over one bond

CH$_4$ 125	H$_2$C=CH$_2$ 157	HC≡CH 250
cyclobutane–H 134	bicyclo[2.2.1] 142	
cyclopropane–H 161	bicyclo–CH 178.5	
bicyclobutane H^1–H^2 (1) 164 (2) 144	cycloheptatriene 131	
H^1, H^2, H^3 (1) 202 (2) 170 (3) 152	142	
H$_5$C$_6$ C$_6$H$_5$ O 190	cyclopropene H 220	
200	cyclobutene H 170	
COOR / ROOC 160	cyclopentene H 162	
H^1 H^1 H^2 H^1 (1) 135.5 (2) 146 (3) 172.5	cyclohexene H 158	
	benzene H 159	
	CH$_3$Cl 151	
	CH$_2$Cl$_2$ 178	
	CHCl$_3$ 209	

4.1.3 ^{13}C,X Coupling constants

Quite a number of magnetic X nuclei lead to line splittings in the ^{1}H-decoupled ^{13}C n.m.r. spectra of inorganic, organic or organometallic molecules, which, depending on the natural abundance of X, gives rise to multiplet structures or satellite lines. In organic chemistry, for instance, numerous ^{13}C,^{19}F, ^{13}C,^{31}P, or ^{13}C,^{15}N coupling constants have been measured which yield interesting structural information. In

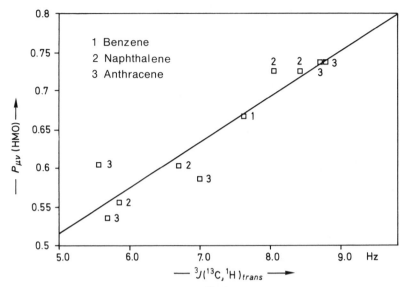

Figure 11.31 Relation between vicinal ^{13}C,^{1}H spin–spin coupling constants and HMO π-bond order of the central C–C bond in the ^{13}C–C–C–^{1}H fragment of benzenoid aromatics (after Ref. 21)

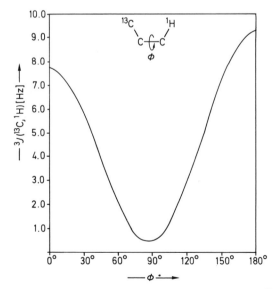

Figure 11.32 Dihedral angle dependence of vicinal ^{13}C,^{1}H spin–spin coupling constants in aliphatic hydrocarbons; one finds $^3J(0°) = 7.7$, $^3J(60°) = 2.0$, $^3J(90°) = 0.5$, $^3J(120°) = 2.9$ and $^3J(180°) = 9.4$ Hz (after Ref. 22)

organometallics, couplings like $J(^{13}C,^{6}Li)$, $J(^{13}C,^{119}Sn)$, or $J(^{13}C,^{199}Hg)$, to name only a few, are additionally observed. For a detailed account on these data the reader is referred to monographs on ^{13}C n.m.r. We note, however, that the one-bond spin–spin interactions between ^{13}C and ^{19}F nuclei are larger than those between ^{13}C and protons. They vary between 170 and 400 Hz and have a negative sign. The sign of the $^{13}C,^{1}H$ coupling constants in contrast, is positive, so that the signs of the other spin–spin interactions can often be correlated with the sign of $^{13}C,^{1}H$ coupling constants and thereby be determined.

5. Carbon-13 spin–lattice relaxation rates

Through the introduction of Fourier transform spectroscopy, *spin–lattice relaxation rates* for nuclei in organic molecules are now readily determined by means of inversion-recovery experiments or similar pulse techniques. We discussed these types of experiments in Chapter 7 and an application for ^{13}C was illustrated in Figure 11.8. It is therefore of interest to consider briefly which information can be derived from such measurements, and how the spin–lattice relaxation rate is related to phenomena of chemical interest.

As was outlined in Chapter 7, spin–lattice relaxation originates from fluctuating magnetic fields that provide an r.f. frequency suitable for an n.m.r. transition. There are several sources of such fluctuating fields, and therefore several mechanisms contribute to the relaxation. Of primary interest to us is the *dipolar relaxation mechanism*, where the fluctuating field results from a modulation of a dipolar spin–spin coupling. The local field induced at a nucleus by a neighbouring magnetic dipole is given by equation (1.12) and its time dependence arises in intramolecular cases through changes in the angle θ, and in intermolecular cases through changes in both θ and the internuclear distance r.

For ^{13}C it turns out that protons bonded to carbon most effectively contribute to the dipolar relaxation of the latter, the modulation of the coupling being provided by the molecular motion in the liquid phase. Following theoretical considerations, the dipolar relaxation rate, $(1/T_1)^{DD} = R_1^{DD}$, can be related to the distance r between the nuclei and the correlation time τ_c through the equation

$$R_1^{DD} = \left(\frac{\mu_0}{4\pi}\right)^2 \hbar^2 \gamma_C^2 \gamma_H^2 r^{-6} \tau_c \qquad (11.20)$$

where the other factors are well known constants. The correlation time characterizes the reorientation of a molecule in a liquid and for nonviscous solutions of samples with a molecular weight below about 500 it is of the order of 10^{-10} s.

Equation (11.20) thus yields the basis for obtaining information on intramolecular distances and molecular dynamics in the liquid state. In order to interpret the experimental results correctly, the extent to which other factors contribute to the observed relaxation rate must be determined. This is most conveniently done by measuring the nuclear Overhauser enhancement factor, η_i, for the particular carbon resonance, since the NOE itself depends on dipolar relaxation. As was pointed out in

Chapter 10, for pure dipolar relaxation the NOE has a maximum value of $\eta = \gamma_H/2\gamma_C = 1.988$. The fractional dipolar relaxation rate is thus given by

$$\text{Dipolar relaxation} = \frac{\eta_i}{1.988} \cdot 100 \ (\%) \tag{11.21}$$

and

$$R_1^{DD} = R_1^{Obs.} \cdot \frac{\eta_1}{1.988} \cdot 100 \tag{11.22}$$

The distance dependence of R_1^{DD} forms the basis for the use of T_1 measurements for assignment purposes that was mentioned on p. 492. It yields considerably different T_1 values for quaternary carbon atoms on one side and protonated carbon atoms of CH-, CH_2-, and CH_3-groups on the other with ranges of 20–100 and 1–20 s, respectively. The long T_1 values of quaternary carbons are responsible for the systematic diminution of intensity observed for such signals during routine measurements. Only if long relaxation delays or relaxation reagents are used, can ^{13}C spectra with correct intensity distributions be expected.

The molecular motion influences R_1^{DD} through the correlation time, τ_c. For rigid, isotropically tumbling systems, such as adamantane, the motion can be described a single τ_c and R_1^{DD} is a function of r or the number of protons present. Thus, for adamantane $R_1^{DD}(CH) = 49$ ms^{-1} and $R_1^{DD}(CH_2) = 88$ ms^{-1}. For anisotropic motion, on the other hand, different correlation times result for different C,H bonds. This is seen from the relaxation rates for the *ortho*- and *meta*-carbon atoms of diphenylacetylene on the one hand (182 ms^{-1}) and the *para*-carbon atom on the other (417 ms^{-1}). The latter is faster because τ_c *(para)* $> \tau_c$ *(ortho, meta)* owing to the preferential rotation of the molecule around its long axis.

R_1^{DD} is thus a probe for molecular dynamics and an interesting topic in this area is segmental motion, that is, localized motion along an aliphatic chain. Since T_1 values can be determined for both carbon atoms and protons, a body of experimental data sufficient to describe certain aspects of the dynamic behaviour of chain-like molecules can be obtained.

Thus, for 1-decanol the following $^{13}C - R_1^{DD}$ values were found:

$$H_3C - CH_2 - CH_2 - CH_2 - (CH_2)_5 - CH_2 - OH$$

$R_1^{DD}(s^{-1})$ 0.32   0.45   0.63   0.91   1.2-1.3   1.54

Their increasing magnitude towards the OH group indicates an increase in the effective correlation time and reduced motion due to hydrogen bonding. A quantitative treatment of such results is by no means straightforward, since sophisticated models are needed to separate the effects of segmental motion from those of overall molecular motion.

6. References

1. *Spectral Catalogue*, Bruker Analytische Messtechnik, Karlsruhe.
2. *Technical Bulletin*, Joel-Kontron Co., München.
3. H. Hausmann, *PhD thesis*, University of Siegen, 1992.
4. W. Bremser, H. P. W. Hill, and R. Freeman, Messtechnik, **78**, 14 (1971).
5. P. Joseph-Nathan, R. L. Santillan, P. Schmitt, and H. Günther, *Org. Magn. Reson.*, **22**, 450 (1984).
6. P. Joseph-Nathan, J. R. Wesener, and H. Günther, *Org. Magn. Reson.*, **22**, 190 (1984).
7. H. Günther, P. Schmitt, H. Fischer, W. Tochtermann, J. Liebe, and Ch. Wolff, *Helv. Chim. Acta*, **68**, 801 (1985).
8. P. Schmitt and H. Günther, *Angew. Chem.*, **95**, 509 (1983); *Angew. Chem. Int. Ed. Engl.* **22**, 499 (1983).
9. H.-E. Mons, and H. Günther, unpublished.
10. P. Schmitt, *PhD thesis*, University of Siegen, 1983.
11. P. Bast, *PhD thesis*, University of Siegen, 1988.
12. U. Ewers, *PhD thesis*, University of Cologne, 1973.
13. H. Günther, and G. Jikeli, *Chem. Ber.*, **106**, 1863 (1973).
14. J. B. Stothers, *Carbon-13 NMR Spectroscopy*, Academic Press, New York 1972.
15. G. A. Olah, and G. P. Mateescu, *J. Am. Chem. Soc.*, **92**, 1430 (1970).
16. H. Spiesecke, and W. G. Schneider, *J. Chem. Phys.*, **35**, 731 (1961).
17. H. Günther, and H. Schmickler, *Pure Appl. Chem.*, **44**, 807 (1975).
18. E. Breitmaier, and K. H. Spohn, *Tetrahedron*, **29**, 1145 (1973).
19. E. Breitmaier, and G. Bauer, ^{13}C-*NMR-Spektroskopie*, Thieme Verlag, Stuttgart 1977.
20. J. T. Clerc, E. Pretsch, and S. Sternhell, ^{13}C-*Kernresonanzspektroskopie*, Akademische Verlagsges., Frankfurt, 1973.
21. H. Günther, and P. Schmitt, unpublished.
22. R. Aydin, and H. Günther, *Magn. Reson. Chem.*, **28**, 448 (1990).

RECOMMENDED READING

Textbooks

H. Friebolin, *One and Two-Dimensional NMR Spectroscopy*, 2nd Ed., VCH Publishers, 1993.
J. K. M. Sanders and B. K. Hunter, *Modern NMR Spectroscopy—A Guide for Chemists*, 2nd Ed., Oxford University Press, Oxford, 1993.
Monographs listed in the Bibliography (p. 549ff.) under (a) and (h).

Review articles

(a) G. A. Morris, Pulsed Methods for Polarization Transfer in ^{13}C NMR, in F6, **4**, 179 (1984).
(b) A. Bax, Two-dimensional NMR-Spectroscopy, in F6, **4**, 197 (1984).
(c) J. Buddrus, and H. Bauer, Direct Identification of the Carbon Skeleton of Organic Compounds Using Double Quantum Coherence ^{13}C-NMR Spectroscopy: The INADEQUATE Pulse Sequence, *Angew. Chem. Int. Ed. Engl.*, **26**, 625 (1987).

(d) C. H. Sotak, C. L. Dumoulin, and G. C. Levy, High-Accuracy Quantitative Analysis by ^{13}C Fourier Transform NMR Spectroscopy, in F6, **4**, 93.

(e) B. E. Mann, Dynamic ^{13}C NMR Spectroscopy, in F5, **11**, 95 (1978).

(f) S. Berger, Chemical Models for Deuterium Isotope Effects in ^{13}C- and ^{19}F-NMR, in F3, **22**, 1 (1989).

(g) G. L. Nelson, and E. A. Williams, Electronic Structure and ^{13}C NMR, *Progr. Phys. Org. Chem.*, **12**, 2290 (1976).

(h) R. N. Young, NMR Spectroscopy of Carbanions and Carbocations, in F5, **12**, 261 (1979).

(i) P. E. Hansen, Carbon–Hydrogen Spin–Spin Coupling Constants, in F5, **14**, 175 (1982).

(j) V. Wray, Carbon–Carbon Coupling Constants: a Compilation of Data and a Practical Guide, in F5, **13**, 177 (1980).

(k) L. B. Krivdin, and G. A. Kalabin, Structural Applications of One-Bond Carbon–Carbon Spin–Spin Coupling Constants, in F5, **21**, 293 (1989).

(l) J. L. Marshall, Carbon–Carbon and Carbon–Proton NMR Couplings: Applications to Organic Stereochemistry and Conformational Analysis, Methods in Stereochemical Analysis, Vol. 2, Verlag Chemie, Weinheim, 1983.

(m) D. A. Wright, D. E. Axelson, and G. C. Levy, Physical Chemical Applications of ^{13}C Spin Relaxation Measurements, in F6, **3**, 104 (1979).

(n) D. J. Craik, and G. C. Levy, Factors Affecting Accuracy in ^{13}C Spin–Lattice Relaxation Measurements, in F6, **4**, 239 (1984)

12 APPENDIX

1. The 'ring current effect' of the benzene nucleus

The effect of the magnetic anisotropy of the benzene nucleus on the resonance frequency of the neighbouring protons is graphically represented in Figure 12.1. Lines of equal shielding or deshielding are represented on a coordinate system the origin of which lies at the centre of the ring. The ordinate z (in units of the ring radius of 0.139 nm) runs along the six-fold rotational axis perpendicular to the plane of the ring and the abscissa ρ (also in units of the ring radius) runs from the centre in the direction of a C–H bond. The contributions, $\Delta\sigma$, to the shielding constant σ are given in parts per million.

2. Tables of proton resonance frequencies and ^{1}H,^{1}H coupling constants

To supplement the exposition concerning the relation between the ^{1}H n.m.r. parameters and the structure of organic molecules that was made in Chapter 4, further illustrative material is presented in Tables 12.1 and 12.2.

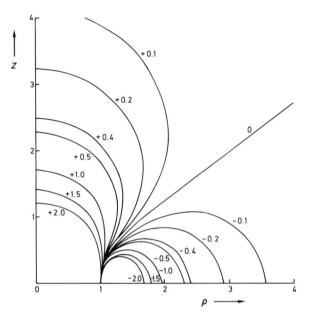

Figure 12.1 Graphic representation of the influence of the magnetic anisotropy of the benzene ring on proton resonance frequencies (after Ref. 1)

Table 12.1 Proton resonance frequencies (δ-scale, ppm) of selected organic compounds

A. Methyl and methylene protons

2.13	1.66	1.93	1.98	0.97
1.81	1.73	1.96	1.94	2.17
2.65	2.46	2.55	2.32	2.37

A. Methyl and methylene protons

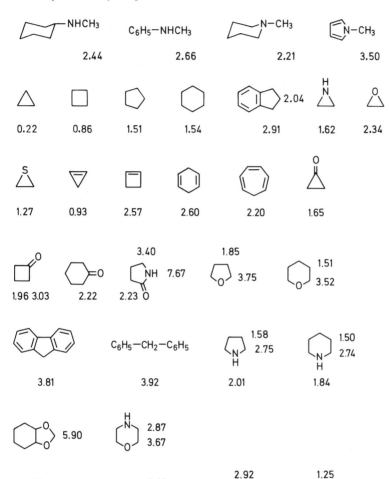

B. Aldehyde protons

H₃C—CHO   H₂C=CH—CHO   ⬡—CHO

   9.72   9.53   10.0

C. Olefinic protons

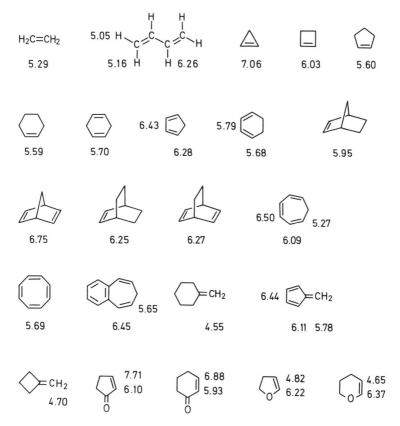

D. Aromatic protons

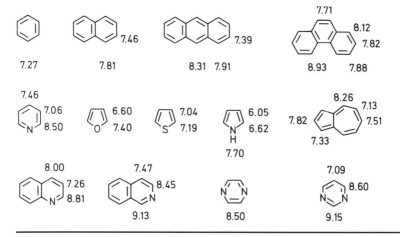

Table 12.2 ^{1}H,^{1}H coupling constants in selected organic molecules

X	J_{cis}	J_{trans}	^2J	Ref.	X	J	Ref.
H	11.6	19.1	2.5	2	Li	8.90	2
Li	19.3	23.9	7.1	2	Si(C$_2$H$_5$)$_3$	8.0	2
COOH	10.2	17.2	1.7	2	H	7.5	2
CN	11.75	17.92	0.91	2	C$_6$H$_5$	7.62	2
C$_6$H$_5$	11.48	18.59	1.08	2	CN	7.60	2
CH$_3$	10.02	16.81	2.08	2	I	7.45	2
OCH$_3$	7.0	14.1	−2.0	2	Br	7.33	2
Cl	1.3	14.6	−1.4	2	CH$_3$	7.26	2
Br	7.1	15.2	−1.8	2	Cl	7.23	2
F	4.65	12.75	−3.2	2	N(C$_2$H$_5$)$_2$	7.13	2
					OC$_2$H$_5$	6.97	2
					$^+$O(C$_2$H$_5$)$_2$	4.7	2

X	$J(1,2)$	$J(1,3)$	$J(2,4)$	$J(3,5)$	$J(2,5)$	$J(2,3)$	Ref.
H	8.97	5.58	8.97	8.97	5.58	−4.34	3
Cl	7.01	3.58	10.26	10.58	7.14	−6.01	4
Br	7.13	3.80	10.16	10.45	7.01	−6.12	4
I	7.51	4.37	9.89	9.97	6.63	−5.94	4
NH$_2$	6.63	3.55	9.65	9.89	6.18	−4.29	4
CN	8.43	5.12	9.18	9.49	7.08	−4.72	4
COOH	8.04	4.57	9.26	9.66	7.14	−4.00	4
COCl	7.88	4.43	9.19	9.99	7.59	−4.46	4
COCH$_3$	7.96	4.55	8.76	9.60	6.94	−3.41	4

X	$J(1,2)$	$J(1,3)$	$J(1,4)$	$J(1,5)$	$J(2,3)$	$J(2,4)$	Ref.
H	7.54	1.37	0.66	1.37	7.54	1.37	5
Li	6.73	1.54	0.77	0.74	1.42	1.29	6
CH$_3$	7.64	1.25	0.60	1.87	7.52	1.51[a]	7
COOCH$_3$	7.86	1.35	0.63	1.79	7.49	1.31	8
I	7.93	1.14	0.47	1.88	7.47	1.75	8
Br	8.05	1.12	0.46	2.1	7.44	1.78	8

X	$J(1,2)$	$J(1,3)$	$J(1,4)$	$J(1,5)$	$J(2,3)$	$J(2, 4)$	Ref.
Cl	8.05	1.13	0.48	2.27	7.51	1.72	8
NH_2	8.02	1.11	0.47	2.53	7.39	1.60	9
$N(CH_3)_2$	8.40	1.01	0.43	2.76	7.29	1.76	9
$N(CH_3)_3$	8.55	0.92	0.48	3.05	7.46	1.69	9
NO_2	8.36	1.18	0.55	2.40	7.47	1.48	9
OH	8.17	1.09	0.49	2.71	7.40	1.74	9
OCH_3	8.30	1.03	0.44	2.94	7.36	1.76	9
F	8.36	1.07	0.43	2.74	7.47	1.82^b	8

[a] $J(1, CH_3)$ −0.75   [b] $J(1, F)$ 8.91
$J(2, CH_3)$ 0.36 $J(2, F)$ 5.69
$J(3, CH_3)$ −0.62 $J(3, F)$ 0.22

Substituent effects $S(J)$ for H,H coupling constants in mono-substituted benzenes (Ref. 10)

J_{ij}	F	Cl	Br	I	NO_2	OCH_3
12	+0.81	+0.61	+0.53	+0.39	+0.77	+0.79
13	−0.34	−0.23	−0.27	−0.25	−0.20	−0.32
14	−0.24	−0.16	−0.20	−0.19	−0.16	−0.22
15	+1.21	+0.87	+0.71	+0.51	+1.02	+1.33
23	−0.04	+0.03	−0.05	−0.04	−0.07	−0.16
24	+0.39	+0.34	+0.36	+0.37	+0.08	+0.38

$J(1,2)$	8.9	8.75	10.14	8.2	8.4
$J(1,3) + J(1,4)$	1.48	1.50	—	1.27	1.38
$J(1,5)$	—	—	−0.25	—	—
$J(1,6)$	—	—	0.22	—	—
$J(2,3)$	5.51	5.59	—	6.97	6.81
$J(2,4)$	0.72	0.54	—	0.73	0.77
$J(2,5)$	0.69	0.76	0.57	0.51	0.34
$J(3,4)$	11.17	10.96	—	8.45	8.43
$J(1,7)$		5.89		3.17	2.75
$J(1,7')$	6.7	—	6.47	8.90	8.70
$J(2,7)$		1.42		−1.07	−1.32
$J(2,7')$	0.4	—	−0.74	0.92	0.94
$J(3,7)$	—	0.32		—	—
$J(7,7')$	−13.0	—	—	−14.05	−14.14
Ref.	11	12	13	14	11

Structure	Coupling Constants			Ref.

Three-membered rings (H O H / H H epoxide, H S H thiirane, H N(H) H aziridine)

Structure	$^3J_{cis}$	$^3J_{trans}$	^2J	Ref.
oxirane (O)	4.45	3.1	5.5	15
thiirane (S)	7.15	5.65	<0.4	15
aziridine (N–H)	6.3	3.8	2.0	15

Structure (H¹–H⁴)	$J(1,2)$	$J(1,3)$	$J(1,4)$	$J(2,3)$	Ref.
	−13.05	3.65	13.12	2.96	16

Structure (H¹, H², H³, H³′)	$J(1,2)$	$J(1,3)$	Ref.
	1.3	1.75	17

Structure (H¹, H², H³, H³′, H⁴, H⁴′)	$J(1,2)$	$J(1,3)$	$J(1,4)$	$J(3,3')$	$J(3,4)$	$J(3,4')$	Ref.
	2.85	−0.35	1.0	−12.00	4.65	1.75	18

Structure (H¹, H², H³, H⁴, H⁵, H⁵′)	$J(1,2)$	$J(1,3)$	$J(1,4)$	$J(2,3)$	$J(1,5)$	$J(2,5)$	Ref.
	5.05	1.09	1.98	1.91	1.33	−1.51	19

Structure (H¹, H², H³, H⁴)	$J(1,2)$	$J(1,3)$	$J(1,4)$	$J(2,3)$	Ref.
	9.64	1.02	1.12	5.04	19

Structure (O; positions 1,2,3,4; R, R′)	$J(1,2)$	$J(1,3)$	$J(1,4)$	$J(2,4)$	Ref.
	10.2	1.9	0.1	3.0	20

Structure (O; R, R; positions 1,2,3,4)	$J(1,2)$	$J(1,3)$	$J(1,4)$	$J(2,3)$	$J(2,4)$	$J(3,4)$	Ref.
	9.76	0.90	0.88	5.78	1.80	9.45	21

Structure	Coupling Constants					Ref.

	$J(1,2)$	$J(1,3)$	$J(1,4)$	$J(1,5)$	$J(1,6)$	
	1.74	10.17	−0.86	0.60	1.30	22
	$J(2,3)$	$J(2,4)$	$J(2,5)$		$J(3,4)$	
	17.05	−0.83	0.60		10.41	

Structure	$J(1,2)$	$J(1,3)$	$J(1,4)$	$J(2,3)$	Ref.
	7.54	1.37	0.66	7.54	5
	8,28	1.24	0.74	6.85	23
	8.55	1.20	0.82	6.59	23
	6.80	0.74	1.08	8.24	24
	1.75	0.85	1.4	3.3	25
	5.00	1.06	2.80	3.50	26
	4.88	1.24	1.00	7.67	27

	$J(1,5)$	$J(2,4)$				Ref.
	−0.13	1.97				27

X	$J(1,2)$	$J(1,3)$	$J(1,4)$	$J(2,3)$	Ref.
CH$_2$	8.97	−0.02	1.46	9.19	28
O	8.77	0.28	1.13	9.28	28
NH	8.82	0.06	1.50	9.31	29

3. The Hamiltonian operator (5.3) in polar coordinates

When polar coordinates are used instead of Cartesian coordinates, the following relations hold (cf. Figure 12.2):

$$x = r \sin \theta \cos \phi \qquad y = r \sin \theta \sin \phi \qquad z = r \cos \theta$$

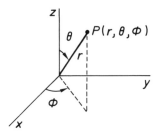

Figure 12.2 Definition of polar coordinates

With the substitution of these values, the Hamiltonian operator (5.3) becomes

$$\mathcal{H} = \frac{-h^2}{8\pi^2 m} \left[\frac{1}{r^2} \frac{\partial}{\partial r} \left(r^2 \frac{\partial}{\partial r} \right) + \frac{1}{r^2 \sin \theta} \frac{\partial}{\partial \theta} \left(\sin \theta \frac{\partial}{\partial \theta} \right) + \frac{1}{r^2 \sin^2 \theta} \frac{\partial^2}{\partial \phi^2} \right] + V$$

$$(12.1)$$

For the model of the electron on a circle, r is a constant and θ is 90°. Under these conditions the first and the second terms in the square brackets are zero and equation (12.1) simplifies to equation (5.4).

4. Commutable operators

We consider two operators $\hat{Q}$ and $\hat{R}$. For $\hat{Q}$ the eigenvalue equation

$$\hat{Q}\Psi_m = q_m \Psi_m \qquad (12.2)$$

with the eigenfunction Ψ_m and the eigenvalue q_m holds. A matrix element $\langle \Psi_n \hat{R} \hat{Q} \Psi_m \rangle$ may then be rewritten as

$$\langle \Psi_n | \hat{R} \hat{Q} | \Psi_m \rangle = q_m \langle \Psi_n | \hat{R} | \Psi_m \rangle \qquad (12.3)$$

Since $\hat{R}$ and $\hat{Q}$ commute it also holds that

$$\langle \Psi_n | \hat{R} \hat{Q} | \Psi_m \rangle = \langle \Psi_n | \hat{Q} \hat{R} | \Psi_m \rangle \qquad (12.4)$$

$\hat{R}$ and $\hat{Q}$ are Hermitian operators so it follows that ****

$$\langle \phi | \hat{R} | \lambda \rangle = \langle \hat{R} \phi | \lambda \rangle \text{ and } \langle \phi | \hat{Q} | \lambda \rangle = \langle \hat{Q} \phi | \lambda \rangle \qquad (12.5)$$

---

* All operators that are associated with real physical properties have this important property.

Then equation (12.4) yields

$$\langle \Psi_n | \hat{Q}\hat{R} | \Psi_m \rangle = \langle \hat{Q}\Psi_n | \hat{R}\Psi_m \rangle = q_n \langle \Psi_n | \hat{R} | \Psi_m \rangle \tag{12.6}$$

With equations (12.3) and (12.4) one obtains

$$q_m \langle \Psi_n | \hat{R} | \Psi_m \rangle = q_n \langle \Psi_n | \hat{R} | \Psi_m \rangle$$

or

$$(q_m - q_n)\langle \Psi_n | \hat{R} | \Psi_m \rangle = 0 \tag{12.7}$$

For $q_n \neq q_m$ this equation holds only if the expression $\langle \Psi_n | \hat{R} | \Psi_m \rangle$ becomes zero. Thus we have proved the expression introduced on p. 151.

5. The F_z operator

The magnetic quantum number of an individual nucleus, i, is $m_I(i)$. The product functions for a spin system with several nuclei are characterized by the total spin m_T, the sum of the magnetic quantum numbers of the individual nuclei:

$$m_T = \sum_i m_I(i) \tag{12.8}$$

By analogy with this, $\hat{I}_z$ is defined as the nuclear spin operator for an individual nucleus and the operator $\hat{F}_z$ is defined as the sum of the nuclear spin operators of the nuclei of the spin system under consideration:

$$\hat{F}_z = \sum_i \hat{I}_z(i) \tag{12.9}$$

All product functions ϕ_k are eigenfunctions of $\hat{F}_z$ the eigenvalues of which are the m_T values. Since the F_z operator commutes with the Hamiltonian operator the Hamiltonian matrix factorizes into submatrices because matrix elements of the type $\langle \phi_k | H | \phi_l \rangle$ vanish when ϕ_k and ϕ_l are product functions with different total spin, i.e. with different eigenvalues of the $\hat{F}_2$ operator.

6. Eigenfunctions of the AB$_2$ system

According to Table 5.2a (p. 166), the basis functions ϕ_2 and ϕ_3 as well as ϕ_4 and ϕ_5, respectively, have the same total spin and consequently mix with one another. Therefore, the variational calculation must be used for the determination of the correct eigenvalues and eigenfunctions. The elements of the secular determinants are obtained by following the rules introduced on p. 157ff. For example, the calculation for the off-diagonal element H_{23} yields:

$$H_{23} = \langle \phi_2 | \mathscr{H} | \phi_3 \rangle$$
$$=< \langle \alpha(\alpha\beta + \beta\alpha)/\sqrt{2} | \mathscr{H} | \beta\alpha\alpha \rangle$$
$$= \frac{1}{\sqrt{2}} (\langle \alpha\alpha\beta | \mathscr{H} | \beta\alpha\alpha \rangle + \langle \alpha\beta\alpha | \mathscr{H} | \beta\alpha\alpha \rangle)$$
$$= \frac{1}{\sqrt{2}} \left(\frac{1}{2} J_{AB} + \frac{1}{2} J_{AB} \right)$$
$$= J_{AB} / \sqrt{2}$$

Because of the symmetry of the Hamiltonian matrix, the identity $H_{23} = H_{32}$ holds and one obtains the same result for $H_{45} = H_{54}$.

The diagonal elements are easily derived from the results for the single nucleus and the A_2 system (pp. 142 and 145, respectively) because the basis functions are product functions of the type $\phi(A)\phi(B)$ and their energies are the sums of the energies E_A and E_B. One then immediately obtains

$$H_{11} = E_1 = \tfrac{1}{2} v_{A} + v_B + \tfrac{1}{2} J_{AB}$$
$$H_{22} = \tfrac{1}{2} v_A$$
$$H_{33} = -\tfrac{1}{2} v_A + v_B - \tfrac{1}{2} J_{AB}$$
$$H_{44} = \tfrac{1}{2} v_A - v_B - \tfrac{1}{2} J_{AB}$$
$$H_{55} = -\tfrac{1}{2} v_A$$
$$H_{66} = E_6 = -\tfrac{1}{2} v_A - v_B + \tfrac{1}{2} J_{AB}$$
$$H_{77} = E_7 = \tfrac{1}{2} v_A - \tfrac{3}{4} J_{BB}$$
$$H_{88} = E_8 = -\tfrac{1}{2} v_A - \tfrac{3}{4} J_{BB}$$

Thus, all of the diagonal elements in the symmetrical set have been reduced by the amount $\tfrac{1}{4} J_{BB}$. The determinants that must be solved for the determination of E_2 to E_5 are

$$\begin{vmatrix} \tfrac{1}{2} v_A - E & J_{AB}/\sqrt{2} \\ J_{AB}/\sqrt{2} & (-\tfrac{1}{2} v_A + v_B - \tfrac{1}{2} J_{AB}) - E \end{vmatrix} = 0 \qquad (12.10)$$

and

$$\begin{vmatrix} (\tfrac{1}{2} v_A - v_B - \tfrac{1}{2} J_{AB}) - E & J_{AB}/\sqrt{2} \\ J_{AB}/\sqrt{2} & -\tfrac{1}{2} v_A - E \end{vmatrix} = 0 \qquad (12.11)$$

From these, one derives

$$(\tfrac{1}{2} v_A - E)(-\tfrac{1}{2} v_A + v_B - \tfrac{1}{2} J_{AB} - E) - \tfrac{1}{2} J_{AB}^2 = 0 \qquad (12.12)$$

and

$$(\tfrac{1}{2} v_A - v_B - \tfrac{1}{2} J_{AB} - E)(-\tfrac{1}{2} v_A - E) - \tfrac{1}{2} J_{AB}^2 = 0 \qquad (12.13)$$

and the desired eigenvalues are

$$E_{2,3} = \tfrac{1}{2}\nu_B - \tfrac{1}{4}J_{AB} \pm \tfrac{1}{2}\sqrt{(\nu_A - \nu_B)^2 + (\nu_A - \nu_B)J_{AB} + \tfrac{9}{4}J_{AB}^2}$$
$$= \tfrac{1}{2}\nu_B - \tfrac{1}{4}J_{AB} \pm C_+$$

$$E_{4,5} = -\tfrac{1}{2}\nu_B - \tfrac{1}{4}J_{AB} \pm \tfrac{1}{2}\sqrt{(\nu_A - \nu_B)^2 - (\nu_A - \nu_B)J_{AB} + \tfrac{9}{4}J_{AB}^2}$$
$$= -\tfrac{1}{2}\nu_B - \tfrac{1}{4}J_{AB} \pm C_-$$

For the transitions represented in the energy level diagram of the AB_2 system (Figure 5.7, p. 167) the following energies are obtained:

A lines:

$$f_1 = E_1 - E_3 = \tfrac{1}{2}(\nu_A + \nu_B) + \tfrac{3}{4}J_{AB} + C_+$$
$$f_2 = E_2 - E_5 = \nu_B + C_+ + C_-$$
$$f_3 = E_7 - E_8 = \nu_A$$
$$f_4 = E_4 - E_6 = \tfrac{1}{2}(\nu_A + \nu_B) - \tfrac{3}{4}J_{AB} + C_-$$

B lines:

$$f_5 = E_2 - E_4 = \nu_B + C_+ - C_-$$
$$f_6 = E_1 - E_2 + \tfrac{1}{2}(\nu_A + \nu_B) + \tfrac{3}{4}J_{AB} - C_+$$
$$f_7 = E_3 - E_5 = \nu_B - C_+ + C_-$$
$$f_8 = E_5 - E_6 = \tfrac{1}{2}(\nu_A + \nu_B) - \tfrac{3}{4}J_{AB} - C_-$$

Combination line:

$$f_9 = E_3 - E_4 = \nu_B - C_+ - C_-$$

The equations given on p. 167 for the direct analysis of the AB_2 system follow from these values. We shall not go into the derivation of the eigenfunctions and their relative intensities. As mentioned earlier, the appearance of an AB_2 spectrum depends only on the ratio $J_{AB}/\nu_0\delta$.

7. Equations for the direct analysis of the AA′BB′ systems

The following equations which we present without proof are used for the direct analysis of AA′BB′ systems:

$$\nu_0\delta = \sqrt{4\,ab}$$
$$= \sqrt{2(ce - df)}$$
$$N = a - b$$
$$K = g + i + k - 2a - l$$
$$= 2b + k - h - j - l$$
$$= b + g + k - a - h - l$$
$$= b + i + k - a - j - l$$

$$M = (ce - df)/v_0\delta$$

$$L = \sqrt{(c - e)^2 - M^2}$$

$$= \sqrt{(d - f)^2 - M^2}$$

The frequencies of the lines a, b, etc., are relative to the centre of the spectrum at $(v_A + v_B)/2$.

In practice, the analysis begins with the assignment of the experimental lines to the specific transitions of the energy level diagram (Figure 5.19, p. 183). For this purpose the following equations that relate certain line separations are helpful:

$$c - d = e - f$$
$$g - h = i - j$$
$$g - a = b - j$$
$$a - i = h - b$$

Difficulties may arise if too many transitions are degenerate. In a number of cases there is also an overlap of the AA' and BB' portion in the centre of the spectrum. In general, however, enough experimental information can be gathered to derive at least an approximate parameter set which can be refined by iterative computer analysis.

Literature: Refs. 30–33.

8. The Bloch equations

According to the classical theory, the behaviour of a magnetic moment $\mathbf{M}$ in a magnetic field B is described by the equation of motion

$$\frac{d\mathbf{M}}{dt} = \gamma[\mathbf{M} \times \mathbf{B}] \tag{12.14}$$

In the n.m.r. experiment the static magnetic field $\mathbf{B}_0$ and the rotating field B_1 contribute to the field $\mathbf{B}$. $\mathbf{B}_0$ coincides with the z-axis in Cartesian coordinates while $\mathbf{B}_1$ rotates with frequency ω clockwise in the x–y plane (cf. Figure 7.1, p. 223). Thus the components of $\mathbf{B}$ are

$$B_x = B_1 \cos \omega t$$
$$B_y = -B_1 \sin \omega t$$
$$B_z = B_0$$

and the vector product in equation (12.14) can be resolved as follows:

$$\frac{dM_x}{dt} = \gamma(M_y B_0 + M_z B_1 \sin \omega t) \tag{12.15a}$$

$$\frac{dM_y}{dt} = \gamma(M_z B_1 \cos \omega t - M_x B_0) \tag{12.15b}$$

$$\frac{dM_z}{dt} = \gamma(-M_x B_1 \sin \omega t - M_y B_1 \cos \omega t) \qquad (12.15c)$$

Here we made use of the general relation for the product of two vectors **A** and **B**:

$$[\mathbf{A} \times \mathbf{B}] = (A_y B_z - A_z B_y)\mathbf{i} + (A_z B_x - A_x B_z)\mathbf{j} + (A_x B_y - A_y B_x)\mathbf{k}$$

where i, j, and k are the coordinate unit vectors.

For the time dependence of the magnetization we must, in addition, consider relaxation effects. These effects were included by Bloch phenomenologically in equation (12.15). The relaxation time T_2 is characteristic for the transverse magnetization in the x–y plane while the variation of the longitudinal magnetization along the z-axis is a function of T_1. The complete Bloch equations thus take the form

$$\frac{dM_x}{dt} = \gamma(M_y B_0 + M_z B_1 \sin \omega t) - \frac{M_x}{T_2} \qquad (12.16a)$$

$$\frac{dM_y}{dt} = \gamma(M_z B_1 \cos \omega t - M_x B_0) - \frac{M_y}{T_2} \qquad (12.16b)$$

$$\frac{dM_z}{dt} = \gamma(-M_x B_1 \sin \omega t - M_y B_1 \cos \omega t) - \frac{M_z - M_0}{T_1} \qquad (12.16c)$$

One now introduces the rotating coordinate system C' in which B_1 is stationary, so that

$$M_x = M_{y'} \cos \omega t - M_{x'} \sin \omega t$$
$$M_y = -M_{y'} \sin \omega t - M_{x'} \cos \omega t$$

In the rotating frame, these equations simplify to

$$\frac{dM_{x'}}{dt} = -(\omega_0 - \omega)M_{y'} - \frac{M_{x'}}{T_2} \qquad (12.17a)$$

$$\frac{dM_{y'}}{dt} = (\omega_0 - \omega)M_{x'} - \gamma B_1 M_z - \frac{M_{y'}}{T_2} \qquad (12.17b)$$

$$\frac{dM_z}{dt} = \gamma B_1 M_{y'} - \frac{M_z - M_0}{T_1} \qquad (12.17c)$$

Here M_0 is the equilibrium magnetization that is present in the magnetic field B_0 at the beginning of the experiment.

Under the conditions of slow adiabatic passage through the resonance — attained by the variation of the frequency ω or the field strength $\mathbf{B}_0$, i.e. the Larmor frequency ω_0 — the time dependence of the components M_z, $M_{x'}$, and $M_{y'}$ can be ignored. Equations (12.17) then have the following solutions:

$$M_{x'} = M_0 \frac{\gamma B_1 T_2^2(\omega_0 - \omega)}{1 + T_2^2(\omega_0 - \omega)^2 + \gamma^2 B_1^2 T_1 T_2} \qquad (12.18a)$$

$$M_{y'} = -M_0 \frac{\gamma B_1 T_2}{1 + T_2^2(\omega_0 - \omega)^2 + \gamma^2 B_1^2 T_1 T_2} \qquad (12.18b)$$

$$M_z = M_0 \frac{1 + T_2^2(\omega_0 - \omega)^2}{1 + T_2^2(\omega_0 - \omega)^2 + \gamma^2 B_1^2 T_1 T_2} \quad (12.18c)$$

Equation (12.18b) is equation (7.4) introduced on p. 227, from which the shape of the absorption signal was derived. Back transformation into the fixed laboratory frame yields for M_x and M_y

$$M_x = \frac{1}{2} M_0 \gamma T_2 \frac{T_2(\omega_0 - \omega) 2B_1 \cos \omega t + 2B_1 \sin \omega t}{1 + T_2^2(\omega_0 - \omega)^2 + \gamma^2 B_1^2 T_1 T_2} \quad (12.19a)$$

$$M_y = \frac{1}{2} M_0 \gamma T_2 \frac{2B_1 \cos \omega t - T_2(\omega_0 - \omega) 2B_1 \sin \omega t}{1 + T_2^2(\omega_0 - \omega)^2 + \gamma^2 B_1^2 T_1 T_2} \quad (12.19b)$$

Literature: Ref. 34.

9. The Bloch equations modified for chemical exchange

For the case in which two or more nuclei, as a result of a fast equilibrium, periodically exchange their chemical environment and thus their Larmor frequencies, the Bloch equations for n.m.r. line shape calculations must be modified. This is done most easily by combining equations (12.17a) and (12.17b) and defining a complex x–y magnetization, G:

$$G = M_{x'} + i M_{y'} \quad (12.20)$$

If one assumes that $M_z \approx M_0$, as is reasonable for weak B_1 fields, it then follows that

$$\frac{dG}{dt} = i(\omega_0 - \omega)G - i\gamma B_1 M_0 - \frac{1}{T_2} G \quad (12.21a)$$

$$\frac{dM_z}{dt} = \gamma B_1 M_{y'} - \frac{M_z - M_0}{T_1} \quad (12.21b)$$

In the absence of exchange for the positions A and B with the resonance frequencies ω_A and ω_B, it also holds that

$$\frac{dG_A}{dt} + \alpha_A G_A = -i\gamma B_1 M_{0A} \quad (12.22a)$$

and

$$\frac{dG_B}{dt} + \alpha_B G_B = -i\gamma B_1 M_{0B} \quad (12.22b)$$

with $\alpha_A = 1/T_{2A} - i(\omega_A - \omega)$ and $\alpha_B = 1/T_{2B} - i(\omega_B - \omega)$.

Through the equilibrium reaction x, y magnetization is transferred from A to B and back again. If one ignores the nuclear precession during the transitions from A and B and from B to A, equation (12.22) can be extended as follows:

$$\frac{dG_A}{dt} + \alpha_A G_A = -i\gamma B_1 M_{0A} + G_B/\tau_B - G_A/\tau_A \quad (12.23a)$$

$$\frac{dG_B}{dt} + \alpha_B G_B = -i\gamma B_1 M_{0B} + G_A/\tau_A - G_B/\tau_B \quad (12.23b)$$

Here the quantities $1/\tau_A = k_A$ and $1/\tau_B = k_B$ denote the probability that a nucleus undergoes a transition from A to B or from B to A.

For adiabatic passage through resonance again a stationary state is assumed, i.e. $dG_A/dt = dG_B/dt = 0$. With $M_{0A} = p_A M_0$ and $M_{0B} = p_B M_0$, where p_A and p_B are the corresponding molar fractions, one obtains

$$G = G_A + G_B$$
$$= -\gamma B_1 M_0 \frac{\tau_A + \tau_B + \tau_A \tau_B(\alpha_A p_B + \alpha_B p_A)}{(1 + \alpha_A p_A)(1 + \alpha_B p_B) - 1} \quad (12.24)$$

This equation contains the real and the imaginary parts of the x, y magnetization. For the calculation of the absorption signal that, according to equation (12.20), corresponds to the imaginary part, the two must be separated. This can be done either by using a computer program or by solving equation (12.24) and separating the imaginary part. In the last case one obtains equation (9.2) introduced on p. 339, in which all quantities were expressed in hertz ($\omega = 2\pi\nu$) and the spectrum was related to $(\nu_A + \nu_B)/2$ as the reference point. In addition, for simplification it was assumed that $1/T_{2A} = 1/T_{2B}$.

Finally, we shall show how the treatment of the Bloch equations set forth above may be extended to deal with exchange phenomena where more than two positions with different Larmor frequencies are involved. Assuming an exchange between three positions that have the Larmor frequencies ω_1, ω_2, and ω_3, one obtains for adiabatic passage:

$$(\alpha_1 + k_{12} + k_{13})G_1 - k_{21}G_2 - k_{31}G_3 = -i\gamma B_1 p_1 M_0 \quad (12.25a)$$
$$-k_{12}G_1 + (\alpha_2 + k_{21} + k_{23})G_2 - k_{32}G_3 = -i\gamma B_1 p_2 M \quad (12.25b)$$
$$-k_{13}G_1 - k_{23}G_2 + (\alpha_3 + k_{31} + k_{32})G_3 = -i\gamma B_1 p_3 M_0 \quad (12.25c)$$

In matrix notation this can be written as

$$\hat{A}\hat{G} = -iC\hat{P} \quad (12.26)$$

where $\hat{A}$ is a quadratic matrix, C is a constant ($= \gamma B_1 M_0$) and $\hat{P}$ and $\hat{G}$ are column vectors. If one multiplies from the left by the inverse of $\hat{A}$ one obtains

$$\hat{G} = -iC\hat{A}^{-1}\hat{P} \quad (12.27)$$

For the summation over all G_i that must be done for the calculation of the total x, y magnetization, G_T, equation (12.27) is multiplied from the left by the row vector $\hat{1}$:

$$G_T = \hat{1}\hat{G} = -iC\hat{1}\hat{A}^{-1}\hat{P} \quad (12.28)$$

The imaginary part of equation (12.28) gives the absorption signal the calculation of which may be accomplished by using an appropriately programmed computer. For our example, equation (12.28) has the form

$$G_T = -iC[1, 1, 1]\hat{A}^{-1}\begin{bmatrix} p_1 \\ p_2 \\ p_3 \end{bmatrix}$$

According to above the matrix $\hat{\mathbf{A}}$ is given by

$$\hat{\mathbf{A}} = 2\pi i(\hat{\mathbf{I}}v - \hat{\mathbf{W}}) + \pi\hat{\Delta} + \hat{\mathbf{X}} \qquad (12.29)$$

where

$\hat{\mathbf{I}}$ = the unit matrix;

v = the variable frequency (in hertz);

$\hat{\mathbf{W}}$ = the diagonal matrix of the Larmor frequencies v_i in the individual positions (in hertz);

$\pi\hat{\Delta}$ = the diagonal matrix of the natural line widths Δ (in hertz) in the individual positions; and

$\hat{\mathbf{X}}$ = the matrix of the reaction rate constants k with the elements

$$x_{ii} = \sum_{j(\neq i)} k_{ij} \quad \text{and} \quad x_{ij} = -k_{ji}$$

The same result is obtained on the basis of the theory of Anderson, Kubo and Sack that was derived from quantum mechanical principles.* Literature: Refs. 35 and 36.

10. The Haigh notation for spin systems

The notation for spin systems described in Chapter 5 as introduced by Pople, Schneider, and Bernstein [37] and modified to accommodate magnetic non-equivalence by Richards and Schaefer [38]. For larger spin systems, where several subsystems are repeated by symmetry, it is sometimes unwieldy and Haigh has proposed a more compact classification [39]. According to Haigh, each subsystem with non-equivalent nuclei is put into square brackets. An index indicates if this system appears again according to symmetry. In this case, apart from the couplings between the nuclei within brackets, couplings between the subsystems exist. This is best explained with an example. The AA'XX' system discussed above is classified after Haigh as an [AX]$_2$ system. The single [AX] system has only the coupling J_{AX}, for the [AX]$_2$ system the additional couplings J_{AA}, J_{XX} and a second A,X coupling, J'_{AX}, in our notation, operate. For smaller spin systems, both notations differ only slightly and the one we used is easier to interpret. For large spin systems the Haigh notation has the advantage of brevity. For example, the ten protons of anthracene are classified after Haigh as an [[AB]$_2$C]$_2$ system, while in our notation an AA'A'¹A'¹'BB'B'¹B'¹'CC' system results.

---

* One should note that the corresponding equations in the references cited are formally different because of a different choice of signs and the arrangement of certain quantities.

11. Chemical shifts of ^{13}C resonances in organic compounds

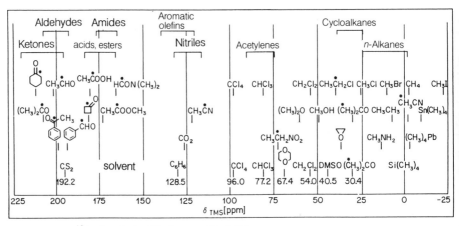

Figure 12.3 ^{13}C chemical shift diagram (Ref. 40)

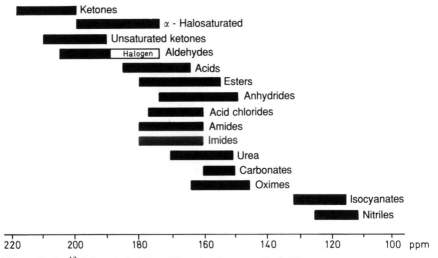

Figure 12.4 ^{13}C chemical shifts of functional groups (Ref. 41)

Table 12.3 ^{13}C resonances of common solvents (after Ref. 42)

	δ_{TMS}(ppm)	
Compound	Hydrogen compound	Perdeuterated compound
Cyclohexane	27.51	26.06
Acetone	30.43	29.22
Dimethyl sulphoxide	40.48	39.56
Dichloromethane	54.02	53.61

Dioxan	67.40	—
Chloroform	77.17	76.91
Carbon tetrachloride	95.99	—
Benzene	128.53	127.96
Acetic acid (^{a}CO)	178.27	—
Carbon disulphide	192.8	
Carbon disulphide*	193.7	

[a] Relative to external TMS at 0.0 ppm.

Table 12.4 Effect of solvent on the ^{13}C resonance of cyclohexane, benzene, and carbon disulphide (after Ref. 42)[a]

Solvent	Cyclohexane	Benzene	Carbon disulphide
Cyclohexane	27.5	128.5	193.0
Acetone	27.4	128.8	193.3
Methanol	27.8	128.3	192.8
Acetic acid	27.6	129.0	193.4
1,4-Dioxan	27.3	128.8	193.2
Chloroform	27.0	128.4	192.6
Carbon tetrachloride	26.8	128.0	192.4
Benzene	27.3	128.5	192.5

[a] Determined for a mixture of the three substances and TMS, concentration 20% (v/v).

Table 12.5 $\delta(^{13}$C) values (ppm) for selected compounds (Refs. 43–46)

Alkanes

Compound	C_1	C_2	C_3	C_4
Methane	−2.3	—	—	—
Ethane	5.7	—	—	—
Propane	15.4	15.9	—	—
Butane	3.0	24.8	—	—
Pentane	13.7	22.6	34.5	—
Hexane	13.9	22.9	32.0	—
Heptane	13.9	23.0	32.4	29.5

Cycloalkanes

Ring size

3	−2.6
4	23.3
5	26.5
6	27.8
7	29.4
8	27.8
9	27.0
10	26.2

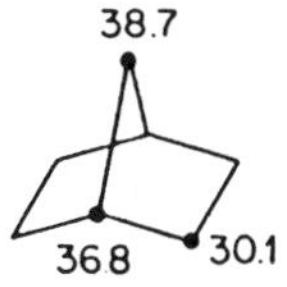

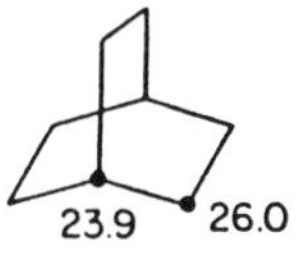

Alkenes

Compound	C_1	C_2
Ethylene	123.3	—
I–	85.3	130.4
Br–	115.5	122.0
Cl–	126.0	117.3
F–		
COOEt–	129.7	130.4
OCH$_3$	153.2	84.1
OCOCH$_3$	141.6	96.3
CH$_2$=CH–	136.9	116.3

Cycloalkenes

Ring size	$-C=-$ D	α-CH$_2$	β-CH$_2$	γ-CH$_2$
3	108.9	2.3	—	—
4	137.2	31.4	—	—
5	130.2	32.3	22.7	—
6	126.9	25.1	22.6	—
7	132.1	29.0	27.3	32.0

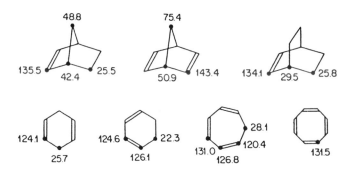

Benzocycloalkenes

Ring size	C_1	C_2	C_3	α-CH$_2$	β-CH$_2$	γ-CH$_2$
3	125.4	114.7	128.8	18.4	—	—
4	145.2	122.1	126.5	29.5	—	—
5	143.3	124.0	125.8	32.7	25.2	—
6	136.4	128.8	125.2	29.3	23.3	—
7	142.7	128.7	125.7	36.6	28.2	32.6
8	140.6	128.7	126.0	32.1	25.8	32.1

Alkynes

HC^1≡C^2 − X	C$_1$	C$_2$
−H	71.9	71.9
−C$_4$H$_9$	66.0	83.0
−C$_6$H$_5$	77.7	83.3
−SCH$_2$CH$_3$	81.4	72.6

Aromatic compounds

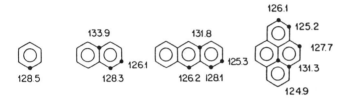

Substituted benzenes
$\delta = 128.5 + S_i(\delta_j)$

Substituent	$S_i(\delta_1)$	$S_i(\delta_\sigma)$	$S_i(\delta_m)$	$S_i(\delta_p)$
−H	0.0	0.0	0.0	0.0
−CH$_3$	9.3	0.6	0.0	−3.1
−CH$_2$CH$_3$	15.7	−0.6	−0.1	−2.8
−CH(CH$_3$)$_2$	20.1	−2.0	0.0	−2.5
−C(CH$_3$)$_3$	22.1	−3.4	−0.4	−3.1
−Cyclopropyl	15.1	−3.3	−0.6	−3.6
−CH$_2$Cl	9.1	0.0	0.2	−0.2
−CH$_2$Br	9.2	0.1	0.4	−0.3
−CF$_3$	2.6	−2.2	0.3	3.2
−CH$_2$OH	13.0	−1.4	0.0	−1.2
−CH=CH$_2$	7.6	−1.8	−1.8	−3.5
−C≡CH	−6.1	3.8	0.4	−0.2
−C$_6$H$_5$	13.0	−1.1	0.5	−1.0
−F	35.1	−14.3	0.9	−4.4
−Cl	6.4	0.2	1.0	−2.0
−Br	−5.4	3.3	2.2	−1.0
−I	−32.3	9.9	2.6	−0.4
−OH	26.9	−12.7	1.4	−7.3
−OCH$_3$	30.2	−14.7	0.9	−8.1

Substituted benzenes
$\delta = 128.5 + S_i(\delta_j)$

Substituent	$S_i(\delta_1)$	$S_i(\delta_0)$	$S_i(\delta_m)$	$S_i(\delta_p)$
$-NH_2$	19.2	-12.4	1.3	-9.5
$-N(CH_3)_2$	22.4	-15.7	0.8	-11.8
$-N(C_6H_5)_2$	19.3	-4.4	0.6	-5.9
$-NO_2$	19.6	-5.3	0.8	6.0
$-CN$	-16.0	3.5	0.7	4.3
$-NCO$	5.7	-3.6	1.2	-2.8
$-SC(CH_3)_3$	4.5	9.0	-0.3	0.0
$-COH$	9.0	1.2	1.2	6.0
$-COCH_3$	9.3	0.2	0.2	4.2
$-COOH$	2.4	1.6	-0.1	4.8
$-COO^-$	7.6	0.8	0.0	2.8
$-COOCH_3$	2.1	1.2	0.0	4.4
$-COCl$	4.6	2.9	0.6	7.0

Heterocyclic Compounds

Compound	C_α	C_β	C_γ
Oxirane	39.5	—	—
Oxetane	72.6	22.9	—
Tetrahydrofuran	68.4	26.5	—
Tetrahydropyran	69.8	27.0	24.2
Dioxan	67.4	—	—
Azirine, N-CH$_3$	28.5	—	—
Azetidine, N-CH$_3$	57.7	17.5	—
Tetrahydropyrrole	56.7	24.4	—
Piperidine	47.9	27.8	25.9
Thietane	18.7	—	—
Tetrahydrothiopyran	27.5	29.7	—
Furan	143.0	109.9	—
Pyrrole	118.7	107.7	—
Thiophen	124.9	126.4	—
Pyridine	149.9	123.8	136.0
Pyridinium ion	141.9	128.5	148.6
Pyridine N-oxide	139.1	126.4	126.4

128.4
144.8
152.0
147.9
154.9

135.0
163.7
159.0
147.8
152.4
152.4

128.5
126.6 124.2
129.4
128.8
136.6
124.7 135.9
129.4

12. The international system (SI) of units (MKSA system)

In the past, physical quantities in chemistry were expressed in general in units of the centimetre-gram-second (CGS) system. Recently, however, an international agreement to adopt the metre-kilogram-second-ampere (MKSA) or SI system was reached and the new convention began to be employed by chemists. Since 1973 the use of the SI system has been required by research journals in chemistry.

Table 12.6 Comparison of the SI and the CGS systems of units

Quantity	Symbol	SI unit	Symbol of the unit	Eqquivalent SI unit	Equivalent unit in the CGS system
Length	l, r	Metre	m	—	10^2 centimetres
Mass	m	Kilogram	kg	—	10^3 grams
Time	t	Second	s	—	—
Force	F	Newton	N	$kg\ m\ s^{-2}$	10^7 dynes
Energy	U	Joule	J	$kg\ m^2 s^{-2}$	10^7 ergs; 0.239 calories
Angle	θ	Radian	rad	—	—
Frequency	ν	Hertz	Hz	—	—
Electric charge	q	Coulomb	C	A s	2.9979×10^9 electrostatic units
Electric current	i	Ampere	A	—	10^{-1} electromagnetic units
Magnetic induction field	B	Tesla	T	$kg\ s^{-2}\ A^{-1}$	10^4 gauss
Magnetic field intensity	H	Ampere metre^{-1}	$A\ m^{-1}$	$A\ m^{-1}$	$4\pi \times 10^{-3}$ oersted
Magnetic dipole moment	μ	Ampere metre^2	m^2	$A\ m^2$	10^3 electromagnetic units
Magnetogyric ratio	γ	Radian tesla^{-1} second^{-1}	$rad\ T^{-1}\ s^{-1}$	rad kg s A	10^{-4} rad gauss^{-1}
Permeability of free space	μ_0	Newton Ampere^{-2}	NA^{-2}	$kgm\ s^{-2}\ A^2$	—

A comparison of the two systems can be made by a reference to Table 12.7, where we have presented the units important to the subject of this text. It is significant for nuclear magnetic resonance that in the SI system the units for magnetic field strength, H, and magnetic induction, B, differ by a factor of $4\pi \times 10^{-7}$ while in the CGS system the two quantities are measured in the same units ($\text{dyn}^{1/2}$ cm^{-1}). Moreover, it has been customary to express field strength incorrectly in gauss rather than in the correct units of oersteds. Since the observed physical properties are functions of B rather than H ($B = H$ only in vacuum), the use of B is correct in the SI system. The two systems also differ significantly in the units in which quantities of energy are expressed. In this case the CGS unit of kcal mol^{-1} has been replaced by kJ mol^{-1} (conversion: 1 kcal = 4.18 kJ).

13. References

1. C. W. Haigh and R. B. Mallion, *Org. Magn. Reson.*, **4**, 203 (1972).
2. A. A. Bothner-By, in F1, **1**, 195 (1965).
3. V. S. Watts and J. H. Goldstein, *J. Chem. Phys.*, **46**, 4165 (1967).
4. L. M. Crecely, V. S. Watt, and J. J. Goldstein, *J. Molecul. Spectr.*, **30**, 184 (1969).
5. J. M. Read Jr., R. E. Mayo, and J. H. Goldstein, *J. Molecul. Spectr.*, **22**, 419 (1967).
6. S. Castellano, private communication.
7. M. P. Williamson, R. Kostelnik, and S. Castellano, *J. Chem. Phys.*, **49**, 2218 (1968).
8. S. Castellano, R. Kostelnik, and C. Sun, *Tetrahedron Lett.*, 4635 (1967).
9. S. Castellano, C. Sun, and R. Kostelnik, *Tetrahedron Lett.*, 5207 (1967).
10. J. M. Read, Jr., R. W. Crecely, R. S. Butler, J. E. Loemker, and J. H. Goldstein, *Tetrahedron Lett.*, 1215 (1968).
11. H. Günther and R. Wenzl, *Z. Naturforschg.*, **22B**, 389 (1967).
12. H. Günther, M. Görlitz, and H.-H. Hinrichs, *Tetrahedron*, **24**, 5665 (1968).
13. M. Görlitz, *Ph. D. thesis*, University of Cologne, 1969.
14. H. Günther, R. Wenzl, and W. Grimme, *J. Am. Chem. Soc.*, **91**, 3808 (1969).
15. F. S. Mortimer, *J. Molecul. Spectr.*, **5**, 199 (1962).
16. E. W. Garbisch, Jr., and M. G. Grith, *J. Am. Chem. Soc.*, **90**, 6543 (1968).
17. J. B. Lambert, A. P. Jovanovich, and W. L. Oliver, Jr., *J. Phys. Chem.*, **74**, 2221 (1970).
18. E. A. Hill and J. D. Roberts, *J. Am. Chem. Soc.*, **89**, 2047 (1967).
19. M. A. Cooper, D. D. Elleman, C. D. Pearce, and S. L. Manatt, *J. Chem., Phys.*, **53**, 2343 (1970).
20. W. Regel and W. v. Philipsborn, *Helv. Chim. Acta*, **52**, 1354 (1969).
21. W. Regel and W. v. Philipsborn, *Helv. Chim. Acta*, **51**, 867 (1968).
22. R. T. Hobgood, Jr. and J. H. Goldstein, *J. Molecul. Spectr.*, **12**, 76 (1964).
23. J. B. Pawliczek and H. Günther, *Tetrahedron*, **26**, 1755 (1970).
24. J. Runsink and H. Günther, *Org. Magn. Reson.*, **13**, 249 (1980).
25. J. M. Read, C. T. Mathis, and J. H. Goldstein, *Spectrochim. Acta*, **21**, 85 (1965).
26. E. W. Garbisch, Jr., *J. Chem. Educ.*, **45**, 492 (1968).
27. S. Castellano, C. Sun, and R. Kostelnik, *J. Chem. Phys.*, **46**, 327 (1967).
28. H. Günther, *Z. Naturforsch.*, **20b**, 948 (1965).
29. H. Günther and H.-H. Hinrichs, *Tetrahedron*, **24**, 7033 (1968).
30. B. Dischler and W. Maier, *Z. Naturforsch.*, **16a**, 318 (1961).
31. B. Dischler and G. Englert, *Z. Naturforsch.*, **16a**, 1180 (1961).
32. B. Dischler, *Z. Naturforsch.*, **20a**, 888 (1965).
33. H. Günther, *Angew. Chem.*, **84**, 907 (1972); *Angew. Chem. Int. Ed. Engl.*, **1**, 861 (1972).

34. Textbook 4
35. G. Binsch, in *Topics in Stereochemistry*, N. L. Allinger and E. L. Eliel, Eds., Vol. 3, Interscience Publ., New York, 1968.
36. R. A. Sack, *Mol. Phys.,* **1**, 163 (1958).
37. J. A. Pople, W. G. Schneider, and H. J. Bernstein, *High-Resolution Nuclear Magnetic Resonance*, McGraw-Hill, New York, 1959.
38. R. E. Richards and T. Schaefer, *Mol. Phys.,* **1**, 163 (1958).
39. C. W. Haigh, *J. Chem. Soc. A.,* 1682 (1970).
40. W. Bremser, *Chemiker Ztg.,* **97**, 248 (1973).
41. G. C. Levy, R. C. Lichter, and G. L. Nelson, *Carbon-13 NMR Spectroscopy,* 2nd Ed., Wiley-Interscience, New York, 1980.
42. Levy, G. C. and J. D. Cargioli, *J. Magn. Reson.,* **6**, 143 (1972).
43. G. C. Levy and G. L. Nelson, *Carbon-13 Nuclear Magnetic Resonance for Organic Chemists*, Wiley-Interscience, New York, 1972.
44. J. B. Stothers, *Carbon-13 NMR Spectroscopy.* Academic Press, New York, 1972.
45. J. T. Clerc, E. Pretsch, and S. Sternhell, ^{13}C-*Kernresonanz-Spektroskopie*, Akadem. Verlagsges., 1973.
46. H.-O. Kalinowski, S. Berger, and S. Braun, ^{13}C-*NMR-Spektroskopie*, Georg Theime Verlag, 1984.

14. Standard definitions of terms, symbols, conventions, and references relating to high-resolution nuclear magnetic resonance (NMR) spectroscopy

 Designation: E 386 – 90*

AMERICAN SOCIETY FOR TESTING AND MATERIALS
1916 Race St , Philadelphia, Pa. 19103
Reprinted from the Annual Book of ASTM Standards. Copyright ASTM
If not listed in the current combined index, will appear in the next edition

Standard Practice for
Data Presentation Relating to
High-Resolution Nuclear Magnetic Resonance (NMR) Spectroscopy[1]

This standard is issued under the fixed designation E 386; the number immediately following the designation indicates the year of original adoption or, in the case of revision, the year of last revision. A number in parentheses indicates the year of last reapproval. A superscript epsilon (ϵ) indicates an editorial change since the last revision or reapproval.

1. Scope

1.1 This standard contains definitions of basic terms, conventions, and recommended practices for data presentation in the area of high-resolution NMR spectroscopy. Some of the basic definitions apply to wide-line NMR or to NMR of metals, but in general it is not intended to cover these latter areas of NMR in this standard. This version does not include definitions pertaining to double resonance nor to rotating frame experiments.

2. Nomenclature and Basic Definitions

2.1 *nuclear magnetic resonance (NMR) spectroscopy*—that form of spectroscopy concerned with radio-frequency-induced transitions between magnetic energy levels of atomic nuclei.

2.2 *NMR apparatus; NMR equipment*—an instrument comprising a magnet, radio-frequency oscillator, sample holder, and a detector that is capable of producing an electrical signal suitable for display on a recorder or an oscilloscope, or which is suitable for input to a computer.

2.3 *high-resolution NMR spectrometer*—an NMR apparatus that is capable of producing, for a given isotope, line widths that are less than the majority of the chemical shifts and coupling constants for that isotope.

NOTE —By this definition, a given spectrometer may be classed as a high-resolution instrument for isotopes with large chemical shifts, but may not be classed as a high-resolution instrument for isotopes with smaller chemical shifts.

2.4 *basic NMR frequency, ν_o*—the frequency, measured in hertz (Hz), of the oscillating magnetic field applied to induce transitions between nuclear magnetic energy levels. The static magnetic field at which the system operates is called H_o (Note 1) and its recommended unit of measurement is the tesla (T) (1 T = 10^4 gauss).

2.4.1 The foregoing quantities are approximately connected by the following relation:

$$\nu_o = \frac{\gamma}{2\pi} H_o$$

where γ = the magnetogyric ratio, a constant for a given nuclide (Note 2). The amplitude of the magnetic component of the radio-frequency field is called H_1. Recommended

units are millitesla and microtesla.

NOTE 1—This quantity is normally referred to as B by physicists. The usage of H to refer to magnetic field strength in chemical applications is so widely accepted that there appears to be no point in attempting to reach a totally consistent nomenclature now.

NOTE 2—This expression is correct only for bare nuclei and will be only approximately true for nuclei in chemical compounds, since the field at the nucleus is in general different from the static magnetic field. The discrepancy amounts to a few parts in 10^6 for protons, but may be of magnitude 1×10^{-3} for the heaviest nuclei.

2.5 *NMR absorption line*—a single transition or a set of degenerate transitions is referred to as a line.

2.6 *NMR absorption band; NMR band*—a region of the spectrum in which a detectable signal exists and passes through one or more maxima.

2.7 *reference compound (NMR)*—a selected material to whose signal the spectrum of a sample may be referred for the measurement of chemical shift (see 2.9).

2.7.1 *internal reference (NMR)*—a reference compound that is dissolved in the same phase as the sample.

2.7.2 *external reference (NMR)*—a reference compound that is not dissolved in the same phase as the sample.

2.8 *lock signal*—the NMR signal used to control the field-frequency ratio of the spectrometer. It may or may not be the same as the reference signal.

2.8.1 *internal lock*—a lock signal which is obtained from a material that is physically within the confines of the sample tube, whether or not the material is in the same phase as the sample (an annulus for the purpose of this definition is considered to be within the sample tube).

2.8.2 *external lock*—a lock signal which is obtained from a material that is physically outside the sample tube. The material supplying the lock signal is usually built into the probe.

NOTE —An external lock, if also used as a reference, is necessarily an external reference. An internal lock, if used as a reference, may be either an internal or an external reference, depending upon the experimental configuration.

2.8.3 *homonuclear lock*—a lock signal which is obtained from the same nuclide that is being observed.

2.8.4 *heteronuclear lock*—a lock signal which is obtained from a different nuclide than the one being observed.

2.9 *chemical shift, δ*—the defining equation for δ is the following:

$$\delta = \frac{\Delta\nu}{\nu_R} \times 10^6$$

where ν_R is the frequency with which the reference substance is in resonance at the magnetic field used in the experiment

---

[1] These definitions are under the jurisdiction of ASTM Committee E-13 on Molecular Spectroscopy and are the direct responsibility of Subcommittee E13.07 on Nuclear Magnetic Resonance Spectroscopy.
  Current edition approved June 29, 1990. Published August 1990. Originally published as E 386 – 69 T. Last previous edition E 386 – 78.

🔷 E 386

and $\Delta\nu$ is the frequency of the subject line minus the frequency of the reference line at constant field. The sign of $\Delta\nu$ is to be chosen such that shifts to the high frequency side of the reference line shall be positive.

· 2.9.1 If the experiment is done at constant frequency (field sweep) the defining equation becomes

$$\delta = \frac{\Delta\nu}{\nu_R} \times \left(1 - \frac{\Delta\nu}{\nu_R}\right) \times 10^6$$

2.9.2 In case the experiment is done by observation of a modulation sideband, the audio upper or lower sideband frequency must be added to or subtracted from the radio frequency.

2.10 *spinning sidebands*—bands, paired symmetrically about a principal band, arising from spinning of the sample in a field (dc or rf) that is inhomogeneous at the sample position. Spinning sidebands occur at frequencies separated from the principal band by integral multiples of the spinning rate. The intensities of bands which are equally spaced above and below the principal band are not necessarily equal.

2.11 *satellites*—additional bands spaced nearly symmetrically about a principal band, arising from the presence of an isotope of non-zero spin which is coupled to the nucleus being observed. An isotope shift is normally observed which causes the center of the satellites to be chemically shifted from the principal band. The intensity of the satellite signal increases with the abundance of the isotope responsible.

2.12 *NMR line width*—the full width, expressed in hertz (Hz), of an observed NMR line at one-half maximum height (FWHM).

2.13 *spin-spin coupling constant (NMR), J*—a measure, expressed in hertz (Hz), of the indirect spin-spin interaction of different magnetic nuclei in a given molecule.

NOTE —The notation $^nJ_{AB}$ is used to represent a coupling over n bonds between nuclei A and B. When it is necessary to specify a particular isotope, a modified notation may be used, such as, $^3J(^{15}NH)$.

3. Types of High-Resolution NMR Spectroscopy

3.1 *sequential excitation NMR; continuous wave (CW) NMR*—a form of high-resolution NMR in which nuclei of different field/frequency ratio at resonance are successively excited by sweeping the magnetic field or the radio frequency.

3.1.1 *rapid scan Fourier transform NMR; correlation spectroscopy*— a form of sequential excitation NMR in which the response of a spin system to a rapid passage excitation is obtained and is converted to a slow-passage spectrum by mathematical correlation with a reference line, or by suitable mathematical procedures including Fourier transformations.

3.2 *broad-band excitation NMR*—a form of high-resolution NMR in which nuclei of the same isotope but possibly different chemical shifts are excited simultaneously rather than sequentially.

3.2.1 *pulse Fourier transform NMR*—a form of broadband excitation NMR in which the sample is irradiated with one or more pulse sequences of radio-frequency power spaced at uniform time intervals, and the averaged free induction decay following the pulse sequences is converted to a frequency domain spectrum by a Fourier transformation.

3.2.1.1 *pulse Fourier difference NMR*—a form of pulse Fourier transform NMR in which the difference frequencies between the sample signals and a strong reference signal are extracted from the sample response prior to Fourier transformation.

3.2.1.2 *synthesized excitation Fourier NMR*—a form of pulse Fourier NMR in which a desired frequency spectrum for the exciting signal is Fourier synthesized and used to modulate the exciting radio frequency.

3.2.2 *stochastic excitation NMR*—a form of broad band excitation NMR in which the nuclei are excited by a range of frequencies produced by random or pseudorandom noise modulation of the carrier, and the frequency spectrum is obtained by Fourier transforming the correlation function between the input and output signals.

3.2.3 *Hadamard transform NMR*—a form of broad band excitation NMR in which the phase of the excitation signal is switched according to a binary pseudorandom sequence, and the correlation of the input and output signals by a Hadamard matrix yields an interference pattern which is then Fourier-transformed.

4. Operational Definitions

4.1 *Definitions Applying to Sequential Excitation (CW) NMR:*

4.1.1 *field sweeping (NMR)*—systematically varying the magnetic field strength, at constant applied radio-frequency field, to bring NMR transitions of different energies successively into resonance, thereby making available an NMR spectrum consisting of signal intensity versus magnetic field strength.

4.1.2 *frequency sweeping (NMR)*—systematically varying the frequency of the applied radio frequency field (or of a modulation sideband, see 4.1.4), at constant magnetic field strength, to bring NMR transitions of different energies successively into resonance, thereby making available an NMR spectrum consisting of signal intensity versus applied radio frequency.

4.1.3 *sweep rate*—the rate, in hertz (Hz) per second at which the applied radio frequency is varied to produce an NMR spectrum. In the case of field sweep, the actual sweep rate in microtesla per second is customarily converted to the equivalent in hertz per second, using the following equation:

$$\frac{\Delta\nu}{\Delta t} = \frac{\gamma}{2\pi} \cdot \frac{\Delta H}{\Delta t}$$

4.1.4 *modulation sidebands*—bands introduced into the NMR spectrum by, for example, modulation of the resonance signals. This may be accomplished by modulation of the static magnetic field, or by either amplitude modulation or frequency modulation of the basic radio frequency.

4.1.5 *NMR spectral resolution*—the width of a single line in the spectrum which is known to be sharp, such as, TMS or benzene (^1H). This definition includes sample factors as well as instrumental factors.

4.1.6 *NMR integral (analog)*—a quantitative measure of the relative intensities of NMR signals, defined by the areas of the spectral lines and usually displayed as a step function in which the heights of the steps are proportional to the areas (intensities) of the resonances.

⊕ E 386

4.2 *Definitions Applying to Multifrequency Excitation (Pulse) NMR:*

4.2.1 *pulse (v)*—to apply for a specified period of time a perturbation (for example, a radio frequency field) whose amplitude envelope is nominally rectangular.

4.2.2 *pulse (n)*—a perturbation applied as described above.

4.2.3 *pulse width*—the duration of a pulse.

4.2.4 *pulse flip angle*—the angle (in degrees or radians) through which the magnetization is rotated by a pulse (such as a 90-deg pulse or $\pi/2$ pulse).

4.2.5 *pulse amplitude*—the radio frequency field, H_1, in tesla.

NOTE —This may be specified indirectly, as described in 8.3.2.

4.2.6 *pulse phase*—the phase of the radio frequency field as measured relative to chosen axes in the rotating coordinate system.[2]

NOTE —The phase may be designated by a subscript, such as, $90°_x$ or $(\pi/2)_x$.

4.2.7 *free induction decay (FID)*—the time response signal following application of an r-f pulse.

4.2.8 *homogeneity spoiling pulse; homo-spoil pulse; inhomogenizing pulse*—a deliberately introduced temporary deterioration of the homogeneity of the magnetic field H.

4.2.9 *filter bandwidth; filter passband*—the frequency range, in hertz, transmitted with less than 3 dB (50 %) attenuation in power by a low-pass filter.

NOTE 1—On some commercial instruments, filter bandwidth is defined in a slightly different manner.

NOTE 2—Other parameters, such as rate of roll-off, width of passband, or width and rejection of center frequency in case of a notch filter, may be required to define filter characteristics adequately.

4.2.10 *data acquisition rate; sampling rate; digitizing rate*—the number of data points recorded per second.

4.2.11 *dwell time*—the time between the beginning of sampling of one data point and the beginning of sampling of the next successive point in the FID.

4.2.11.1 *aperture time*—the time interval during which the sample-and-hold device is receptive to signal information. In most applications of pulse NMR, the aperture time is a small fraction of the dwell time.

NOTE —*Sampling Time* has been used with both of the above meanings. Since the use of this term may be ambiguous, it is to be discouraged.

4.2.12 *detection method*—a specification of the method of detection.

4.2.12.1 *single-phase detection*—a method of operation in which a single phase-sensitive detector is used to extract signal information from a FID.

4.2.12.2 *quadrature detection*—a method of operation in which dual phase-sensitive detection is used to extract a pair of FID's which differ in phase by 90°.

4.2.13 *spectral width*—the frequency range represented without foldover. (Spectral width is equal to one half the data acquisition rate in the case of single-phase detection; but is

equal to the full data acquisition rate if quadrature detection is used.)

4.2.14 *foldover; foldback*—the appearance of spurious lines in the spectrum arising from either *(a)* limitations in data acquisition rate or *(b)* the inability of the spectrometer detector to distinguish frequencies above the carrier frequency from those below it.

NOTE —These two meanings of *foldover* are in common use. Type *(a)* is often termed "aliasing." Type *(b)* foldover is obviated by the use of quadrature detection.

4.2.15 *data acquisition time*—the period of time during which data are acquired and digitized; equal numerically to the product of the dwell time and the number of data points acquired.

4.2.16 *computer-limited spectral resolution*—the spectral width divided by the number of data points.

NOTE—This will be a measure of the observed line width only when it is much greater than the spectral resolution defined in 4.1.5.

4.2.17 *pulse sequence*—a set of defined pulses and time spacings between these pulses.

NOTE —There may be more than one way of expressing a sequence, for example, a series $(90°, \tau)_n$ may be one sequence of n pulses or n sequences each of the form $(90°, \tau)$.

4.2.18 *pulse interval*—the time between two pulses of a sequence.

4.2.19 *waiting time*—the time between the end of data acquisition after the last pulse of a sequence and the initiation of a new sequence.

NOTE —To ensure equilibrium at the beginning of the first sequence, the software in some NMR systems places the waiting time prior to the initiation of the first pulse of the sequence.

4.2.20 *acquisition delay time*—the time between the end of a pulse and the beginning of data acquisition.

4.2.21 *sequence delay time; recovery interval*—the time between the last pulse of a pulse sequence and the beginning of the succeeding (identical) pulse sequence. It is the time allowed for the nuclear spin system to recover its magnetization, and it is equal to the sum of the acquisition delay time, data acquisition time, and the waiting time.

4.2.22 *sequence repetition time*—the period of time between the beginning of a pulse sequence and the beginning of the succeeding (identical) pulse sequence.

4.2.23 *pulse repetition time*—the period of time between one r-f pulse and the succeeding (identical) pulse; used instead of *sequence repetition time* when the "sequence" consists of a single pulse.

4.2.24 *inversion-recovery sequence*—a sequence that inverts the nuclear magnetization and monitors its recovery, such as $(180°, \tau, 90°)$, where τ is the pulse interval.

4.2.25 *saturation-recovery sequence*—a sequence that saturates the nuclear magnetization and monitors its recovery, such as the sequence (90°, homogeneity-spoiling pulse, τ, 90°, T, homogeneity-spoiling pulse) or the sequence $(90°)_n$, τ, 90°, T, where $(90°)_n$ represents a rapid burst of 90° pulses.

4.2.26 *progressive saturation sequence*—the sequence 90°, $(\tau, 90°)_n$, where n may be a large number, and data acquisition normally occurs after each pulse (except possibly the first three or four pulses).

4.2.27 *spin-echo sequence*—the sequence 90°, τ, 180°

---

[2] For a discussion of the rotating coordinate system, see Abragam, "Principles of Nuclear Magnetism," Oxford, 1961, pp. 19ff.

⬡ E 386

4.2.28 *Carr-Purcell (CP) sequence*—the sequence 90°, τ, 180°, (2τ, 180°)ₙ, where *n* can be a large number.

4.2.29 *Carr-Purcell time*—the pulse interval 2τ between successive 180° pulses in the Carr-Purcell sequence.

4.2.30 *Meiboom-Gill sequence; CPMG sequence*—the sequence 90°ₓ, τ, 180°ᵧ, (2τ, 180°ᵧ)ₙ.

4.2.31 *spin-locking sequence*—the sequence 90°ₓ, (SL)ᵧ, where SL denotes a "long" pulse (often measured in milliseconds or seconds, rather than microseconds) and *H* (lock) ≫ *H* (local).

4.2.32 *zero filling*—supplementing the number of data points in the time response signal with trailing zeroes before Fourier transformation.

4.2.33 *partially relaxed Fourier transform (PRFT) NMR*—a set of multiline FT spectra obtained from an inversion-recovery sequence and designed to provide information on spin-lattice relaxation times.

4.2.34 *NMR integral (digital)*—the integrals (see 4.1.6) of pulse-Fourier transform spectra or of digitized CW spectra, obtained by summing the amplitudes of the digital data points that define the envelope of each NMR band. The results of these summations are usually displayed either as a normalized total number of digital counts for each band, or as a step function (running total of digital counts) superimposed on the spectrum.

5. NMR Conventions

5.1 The dimensionless scale used for chemical shifts for any nucleus shall be termed the δ scale. The correct usage is δ = 5.00 or δ 5.00. Alternative forms, such as δ = 5.00 ppm or shift = 5.00 δ shall not be used.

5.2 The unit used for line positions should be hertz.

5.3 The dimensionless and frequency scales should have a common origin.

5.4 The standard sweep direction should be from high to low radio frequency (low to high applied magnetic field).

5.5 The standard orientation of spectra should be with low radio frequency (high field) to the right.

5.6 Absorption mode peaks should point up.

6. Referencing Procedures and Substances

6.1 *General:*

6.1.1 Whenever possible, in the case of proton and carbon-13 spectra, the chemical shift scale should be tied to an *internal* reference.

6.1.2 In case an external reference is used, either a coaxial tube or a capillary tube is generally adequate.

6.1.3 For nuclei other than protons or ^{13}C, for which generally agreed-upon reference substances do not yet exist, it is particularly important to report the reference material and referencing procedure fully, including separations in hertz and the spectrometer radio frequency when it is known.

6.2 *NMR Reference Substances for Proton Spectra:*

6.2.1 The primary internal reference for proton spectra in nonaqueous solution shall be tetramethylsilane (TMS). A concentration of 1 % or less is preferred.

6.2.2 The position of the tetramethylsilane resonance is defined as exactly zero.

6.2.3 The recommended internal reference for proton spectra in aqueous solutions is the sodium salt of 2,2,3,3-tetradeutero-4,4-dimethyl-4-silapentanoic acid (TSP-d_4). Its chemical shift is assigned the value zero.

6.2.4 The numbers on the dimensionless (shift) scale to high frequency (low field) of TMS shall be regarded as positive.

6.3 *NMR Reference Substances for Nuclei Other than Protons:*

6.3.1 For all nuclei the numbers on the dimensionless (shift) scale to high frequency (low field) from the reference substance shall be positive. In the interim, until this proposal has been fully adopted, the sign convention used should be explicitly given.

NOTE —The existing literature on NMR contains examples of both the sign convention given above and its opposite. It seems desirable to adopt a uniform convention for all nuclei, and the convention recommended herein is already widely used in both proton and ^{13}C NMR. The recommended convention will result in assigning the most positive numerical value to the transition of highest energy.

6.3.2 The primary internal reference for ^{13}C spectra of nonaqueous solutions shall be tetramethylsilane (TMS). For aqueous solutions, secondary standards such as dioxane have been found satisfactory. When such standards are used the line positions and chemical shifts should be reported with reference to TMS, and the conversion factor should be stated explicitly.

6.3.3 The primary external reference for boron spectra (^{10}B and ^{11}B) shall be boron trifluoride-diethyletherate [$(C_2H_5)_2O:BF_3$].

6.3.4 The primary external reference for ^{31}P spectra shall be phosphorus trioxide (P_4O_6).

6.3.5 Specific recommendations for nuclei other than those mentioned above are not offered here. The following guidelines should be used: If previous work on the nucleus under study exists, any earlier reference should be used unless there are compelling reasons to choose a new reference. A reference substance should have a sharp line spectrum if possible. A singlet spectrum is preferred. A reference substance should be chosen to have a resonance at low frequency (high field) so far as possible, in order that the majority of chemical shifts will be of positive sign. Internal references should be avoided unless it is possible to include a study of solvent effects on chemical shift.

7. Recommended Practice for Signal-to-Noise Determination in Fourier Transform NMR

7.1 *General*—This section gives the recommended practice for signal-to-noise ratio (S/N) determination in three specific situations: ( ) proton single pulse mode; (*b*) carbon-13 single pulse mode; and (*c*) carbon-13 multiple pulse mode.

NOTE 1—Some of the materials recommended for use in this section are known to present health hazards if used improperly. Anyone making up solutions containing benzene, dioxane, or chloroform should consult and abide by OSHA regulations 29CFR 1910.1000 (solvents) and 29CFR 1910.1028 (benzene).

7.2 *Proton Single Pulse Mode:*

7.2.1 *Sample*—Dilute ethylbenzene in CDCl₃.

7.2.2 *Measurement*—Proton signal-to-noise ratio is measured using a single pulse of radio-frequency power applied to a dilute solution of ethylbenzene in CDCl₃. Choose the concentration of ethylbenzene appropriate to the sensitivity

⬡ E 386

of the instrument under test, such that the S/N as measured on the methylene quartet is 25 : 1. State the determined S/N as "equivalent one percent ethylbenzene sensitivity." Carry out the measurement using the following conditions:

Spectral width	0 to 10 ppm ($\delta^1H_{MS} \equiv 0$)
Data acquisition time	≥0.4 s
Flip angle	90°
Analog filter	appropriate for method of detection
Detection method	specify (for example, single phase, SSB, QPD)
Equilibration delay	60 s

Following the data acquisition, multiply the data by a decaying exponential function of the form $e^{-t/A}$, where A is equivalent to a T_2 contribution. A may be expressed as a time constant in units of seconds, or, alternatively, the line broadening (LB) resulting from the exponential multiplication may be expressed in units of hertz (Hz). For the measurement, $A = 0.3$ or LB = 1 Hz. Perform no data smoothing after transformation. Plot the resulting absorption mode spectrum over the full 0 to 10 ppm. Measure S/N on a plot expansion covering the range of 2 to 6 ppm, in which the methylene quartet is plotted to fill the chart paper as closely as practical. Use sufficient vertical amplitude to obtain a peak-to-peak noise measurement greater than 2 cm. Measure peak-to-peak noise over the 4 to 6 ppm region on the same trace or calculate rms noise by computer (see Note 2). The S/N is then calculated on the strongest line in the quartet as follows (see Fig. 1):

[(signal intensity)/(peak-to-peak noise)] × 2.5 = S/N

NOTE 2—The true rms noise can be calculated by computer and used in the S/N determination. Since peak-to-peak noise is approximately five times rms noise, rather than 2.5 times, the rms noise must be doubled to obtain a comparable S/N. When this is done, it is felt that the S/N determined by computer should be reliable and less subject to human error than the alternate method of estimating peak-to-peak noise from a chart recording. The computer program should do the following:

(*a*) Select the region in which noise is to be measured as specified in the above test.

(*b*) Obtain the algebraic mean of all the observed points in this region, and subtract the mean from each point (zero-order correction).

(*c*) If the base line slopes, a first order correction may be made by using a standard least-squares method to obtain the slope and intercept of the baseline, then subtracting each calculated point from the corresponding observed point.

(*d*) Corrections calculated on the noise in the specified region of the spectrum should be applied to that region and also to the spectral region containing the signal.

(*e*) Form the sum of the squares of each amplitude (point), corrected as described previously, divide by one less than the number of points in the region, and take the square root. This is the rms noise.

$$\text{rms noise} = [(\Sigma[\text{amplitude}]^2)/(N-1)]^{1/2}$$

No other processing should be done; in particular, points that appear to be extreme should not be deleted. S/N becomes simply (signal intensity/2)/(rms noise).

7.2.3 *Discussion*—The 1 % ethylbenzene S/N measurement is a widely used method for ^1H S/N both in CW and FT NMR. Although presenting few difficulties in CW work, the typical samples used in FT NMR do present some problems which we hope to avoid using this procedure.

7.2.3.1 The 1 % concentration traditionally employed generates a very high S/N on modern FT spectrometers, particularly at very high magnetic field strengths.

7.2.3.2 TMS is usually present in standard samples at the 1 % level. This causes a very strong signal which can lead to an erroneous S/N measurement.

7.2.3.3 The variety of sample tube sizes and S/N values has made it inconvenient to use a uniform concentration. The solution(s) should be made up by volume composition at 25°C using good volumetric practice. Suggested solutions:

No.	Ethylbenzene, %	TMS, % (Note 3)
1	3.0	0.3
2	1.0	0.1
3	1.0	1.0 (also valuable for CW TMS-locked spectrometers)
4	0.33	0.03
5	0.10	0.01
6	0.033	0.003
7	0.010	0.001

NOTE 3—The TMS is added for a reference material.

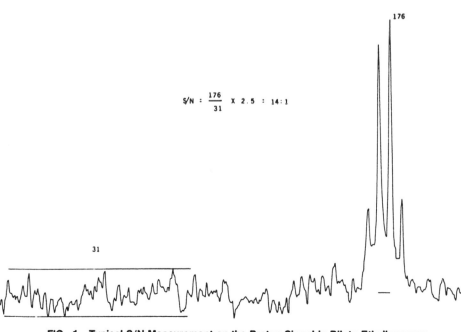

$$S/N : \frac{176}{31} \times 2.5 : 14 : 1$$

FIG. 1   Typical S/N Measurement on the Proton Signal in Dilute Ethylbenzene

E 386

7.3 Carbon-13 Single Pulse Mode:

7.3.1 *Sample*—60 %C_6D_6(>98atom %D),40 % p-dioxane (v/v).

7.3.2 *Measurement*—Measure carbon-13 signal-to-noise ratio on the benzene carbon signal in a solution of 60 % perdeuterobenzene – 40 % p-dioxane, with the spectrometer locked to the deuterium in the sample, using the following conditions:

Spectral width	0 to 200 ppm ($\delta \, _{TMS}^{13} \equiv 0$ ppm)
Data acquisition time	≥0.4 s
Flip angle	90°
Analog filter	appropriate for method of detection
Detection method	specify (for example, single phase, SSB, QPD)
Equilibration delay	300 s
Decoupler	off

Following the data acquisition, multiply the data by a decaying exponential function of the form $e^{-t/A}$, where A is equivalent to a T_2 contribution. A may be expressed as a time constant in units of seconds, or, alternatively, the line broadening (LB) resulting from exponential multiplication may be expressed in units of hertz (Hz). For the measurement, $A = 0.3$ or LB = 1 Hz. Perform no data smoothing after transformation. Plot the resulting absorption mode spectrum over the full 0 to 200 ppm chemical shift range. Plot the C_6D_6 triplet to fill the vertical range of the chart paper as closely as practical. Use sufficient vertical amplitude to obtain a peak-to-peak noise measurement greater than 2 cm. Signal-to-noise is to be measured as:

[(average triplet intensity)/(peak-to-peak noise)]× 2.5 = S/N

Measure the peak-to-peak noise between the C_6D_6 and dioxane triplets, specifically between and inclusive of 80 and 120 ppm on the ^{13}C chemical shift scale, or calculate rms noise by computer (see Note 2 and Fig. 2).

7.3.3 Characteristics of the Proposed Standard:

7.3.3.1 The S/N of the C_6D_6 triplet is low enough to permit a plot from which both signal and noise may be measured. For a full scale vertical display of the C_6D_6 triplet, the peak-to-peak noise amplitude should be adequately measured and have two significant figures. (For those spec-

trometers with very high sensitivity, noise would still have to be blown up to at least 2 cm peak-to-peak in a separate trace of the same transformed data.)

7.3.3.2 The C_6D_6 triplet has linewidth of 14 Hz under these conditions, reasonably independent of magnet resolution, permitting easy tune up and small 4 K data table for the measurement.

7.3.3.3 The C_6D_6 S/N can be measured in the presence of or absence of high power proton decoupling facilitating servicing diagnostic procedures. It is particularly valuable in diagnosing decoupler-caused noise contributions.

7.3.3.4 The broad lines of the C_6D_6 result from long-range ^{13}C-^2H coupling and thus the linewidth is not field-dependent.

7.3.3.5 C_6D_6 has no nuclear Overhauser enhancement (NOE).

7.3.3.6 The reference material is widely available and can serve as an internal ^2H lock.

7.3.3.7 The C_6D_6 S/N is independent of applied lock power in normal locking power range up to and beyond saturation of the deuterium signal.

7.3.3.8 The C_6D_6 S/N is temperature independent over normal working temperatures.

7.3.3.9 The dioxane serves several purposes: ready reference to prior data; a conveniently short T_1 (<10 s); under decoupled conditions it possesses a strong signal serving for $\gamma H_1/2\pi$ measurement by means of a 90° pulse determination; under off-resonance conditions its residual ^{13}C-^1H coupling can serve to measure $\gamma H_2/2\pi$; the decoupled singlet can be used to measure resolution in terms of full linewidth at half-height, also line shape and spinning sidebands; and under coupled conditions and longer acquisition times, it can provide a coupled spectrum with long-range couplings. The strong signal available from decoupled dioxane permits facile tests of decoupler gating through measurement of the NOE via "Suppressed Overhauser" gating schemes vs use of coupled dioxane as the base point for calculating the NOE. The short T_1 of dioxane allows routine check of automatic T_1 programs and calculations.

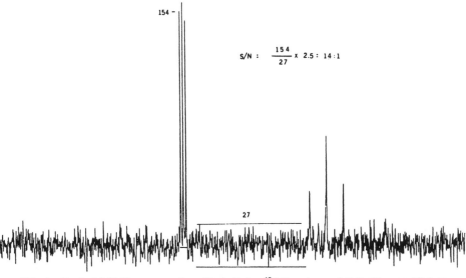

154 –

S/N : $\dfrac{154}{27}$ x 2.5 : 14 : 1

27

FIG. 2 Typical S/N Measurement on Single Pulse ^{13}C Spectrum of C_6D_6-Dioxane Mixture

E 386

7.3.4 *Discussion*—The proposed measurement is possible and convenient on any modern FT instrument. This method ensures that the maximum available S/N is obtained, thus preventing confusion in parameter choice, particularly in the case of the exponential weighting. A new standard is necessary in view of the difficulty in widespread reliable use of the 90 % ethylbenzene sample previously used. The natural linewidths of the ethylbenzene lines are less than 0.1 Hz requiring exacting field homogeneity to obtain maximum resolution. The narrow lines also demand long data acquisition times in each FID to define the lines adequately. Since ethylbenzene S/N is measured on a decoupled protonated carbon signal, decoupler power, modulation efficiency, and offset are all factors in determining S/N. The S/N for most spectrometers is >100:1 for 90 % ethylbenzene making noise measurements the primary factor in the derived S/N.

7.3.4.1 Dioxane has been proposed for the S/N sample but it has some serious drawbacks in addition to several advantages shared with deuterobenzene. Its T_1 is dipole-dipole dominated and has full NOE in the decoupled experiment. It is easily possible to have residual NOE in a *coupled* spectrum by not waiting long enough for the NOE to decay away prior to the sampling pulse. Although deuterobenzene has the common requirement of sufficient equilibration delay the error is *always* on the side of lower S/N, whereas dioxane's apparent S/N can be up to a factor of three greater than that assumed by simple inspection of the spectrum. This makes comparison of intrinsic S/N susceptible to error. The addition of dioxane to the 40 % level provides all the advantages listed above for routine tuning up and quick S/N checking, while the C_6D_6 permits an absolute measurement. The other major disadvantage of dioxane is the dependence of the character of the spectrum on acquisition time and weighting function. If more than 0.5-s acquisition is used with a less severe weighting function than above, the fine structure from the long-range coupling becomes visible. While no problem for the experienced spectroscopist, this can be and has been confusing to inexperienced users.

7.3.4.2 In summary, the sample in 7.3 for S/N measurement is recommended particularly when comparing instruments in different laboratories. For use within a laboratory by knowledgeable operators, ethylbenzene still offers a practical sample for simultaneous checking of S/N, resolution and decoupling efficiency. The adoption of an intrinsic S/N sample such as that described above also identifies the need for separate measurement of resolution and $\gamma H_2/2\pi$ to more completely characterize the performance of an FT spectrometer on ^{13}C. In addition, this measurement is understood to measure only intrinsic sensitivity and not the sensitivity of a time-averaged spectrum on a "routine" sample.

7.4 *Carbon-13 Multiple Pulse Mode:*

7.4.1 *Sample*—0.1 M Sucrose in D_2O equilibrated with toluene. Dissolve 3.423 g of sucrose (stored at a relative humidity of 50 % or less; NBS SRM sucrose is satisfactory) in about 90 cc of D_2O in a 100-cc volumetric flask, then dilute to the mark at 25°C with D_2O after all the sucrose is dissolved. Add 0.05 ml of toluene as a preservative.

7.4.2 *Measurement*—Carry out the measurement in the multiple-pulsed mode locked to the internal D_2O using the following conditions:

Spectral width	0 to 200 ppm ($\delta_{\mathrm{TMS}}^{13C} \equiv 0$)
Data acquisition time	≥0.4 s
Flip angle	90°
Analog filter	appropriate for method of detection
Detection method	specify (for example, single phase, SSB, QPD)
Pulse repetition rate	1 pulse/s
^{1}H decoupler	broadband
^{1}H decoupler frequency	centered at 5 ± 1 ppm in the ^{1}H spectrum
^{1}H decoupler modulation mode	specify (for example, noise, square wave, etc.)
^{1}H decoupler modulation frequency	specify
Number of transients	4000 for 5-mm sample size
	1000 for 10 to 12-mm sample size
	100 for >12-mm sample size
Operating temperature	specify

Following the data acquisition, multiply the data by a decaying exponential function of the form $e^{-t/A}$, where A is equivalent to a T_2 contribution. A may be expressed as a time constant in units of seconds, or, alternatively, the line broadening (LB) resulting from the exponential multiplication may be expressed in units of Hz. For the measurement, $A = 0.3$ or LB = 1.0 Hz. Perform no data smoothing after transformation. Plot the resulting absorption mode spectrum over the full 200 ppm chemical shift range. Plot the spectrum to fill the vertical range of the chart paper as closely as practical. Measure the peak-to-peak noise between 120 and 140 ppm of the spectral window or calculate rms noise by computer (see Note 2). For those spectrometers with very high sensitivity, noise may have to be blown up to at least 2 cm peak-to-peak in a separate trace of the same transformed data. Measure signals Nos. 2, 3, 9, and 12 (identified on Fig. 3) and calculate S/N as follows:

$$[(2 + 3 + 9 + 12)/(\text{peak-to-peak noise})] \times 0.625 = S/N$$

7.4.3 *Discussion*—This measurement permits evaluation of sensitivity under "typical" conditions; that is, the decoupler is on and many transients are obtained. In addition to a knowledge of the basic, or intrinsic, ^{13}C sensitivity as measured in the C_6D_6 test, it is extremely important to evaluate the long term sensitivity as reflected in a proton-decoupled, time-averaged spectrum. The type and quality of the decoupling, as well as long term and short term instabilities in any instrument element, can profoundly affect sensitivity. This test is designed to monitor this performance.

7.4.3.1 Sucrose is chosen because of its widespread availability, purity, low cost, stability (in toluene equilibrated water) and spectral characteristics. Among these are the reasonable (1 Hz) linewidths, short T_1s, and full NOE. The number of transients is chosen to provide a reasonable total experimental time, typically 20 min, while still running long enough to simulate normal experiments adequately.

7.4.3.2 Decoupling efficiency is another highly variable element in "routine sensitivity." It certainly determines the ultimate sensitivity in the 90 % ethylbenzene sensitivity test (magnet homogeneity permitting). For this reason ethylbenzene is unsuitable for an absolute sensitivity determination. Yet, it is necessary to include the decoupler in sensitivity considerations since a poorly operating decoupler can be the main determinant in apparent sensitivity. Thus, proper consideration must be given not only to intrinsic sensitivity but also to "routine" sensitivity in characterizing spectrometer performance.

⊕ E 386

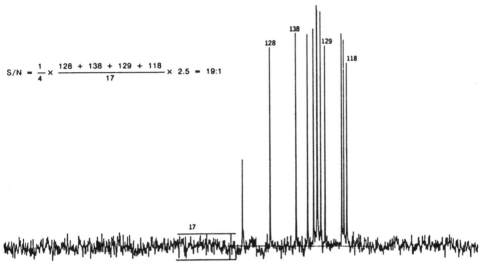

$$S/N = \frac{1}{4} \times \frac{128 + 138 + 129 + 118}{17} \times 2.5 = 19:1$$

FIG. 3   Typical S/N Measurement on Accumulated ^{13}C Spectrum of 0.1 M Sucrose in D_2O

8. Presentation of NMR Data and Spectrometer Parameters

8.1 *General*—The following should be specified whenever NMR data are published:

8.1.1 Nucleus observed. In cases where possible ambiguity exists, the isotope must be specified, for example, ^{14}N, ^{11}B. In other cases the isotope may be specified, even though superfluous, such as, ^{19}F, ^{31}P.

8.1.2 Name of solvent and concentration of solution.

8.1.3 Name of external reference, or name and concentration of internal reference, as applicable.

8.1.4 Temperature of sample and how measured.

8.1.5 Procedure used for measuring peak positions.

8.1.6 Radio frequency at which measurements were made.

8.1.7 Magnitude of radio frequency field (see 2.4), or assurance that saturation of the signal has not occurred (in the case of CW spectra), or both.

8.1.8 Mathematical operations used to analyze the spectrum. In cases where a computer program has been used to assist in the analysis of the spectrum, the following information should be included: Identification/source of program, number of lines fitted, identity of parameters varied, rms deviation of all lines, estimated precision of fitted parameters, and maximum deviation of worst line.

8.1.9 Numbers on the frequency scale (if used). They should increase from low to high frequency (high to low applied field if field sweep is used).

8.2 When CW spectra are published the following information should be included:

8.2.1 Sweep rate.

8.2.2 Values of both r-f fields when spin decoupling or double resonance is employed.

8.2.3 The shifts and couplings obtained from the spectra should be reported when available, the former in dimensionless units (ppm) and the latter in frequency units (hertz).

8.3 *Pulse-Fourier Transform Spectra*—For high-resolution pulse-Fourier transform experiments, all of the following that are applicable should be specified:

8.3.1 Pulse flip angle used.

8.3.2 90° pulse width, or pulse amplitude.

NOTE —Both 8.3.1 and 8.3.2 must always be specified. They may be given indirectly, for example, as pulse width used *and* as pulse width for a 90° pulse for the nucleus being studied.

8.3.3 Bandwidth and rolloff characteristics of all limiting filters (low-pass and crystal filters). Usually given as bandwidth (see 4.2.9) and type (such as, a 4-pole Butterworth).

8.3.4 Spectral width (or data acquisition rate or dwell time).

8.3.5 Data acquisition time (and acquisition delay time if relevant).

8.3.6 Pulse repetition time and number of pulses if the "sequence" consists of a single pulse.

8.3.7 Description of pulse sequence including *(a)* common name or details of pulses and phases, *(b)* sequence repetition time, *(c)* pulse intervals, *(d)* waiting time, *(e)* number of sequences, and *(f)* the specific pulse intervals during which data are acquired.

8.3.8 Quadrature phase detection, if used.

8.3.9 Number of data points Fourier transformed (it is desirable to indicate specifically whether zero filling is used).

8.3.10 The time constant of exponential weighting function (exponential filter), if used.

8.3.11 Details of apodization or other weighting of the time response signal.

8.3.12 Details of any other data processing such as spectral smoothing, baseline corrections, etc.

8.3.13 Details of systematic noise reduction, if used.

8.3.14 Relation of pulse frequency to observed frequencies.

🎄 E 386

BIBLIOGRAPHY

There exists extensive literature on n.m.r. spectroscopy that covers both the physical principles and the applications to various research fields in physics, chemistry, and biology. Here we present listings of comprehensive texts, monographs, data compilations, and serial progress reports, each section arranged in alphabetical order by author.

Since today it is impossible to present all aspects of the subject in a single work, monographs covering special topics and review articles dealing with new developments have become indispensable. Six serial publications devoted to n.m.r. spectroscopy are published, and many of the aspects touched upon in our text only briefly are treated there with greater depth.

The current research literature is covered by two *Chemical Abstracts Selects* titles (*¹H NMR, Carbon and Heteroatom NMR*) and by the yearly *Specialist Periodical Report—Nuclear Magnetic Resonance* of the Royal Society of Chemistry, London.

COMPREHENSIVE TEXTBOOKS

(a) Physics-oriented:

1. A. Abragam, *The Principles of Nuclear Magnetism*, Clarendon Press, Oxford, 1961, 599 pp.
2. A. Carrington and A. D. McLachlan, *Introduction to Magnetic Resonance*, Chapman and Hall, London, 1979, 266 pp.
3. N. Chandrakumar and S. Subramanian, *Modern Techniques in High-Resolution FT-NMR*, Springer-Verlag, New York, 1987, 388 pp.
4. J. W. Emsley, J. Feeney, and L. H. Sutcliffe, *High-resolution Nuclear Magnetic Resonance Spectroscopy*, Pergamon Press, Oxford, 2 Vols., 1965, 1154 pp.
5. R. R. Ernst, G. Bodenhausen, and A. Wokaun, *Principles of Nuclear Magnetic Resonance in One and Two Dimensions*, Clarendon Press, Oxford, 1987, 610 pp.
6. M. Goldman, *Quantum Description of High-Resolution NMR in Liquids*, Clarendon Press, Oxford, 1988, 268 pp.
7. R. K. Harris, *Nuclear Magnetic Resonance Spectroscopy*, Longman, Harlow, 1986, 260 pp.
8. J. W. Hennel and J. Klinowski, *Fundamentals of Nuclear Magnetic Resonance*, Longman, 1993, 288 pp.
9. M. Munowitz, *Coherence and NMR*, John Wiley, New York, 1988, 289 pp.
10. C. P. Slichter, *Principles of Magnetic Resonance*, 3rd Ed., Springer-Verlag, Berlin, 1990, 397 pp.

(b) Chemistry-oriented:

11. Atta-ur-Rahman, *One- and Two-Dimensional NMR Spectroscopy*, Springer-Verlag, New York, 1989.
12. R. Abraham, J. Fisher, and P. Loftus, *Introduction to NMR Spectroscopy*, Wiley, Chichester, 1988, 271 pp.
13. E. D. Becker, *High Resolution NMR*, 2nd Ed., Academic Press, New York, 1980, 354 pp.
14. F. A. Bovey, *Nuclear Magnetic Resonance Spectroscopy*, Academic Press, New York, 1969, 396 pp.
15. H. Friebolin, *One and Two-Dimensional NMR Spectroscopy*, 2nd Ed., VCH Publishers, 1993, 317 pp.
16. V. M. S. Gil and C. F. G. C. Geraldes, *Ressonância Magnética Nuclear–Fundamentos, Métodos e Aplicações*, Fundação Calouste Gulbenkian, Lissabon, 1987, 1088 pp.
17. L. M. Jackman and S. Sternhell, *Application of Nuclear Magnetic Resonance Spectroscopy in Organic Chemistry*, Pergamon Press, Oxford, 1969, 456 pp.
18. J. K. M. Sanders and B. K. Hunter, *Modern NMR Spectroscopy—A Guide for Chemists*, 2nd Ed., Oxford University Press, Oxford, 1993, 314 pp.

(c) Workbooks

E. Breitmaier, *Structure Elucidation by NMR in Organic Chemistry. A Practical Guide*, Wiley, Chichester, 1993, 230 pp.
H. Duddeck and W. Dietrich, *Structure Elucidation by Modern NMR*, Springer, New York, 1988, 197 pp.
J. K. M. Sanders, E. C. Constable and B. K. Hunter, *Modern NMR-Spectroscopy—A Workbook of Chemical Problems*, Oxford University Press, Oxford, 1989.

MONOGRAPHS DEALING WITH SPECIAL TOPICS

(a) New 1D and 2D Techniques, Experimental Aspects

A. D. Bax, *Two-Dimensional NMR in Liquids*, Reidel, Amsterdam, 1982, 208 pp.
W. S. Brey, Editor, *Pulse Methods in 1D and 2D Liquid-Phase NMR*, Academic Press, New York, 1988, 561 pp.
W. R. Croasmun and R. M. K. Carlson, Editors, *Two-Dimensional NMR Spectroscopy, Methods in Stereochemical Analysis*, Vol. 9, VCH Publishers, Weinheim, 2nd Ed., 1994, 958 pp.
A. E. Derome, *Modern NMR Techniques for Chemistry Research*, Pergamon Press, Oxford, 1987, 280 pp.
R. Freeman, *A Handbook of Nuclear Magnetic Resonance*, Longman, Harlow, 1987, 312 pp.
E. Fukushima and S. B. W. Roeder, *Experimental Pulse NMR: A Nuts and Bolts Approach*, Addison-Wesley, London, 1981, 539 pp.
S. W. Homans, *A Dictionary of Concepts in NMR*, Clarendon Press, Oxford, 1989, 343 pp.
G. E. Martin and A. S. Zektzer, *Two-Dimensional NMR Methods for Establishing Molecular Connectivity—A Chemist's Guide to Experiment Selection, Performance, and Interpretation*, VCH Publishers, Weinheim, 1988, 508 pp.
M. L. Martin, J.-J. Delpuech, and G. J. Martin, *Practical NMR-Spectroscopy*, Heyden, London, 1980, 460 pp.
N. J. Oppenheimer and T. L. James, Editors, *Methods in Enzymology, Nuclear Magnetic Resonance*, Vol. 176, Part A: Spectral Techniques and Dynamics, Vol. 177, Part B: Structure and Mechanism, 1989.
J. Schraml and J. M. Bellama, *Two-Dimensional NMR Spectroscopy*, Wiley, New York, 1988,

220 pp.
D. A. W. Wendisch, *Acronyms and Abbreviations in Molecular Spectroscopy—An Encyclopedic Dictionary*, Springer-Verlag, Berlin, 1990, 315 pp.

(b) Spectral Analysis, Theory, Relaxation

R. J. Abraham, *Analysis of High Resolution NMR Spectra*, Elsevier, Amsterdam, 1971, 324 pp.
I. Ando and G. A. Webb, *Theory of NMR Parameters*, Academic Press, London, 1983, 217 pp.
L. Bauci, I. Bertini, and C. Luchinat, *Nuclear and Electron Relaxation*, VCH Publishers, Weinheim, 1991, 216 pp.
N. Bloembergen, *Nuclear Magnetic Relaxation*, W. A. Benjamin, New York, 1961, 178 pp.
P. L. Corio, *Structure of High-Resolution NMR Spectra*, Academic Press, New York, 1966, 548 pp.
W. T. Dixon, *Theory and Interpretation of Magnetic Resonance Spectra*, Plenum Press, London, 1971, 164 pp.
J. D. Memory, *Quantum Theory of Magnetic Resonance Parameters*, McGraw-Hill, New York, 1968, 192 pp.
D. P. Poole and H. Farach, *Relaxation in Magnetic Resonance*, Academic Press, New York, 1971, 392 p.

(c) Pulse and Fourier Transform Spectroscopy

T. C. Farrar and E. D. Becker, *Pulse and Fourier Transform NMR*, Academic Press, New York, 1971, 115 pp.
K. Müllen and P. S. Pregosin, *Fourier Transform NMR Techniques: A Practical Approach*, Academic Press, London, 1976, 149 pp.
D. Shaw, *Fourier Transform NMR Spectroscopy*, Elsevier, Amsterdam, 2nd Ed., 1984, 304 pp.
D. Ziessow, *One-line Rechner in der Chemie, Grundlagen und Anwendung in der Fourierspektroskopie*, Walter de Gruyter, Berlin, 1973, 376 pp.

(d) Dynamic Phenomena, Analytical Applications

L. D. Field and S. Sternhell (Eds.), *Analytical NMR*, Wiley, Chichester, 1989.
L. M. Jackman and F. A. Cotton (Eds.), *Dynamic NMR Spectroscopy*, Academic Press, New York, 1975, 660 pp.
J. I. Kaplan and G. Fraenkel, *NMR of Chemically Exchanging Systems*, Academic Press, New York, 1980, 165 pp.
D. E. Leyden and R. H. Cox, *Analytical Applications of NMR*, Wiley, New York, 1977, 456 pp.
J. Sandström, *Dynamic NMR Spectroscopy*, Academic Press, London, 1982, 226 pp.

(e) Nuclear Overhauser Effect, CIDNP, Shift Reagents

G. N. LaMar, W. D. Horroks, and R. H. Holm, (Eds.), *NMR of Paramagnetic Molecules*, Academic Press, New York, 1973, 678 pp.
A. R. Lepley and G. L. Closs, *Chemically Induced Magnetic Polarization*, Wiley, New York, 1973, 416 pp.
D. Neuhaus and M. Williamson, *The Nuclear Overhauser Effect in Structural and Conformational Analysis*, VCH Publishers, Weinheim, 1989, 522 pp.

R. E. Sievers, *Nuclear Magnetic Resonance Shift Reagents*, Academic Press, New York, London, 1973, 410 pp.

(f) Macromolecules, Liquid Crystals, Solids

F. A. Bovey, *High Resolution NMR of Macromolecules*, Academic Press, New York, 1972.
J. W. Emsley and J. C. Lindon, *NMR Spectroscopy Using Liquid Crystal Solvents*, Pergamon Press, Oxford, 1975, 367 pp.
C. A. Fyfe, *Solid State NMR for Chemists*, C.F.C. Press, Ontario, 1983, 593 pp.
M. Mehring, *Principles of High Resolution NMR in Solids*, Springer-Verlag, Berlin, 1983, 342 pp.

(g) Biological Applications, Medicine, Imaging

B. Blümich and W. Kuhn, *Magnetic Resonance Microscopy—Methods and Applications in Materials Science and Biomedicine*, VCH Publishers, weinheim, 1992, 604 pp.
E. M. Bradbury and C. Nicolini, *NMR in the Life Sciences*, NATO Asi Series, A, Vol. 107, Plenum Press, New York, 1985, 237 pp.
A. F. Casy, *NMR Spectroscopy in Medicinal and Biological Chemistry*, Academic Press, London, 1971, 425 pp.
R. A. Dwek, *Nuclear Magnetic Resonance in Biochemistry: Applications to Enzyme Systems*, Clarendon Press, Oxford, 1973, 395 pp.
R. A. Dwek, I. D. Campbell, R. E. Richards, and R. J. P. Williams, *NMR in Biology*, Academic Press, London, 1977, 381 pp.
D. G. Gadian, *NMR and its Applications to Living Systems*, Oxford University Press, Oxford, 1982, 216 pp.
K. H. Hausser and H. R. Kalbitzer, *NMR für Mediziner und Biologen*, Springer-Verlag, Berlin, 1989, 221 pp.
O. Jardetzky and G. C. K. Roberts, *NMR in Molecular Biology*, Academic Press, New York, 1981, 681 pp.
K. Roth, *NMR Tomographie und Spektroskopie in der Medizin. Eine Einführung*, Springer-Verlag, Berlin, 1984, 128 pp.
K. Wüthrich, *NMR in Biological Research: Peptides and Proteins*, Elsevier, Amsterdam, 1976, 379 pp.
K. Wüthrich, *NMR of Proteins and Nucleic Acids*, Wiley, New York, 1986.

(h) Carbon-13 NMR

E. Breitmaier and G. Bauer, *^{13}C-NMR-Spektroskopie*, George Thieme Verlag, Stuttgart, 1977, 401 pp.
E. Breitmaier and W. Voelter, *^{13}C-NMR Spectroscopy*, (Monographs in Modern Chemistry, 5), VCH Publishers, Weinheim, 1978, 334 pp.
H.-O. Kalinowski, S. Berger and S. Braun, *^{13}C-NMR-Spektroskopie*, Georg Thieme Verlag, Stuttgart, 1984, 685 pp.
H.-O. Kalinowski, S. Berger and S. Braun, *Carbon-13 NMR Spectroscopy*, Wiley, Chichester, 1988, 792 pp.
G. C. Levy, R. C. Lichter, and G. L. Nelson, *Carbon-13 NMR Spectroscopy*, 2nd Ed., Wiley-Interscience, New York, 1980, 338 pp.
J. B. Stothers, *Carbon-13 NMR Spectroscopy*, Academic Press, New York, 1972, 559 pp.
F. W. Wehrli, A. P. Marchand, and T. Wirthlin, *Interpretation of Carbon-13 NMR Spectra*, Wiley, Chichester, 1988, 484 pp.

(i) Nitrogen-15 NMR

G. C. Levy and R. L. Lichter, *Nitrogen-15 NMR-Spectroscopy*, Wiley, New York, 1979, 221 pp.
G. J. Martin, M. L. Martin, and J.-P. Gouesnard, *^{15}N-NMR Spectroscopy*, NMR-Basic Principles and Progress, Vol. 18, Springer-Verlag, Berlin, 1981, 382 pp.
M. Witanowski and G. A. Webb (Eds.), *Nitrogen NMR*, Plenum Press, London, 1973, 403 pp.

(j) Multinuclear MR

J. W. Akitt, *NMR and Chemistry. An Introduction to the FT Multinuclear Era*, Methuen, London, 1983, 224 pp.
T. Axenrod and G. A. Webb, (Eds.), *Nuclear Magnetic Resonance Spectroscopy of Nuclei Other than Protons*, Wiley, New York, 1974, 407 pp.
C. Brevard and P. Granger, *Handbook of High Resolution Multinuclear NMR*, John Wiley, New York, 1981, 229 pp.
E. A. Evans, D. C. Warrell, J. A. Elvidge and J. R. Jones, *Handbook of Tritium NMR Spectroscopy and Applications*, Wiley, Chichester, 1985, 249 pp.
R. K. Harris and B. E. Mann, (Eds.) *NMR and the Periodic Table*, Academic Press, London, 1978, 459 pp.
P. Granger and R. K. Harris, (Eds.), *Multinuclear Magnetic Resonance in Liquids and Solids—Chemical Applications*, Kluwer, Dordrecht, 1990.
J. B. Lambert and F. G. Riddell (Eds.), *The Multinuclear Approach to NMR Spectroscopy*, D. Reidel Publishing, Dordrecht, 1983, 548 pp.
P. Laszlo (Ed.), *NMR of Newly Accessible Nuclei*, 2 Vols., Academic Press, New York, 1983.
J. Mason (Ed.), *Multinuclear NMR*, Plenum Press, New York, 1987, 639 pp.

DATA COMPILATIONS

D1. A. Ault and M. R. Ault, *A Handy and Systematic Catalogue of NMR Spectra ($^1H,^{13}C$)*, University Science Books, Mill Valley, Calif., 1980, 437 pp.
D2. F. A. Bovey, *NMR Data Tables of Organic Compounds*, Wiley-Interscience, New York, 1967.
D3. E. Breitmaier, G. Haas, and W. Voelter, *Atlas of Carbon-13 NMR Data*, 2 Vols., Heyden, London, 1975, 1979.
D4. W. Brügel, *Handbook of NMR Spectral Parameters*, Heyden, London, 1979, 1016 pp.
D5. W. Bremser, B. Franke, and H. Wagner, *Chemical Shift Ranges in Carbon-13 NMR Spectroscopy*, Verlag Chemie, Weinheim, 1981; W. Bremser, L. Ernst, B. Franke, R. Gerhards, and A. Hardt, *Carbon-13 NMR Spectral Data*, 3rd Ed., Verlag Chemie, Weinheim, 1981.

PROGRESS REPORTS

F1. *Advances in Magnetic Resonance*, J. S. Waugh and W. S. Warren, (Eds.), Academic Press, New York, 1965 ff.
F2. *Annual Reports on NMR Spectroscopy*, E. F. Mooney and G. A. Webb (Eds.), Academic Press, New York, 1968 ff.
F3. *NMR-Basic Principles and Progress, (NMR—Grundlagen und Fortschritt)*, P. Diehl, E. Fluck, R. Kosfeld, H. Günther and J. Seelig (Eds.), Springer-Verlag, Berlin, 1969 ff.
F4. *Nuclear Magnetic Resonance*, R. K. Harris, R. J. Abraham, and G. A. Webb (Eds.), Specialist Periodical Report, Chemical Society, London, 1972 ff.

F5. *Progress in Nuclear Magnetic Resonance Spectroscopy*, J. M. Emsley, J. Feeney, and L. H. Sutcliff, (Eds.), Pergamon Press, Oxford, 1966 ff.

F6. *Topics in Carbon-13 NMR-Spectroscopy*, G. C. Levy (Ed.), Wiley, New York, 1974 ff.

SOLUTIONS TO EXERCISES

2.1 The average area ratio for signals A and C is 28.8:64.4. Per proton this corresponds to 14.4:21.47 and a molar ratio of 1:1.491.

2.2 Firstly the correct assignments of the resonance signals must be made. According to Table 2.1 the signals at δ 7.27, δ 5.30, and δ 2.17 arise from chloroform, methylene chloride, and acetone, respectively. Considering the number of protons in each compound the integration indicates a molar ratio of 10:9:6 or 40, 36 and 24 mole-% in the above order.

2.3 Toluene:methylene chloride:benzene = 38.6:48.9:12.5.

2.4 The difference in the chemical shifts of the signals is 4.78 ppm. This corresponds to 286.8 Hz at 60 MHz. Reference to Table 2.1 and the molecular formula indicates that the substance is toluene.

2.5 (a) $\sim \delta$ 0.9, 3.5, 7.2 (9:2:5); (a′) $\sim \delta$ 1.2, 2.2, 3.5, 7.2 (6:3:3:4); (b) $\sim \delta$ 0.9, 2.0 (3:1); (b′) $\sim \delta$ 0.9, 2.7, 3.0 (9:1:2).

2.6 (a) 1,2-Dichloroethane; (b) 1,3,5-trimethylbenzene; (c) 1,2-dimethoxyethane; (d) 1,2,5,6-dibenzocycloheptadiene; (e) diphenylmethane; (f) 2,5-dimethylfuran; (g) chloroacetic acid methyl ester; (h) 1,2,4,5-tetramethylbenzene; (i) 9,10-dihydroanthracene; (j) terephthaldehyde.

2.7 (1) a, Doublet (1:1); b, quartet (1:3:3:1).
(2) a, Singlet; b, singlet for $J = 0$.
a, Triplet (1:2:1); b, 10-line multiplet (1:9:36:84:126:126:84:36:9:1) for $J > 0$.
(3) a, Doublet (1:1); b, quartet (1:3:3:1) of triplets (1:2:1); c, doublet (1:1); d, singlet.
(4) a, Doublet (1:1) of doublets (1:1); b, doublet (1:1) of quartets (1:3:3:1) of quartets (1:3:3:1); c, as b; d, doublet (1:1) of doublets (1:1).
(5) a, Doublet (1:1); b, septet (1:6:15:20:15:6:1); c, quartet (1:3:3:1); d, triplet (1:2:1).
(6) a, Triplet (1:1:1); b, doublet (1:1).

2.8 (a) AMX; $J_{AM} = 4$, $J_{AX} = 2$, and $J_{MX} = 0$ Hz; (b) AM₂X; $J_{AM} = 1$, $J_{AX} = 5$, and $J_{MX} = 0$ Hz; (c) AM₂X; $J_{AM} = 3$, $J_{AX} = 4$, and $J_{MX} = 1$ Hz; (d) AMX₃; $J_{AM} = 5$, $J_{AX} = 2$, and $J_{MX} = 1$ Hz.

2.9 (a) Na[^{10}BH₄] and Na[^{11}BH₄]; (b) C₆H₅CHD₂; (c) ^{15}NH₄$^+$; (d) ^{14}NH₄$^+$; cf. Table 1.1 (p. 11).

2.10 (1) (a) Isopropyl chloride; (b) C₆H₅CH₂SH; (c) diethyl phthalate; (d) CH₃CH(NO₂)-CO₂C₂H₅.
(2) In the δ-scale 1 mm = 4.2 Hz; thus in the expanded spectrum 1 mm = 2.1 Hz. The absorption at lowest field (δ 7.1) is split by 6.3 and 14.7 Hz that must arise from *cis* and *trans* coupling across a double bond. This absorption is therefore due to the proton adjacent to the ester function. Splittings of 14.7 and 6.3 Hz are found at δ 4.7 and 4.4 so that these δ-values arise from the protons *cis* and *trans*, respectively, to the ester group. These protons are coupled to one another with a geminal coupling constant of about 1.5 Hz.
(3) The spectrum shows two absorptions separated by about 50 Hz. The one at lower field can be recognized as two partially superimposed quartets, the separation and

splitting of which leads to coupling constants of 15 and 6.6 Hz. At higher field four quartets are observed with coupling constants of 15, 7.8, and 1.5 Hz. Therefore, the proton adjacent to the methyl group absorbs at the lower field [J(H,H), J(CH$_3$,H)] and the proton adjacent to the aldehyde group absorbs at the higher field [J(H,H), J(H,CHO), J(CH$_3$,H)].

2.11 δ(a) 7.36, δ(b) 8.77 ppm; J(a,b) 5.0, J(a,c) 1.5 Hz.

2.12   See H. Günther, M. Görlitz, and H. Meisenheimer, *Org. Magn. Resonance*, **6**, 388 (1974).

2.13   F. Bottino and S. Pappalardo, *Org. Magn. Reson.*, **16**, 1 (1981).

4.1   (a) δ_o 7.06; δ_m 7.27, δ_p 7.08; (b) δ_o 7.81, δ_m 7.91, δ_p 9.30; with the proportionality factor 12.7 one obtains δ_o 7.02, δ_m 7.27, δ_p 7.04. Compared with the empirical constants of Table 4.6 the calculated $\Delta\delta$ values are too small.

4.2   The chemical shifts for compounds **b** and **e** show the charge polarization induced by the electron donor atom (S, O); for compounds **c** and **f** the electron acceptor properties of the carbonyl group dominate.

4.3   Compared with the protons in benzene H(4) and H(9) in phenanthrene are exposed to the additional ring current effects of rings A and B and A and C, respectively. Because of the planarity of the compound $\theta = 90°$ and equation (4.6) simplifies to $\Delta\sigma = \Delta\chi/12\pi R^3$. With the indicated $\Delta\chi$ value of -630×10^{-36} m^3/molecule it follows further that $\Delta\sigma = -16.67/R^3$ ppm when R is expressed in nm. Now the distances of the protons from the centres of their respective rings must be determined trigonometrically. Using 0.140 and 0.110 nm for the C–C and C–H bond lengths, respectively, and a uniform bond angle of 120° we obtain:

$$H(4): \quad R_B = 0.348 \text{ nm and } \Delta\sigma_B = -0.40 \text{ ppm};$$
$$R_A = 0.366 \text{ nm and } \Delta\sigma_A = -0.34 \text{ ppm};$$
$$H(9); \quad R_C = 0.475 \text{ nm and } \Delta\sigma_C = -0.16 \text{ ppm.}$$
$$R_A = 0.348 \text{ nm and } \Delta\sigma_A = -0.40 \text{ ppm}$$

Total shielding contributions of $\Delta\sigma(4) = -0.74$ and $\Delta\sigma(9) = -0.56$ result in δ values of 8.01 and 7.83 for H(4) and H(9), respectively, relative to benzene (δ 7.27).

Experimentally determined values are δ(4) 8.93 and δ(9) 7.71. The deviation in the case of H(9) is within the limits of error for such calculations but the large discrepancy for H(4) is indicative of an additional shielding effect. This is the van-der-Waals effect (cf. p. 37), which has its origin in the steric interaction between H(4) and H(5).

4.4 According to the π charge density, H(1) should resonate at higher field (δ 1.27). The high-field shifts measured amount to $\Delta\delta H(1) = 6.54$ ppm and $\Delta\delta H(2) = 4.37$ ppm. The charge density contributions to these shifts, following equation (4.2), are 3.61 and 1.38 ppm, respectively. The paramagnetic ring current contribution for H(1) thus amounts to 2.93 ppm and for H(2) to 2.99 ppm (see R. Benken and H. Günther, *Helv. Chim. Acta*, **71**, 694 (1988)).

4.5 The comparison of the partial charges in pyridine and the pyridinium cation leads to the following $\Delta\rho_i$ values:

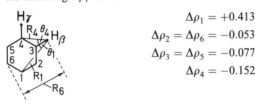

$$\Delta\rho_1 = +0.413$$
$$\Delta\rho_2 = \Delta\rho_6 = -0.053$$
$$\Delta\rho_3 = \Delta\rho_5 = -0.077$$
$$\Delta\rho_4 = -0.152$$

On the basis of trigonometric analysis it follows that for H_β

$$R_1 = R_5 = 0.341 \text{ nm}; \quad \theta_1 = \theta_5 = 20°44'$$
$$R_2 = R_4 = 0.217 \text{ nm}; \quad \theta_2 = \theta_4 = 33°53'$$
$$R_3 = 0.110 \text{ nm}; \quad \theta_3 = 0°$$
$$R_6 = 0.390 \text{ nm}; \quad \theta_6 = 0°$$

and for H_γ

$$R_1 = 0.390 \text{ nm}; \quad \theta_1 = 0°$$
$$R_2 = R_6 = 0.341 \text{ nm}; \quad \theta_2 = \theta_6 = 20°44'$$
$$R_3 = R_5 = 0.217 \text{ nm}; \quad \theta_3 = \theta_5 = 33°53'$$
$$R_4 = 0.110 \text{ nm}; \quad \theta_4 = 0°$$

With these we obtain from equation 4.16

$$\Delta\sigma_\beta = 12.5\left(\frac{0.413}{11.62}0.935 - \frac{0.053}{4.71}0.830 - \frac{0.077}{1.21} - \frac{0.152}{4.71}0.830\right.$$
$$\left. -\frac{0.077}{11.62}0.935 - \frac{0.053}{15.21}\right) - 17.0\left(\frac{0.413}{11.62} - \frac{0.053}{4.71} - \frac{0.077}{1.21}\right.$$
$$\left. -\frac{0.152}{4.71} - \frac{0.077}{11.62} - \frac{0.052}{15.21}\right)^2$$
$$\Delta\sigma_\beta = -0.95 - 0.11 = -1.06 \text{ ppm.}$$
$$\Delta\sigma_\gamma = 12.5\left(\frac{0.413}{15.21} - 2\frac{0.053}{11.62}0.935 - 2\frac{0.077}{4.71}0.830 - \frac{0.152}{1.21}\right)$$
$$-17.0\left(\frac{0.413}{15.21} - 2\frac{0.053}{11.62} - 2\frac{0.077}{4.71} - \frac{0.152}{1.21}\right)^2$$
$$\Delta\sigma_\gamma = -1.68 - 0.33 = -2.01 \text{ ppm.}$$

The experimental values are $\Delta\sigma_\beta = -1.07$ and $\Delta\sigma_\gamma = -1.22$ ppm. The agreement of the experimental and theoretical values for the β position is doubtlessly fortuitous. The deviation that is found for the resonance frequency of H_γ is rather typical for this type of approximate calculation. Certainly one can assume that in addition to the charge density effect still other factors are responsible for the variation of the resonance frequencies. Thus, the nitrogen in pyridine has a magnetic anisotropy ($\Delta\chi = -91 \times 10^{-36}$ m^3 per molecule perpendicular to the plane of the ring) that vanishes upon protonation. Also, medium effects may not be neglected in this case. Nevertheless, it is often possible by consideration of all factors to obtain good results (compare, for example, S. Castellano, H. Günther, and S. Ebersole, *J. Phys. Chem.,* **69**, 4166 (1965); H. Günther and S. Castellano, *Ber. Bunsenges. Phys. Chem.,* **70**, 913 (1966)).

4.6   The integration curve indicates that the resonance signals of the four olefinic protons begin with the signal 0 Hz. One finds 1 H (0, 11 Hz), 2 H (26–52 Hz) and 1 H (72–84 Hz). Taking only vicinal coupling constants into consideration it is advantageous to begin the analysis with the doublet at the lowest field ($J = 11$ Hz) that must be assigned to H_a or H_d. There is a separation of 11 Hz between lines 35 and 46 and lines 41 and 52, so these signals probably arise from the neighbouring proton. This conclusion is supported by the roof effect (35 and 41 are more intense than 46 and 52). In addition, this proton interacts, as indicated by the roof effect, with the proton at the highest field with a coupling constant of 6 Hz. The latter proton is also coupled with a coupling constant of 6 Hz with the fourth proton, the resonance signals of which are found at 26 and 32 Hz. In order to make an assignment it must be decided whether $\delta(H_a) > \delta(H_d$ or $\delta(H_a) < \delta(H_d)$. Both protons are deshielded, H_a by the effect of the benzene ring and H_d by the effect of the adjacent oxygen atom. It is difficult to predict which effect is more important. Thus, on the basis of chemical shifts alone the following assignments are possible:

$$\delta(H_a) > \delta(H_d) > \delta(H_b) > \delta(H_c) \text{ and } \delta(H_d) > \delta(H_a) > \delta(H_c) > \delta(H_b)$$

A decision between these possibilities can be made on the basis of the coupling constants: the electronegativity effect of the oxygen works to reduce J_{cd} so that J_{cd} must be smaller than J_{ab}. The first alternative is therefore correct and the coupling constants are

$$J_{ab} = 11, \; J_{bc} = J_{cd} = 6 \text{ Hz}$$

Long-range coupling is observed in this system only if the spectrum is expanded or if it is recorded under conditions of higher resolution.

4.7   This is a mixture of the *cis* and *trans* isomers in the ratio 7:3. The coupling constants are:

$$^3J_{trans} = 19.6 \text{ Hz}; \; ^3J_{cis} = 14.2 \text{ Hz}; \; ^3J(H, CH_2)_{trans} = 6.5 \text{ Hz}; \; ^3J(H, CH_2)_{cis}$$
$$= 8.0 \text{ Hz}; \; |^4J(H, CH_2)_{trans}| = 1.4 \text{ Hz and } |^4J(H, CH_2)_{cis}| = 1.2 \text{ Hz}$$

The difference in the couplings between the olefinic protons and the methylene protons in the two isomers are functions of conformation. In the *cis* isomer the conformation with the methylene chain *gauche* is more stable than that conformation in the *trans* isomer because of the large silyl group (cf. **60**, p. 126). The vicinal coupling $^3J(H,CH_2)$ therefore has a larger '*trans*-component', i.e. its magnitude increases according to the Karplus curve. Simultaneously the π contribution to the allylic coupling $^4J(H,CH_2)$ decreases and the magnitude of this coupling becomes smaller.

4.8   The negative allylic coupling shows that the π contribution dominates. Thus the more favourable conformation is **71b** in which the C–H^7 bond is aligned nearly parallel to the $2p_z$ orbitals at C–1 and C–6. This is also the conformation that is necessary for the observation of a measurable homoallylic coupling.

4.9 A value of 13.1 Hz is typical for J_{aa} in cyclohexane. Therefore **72** is the correct conformation.

4.10 Only in conformation **74** is a coplanar zig-zag arrangement of σ-bonds, a prerequisite for the existence of a large ^4J coupling *via* the M mechanism, possible. The methyl protons couple with the axial hydrogens in the 2- and 6-positions.

4.11 The small value of the vicinal coupling is compatible only with structure **77** (cf. p. 118).

5.1

(a) AA'BB'	(e) A_6X	(i) A_3B_2	(m) A_3KMX
(b) AA'BB'C	(f) AA'MM'XY	(j) A_2X_6	
(c) ABC	(g) AA'BB'	(k) ABCD	
(d) A_3B	(h) AA'BB'	(l) ABC_2	

5.2

$$-\frac{h^2}{8\pi^2 m}\frac{\partial^2}{\partial x^2} N \sin ax = EN \sin ax$$

$$\frac{h^2}{8\pi^2 m} a^2 N \sin ax = EN \sin ax$$

From this it follows that $E = a^2 \dfrac{h^2}{8\pi^2 m}$.

Now we have to determine whether all values of a in our trial function are allowed. If the electron is moving in a one-dimensional box of length L this is not the case. Under these conditions $\Psi(0)$ and $\Psi(L)$ must equal zero, i.e., the wavefunction must vanish at $x = 0$ and $x = L$. Thus:

$$\Psi(0) = N \sin a0 = 0 \qquad \Psi(L) = N \sin aL = 0$$

The second condition is met only if $aL = q\pi$, where q is an integer. It follows that $a = q\pi L$ and one obtains the eigenvalues

$$E_q = \frac{h^2}{8mL^2} Q^2 \text{ with } q = 1, 2, 3, \ldots, n$$

The eigenfunctions are then

$$\Psi_q = N \sin q\pi x/L$$

with $N\sqrt{2/L}$.

5.3 The basis functions are given on p. 160. The Hamiltonian operator is $\hat{H} = \hat{I}_z(A) + \hat{I}_z(B) + \hat{I}_z(C)$. From this one obtains

$$
\begin{aligned}
E_1 &= \tfrac{1}{2}\left(v_A + v_B + v_C\right) & E_5 &= \tfrac{1}{2}\left(v_A - v_B - v_C\right) \\
E_2 &= \tfrac{1}{2}\left(v_A + v_B - v_C\right) & E_6 &= \tfrac{1}{2}\left(-v_A + v_B - v_C\right) \\
E_3 &= \tfrac{1}{2}\left(v_A - v_B + v_C\right) & E_7 &= \tfrac{1}{2}\left(-v_A - v_B + v_C\right) \\
E_4 &= \tfrac{1}{2}\left(-v_A + v_B + v_C\right) & E_8 &= \tfrac{1}{2}\left(-v_A - v_B - v_C\right)
\end{aligned}
$$

5.4

$$
\begin{aligned}
I(f_1) : \langle\beta\alpha|\hat{I}_x(A) + \hat{I}_x(B)|\alpha\alpha\rangle^2 &= [\langle\beta\alpha|\hat{I}_x(A)|\alpha\alpha\rangle + \langle\beta\alpha|\hat{I}_x(B)|\alpha\alpha\rangle]^2 \\
&= [\langle\beta\alpha|\tfrac{1}{2}\beta\alpha\rangle + \langle\beta\alpha|\tfrac{1}{2}\alpha\beta\rangle]^2 \\
&= \tfrac{1}{4}
\end{aligned}
$$

$$
\begin{aligned}
I(f_2) : \langle\beta\beta|\hat{I}_x(A) + \hat{I}_x(B)|\alpha\beta\rangle^2 &= [\langle\beta\beta|\hat{I}_x(A)|\alpha\beta\rangle + \langle\beta\beta|\hat{I}_x(B)|\alpha\beta]^2 \\
&= [\langle\beta\beta|\tfrac{1}{2}\beta\beta\rangle + \langle\beta\beta|\tfrac{1}{2}\alpha\alpha\rangle]^2 \\
&= \tfrac{1}{4}
\end{aligned}
$$

One obtains $I(f_3)$ and $I(f_4)$ analogously.

5.6
$$C = \tfrac{1}{2}\sqrt{v_0\delta^2 + J^2} = \tfrac{1}{2}\sqrt{400 + 225} = 12.5;$$
$$\sin 2\theta = J/2C = 15/25 = 0.6;$$
$$f_1 = 7.5 + 12.5 = 20.0; \quad I_1 = 0.4$$
$$f_2 = -7.5 + 12.5 = 5.0; \quad I_2 = 1.6$$
$$f_3 = 7.5 - 12.5 = -5.0; \quad I_3 = 1.6$$
$$f_4 = -7.5 - 12.5 = -20.0; \quad I_4 = 1.4$$

5.7
$$\begin{vmatrix} H_{22} - E & H_{23} & H_{24} \\ H_{32} & H_{33} - E & H_{34} \\ H_{42} & H_{43} & H_{44} - E \end{vmatrix}$$
$$H_{22} = \tfrac{1}{2}(v_A + v_B - v_C) + \tfrac{1}{4}(J_{AB} - J_{AC} - J_{BC})$$
$$H_{33} = \tfrac{1}{2}(v_A - v_B + v_C) + \tfrac{1}{4}(-J_{AB} + J_{AC} - J_{BC})$$
$$H_{44} = \tfrac{1}{2}(-v_A + v_B + v_C) + \tfrac{1}{4}(-J_{AB} - J_{AC} + J_{BC})$$
$$H_{23} = H_{32} = \tfrac{1}{2}J_{BC}$$
$$H_{24} = H_{42} = \tfrac{1}{2}J_{AC}$$
$$H_{34} = H_{43} = \tfrac{1}{2}J_{AB}$$

5.8   (a) $1/\sqrt{3}$; (b) $1:3:3:1$; (c) variation of the spectrometer frequency or solvent effects can influence $v_0\delta$ in an AB system and, as a result, the separation $f_2 - f_3$ and the intensity ratio are variable.

5.9 $v_A = 31.55$ Hz; $v_B = 16.95$ Hz; $J = 10.5$ Hz; $I_1 = I_4 = 0.42$; $I_2 = I_3 = 1.58$.

5.10 $v_A = 95.00$ Hz; $v_B = 105.00$ Hz; $J_{AB} = 6.00$ Hz.

5.11   Solution 1, $I_{11} = 0.993$; solution 2, $I_{11} = 0.595$.

5.12   (1) ab subspectrum: 43.29 45.78 47.75 50.29 Hz.
(2) ab subspectrum: 45.21 47.75 54.03 56.52 Hz.
The analyses of the ab subspectra give an average value of 2.52 Hz for $|J_{AB}|$. We also find that:

Solution 1:

$$v_A = 45.76 \text{ Hz}; \quad J_{AX} = 1.67 \text{ Hz};$$
$$v_B = 51.86 \text{ Hz}; \quad J_{BX} = 6.47 \text{ Hz}.$$

Solution 2:

$$v_A = 50.00 \text{ Hz}; \quad J_{AX} = 10.17 \text{ Hz};$$
$$v_B = 47.61 \text{ Hz}; \quad J_{BX} = -2.03 \text{ Hz}.$$

Here X is the fluorine nucleus. In order to determine which solution is correct the ^{19}F spectrum of the compound must be measured. When this is done one finds the following resonance frequencies (in hertz relative to v_X) for the lines f_9, f_{10} and f_{11} where the intensities are indicated in parentheses: 6.66 (0.822), 4.08 (1.000), and 2.15 (0.178). It follows then from equation (5.25) that solution 2 is correct. That is, *ortho* and *para* H,F couplings have opposite signs (the combination -10.17 and 2.03 is also possible). J_{AB}, as the *meta* H,H coupling, is assumed to be positive. For the experimental spectrum see Corio, Monographs (b), p. 338.

5.13   Since the inner lines of one of the ab subspectra are not resolved and the positions of the four signals must be estimated, only an approximate analysis is possible. The relative chemical shifts in this subspectrum can be calculated only if the separation

$f_2 - f_3$ can be evaluated from the line width. In this case a value of 0.25 Hz seems reasonable. One then obtains:

ab subspectrum 1: 0.0 9.78 10.02 (19.8) Hz
ab subspectrum 2: 6.2 16.2 20.4 30.8 Hz

From subspectrum 2 an average value for $|J_{AB}|$ of 10.2 Hz can be determined. In addition the analysis yields:

Solution 1:

$$v_A = 10.55 \text{ Hz}; \quad J_{AX} = 3.5 \text{ Hz}; \quad I_{11} = 0.92;$$
$$v_B = 16.75 \text{ Hz}; \quad J_{BX} = 11.5 \text{ Hz}.$$

Solution 2:

$$v_A = 15.7 \text{ Hz}; \quad J_{AX} = 13.7 \text{ Hz}; \quad I_{11} = 0.78;$$
$$v_B = 11.7 \text{ Hz}; \quad J_{BX} = 1.3 \text{ Hz}.$$

The experimental X portion indicates that solution 1 is the correct one. For the assignment of these parameters to the protons of L-asparagine it can be assumed that the AB portion is formed from the methylene protons. J_{AB}, as a geminal coupling constant, would then be negative (cf. p. 109). The X proton shows a large *trans* coupling with H_B and a small *gauche* coupling with H_A so that the following conformation for the compound can be assumed:

5.15 After determining that $N = 9.0$ Hz analysis of the ab subspectrum gives $L = 7.0$, $K = 11.0$ (7.0), and $M = 7.0$ (11.0) Hz. One then derives the following coupling constants: $J = 8.0$, $J' = 1.0$, $J_A = 9.0$ (2.0), and $J_X = 2.0$ (9.0) Hz.

5.16 We find that $N = 9.67$ Hz and that the ab subspectra consist of the lines:
(1) 0.0 7.7 10.6 18.7 Hz;
(2) 1.7 7.4 11.4 16.9 Hz.
Thus $L = 7.4$ or 7.8 Hz or an average of 7.6 Hz, $K = 7.9$, and $M = 5.6$ Hz. Another assignment of the lines to the ab subspectra leads to very different values of L and thus can be eliminated from consideration. The coupling constants are $J = 8.6$, $J' = 1.1$, $J_A = 6.8$, and $J_X = 1.1$ Hz. On the basis of the structure J_A must be J_{23}. The results of an AA'BB' analysis are, in the same order, 8.65, 1.04, 6.79, and 1.03 Hz, so the X approximation can be tolerated here.

6.1 (a) AA'BB'B''B'''X$_2$ (d) A$_4$
(b) ABX$_2$ (e) ABCD
(c) AA'XX'   (f) AA'BB'

6.2 2-Acetylpyridine is an example of an ABCD system that produces a multitude of lines (V. J. Kowalewski and D. G. de Kowalewski, *J. Chem. Phys.*, **37**, 2603 (1962)). The 3-isomer produces a spectrum of nearly first-order simplicity (V. J. Kowalewski and D. G. Kowalewski, *J. Chem. Phys.*, **36**, 266 (1962)) while the 4-isomer can be recognized by the characteristically symmetric appearance of the AA'BB' or AA'XX' system.

6.3   Only (b) possesses the molecular symmetry required for an AA′BB′ system.

6.4   Spectrum 1: **e**: spectrum 2: **d**.

6.5   Enantiotopic = **e**; diastereotopic = **d**.

(a) **d**; (b) **e**; (c) **e**; (d) **d**; (e) **e**; (f) **e**; (g) **d**; (h) **e**; (i) **e**; (j) **d**.

6.6   (a) AX; (c) AX_2; (e) AX_2Y;

(b) AX; (d) AA′XX′; (f) A_2X_6.

8.1

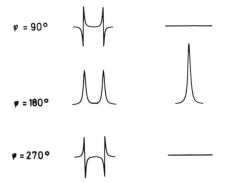

8.2

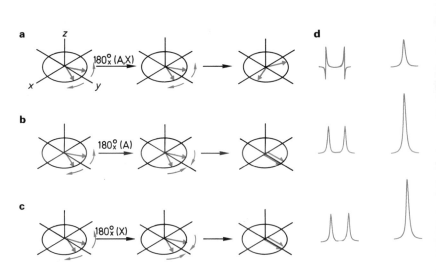

8.3

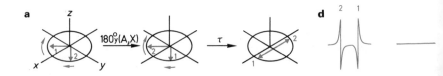

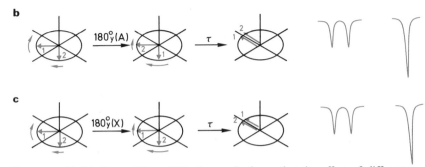

If compared with Figure 8.7, p. 281, the result shows that the effect of different Larmor frequencies is eliminated by the spin echo experiment.

8.4

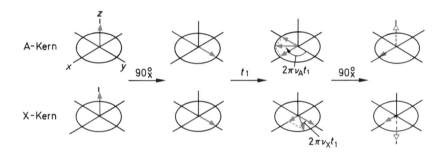

At the end of the evolution time t_1 the vectors have progressed differently because of their Larmor frequency. The second 90°_x pulse leads to a situation where only the x-component is retained as transverse magnetization and finally as t_2 signal. Its amplitude depends on t_1 and ν_A and ν_X, respectively.

8.5

During the evolution time t_1, z-magnetization is formed by relaxation which yields,

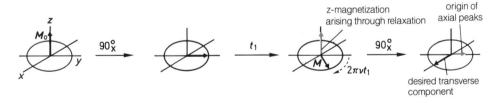

after the second 90°_x pulse, the unwanted transverse component. This signal increases with increasing t_1. It is, however, not frequency-modulated.

8.6   The t_1 increment corresponds to the dwell time in the F_1 dimension. A prolongation would reduce the spectral window (see Chapter 7). For the COSY experiment, however, the condition $F_1 = F_2$ must be met.

8.7 $H_a, H_b = 1,2; H_c = 5; H_d = 6; H_e = 4; H_f = 3.$

8.8 $\hat{I}_x \xrightarrow{90^\circ[\hat{I}_y]} \hat{I}_z; \ \hat{I}_y \xrightarrow{90^\circ[\hat{I}_y]} \hat{I}_y; \hat{I}_x \xrightarrow{90^\circ[-\hat{I}_y]} -\hat{I}_z; \ \hat{I}_y \xrightarrow{90^\circ[-\hat{I}_y]} \hat{I}_y$

8.9 $2\hat{I}_z(X)\hat{I}_y(A)\sin X1\sin J1 \xrightarrow{\text{Larmor precession, spin–spin coupling}}$

$2\hat{I}_z(X)\hat{I}_y(A)\sin X1\sin J1\cos A2\cos J2 + \hat{I}_x(A)\sin X1\sin J1\cos A2\sin J2$

$\quad + 2\hat{I}_z(X)\hat{I}_x(A)\sin X1\sin J1\sin X2\cos J2 - \hat{I}_y(A)\sin X1\sin J1\sin X2\sin J2$

8.10   For the diagonal signals the expressions 3, 4, 5, 6, 9, 10 and 12 are responsible. The cross peaks result from 13 and 14. For $\beta = 0°$, all expressions which contain $\sin\beta$ are eliminated and only diagonal peaks are detected. For $\beta = 90°$, all expressions which contain $\cos\beta$ are eliminated and the diagonal peaks arise from 3 and 4, the cross peaks from 13 and 14. For $\beta = 45°$, $\sin\beta = 0.85$ and $\cos\beta = 0.53$. The diagonal peaks are thus more strongly reduced than the cross peaks.

8.11   Coherence A is cancelled, B is detected.

8.12   For one pathway we have $\Delta p = +1, -2, +2$, for the other $\Delta p = -1, +2, 0$. With $\Delta\theta(P3) = 0$ the following signal phases result: $(0\ 3\ 2\ 1)$ and $(0\ 1\ 2\ 3)$, respectively. Both arrangements lead to cancellation.

8.13   From the coherence pathways drawn in Fig 8.26, the following table can be set up for the coherence orders, p, and the corresponding gradient effects, Φ:

(a) p: $+1, +2, +1$ $\Phi = (+1 + 2 + 3) = 6$
             $+1, -2, +1$ $\Phi = (+1 - 2 + 3) = 2$
             $-1, +2, +1$ $\Phi = (-1 + 2 + 3) = 4$
             $\boldsymbol{-1, -2, +1}$ $\boldsymbol{\Phi = (-1 - 2 + 3) = 0}$

(b) p: $+1, +1, +1$ $\Phi = (+1 + 1 + 3) = 5$
             $+1, \ \ 0, +1$ $\Phi = (+1 + 0 + 3) = 4$
             $-1, \ \ 0, +1$ $\Phi = (-1 + 0 + 3) = 2$
             $-1, -1, +1$ $\Phi = (-1 - 1 + 3) = 1$

Refocusing is thus expected only for the pathway shown in Fig 8.27 (italics above).

8.14   If we use $1/2J_{AX}$ for the t_1 delay, we obtain for $\cos(\pi J_{AX}t_1)$ the expression $\cos(\pi/2)$ and for $\sin(\pi J_{AX}t_1)$ the expression $\sin(\pi/2)$. Consequently, $\hat{I}_y(A)$ yields $2\hat{I}_x(A)\hat{I}_z(X)$ and $\hat{I}_y(X)$ correspondingly $2\hat{I}_x(X)\hat{I}_z(A)\ [\ = 2\hat{I}_z(A)\hat{I}_x(X))]$, respectively.

10.1   See Figure 10.12, p. 410

11.1

a   b   c

For an excitation pulse with the angle $\beta < 90°$, part of the z-magnetization is retained (a). This is transformed into negative z-magnetization by the first $180°$ pulse (b) which would be transformed into transverse magnetization by the excitation pulse of the second experiment. The $180°$ pulse of the additional spin echo sequence directs this magnetization again along the positive z-axis (c) and for the following expressions complete z-magnetization is available.

11.3   The important aspect of this pulse sequence is the magnetization transfer from the sensitive A-nucleus (^{1}H) to the insensitive X-nucleus (^{13}C). In order to simplify our treatment, we neglect the evolution of the Larmor frequencies in the t_1 period.

Sequence (a): After the excitation pulse, transverse A-magnetization $\hat{I}_y(A)$ is obtained. The components of the chemically different nuclei A_i are labelled during the evolution time with their Larmor frequencies ω_A^i. The $180°(X)$ pulse leads to A,X decoupling. During the delay Δ_1, antiphase A-magnetization, $2\hat{I}_x(A)\hat{I}_z(X)$, develops under the action of the propagator $\pi J\Delta_1 2\hat{I}_z(A)\hat{I}_z(X)$ which, because of $\Delta_1 = 1/2J_{AX}$ (equation (8.23a,b), p. 308), has the simple form $(\pi/2)2\hat{I}_z(A)\hat{I}_z(X)$.

Polarization transfer is then introduced by the $90°$ pulse pair at the A and X channel:

$$2\hat{I}_x(A)\hat{I}_z(X) \xrightarrow{90°\hat{I}_y(A)} 2\hat{I}_z(A)\hat{I}_z(X) \xrightarrow{90°\hat{I}_z(X)} 2\hat{I}_z(A)\hat{I}_y(X)$$

Antiphase A-magnetization is, therefore, transformed into antiphase X-magnetization which is now modulated during t_2 with ω_X^i. This yields, after double Fourier transformation, cross peaks at ω_A^i, ω_X^i. A refocusing delay before detection with $\Delta_2 = 1/2J_{AX}$ leads for an AX system (CH group) to in-phase X-magnetization $\hat{I}_x(X)$. In this case A-decoupling can be applied.

Sequence (b): The experiment starts after the excitation pulse with a delay $\Delta_1 = 1/2J_{AX}$, at the end of which antiphase A-magnetization is formed. The $90°$ pulse transforms this to double and zero quantum magnetization:

$$2\hat{I}_x(A)\hat{I}_z(X) \xrightarrow{90°\hat{I}_x(X)} 2\hat{I}_x(A)\hat{I}_y(X).$$

These develop during the following evolution time t_1 with the sum and the difference of the Larmor frequencies ω_A and ω_X, respectively. Again a $180°$ pulse is used for A,X decoupling. The second $90°(X)$ pulse, which is phase cycled in order to select double quantum magnetization, yields antiphase A-magnetization which is now modulated in t_2 with ω_A leading to cross peaks:

$$2\hat{I}_x(A)\hat{I}_y(X) \xrightarrow{90°\hat{I}_x(X)} -2\hat{I}_x(A)\hat{I}_z(X).$$

11.4 Main signal magnetization S_0, satellite magnetization $s_{\pm1}$ (red), the $180°$ pulse is neglected:

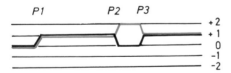

Starting with situation (e) in Figure 11.19 (p. 487), the selection of $s_{\pm1}$ is possible on the basis of the phase rule $\Delta\phi = \Delta p \times \Delta\theta$ [equation (8.37), p. 315] with a phase cycle for the last $90°$ pulse $P3$. From the phase diagram we have for $s_{\pm1}$ $\Delta p = -1$ and $+3$, while the z-magnetization S_0 follows the classical Bloch vector rules with $\Delta p = +1$.

This yields:

experiment	$\theta(P3)$	S_0	s_{+1}	s_{-1}	detector phase
1	+x	−y	+x	−x	+x
2	+y	+x	−y	+y	−y
3	−x	+y	−x	+x	−x
4	−y	−x	+y	−y	+y

The detector phase follows the phase of the satellite magnetization. One detects the following signals (cf. p. 248) and addition leads to elimination of S_0 while the satellite magnetization $s_{\pm 1}$ is retained:

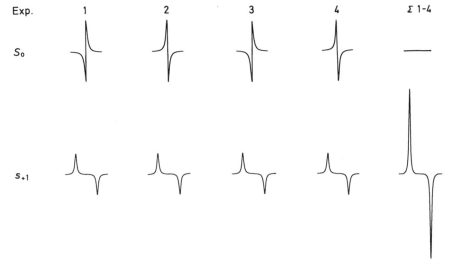

Exp.   1   2   3   4 Σ 1-4

S_0

s_{+1}

According to the coherence level diagram shown above, a first gradient pulse ΔG_z after the pulse $P2$ and a second one with amplitude $-2\Delta G_z$ after the pulse $P3$ refocuses selectively the satellite magnetization [see W. Willker and D. Leibfritz, *Magn. Reson. Chem.* **32**, 665 (1994)].

11.5
$$\overset{1}{H_3C}-\overset{2}{C}H_2-\overset{3}{C}H-\overset{4}{C}H_2-\overset{5}{C}H_2-\overset{6}{C}H_2-\overset{7}{C}H_3$$

$$\overset{8}{|}\\ CH_3$$

	1	2	3	4	5	6	7	8
δ_{exp}	11.3	29.7	34.7	36.5	29.7	23.3	14.1	19.3
δ_{ber}	11.4	29.9	33.9	36.3	29.7	23.3	14.1	19.7

For C(2) as example: $\delta_{calc} = -2.6 + 2 \times 9.1 - 2.5 + 0.3 + 0.2 - 2.5 = 29.9$

11.6 We start at C(3) and use equation (11.16). This yields s(i) = 32.2% and for the CH-bonds sp$^{2.11}$-hybrids. For the C(3)-B C(2) bond 35.6% s-character remains, which leads to an sp$^{1.81}$-hybrid. Equation (11.15) yields then $95.2 = 550 \times 0.356 \times x$ and $x = 0.486$. The C(2)-C(3)-hybrid, therefore, has 48.6% s-character and is a sp$^{1.06}$-hybrid. The residual s-character at C(2) is distributed between the two bonds in the three membered ring which receive 25.7% each. Within the framework of the Walsh model a sp$^{0.95}$-hybrid is directed towards the ring. Equation 11.15 yields $23.2 = 550 \times 0.257 \times x$ and $x = 0.164$. C(1) has, therefore, 16.4% s-character for the C(1)–C(2) bond. If one assumes the same s-character for the C(1)–C(1′) bond, 67.8% remain for the other CH bonds a C(1), that is 33.9% for each bond. These are then formed by sp$^{1.98}$-hybrids. The ^{13}C,^{1}H coupling at the three-membered ring should then amount to 168.4 Hz. The difference found to the experimental result has its origin in the unknown coupling C(1)-C(1′) and in an accumulation of the errors in the last equation.

INDEX